Biophysics of Photoreceptors and Photomovements in Microorganisms

NATO ASI Series

Advanced Science Institutes Series

A series presenting the results of activities sponsored by the NATO Science Committee, which aims at the dissemination of advanced scientific and technological knowledge, with a view to strengthening links between scientific communities.

The series is published by an international board of publishers in conjunction with the NATO Scientific Affairs Division

A	**Life Sciences**	Plenum Publishing Corporation
B	**Physics**	New York and London
C	**Mathematical and Physical Sciences**	Kluwer Academic Publishers
D	**Behavioral and Social Sciences**	Dordrecht, Boston, and London
E	**Applied Sciences**	
F	**Computer and Systems Sciences**	Springer-Verlag
G	**Ecological Sciences**	Berlin, Heidelberg, New York, London,
H	**Cell Biology**	Paris, Tokyo, Hong Kong, and Barcelona
I	**Global Environmental Change**	

Recent Volumes in this Series

Volume 205—Developmental Patterning of the Vertebrate Limb
edited by J. Richard Hinchliffe, Juan M. Hurle, and Dennis Summerbell

Volume 206—Alcoholism: A Molecular Perspective
edited by T. Norman Palmer

Volume 207—Bioorganic Chemistry in Healthcare and Technology
edited by Upendra K. Pandit and Frank C. Alderweireldt

Volume 208—Vascular Endothelium: Physiological Basis of Clinical Problems
edited by John D. Catravas, Allan D. Callow, C. Norman Gillis, and Una S. Ryan

Volume 209—Molecular Basis of Human Cancer
edited by Claudio Nicolini

Volume 210—Woody Plant Biotechnology
edited by M. R. Ahuja

Volume 211—Biophysics of Photoreceptors and Photomovements in Microorganisms
edited by F. Lenci, F. Ghetti, G. Colombetti, D.-P. Häder, and Pill-Soon Song

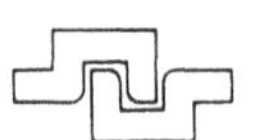

Series A: Life Sciences

Biophysics of Photoreceptors and Photomovements in Microorganisms

Edited by

F. Lenci, F. Ghetti, and G. Colombetti

CNR-Istituto di Biofisica
Pisa, Italy

D.-P. Häder

Friedrich-Alexander-Universität
Erlangen, Germany

and

Pill-Soon Song

University of Nebraska-Lincoln
Lincoln, Nebraska

Springer Science+Business Media, LLC

Proceedings of a NATO Advanced Study Institute on
Biophysics of Photoreceptors and Photomovements in Microorganisms,
held September 16-28, 1990,
in Tirrenia (Pisa), Italy

Library of Congress Cataloging-in-Publication Data

NATO Advanced Study Institute on Biophysics of Photoreceptors and
Photomovements in Microorganisms (1990 : Tirrenia, Italy)
Biophysics of photoreceptors and photomovements in microorganisms
/ edited by F. Lenci ... [et al.].
p. cm. -- (NATO ASI series. Series A, Life sciences ; v.
211)
"Proceedings of a NATO Advanced Study Institute on Biophysics of
Photoreceptors and Photomovements in Microorganisms, held September
16-28, 1990, in Tirrenia (Pisa), Italy"--T.p. verso.
"Published in cooperation with NATO Scientific Affairs Division."
Includes bibliographical references and index.

1. Microorganisms--Motility--Congresses. 2. Phototropism-
-Congresses. 3. Photoreceptors--Congresses. I. Lenci, Francesco.
II. North Atlantic Treaty Organization. Scientific Affairs
Division. III. Title. IV. Series.
QR96.N37 1990
576'.118--dc20 91-24636
CIP

DOI 10.1007/978-1-4684-5988-3

Originally published by Plenum Press New York in 1991
MyCopy version of the original edition 1991

Dedicated to
Professor Dr. Alessandro Checcucci
and
Professor Dr. Wolfgang Haupt

Preface

This volume contains the lectures given at the NATO Advanced Study Institute on "Biophysics of Photoreceptors and Photomovements in Microorganisms" held in Tirrenia (Pisa), Italy, in September 1990. The Institute was sponsored and mainly funded by the Scientific Affairs Division of NATO; the Physical Science Committee and the Institute of Biophysics of National Research Council of Italy also supported the School and substantially contributed to its success. It is our pleasant duty to thank these institutions.

Scientists from very different backgrounds contributed to the understanding of this fast developing field of research, which has seen considerable progress during the last years. The areas of expertise ranged from behavioral sciences, supported by sophisticated techniques such as image analysis or laser light scattering, to spectroscopy, applied, in different time domains, to the study of the primary photoreactions, to electrophysiology, biochemistry or molecular biology, with the aim of analyzing the various steps of the transduction chains and how they control the motor apparatus of the cells. The organisms studied covered a wide range, from bacteria to algae, fungi and other eukaryotes.

Thus, the ASI represented a successful opportunity for carrying on and implementing an interdisciplinary approach to the study of the biophysical basis of photoreception and photosensory transduction in aneural organisms, with special attention to the basic phenomena and the underlying molecular events. We hope that this book has caught the spirit in which the ASI was conceived.

The first part of the book intends giving a general view on the subject of photomotile reactions in microorganisms and on the various experimental approaches. In particular the first chapters introduce the phenomenon of photoreception and phototransduction, together with a discussion of the molecular properties of biological light sensors, and the main behavioral aspects of the photosensory responses in different classes of microorganisms. Subsequent contributions present methods both for investigating the physical parameters of cell motion and their alteration due to the effect of light and for analyzing the spectral efficiency of the light stimulus. Absorption, steady-state and time-resolved emission spectroscopy and laser flash photolysis, as powerful methods for studying photoreceptor properties and the primary photoreactions they undergo, are then discussed in detail. Finally, the subject of environmental photobiology is briefly introduced.

The second part of the book deals with the most recent experimental results in some cases in detail, such as the flagellated algae *Euglena gracilis, Chlamydomonas reinhardtii* and *Haematococcus pluvialis,* the archaebacterium *Halobacterium halobium* and the ciliate protozoa *Stentor coeruleus* and *Blepharisma japonicum*.

During the ASI many students gave talks on their research activity which substantially contributed to giving the audience deeper insight into the problems presented in the main lectures. It is impossible to report their results in this book, but we hope to read them in the literature in the next few years.

As a final remark, we want to stress that the international scientific community has always been, even during the worst periods of cold war, an open community, based on cooperation and the free exchange of information.

We hope this ASI has served not only to advance our understanding of the biophysics of photoreceptors and photomovements of microorganisms, but also to favor cooperation, mutual confidence and trust among us and among the many different countries we come from. This can be our tiny contribution towards making our world closer to the open world Niels Bohr called for.

Francesco Lenci
Giuliano Colombetti
Francesco Ghetti
Donat-P. Häder
Pill-Soon Song

Contents

Survey of Photomotile Responses in Microorganisms

Wilhelm Nultsch

Lehrstuhl für Botanik
Fachbereich Biologie - Botanik
Karl-von-Frisch-Straße
D-3550 Marburg 1
Germany

Introduction

As light is used as energy source in photosynthesis, plants and photosynthetic microorganisms have developed several different photosensing and photomotile systems, respectively. They enable them to find places of proper irradiances and to stay there, or to bring themselves or their organs into a position in which they can capture an optimal number of photons. Moreover, the same or other photosensing systems can be used to avoid places of too high irradiances and/or UV, this way preventing the cells from photodamage. Even some non-photosynthetic organisms can respond to light stimuli, e.g., the sporangiophores of some fungi as well as some colorless flagellates and other protozoa. In the latter case photomovement may have also a photoprotective function, whereas in fungi phototropic responses facilitate the propagation of spores whose germination is light-dependent.

Unlike animals in which movements are mostly brought about with the aid of the acto-myosin-system, and as photoreceptors preferably rhodopsin-like pigments are used, there is a great variety of photomovement phenomena in plants and microorganisms as well, concerning the phenomena, the mechanisms, the photoreceptors, the mode of light action and, as far as we know, the sensory transduction.

As movements we denote locomotive movements of freely motile organisms, e.g., unicellular organisms, colonies, cell aggregates and filaments as well as changes in the position of organs or even parts of organs. Finally, even intracellular movements of cytoplasm and organelles such as chloroplasts occur.

Tropic bending can be the result of either a differential growth or differential changes of the tugor pressure on the opposite flanks of an organ. Flagellar and ciliary movements are brought about by the microtubuli system, ameboid and intracellular movements by the acto-myosin system. Gliding movements are either also caused by the acto-myosin system as probably in diatoms, or by other contractile protein filaments, as in cyanobacteria. Finally, movements can be the result of a unilateral secretion of mucilage, as in desmids and in the red alga *Porphyridium*.

Several different photoreceptors are involved in photomotile responses such as flavoproteins, rhodopsin, phytochrome, stentorin and photosynthetic pigments. Concerning the mode of response to light we distinguish light-oriented movements, tran-

Biophysics of Photoreceptors and Photomovements in Microorganisms
Edited by F. Lenci *et al.*, Plenum Press, New York, 1991

sient changes of movement velocity and direction, effects on the steady state velocity of movement and, as an indirect effect, setting the clock of circadian movements.

For the classification of photomovements mostly the mode of responses to light is used, though this system is rather descriptive and not sufficient to comprehend all the different photoresponses observed so far, as will be shown later.

Light-oriented movements

The signal perceived is the direction of light.

1. **Phototaxis** is defined as an oriented movement of locomotive microrganism with respect to the light direction. It results in a movement either toward the light source (positive phototaxis) or away from it (negative phototaxis, Nultsch and Häder, 1979, 1988). According to this definition it seems to be relatively easy to distinguish phototaxis from other photic reaction types. This, however, is true only in some cases, e.g., in the cyanobacterium *Anabaena* (Nultsch and Schuchart, 1985; Nultsch, 1985).

Anabaena: As the trichomes of Nostocaceae do not rotate during movement, they have an irradiated and a shaded flank if unilateral illumination impinges from the side, so that they bend toward the light source in dim and away from it in strong light, resulting in a movement toward the light source or away from it. In this case the light direction is perceived by a true "one instant mechanism", i.e., by simultaneous measurement on both flanks of the trichome by two or more photoreceptors.

Porphyridium: In the unicellular red alga *Porphyridium* the directness of movement toward the light source under lateral illumination can be easily demonstrated by staining the traces of mucilage (Nultsch and Schuchart, 1980). In diffuse light, the cells move in any direction and often describe a circle. Thus one should expect that cells moving toward the light source should curve into the new direction when they are illuminated from the side perpendicular to the initial light direction. In fact, this is often the case. Analyzing some movies, however, we could also observe that some cells did not turn but that the illuminated side now became the anterior end. This can be observed more often when the direction of illumination is changed by an angle of 180°. In this case most of the cells stop and resume their movement in the opposite direction without turning.

In *Phormidium* and other Oscillatoriaceae the trichomes rotate around their long axis so that the light direction cannot be measured by a one instant mechanism. In these organisms phototaxis is the result of a change of the autonomous rhythm of forward and backward movement (Nultsch, 1975). Under unilateral illumination only in those individuals being in a more or less parallel position to the light beam the movement toward the light source is prolonged, while movement away from it is shortened, during positive phototaxis and *vice versa* in negative phototaxis. However, they cannot change the movement direction actively. Consequently, a true steering is not possible, and the trichomes approach the light source rather by chance. With time the spreading area of a population is also shifted toward the light source and, finally, the trichomes accumulate on the side of the Petri dish facing the light, as in the case a true steering mechanism. Though the result is the same, it is brought about by a completely different mechanism.

Euglena: In the flagellate *Euglena* phototaxis is the result of repetitive photophobic responses. According to the classic shading hypothesis (Jennings, 1906; Mast, 1911) these responses occur under unilateral illumination when during rotation of the cells the stigma casts a shadow on the PFB (paraflagellar body) which is the presumptive photoreceptor site. More recently, Häder et al. (1987) have shown in experiments with polarized light that the photoreceptor pigments of *Euglena* are dichroically oriented. Thus, the stepwise orientation to the light direction seems to be the result of periodic dichroism rather than of aperiodic shading.

Chlamydomonas: In *Chlamydomonas* cells the light direction is perceived by modulation. The photoreceptor area which, according to Melkonian and Robenek (1980), differs in its ultrastructure from other parts of the cytoplasmic membrane is supposed to be located in the cytoplasmic membrane overlying the stigma, as already shown by Walne and Arnott (1967). During rotation (1-2 times/s) it is periodically irradiated and shaded. Therefore, it has been suggested that the orientation of the cell to the light direction may be brought about by a change of the synchrony of the flagellar beat (Nultsch, 1983), i.e., that in case of positive phototaxis the flagella beat faster while passing the shaded side and *vice versa*. Studying single cells held on micropipettes Rüffer and Nultsch (1985, 1987) observed with the aid of high speed cinematography that generally the flagella of *Chlamydomonas* beat synchronously, though asymmetrically. In a more recent paper (Rüffer and Nultsch, 1990) it has been shown that both step-up and step-down stimuli can cause changes in beat frequency, both an increase and a decrease. However, and this is important, independently whether the frequency was increased or decreased, both the flagella responded in the same way, i.e., no asynchrony was induced by the laterally impinging light. Apparently asymmetric beat frequency changes can be ruled out as orientation mechanism in *Chlamydomonas*.

Most recently we filmed the effects of unilateral irradiation impinging on both the *cis*-(stigma)- and the *trans*-side of the cell and analyzed 85 films frame by frame. We found that by both step-up and step-down stimuli two changes of the beat pattern can be caused: either the flagellum enlarges the front amplitude by further advancing the recovery bend and unrolling the flagellum in front of the cell, or the front amplitude is flatter than before stimulation so that the effective stroke starts already with a recovery bend. These two kinds are combined in various ways: *cis* changes always opposite to *trans*, and step-down opposite to step-up. Thus, phototactic orientation seems to be the result of both periodic irradiation and periodic shading, causing either step-up or step-down responses on the one side and the opposite on the other.

These examples clearly demonstrate that movement towards the light source or away from it can be brought about by different types of responses, all called phototaxis, but being different in the modes of light perception and motor responses as well. Thus the terms one and two instant mechanisms are not sufficient to define all these responses.

2. Phototropism is defined as an oriented movement of sessile organisms or their organs with respect to the light direction, brought about mostly by a differential growth or, in some cases, by differential changes in turgor pressure on the two opposite flanks of the organ. The result is a bending of the organs toward the light source (positive phototropism) or away from it (negative phototropism). Phototropism is widespread among higher plants, but occurs also in microorganisms, if we include the lower fungi in this group (Galland, 1990). At lower fluence rates the normally vertical growing sporangiophores bend toward the light source.

Light-induced transient changes in movements

In this category of photomovements changes in the photon fluence rate are perceived, either a decrease or an increase. They occur independently of the light direction.

1. Photo-phobic responses are transient changes in locomotion of microorganisms caused by a sudden increase (step up) oder decrease (step down) of the fluence rate, resulting in either a tumbling or turning of the cells, in a reversal of the movement direction or simply in a stop (Nultsch and Häder, 1979, 1988). In *Chlamydomonas* the photophobic response is caused by a snap over of the flagella that show an undulating movement for a while, but then the cells resume the forward movent in another direction. This behavior is mostly called stop-response. However, as shown by Rüffer and Nultsch (1990), photophobic responses in *Chlamydomonas* can consist also in an increase or decrease of beat frequency. Photophobic responses can be brought about either by a temporal change, e.g., when the fluence rate is suddenly changed, or by a spatial gradient, e.g., when an organism moves from an area of given fluence rate to another with a higher or lower one which is perceived by moving organisms as a temporal change as well. As a result of a step down response, they accumulate in a light field, but leave it in case of step up. Again we have a similar situation as in phototaxis. Though the movement mechanisms and, hence, the motor responses vary from organism to organism, the results are the same. The cells accumulate in areas irradiated by light of moderate fluence rates, favorable for photosynthesis, but avoid areas of high fluence rates eventually causing photodamage.

2. Photonasty denotes photomovements of organs of sessile plants caused by changes in the photon fluence rate, independent of the light direction. However, as they do not occur in microorganisms, they are out of the scope of this talk.

Effect of light on the steady state velocity of movement

In this case light can also function as a trigger, but in photosynthetic organisms it serves mostly as an energy source for movement.

1. Photokinesis is the effect of light on steady state velocity of locomotive microorganisms. If movement is initiated or accelerated, photokinesis is called positive, if it is decelerated or stopped, it is called negative. Photokinesis is independent of the light direction. The signal perceived is the fluence rate (Nultsch and Häder, 1979, 1988). As far as investigated, the action spectra indicate that the photokinetically active light is absorbed by the photosynthetic apparatus. Obviously light acts via photophosphorylation, irrespective if ATP or the proton motive force itself serve as energy source.

2. Photodinesis denotes the initiation of a protoplasmic streaming or an effect on its speed. However, it has not yet been observed in microorganisms.

References

Galland, P., 1990, Phototropism of the *Phycomyces* sporangiophore: a comparison with higher plants, *Photochem. Photobiol.*, 52, 233.

Häder, D.-P., Lebert, M., and DiLena, M. R., 1987, New evidence for mechanism of phototactic orientation of *Euglena gracilis*, *Curr. Microbiol.*, 14, 157.

Jennings, H. S., 1906, *Behavior of the Lower Organisms*, Indiana University Press. Bloomington, 1976 reprint.

Mast, S. O., 1911, *Light and Behavior of Organisms*, Wiley, New York.

Melkonian, M., and Robenek, H., 1980, Eyespot membranes of *Chlamydomonas reinhardtii*: a freeze-fracture study, *J. Ultrastruct. Res.*, 72, 90.

Nultsch, W., 1975, Phototaxis and photokinesis. *In* "Primitive Sensory and Communication Systems," Carlile, M. J., ed., Academic Press, London, pp. 29.

Nultsch, W., 1983, The photocontrol of movement of *Chlamydomonas*. *In* "The Society for Experimental Biology Symposium XXXVI," Cosens, D. J., and Vince-Prue, D., eds., Society for Experimental Biology, Great Britain, pp. 521.

Nultsch, W., 1985 , Photosensing in cyanobacteria. *In* "Sensory Perception and Transduction in Aneural Organisms," Colombetti, G., Lenci, F., and Song, P.-S., Plenum Press, New York, pp. 147.

Nultsch, W., and Häder, D.-P., 1979, Photomovement of motile microorganisms, *Photochem. Photobiol.*, 29, 423.

Nultsch, W., and Häder, D.-P., 1988, Photomovement in motile microorganisms-II, *Photochem. Photobiol.*, 47, 837.

Nultsch, W., and Schuchart, H., 1980, Photomovement of red alga *Porphyridium cruentum* (Ag., Naegeli.) II. Phototaxis, *Arch. Microbiol.*, 125, 181.

Nultsch, W., and Schuchart, H., 1985, A model of the phototactic reaction chain of the cyanobacterium *Anabaena variabilis*, *Arch. Microbiol.*, 142, 180.

Rüffer U., and Nultsch, W., 1985, High-speed cinematographic analysis of the movement of *Chlamydomonas*, *Cell Motility Cytoskeleton*, 5, 251.

Rüffer, U., and Nultsch, W., 1987, Comparison of the beating of *cis* and *trans* flagella of *Chlamydomonas* cells held on micropipettes, *Cell Motility Cytoskeleton*, 7, 87.

Rüffer, U., and Nultsch, W., 1990, Flagellar photoresponses of *Chlamydomonas* cells held on micropipettes: I. Change in flagellar beat frecency, *Cell Motility Cytoskeleton*, 15, 162.

Walne, P. L., and Arnott, J. J., 1967, The comparative ultrastructure and possible function of eyespots: *Euglena granulata* and *Chlamydomonas eugametos*, *Planta*, 77, 289.

Introduction to Photosensory Transduction Chains

Wolfgang Haupt

Institut für Botanik und Pharmazeutische Biologie
Universität Erlangen-Nürnberg
Germany

Introduction

All life on earth is eventually based on sunlight as the source of energy. This is a very trivial statement. Besides, however, light has many effects on organisms that cannot simply be explained by a transformation of light energy into chemical energy for the metabolism or for mechanical work. A classical example is an etiolated plant, which only after a short pulse of light starts to synthesize chlorophyll, anthocyanin, many new enzymes, and which develops new organs. A selection of *photoresponses* is listed in Table 1.

In many of these cases it can easily be shown that the response requires much more energy than was provided by the short irradiation, and this is most obvious in the case of photomorphogenesis of an etiolated seedling as mentioned before. Thus, in such a case light does not provide *exogenous energy* for the metabolism, but it acts as a *signal*, which in some way controls the metabolism of *endogenous energy*, i.e., it redistributes metabolic energy; this finally results in the control of biosyntheses, development and behavior.

Table 1. Photoresponses of plants and microorganisms making use of light as a signal rather than as a source of energy for metabolism

photomorphogenesis, e.g.
deetiolation
phototropism
photomovement, e.g.
photophobic response
phototaxis
chloroplast orientation
photoreactivation
photoperiodic responses, e.g.
induction of flowering
induction of or release from dormancy

Biophysics of Photoreceptors and Photomovements in Microorganisms
Edited by F. Lenci *et al.*, Plenum Press, New York, 1991

From these considerations it becomes evident that in terms of energy there is a substantial amplification from light perception to response, and this amplification may easily amount to a factor of thousand or more. Thus, there must be a connecting link between *perception* of the signal, i.e., the *input*, and the terminal *response*, i.e., the *output*. This connection is called *sensory transduction* (Fig. 1); it is usually thought to be a complex of reactions, comprising several steps in sequence, and is therefore called *transduction chain*.

For most of the responses of microorganisms or plants to light, the single steps of the transduction chain and their interdependence are still unknown, and it is the aim of the present school to discuss typical examples where some progress appears promising to identify and to characterize the contents of this *black box*. This introductory lecture will present two examples of photoresponse in unicellular systems, which may serve to point out main features of transduction chains, methods of approach, but also problems and limitations, although the second example does not deal with a movement. This may help, then, in analyzing transduction chains in other systems.

Chloroplast Orientation Toward Light

The first example to be discussed is the orientation movement of a chloroplast with respect to the light direction (for details, cf. references in Haupt and Wagner, 1984; Haupt and Scheuerlein, 1990). The green alga *Mougeotia* has its cells arranged in series, forming a filament; each cylindrical cell contains one single ribbon-shaped chloroplast, which orients its face to low-intensity light or its profile to high-intensity light (Fig. 2 A, B). Whenever intensity and/or direction of light change, the chloroplast reorients by a proper movement. For the present example, the reorientation to the face position will be considered (Fig. 2 C).

In the laboratory, reorientation can be induced already if light is applied as a short flash only, e.g., in the microsecond range. Movement, then, starts a few minutes thereafter and can be observed during the next 30 min. Thus, the terminal response proceeds some time after the light signal has acted (cf. also the second example below, Figs. 7 and 9); i.e., input and output are separated in time, and it is an important characteristic of the presumed transduction chain to *connect perception and response in time*.

To characterize *perception*, the primary question concerns the *photoreceptor pigment*, which in photobiology is tried to be answered by running an action spectrum. In *Mougeotia*, its peak is in the red region around 660 nm; in addition, a red flash or pulse can be made completely ineffective by a subsequent irradiation with far red, i.e., light above 700 nm. In combination this is proof for phytochrome as the photoreceptor pigment.

Phytochrome can exist in two forms, the red light absorbing Pr and the far red-light absorbing Pfr (Fig. 3). Either form is reversibly phototransformed into the other one by absorption of the corresponding light quality. Moreover, Pr is considered as

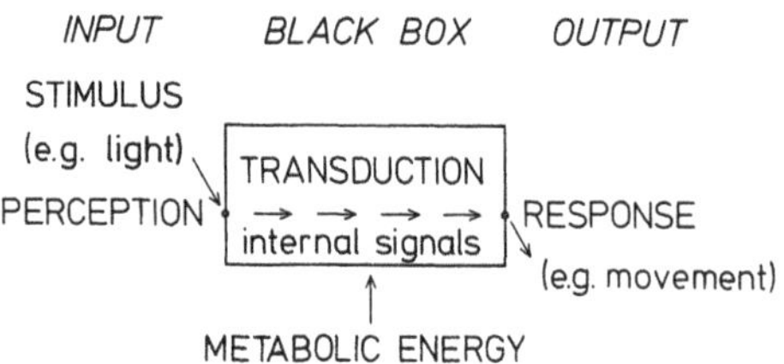

Fig. 1. General and simplified scheme of a sensory transduction chain.

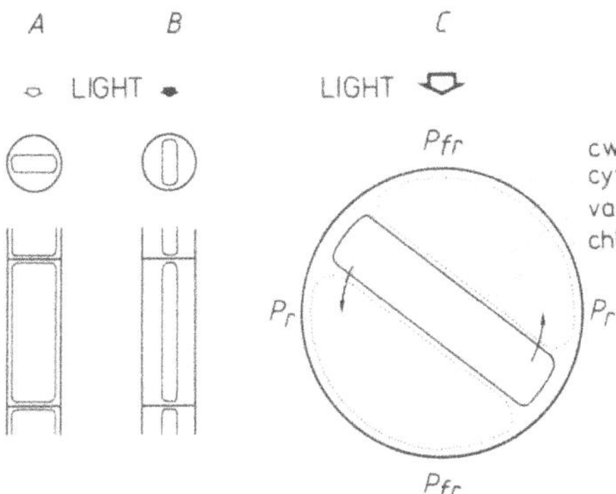

Fig. 2. Chloroplast orientation in *Mougeotia*. A. and B. Cell in cross section (above) and in surface view (below), with the chloroplast in low-intensity (A) and high-intensity position (B), respectively. C. Simplified cytomorphology of the cell in cross section with the chloroplast moving to the low-intensity position in a gradient of phytochrome Pfr. The prevailing forms of phytochrome due to the unidirectional light signal are indicated for some regions. cw = cell wall; cyt = cytoplasm; va = vacuole; chl = chloroplast.

being physiologically inactive, but Pfr as being active. In other - more trivial - words: phytochrome is a physiological switch, which can be "off" (Pr) or "on" (Pfr), depending on the light signal. Thus, we now can adequately characterize the perception process as the phototransformation of phytochrome molecules from the "off" form to the "on" form. It is important to note that the "on" form Pfr is relatively stable, in *Mougeotia* even for several hours (Kraml et al, 1987). Thus, ihe information about the light pulse is stored in Pfr for its further action, and Pfr can be considered as an *internal signal* that starts the transduction chain.

However, for *reorientation* to light, the cell has to know not only that there was a light signal a few minutes ago, but also what its direction was. Thus, characterization of the perception process has also to consider the directional aspect. Without presenting the underlying experiments, it should simply be stated that in *Mougeotia* phytochrome is localized close to the cell surface and that by physical reasons unidirectional light transforms more phytochrome to the Pfr form at the proximal and the distal surface than at the flanks (cf. Fig. 2 C); this gradient of Pfr stores the information about the light direction, necessary for the reorientation movement.

Next, the output has to be characterized, i.e., the *terminal response*. It can be shown that the edge of the chloroplast, merging into the peripheral cytoplasm, moves with respect to the outermost layer of cytoplasm. Thus, the motive force must be exerted there (Fig. 4). From various inhibitor experiments it has been concluded that an interaction of actin and myosin is responsible for generating the motive force. Moreover, actin microfilaments have been found just in those parts of cytoplasm that are close to the chloroplast's edge. As the chloroplast moves in a given direction, there must be an *asymmetry* in the activity of actin (Fig. 4). For full understanding of the mechanism of response, we have to know what this activity means (see below). No final answer can be given yet.

With these pieces of knowledge we can now address the proper questions for the *transduction chain*: How is the information, residing in the Pfr gradient, transformed into a gradient of actin activity? One observation is important at the beginning: If during an ongoing movement phytochrome is "switched off", i.e., a far red pulse is given

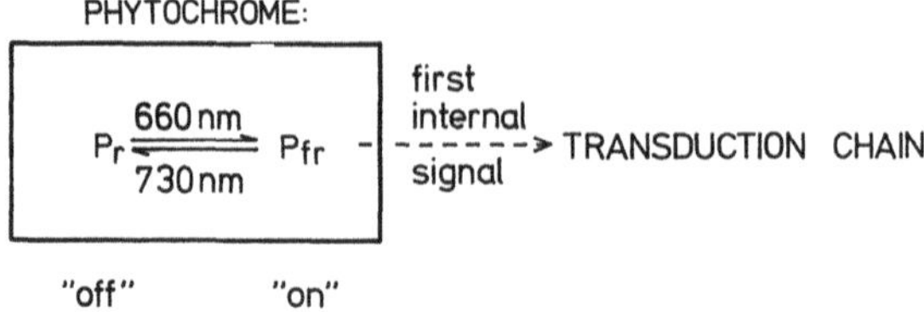

Fig. 3. Phytochrome as a molecular switch. Pr and Pfr denote the red-absorbing and the far red-absorbing form of phytochrome, respectively. The phototransformations are obtained with red = 660 nm and far red = 730 nm.

to retransform Pfr to Pr, the movement stops immediately, i.e., in less than a minute (Haupt and Übel, 1975). Thus, the information is stored only in Pfr, no site of storage can be assumed in the transduction chain behind Pfr. This allows the conclusion that the transduction chain is short and/or its single steps are fast processes.

From our knowledge about actin-myosin interaction in animals and its control by calcium, it is reasonable to speculate about calcium as controlling actin activity also in the *Mougeotia* cell, and to ask for a possible effect of phytochrome on the activity of calcium. Indeed, there is good experimental evidence that calcium is involved in sensory transduction in *Mougeotia*, and that it becomes effective by its interaction with the calcium-binding protein calmodulin. Thus, a minimum model of the transduction chain can be established:

Pfr → calcium → calmodulin → actin activity.

This model, however, is nothing but a basis to ask good questions for further research. e.g.:

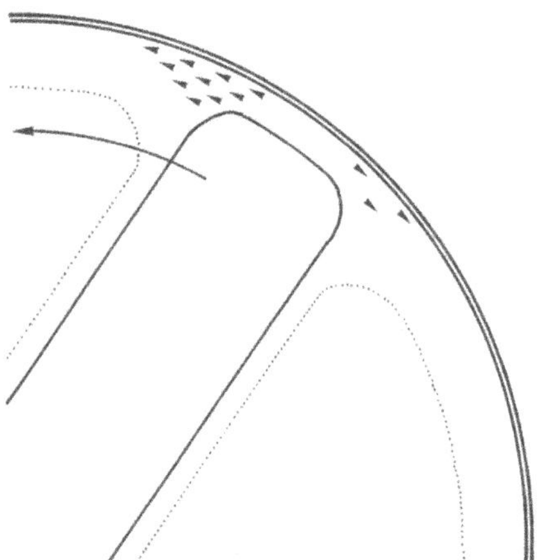

Fig. 4. Part of a cross section through a *Mougeotia* cell, schematic (cf. Fig. 2), showing the chloroplast movement (large arrow) and the postulated asymmetric action of actin microfilaments (small arrowheads) as underlying this movement.

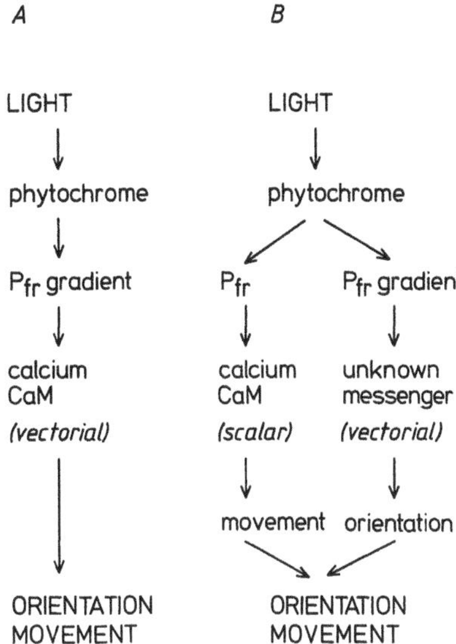

Fig. 5. Simple (A) and complicated (B) model for the transduction chain of light-oriented movement of the *Mougeotia* chloroplast. In A, the light signal and its direction is transduced by one series of steps only. In B, the signal in general (scalar) and its direction (vectorial) are transduced by two different series of steps, with their integration resulting in the oriented movement. CaM = calmodulin.

i) What kind of actin activity is controlled by calcium/calmodulin? Interaction of actin with myosin? Anchorage of actin to its reference cell structures? Stage of polymerization of G-actin to F-actin?
ii) How is the calcium activity controlled by Pfr? By controlling uptake from the external medium? By controlling release from internal calcium vesicles? By regulating calcium pumps? By controlling its interaction with calmodulin?
iii) Is calcium an internal messenger only for the light signal in general, or also for its directionality (Fig. 5)? In other words: Is there any need to assume a directional messenger, in addition to calcium as a purely scalar messenger? Interestingly, this latter alternative would mean that the transduction chain is not a single linear sequence (a "chain") of steps, but is branched into two chains, one containing scalar information only, the other one the vectorial information (Fig. 5 B).

These problems or questions become still more complex if we add another observation: The response can be induced also by blue light, which acts via a separate photoreceptor pigment, probably a flavoprotein. The main question is: What parts of the transduction chains are common to both these different inputs - or: where do the two transduction chains fuse (Fig. 6)? A more specific question concerns the storage of the information, which is also found after a blue light pulse.As in chloroplast reorientation there is no indication for a long-living reduced form of flavin resulting from a blue light pulse, the photoreceptor molecule is assumed to return to its ground state after excitation within very short fractions of a second. Thus, in contrast to phytochrome with its metastable "active form" Pfr, the information is not stored in the photoreceptor molecule proper, but the blue light-started transduction chain is postulated to contain an early step ("X") which acts as the "memory". This step is not found in the red light-started transduction chain, as there was no indication for storage of information behind Pfr (see p. 10). For further information, cf. Haupt (1991).

Spore Germination in Ferns

In our second example, the induction of fern-spore germination by light, the transduction chain proceeds much more slowly and thus allows for time-resolving experiments. This system is less complicated insofar as it does not require directional information and as it makes use of one photoreceptor pigment only; but it is more complicated insofar as it involves, in addition, the concept of "competence", as will be shown below.

For a dormant spore to start germination, it is not sufficient to be imbibed in water, but it must receive a light signal (cf. Furuya, 1983; Raghavan, 1989). As in our first example, a very short pulse or flash can saturate the response already; phytochrome is the photoreceptor pigment as well (cf. Fig. 3), it is probably located in the peripheral cytoplasm, and Pfr is the first internal signal.

The terminal response is germination, but this output is rather complex. In classical work (cf. Mohr, 1956), three parameters are used to characterize a spore as germinating: swelling and rounding of the spore by uptake of water, greening of the etioplasts, and breaking of the exospore. Other parameters can be added, as, e.g., mitosis, initiation and growth of rhizoids (cf. Scheuerlein et al., 1988; Dürr and Scheuerlein, 1990). It is evident that at least for some of these parameters the site of response is located apart from the perception site; thus, the transduction chain has to *connect* perception and response not only in time, but also *in space*, and this can be taken as a widespread general characteristic of transduction chains.

As a complication, it has been found that not all of these parameters are always combined, and thus the transduction chains are not identical (cf. Fig. 9). They therefore are assumed to branch somewhere after Pfr (Dürr, personal communication).
In order to avoid this complication, we concentrate on one parameter, i.e., chlorophyll formation in the etioplasts, which can be detected very early by the conspicuous chlorophyll fluorescence (Scheuerlein et al., 1988). We apply a pulse of red light, thus

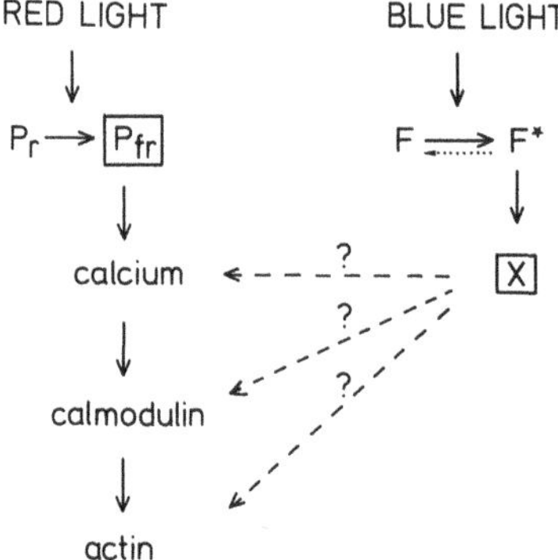

Fig. 6. Postulated transduction chains for red and blue light as the signal for the chloroplast orientation in *Mougeotia*. P = phytochrome (cf. Fig. 3), F = flavin, F^* = excited flavin, X = unknown substance. Letters in the rectangles denote the "storage" forms at the beginning of the transduction chains; possible transient steps are omitted, as well as the excited state in phytochrome phototransformation.

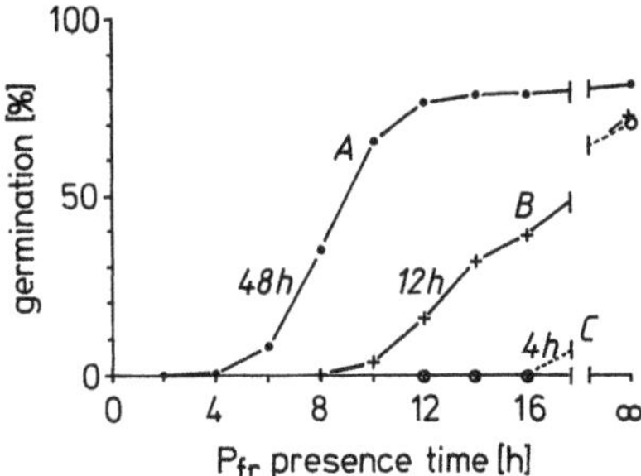

Fig. 7. Pfr-escape curves for the spore germination of *Dryopteris filix-mas*. A saturating red pulse (662 nm) is given 4, 12 or 48 h after sowing, as indicated at the curves, and the presence of Pfr is terminated by a saturating far red pulse after various periods (abscissa: interval between red and far red). After Haupt and Psaras (1989).

establishing Pfr, and after various intervals Pfr action is terminated by far red. Fig. 7 (curve A) shows that Pfr has to be present during about 12 h in order to enable chlorophyll synthesis in most of the spores of the population. Thus, during this period the transduction chain is under the control of Pfr at least for part of the spore population, but thereafter it has escaped Pfr control in the whole population. Obviously, a first step of the transduction chain has passed, and this step can be characterized as *coupling of Pfr to the transduction chain*. But thereafter it takes about another two days until chlorophyll formation becomes visible.

Thus, we can distinguish between a *coupling phase* and a *postcoupling phase*, the latter being independent of the presence of Pfr. It might be added that in these experiments red light has been applied not before one or two days have elapsed after sowing, thus allowing for a *preinduction phase*, too. It is a challenge to analyze the pro-

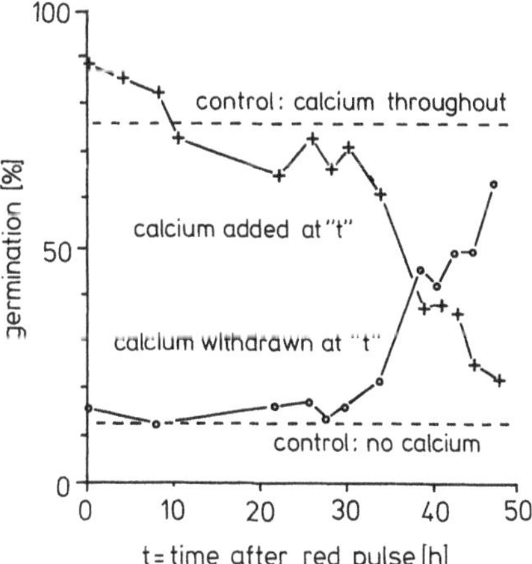

Fig. 8. Kinetics of the calcium effect on the light-induced spore germination of *Dryopteris paleacea*. Spores were sown in medium with and without calcium (controls). Germination was induced by a saturating pulse of red light (662 nm), and at various times thereafter (t, abscissa) calcium was added or withdrawn. The calcium concentration was 10^{-3} M. After Dürr and Scheuerlein (1990).

cesses occuring during these periods. A useful tool has been found by the observation that several external factors can strongly modify the effectivity of the transduction chain. Among them, the composition of the imbibition medium is important.

Modifying Effects of Nitrate and Calcium

Without nitrate, the response is strongly reduced, i.e., a smaller percentage of the spores will germinate (Haas and Scheuerlein, 1990); for a good effect, micromolar concentrations are already sufficient. When nitrate is applied or omitted during one of the phases only, its effect can be located mainly in the coupling phase, but it extends a few hours into the postcoupling phase. Thus, it can be assumed that the nitrate effect is closely related to the primary reaction of Pfr, but not restricted to the coupling process proper (see below; cf. also Fig. 9).

Another effective ion is calcium. For most of the spores, presence of calcium in micro- to millimolar concentrations is an absolute requirement for the light induction to become effective (Dürr and Scheuerlein, 1990). Remarkably, however, calcium is completely ineffective during the coupling phase (and during the preinduction phase as well); moreover, it is ineffective during 30 h after the red pulse, in which period the intracellular calcium concentration depends quantitatively on the calcium concentration of the culture medium, as determined with the fura-2-staining method (Scheuerlein et al., 1991). Instead, the action of calcium is limited to the period between 30 and 50 h after red (Fig. 8), it is centered around 20 h after coupling has occured. There is some evidence that for a single spore a period of only one hour or below exists in which calcium is required and can act. The start of this period is determined by the red pulse, more precisely by switching on the action of Pfr; and if the sensitive period is passed without calcium being present, the possibility of the calcium effect in question is irreversibly lost (Dürr and Scheuerlein, 1990).

Apart from detailed interpretations, based on sophisticated experiments of Scheuerlein's group, two conclusions are immediately obvious:

i) In contrast to recent hypotheses on the mechanism of Pfr action (e.g., Roux et al., 1986), calcium is not a second messenger located immediately after Pfr and thus not *starting* the transduction chain (cf. also Iino et al., 1989; Scheuerlein et al., 1989).
ii) The time course of calcium action allows to *subdivide* the postcoupling phase into a) an early, calcium-independent phase; b) a rather short calcium-dependent phase; and c) a late phase that appears again independent of calcium (Fig. 9). These results might be a promising tool to further analyze and characterize the obviously complex postcoupling phase.

So far, the *all-or-none* effect of calcium concerns two light-induced parameters, viz., appearance of chlorophyll in the etioplasts and initiation of rhizoid formation. In nature, however, "greening" involves not only chlorophyll formation in the already existing etioplasts, but also formation of new chloroplasts with their photosynthetic pigments. Provided Pfr has started the postcoupling processes and calcium in the sensitive phase has allowed chlorophyll formation in the etioplasts, the formation of additional chloroplasts requires calcium once more. Interestingly, this developmental process depends on the calcium concentration in a *quantitative* manner, and the action of calcium is not restricted to a short period, but can become effective at any time in the later postcoupling phase (Dürr and Scheuerlein, 1990). A similar situation appears to hold for the growth rate of the rhizoid after it has been initiated as an all-or-none response (Scheuerlein, personal communication).

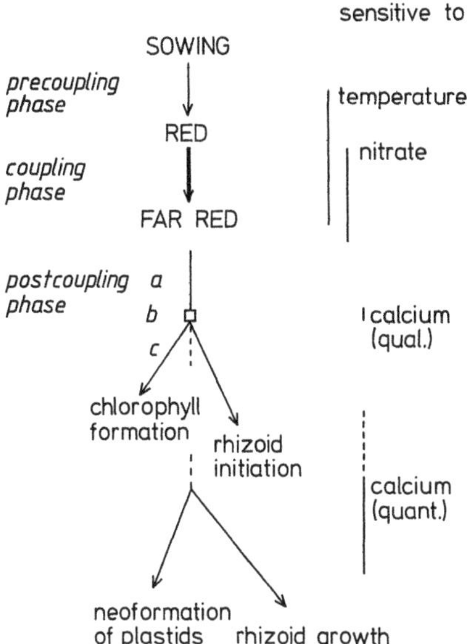

Fig. 9. Phases in the transduction chain of fern-spore germination and their sensitivities to elevated temperature, nitrate and calcium. For the latter, the qualitative and quantitative effects are distinguished. Presence of Pfr shown by the heavy arrow. The open square denotes the short qualitative sensitivity to calcium, dividing the postcoupling phase into the subphases a, b and c. After Dürr and Scheuerlein (1990), Haas and Scheuerlein (1990), Haupt (submitted).

This dual role of calcium clearly shows a fundamental problem: the analysis of a sensory transduction chain strongly depends on the parameter chosen as the terminal response, and conclusions drawn from experiments with ill-defined responses should be considered with care. We will come back to this problem below (p. 17).

Modifying Effect of Temperature, and the Concept of Competence

Besides nitrate and calcium, the temperature can modulate the light-induced germination. But before this temperature effect will be discussed, the concept of *competence* has to be introduced. In the basic experiments as presented in Fig. 7 A, the light pulse was given 48 h after sowing. If, instead, this pulse is given at various times shortly after sowing, and Pfr allowed to act infinitely, full effectivity is achieved within 5 minutes after sowing. Thus, phytochrome becomes competent to *perceive* the light signal almost instantly. But can *transduction* start that early, too?

To answer this question, Pfr-escape curves are run at various times after sowing. Under standard conditions, i.e., two days after sowing, the most sensitive spores require about 6 hours for coupling, but the least sensitive spores have coupled not before 12 h after the light pulse (see above, Fig. 7, curve A). If, however, the same experiment is performed after a shorter preinduction phase, there is an extended lag phase (curve B for the interval of 12 h), and this amounts to more than 16 h if the light is given 4 h after sowing already (curve C). The whole escape curve is shifted accordingly. This means, although Pfr is present very early, it starts coupling at a later time. At least during the first 16 h the spore is not yet able to couple, it is not *competent* for the action of Pfr (or, in other words, competent to transduce the signal stored in Pfr). Thus, the *preinduction phase* or, more adequately, *precoupling phase* comprises time-consuming processes; this may be interpreted as showing that something has to be produced that is required for the process of coupling. Further characterization of this early phase might be possible if the time-resolved experiments on the action of external factors are extended to it, although strictly speaking it is not part of the transduction chain, but it precedes the latter.

As has been mentioned before, there is no effect of nitrate and calcium during the precoupling phase. There is, however, an inhibiting effect of an elevated temperature, viz. 27° C or 32° C instead of 22° C, which is well pronounced in the precoupling phase (Fig. 10 A, cf. Haupt, in preparation). Moreover, if the temperature treatment is restricted to part of a precoupling phase of 48 h, its effect is most obvious in the second 24 h subperiod (Fig. 10 B). This means that something is destroyed or inactivated that is required for coupling, assumedly the reaction partner of Pfr.

Interestingly, the temperature effect is found also during coupling, but there is little effect thereafter - if at all (Fig. 10 A; cf. also Fig. 9). This is consistent with the above interpretation: even during coupling the reaction partner could well be sensitive to be inactivated by the high temperature.

Alternatively, this extension of the effect from one phase to the next can be interpreted by an overlapping of phases in the population. We remember from Fig. 7 that some of the spores have completed coupling after a few hours already, but that others couple much later. Thus, in the early hours after the red pulse the population comprises a mixture of spores being in the coupling phase and others still in the precoupling phase (Fig. 11). In consequence, a factor that is only effective in the precoupling phase of the individual spore, may appear to affect also the coupling process if considered on the population level. Thus, the above-mentioned additional temperature effect during presence of Pfr is not proof for its affecting the coupling process proper. On the other hand, in the later hours after red light, it can be predicted that there is a corresponding mixture of spores in the coupling phase and in the postcoupling phase. For the effect of nitrate, as discussed above, this could mean that in fact this ion

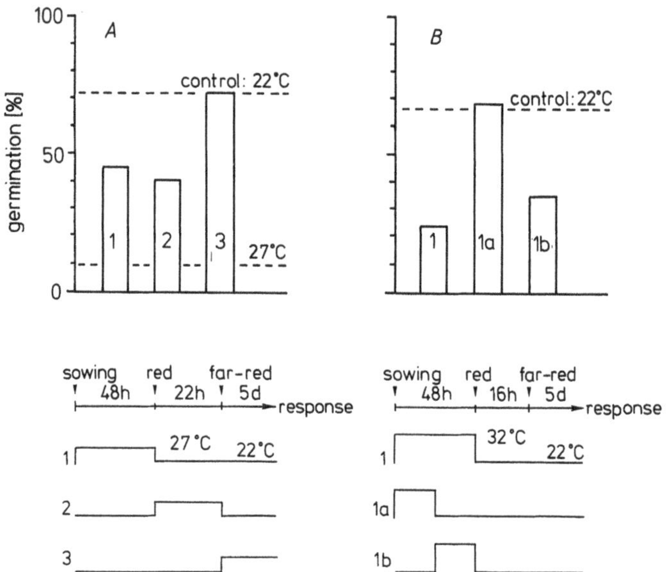

Fig. 10. Inhibiting effect of elevated temperatures on the light-induced spore germination of *D. filix-mas*. In the controls, the spores were kept at the indicated temperature all the time; the columns show treatments with the elevated temperature for limited periods as explained by the lower part of the figure. The red pulse was given at 688 nm, thus establishing a Pfr level that is not fully saturating for germination.

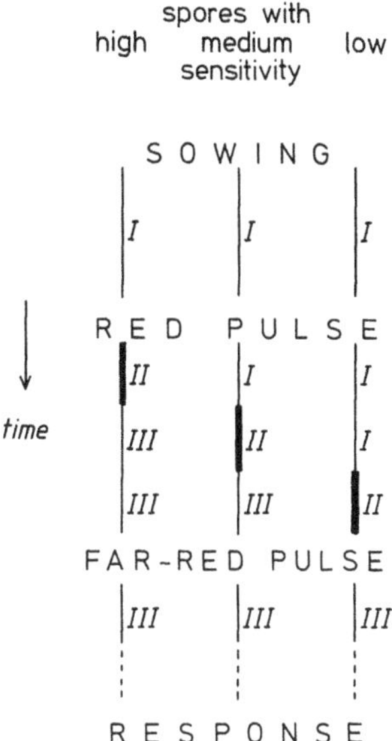

Fig. 11. Schematic comparison of the time course of the various phases within a population of *Dryopteris* spores. I, II, and III denote precoupling, coupling and postcoupling phase, respectively, for individual spores with various sensitivities (or responsivities). Coupling phase indicated by the heavy line. Notice that for the individual spore phases I and III can extend into the coupling phase of the population.

does not act on the coupling process, but shortly thereafter. As long as it is necessary to work with populations, these complications cannot be completely circumvented, but they may be strongly reduced by shortening the phase of presence of Pfr so as to work with a sensitive fraction of the population only and thus to make overlapping as small as possible.

Final Remarks

We have discussed two systems that appeared relatively simple. Yet, it turned out that a transduction chain and/or its investigation is much more complicated than one might have anticipated. Among the problems were, e.g.: two separate input systems (red vs. blue light receptors), which somewhere fuse in the transduction chain (Fig. 6); vice versa, branching of the chain to separate output systems (chlorophyll formation vs. rhizoid initiation, Fig. 9); as a result, dependence of the conclusions on the parameter chosen as response; separate transduction of diverse parameters of the external signal (light quality to operate the switch, vs. light direction, Fig. 5); more than one point of attack for a modifying factor (calcium, Fig. 9); overlapping of phases in a population (Fig. 11).

If the concept of perception-transduction-response is applied to more complicated systems, the problems and difficulties may strongly increase. As an example, phototropism of etiolated grass coleoptiles may be mentioned (cf. Firn, 1990) with a rather trivial problem: What is the terminal response after a unilateral light pulse? It is, of course, the curvature of the organ towards the light - but is it really? The curvature is brought about by differential growth of opposite flanks as the latest step in the transduction chain. Theoretically, this differential growth can be due to increase and/or decrease of the growth rate, and accordingly it has recently been concluded that measurement of these growth rates is a much more adequate and much more reliable parameter to characterize the response than is just the angle of curvature. Modulation of growth rate is generally assumed to be the result of unequal distribution of auxin. Thus, a hormone physiologist might be interested mainly in that redistribution, and we have again a different parameter for terminal response. But this may lead to a completely wrong direction of research, as recently there is increasing doubt whether indeed auxin plays a key role in phototropism.

This finally results in the fundamental doubt as to whether the concept of perception-transduction-response can be a useful tool at all for investigating this phototropism rather than to impede progress (Firn, 1990). This skeptic view has to be taken very seriously for other photoresponses, too, as a warning: the concept must not be applied without carefully considering its limitations. However, there is no reason to abandon the concept; rather, the skepticism of Firn should be taken as a challenge to always be aware of the problems involved in the individual case and of the importance of properly select the parameters and the terminology. Under these conditions, consequent application of the perception-transduction-response concept can become very useful to ask good questions; its usefulness for further analyzing the causality of photoresponses should have become obvious from the two examples in this chapter.

References

Dürr, S., and Scheuerlein, R., 1990, Characterization of a calcium-requiring phase during phytochrome-mediated fern-spore germination of *Dryopteris paleacea* Sw., *Photochem. Photobiol.*, 52:73.

Firn, R. D., 1990, Phototropism - the need for a sense of direction?, *Photochem. Photobiol.*, 51:255.

Furuya, M., 1983, Photomorphogenesis in ferns, *in*: "Encyclopedia of Plant Physiology," N. S., vol. 16, Shropshire, W., jr., and Mohr, H., eds., Springer, Berlin Heidelberg, New York, Tokyo, p. 569.

Haas, C. J., and Scheuerlein, R., 1990, Phase-specific effect of nitrate on phytochrome-mediated germination in spores of *Dryopteris filix-mas* L., *Photochem. Photobiol.*, 52:67.

Haupt, W., 1991, Phytochrome and cryptochrome: coaction or interaction in the control of chloroplast orientation, *in*: "Photobiology: The Science and its Application", E. Riklis, ed., Plenum Press, New York, London, p. 479.

Haupt, W., and Psaras, G. K., 1989, Phytochrome-controlled fern-spore germination: Kinetics of Pfr action, *J. Plant Physiol.*, 135:31.

Haupt, W., and Scheuerlein, R., 1990, Chloroplast movement, *Plant, Cell Environm.*, 13:595.

Haupt, W., and Übel, H., 1975, Zum Mechanismus der Phytochromwirkung bei der Chloroplastenbewegung von *Mougeotia*, *Zeitschr. Pflanzenphysiol.*, 75:165.

Haupt, W., and Wagner, G., 1984, Chloroplast movement, *in*: "Membranes and Sensory Transduction," Colombetti, G., and Lenci, F., eds., Plenum Press, New York, London, p. 331.

Iino, M., Endo, M., and Wada, M., 1989, The occurence of a Ca^{2+}-dependent period in the red-light-induced late G1 phase of *Adiantum* spores, *Plant Physiol.*, 91:610.

Kraml, M., Leopold, K., and Winkler, B., 1987, Long-lasting activity of Pfr and Pfr gradients in *Mougeotia* chloroplast movement?, *Acta Physiol. Plant.*, 9:189.

Mohr, H., 1956, Die Beeinflussung der Keimung von Farnsporen durch Licht und andere Faktoren, *Planta*, 46:534.

Raghavan, V., 1989, "Developmental Biology of Fern Gametophytes," Cambridge Univ. Press, Cambridge, New York, Port Chester, Melbourne, Sydney.

Roux, S. J., Wayne, R. O., and Datta, N., 1986, Role of calcium ions in phytochrome response: an update, *Physiol. Plant.*, 66:344.

Scheuerlein, R., Wayne, R., and Roux, S. J., 1988, Early quantitative method for measuring germination in non-green spores of *Dryopteris paleacea* using an epifluorescence-microscope technique, *Physiol. Plant.*, 73:505.

Scheuerlein, R., Wayne, R., and Roux, S. J., 1989, Calcium requirement of phytochrome-mediated fern-spore germination: No direct phytochrome-calcium interaction in the phytochrome-initiated transduction chain, *Planta*, 178:25.

Scheuerlein, R., Schmidt, K., Poenie, M., and Roux, S. J., 1991, Determination of cytoplasmic calcium concentration in *Dryopteris* spores: A developmentally non-disruptive technique for the loading of the calcium indicator fura-2, *Planta*, in press.

Addendum

An important sentence had been lost when the author rearranged the text referring to the model (Fig. 5) for *Mougeotia*:

Although the obligatory involvement of calcium and calmodulin in the transduction chain had carefully been worked out by Wagner, some inconsistencies with results of Schönbohm have still to be resolved; for details see Haupt and Scheuerlein (1990) and references therein.

Molecular Properties of Biological Light Sensors

Pill-Soon Song

Institute for Cellular and Molecular Photobiology
Department of Chemistry
University of Nebraska
Lincoln, NE 68588
USA

Satoshi Suzuki

Department of Industrial Chemistry
Shinshu University
Nagano 380
Japan

Il-Doo Kim

Department of Chemistry
Chosun University
Kwangju
Korea

Ja Hong Kim

Department of Chemistry
Cheonbuk National University
Cheonju
Korea

Introduction

As the source of energy and as an environmental factor, light has played a crucial role in selection and adaptation processes in chemical and organismic evolution. Organisms ranging from prokaryotic bacteria to eukaryotic mammals directly absorb light of varying wavelengths for energy supply, for survival and/or light sensory signal transduction. Efficient absorption of a specific wavelength of light by the photoreceptor/light sensor molecules triggers a variety of photobiological responses in different organisms. Figure 1 is an attempt to demonstrate the diversity of photosensor molecules and their corresponding light absorbance characteristics, particularly specific wavelength light for absorbance maximum. Some organisms such as the firefly are capable of converting chemical energy to light energy (bioluminescence). This lighting phenomenon is also included in what we might call the "photobiological spectrum" (Fig. 1). Figure 2 illustrates the simplest route for light absorption by a photosensor molecule and for the resulting excitation of the molecule which initiates a sensory transduction chain in photobiological responses of organisms.

In this chapter, several factors that determine light absorption properties of photoreceptors, specifically absorbance wavelength maximum, absorption intensity (molar extinction coefficient or oscillator strength) and polarization, will be discussed with the

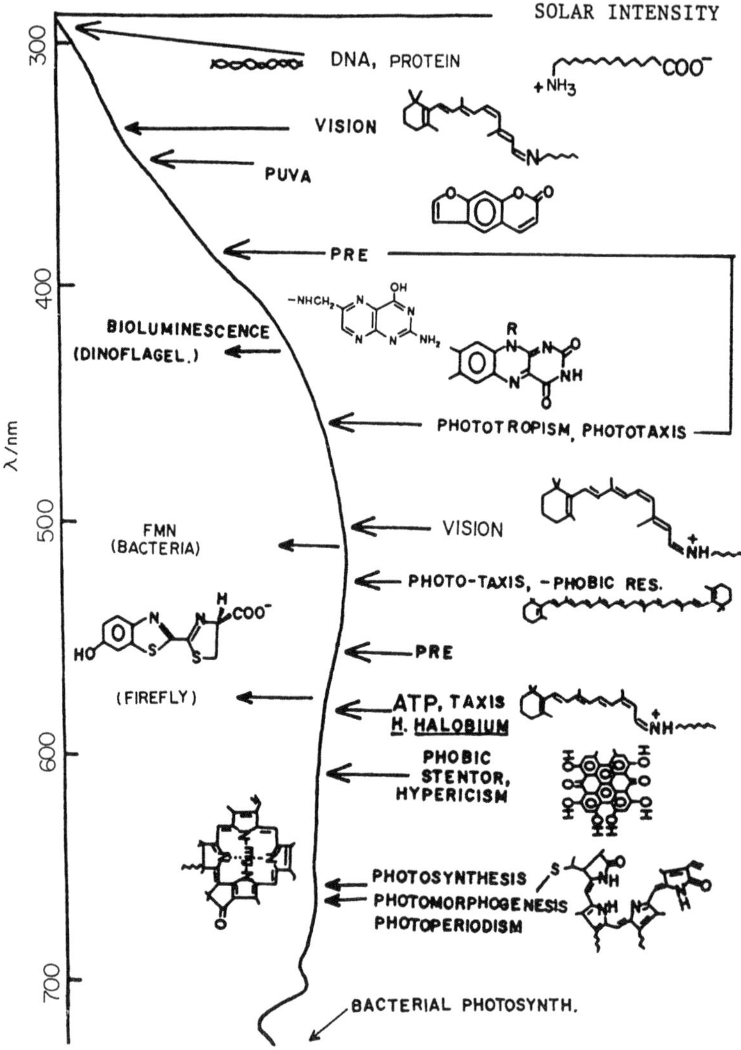

Fig. 1. A photobiological spectrum showing solar radiation spectrum (solid line) and photoresponses at the wavelength maximum of photosensitivity (arrows), corresponding to the photobiological action spectral maximum. Arrows below the solar spectrum represent bioluminescence. The chemical structure of a possible photosensor molecule/chromophore for each photoresponse reaction is indicated, corresponding to the photobiological action spectral maxima (Modified from Song, 1983a).

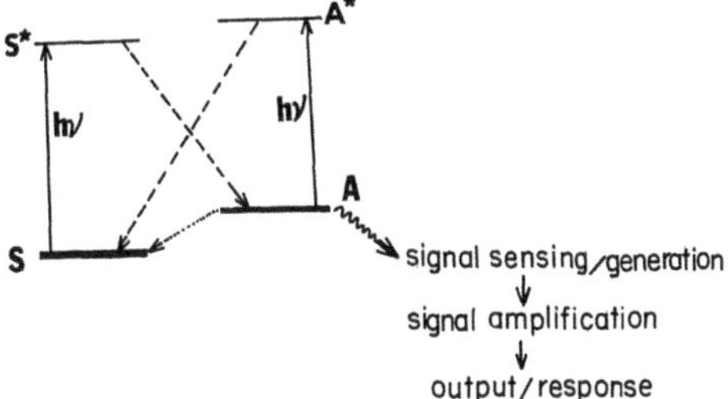

Fig. 2. A simplest scheme for the electronic excitation of photosensor molecules and for the energy routing of excitation energy to the sensory signal transducing chain in light-sensitive organisms. S; primary sensor molecule. A; activated/metastable intermediate or product which can be cycled back to "S" via photoreversion and/or dark reversion (dotted line). Both rhodopsins including sensory and phytochrome work according to this scheme.

aim of reconciling the wide spectral distribution of photobiological sensor molecules covering the entire UV and visible spectrum of solar radiation as shown in Figure 1. Absorption spectra give us information about the transition energy (cf. Fig. 2), transition probability (intensity) and direction of transition dipole (polarization) in a given photobiological sensor molecule.

UV-absorbing chromophores including DNA, RNA and aromatic amino acid residues will not be discussed in this chapter. The student is referred to a previous review on the spectroscopy of the UV-absorbing chromophores and their photobiological implications (Song, 1983a).

Electronic Structure and $\pi \rightarrow \pi^*$ Transitions

From the photobiological spectrum shown in Figure 1, it is clear that the size and extent of the π-electron conjugated system of the chromophore molecules increases with the photobiological action spectral maximum for each photoresponse indicated by an arrow. Although this trend of red shift of the absorbance/action maximum (bathochromism) with increase in the size and extent of π-conjugation of the molecule is well recognized by the student of photobiology, he/she may not be well aware that a parallel increase in absorption intensity (hyperchromism) with red shift of the absorbance maximum is not necessarily obeyed. The latter has to do with the shape of absorption spectrum. Thus, linear conjugated molecules such as all-*trans* retinal and all-*trans* ß-carotene show a maximum absorbance band in the longest wavelength (corresponding to the $S_0 \rightarrow S_1$ transition), whereas cyclic conjugated molecules such as porphyrins and chlorophylls show stronger absorption in the near UV and blue spectral region ("Soret" band), compared to the absorption band in the red spectral region. To illustrate these two extremes, linear vs. cyclic, the absorption spectra of all-*trans* ß-carotene and protoporphyrin IX, respectively, are shown in Figure 3.

Any deviation from the linearity and cyclicity of a π-conjugated system of the photoreceptor chromophore results in a profound change in spectral shape of the absorption spectrum. For example, when *cis*-isomerization of the central double bond perturbs the near perfect linearity of all-*trans* ß-carotene, the longest wavelength absorbance band at 450 nm loses its intensity (hypochromism) while the second absorption band at 340 nm gains in intensity ("*cis*" peak; hyperchromicity). The *cis* peak hyperchromicity occurs because the dipole forbidden band with its transition dipole perpen-

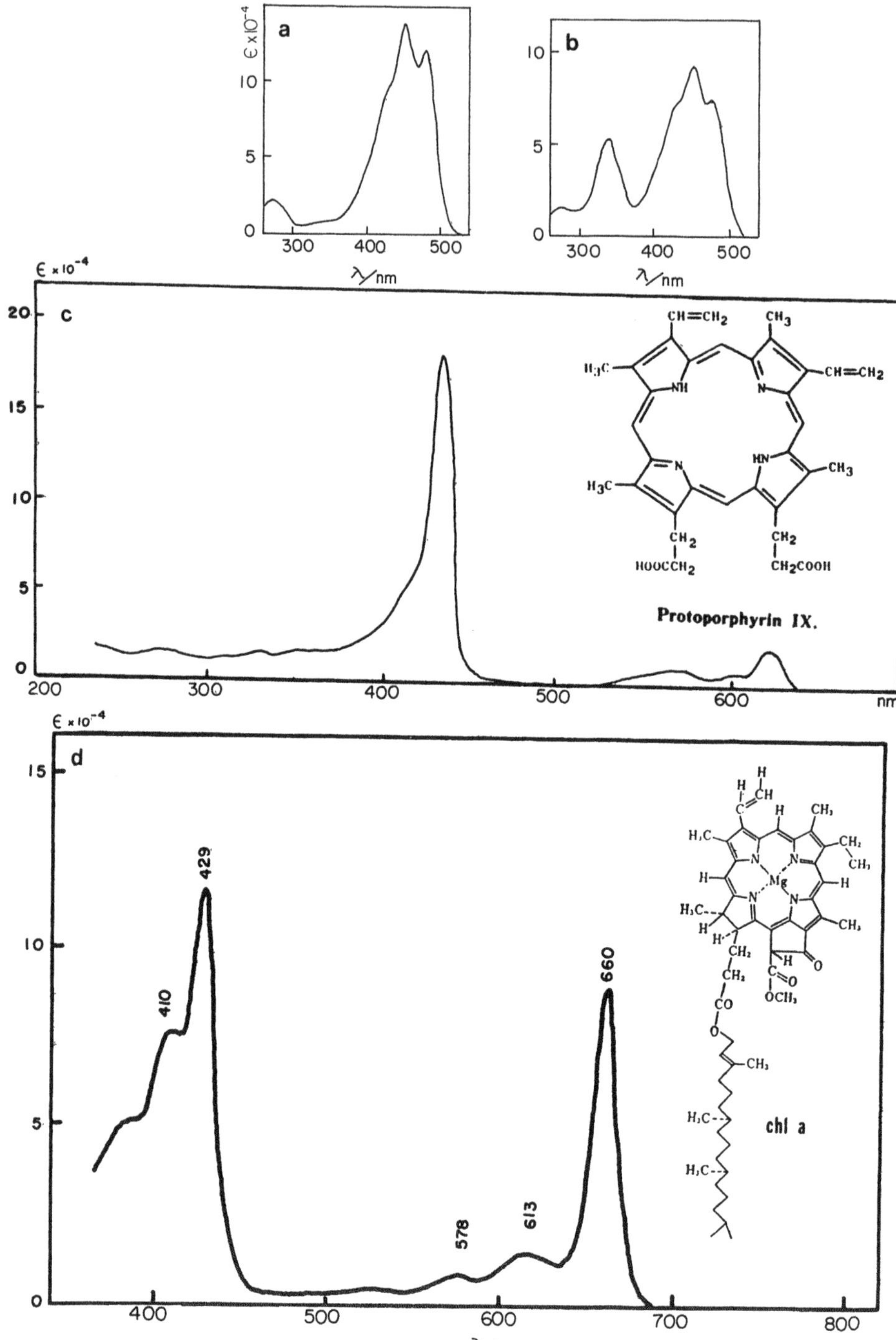

Fig. 3. Absorption spectra of (a) all-*trans* and (b) 15,15'-mono-*cis* ß-carotene (redrawn from Zechmeister, 1962), (c) protoporphyrin IX and (d) chlorophyll a (Redrawn from Clayton, 1965) at room temperature.

dicular to the long molecular axis becomes allowed as the result of a bend in the molecule which introduces a non-zero component of the transition dipole moment perpendicular to that of the blue absorbance band. The transition dipole of the latter along the long molecular axis of the molecule is shortened in the *cis*-isomer. In fact, when additional *cis-s-cis* isomerization is introduced at the 9,10- and 9',10'-double bond the left and right halves of the mono-*cis* ß-carotene, making the π-conjugation network approximately square in shape, the first absorption band is expected to be weaker than the second band. Thus, such a tri-*cis* ß-carotene resembles a cyclic conjugated molecule, producing an absorption spectrum that is mutually similar. Chlorophyll *a* can be thought of as a disrupted cyclic polyene as the result of hydrogenation/saturation at one of the porphyrin double bonds. Thus, its absorption spectrum in the red wavelength region (660 nm) exhibits a significantly higher intensity than does the corresponding band in protoporphyrin IX (Fig. 3). Additional hydrogenation of the chlorophyll *a* conjugated system gives rise to a strongly red shifted spectrum for bacteriochlorophyll *a*, with an absorbance maximum at 770 nm. It is interesting that in the evolution of the photosynthetic pigments higher plants developed the hydrogenation system (protochlorophyllide reductase) to harness more energetic photons for water splitting.

The student is referred to a chapter in the previous NATO ASI volume for a theoretical explanation for the dependence of absorbance spectral shape on the linearity and cyclicity of π-conjugated systems (Song, 1985). The nature of the lowest excited singlet state of carotenoids (^1A_g state?) will be discussed in Section IV, in connection with the role of carotenoids in photobiology.

Wavelength Modulation by Proteins and Membrane

In general, photosensor molecules in photobiologically responsive systems are associated with an apoprotein and/or a membrane. Association with the protein can be either covalent or non-covalent. Arguments as to why most of the well known photoreceptor chromophores are covalently linked to the apoproteins have been presented elsewhere and will not be repeated here (Song, 1987). In this section, we will discuss the role of apoproteins/membrane in their photobiologically native state of spectral (absorbance spectral shape and wavelength) and photochemical integrity. Through the process of evolution and supramolecular architecture, the photoreceptor apparatus can be said to have been tailor-made from the free chromophores and apoproteins/membrane to achieve optimal spectral and photochemical efficiencies for photobiological responses of organisms to light.

The effect of apoprotein and/or membrane on the spectral properties of the photosensor molecule is discussed here in terms of the phenomenon of λ_{max} modulation by apoprotein and/or membrane. The wavelength modulation is predetermined by the amino acid sequence, configuration and conformation, and other properties of the apoprotein, which evolved through molecular evolutionary determinants.

Covalent-Bond Formation

The chromophores of photoreceptors are frequently covalently bound to the apoproteins (Song, 1987). The covalent linkage between the chromophore and the apoprotein affects the electronic states of the chromophore, thus causing spectral shift/wavelength modulation. Phytochrome is a good example. The tetrapyrrole chromophore in phytochrome is covalently bound through a thioether bond with a cystein residue of the apoprotein (Lagarias and Rapoport, 1980), thus reducing the number of conjugated double bonds in the chromophore by one. As expected, this type of covalent linkage brings about a significant blue shift of the absorbance maximum wavelength. A molecular orbital theory predicts a 40-50 nm blue shift for the red wavelength absorbance maximum as the result of the covalent addition of a sulfhydryl group to the vinyl double bond of the phytochrome chromophore (Song and Chae, 1979).

Steric and Conformational Effects

A chromophore bound to the apoprotein may form a unique conformation due to the steric requirement around the chromophore binding site on the apoprotein. As a result of the covalent bond formation, chromophores and specific binding site residues can be held close together within the chromophore binding site or pocket, thus inducing a spectral shift. As will be discussed later in more detail, phytochrome serves as a useful example for the wavelength modulation by the apoprotein. The tetrapyrrole chromophore in the free form or in denatured phytochrome assumes a thermodynamically stable cyclic conformation, whereas the chromophore in native phytochrome is maintained in its extended, semi-cyclic conformation (for review, Song and Chae, 1979; Song, 1988). This is illustrated in Figure 4. This type of microenvironmental effects on the spectral shape and the absorbance maximum is also present in both covalently and non-covalently bound chromophores. In the following paragraphs, we briefly discuss several factors that contribute to the wavelength modulation induced by the chromophore binding site or pocket.

Hydrogen Bonding

Hydrogen bonding between the chromophore and the apoprotein may contribute to the origin of wavelength modulation in photobiological sensors, although the effect of hydrogen bonding on the absorbance maximum is significantly less than the effects mentioned in paragraphs (a) and (b) above. However, multiple hydrogen bonding is possible between the chromophore and the apoprotein, which can produce a marked spectral shift by exerting the combined effects of hydrogen bonding itself on the absorption spectrum and conformation fixation of the chromophore by hydrogen bon-

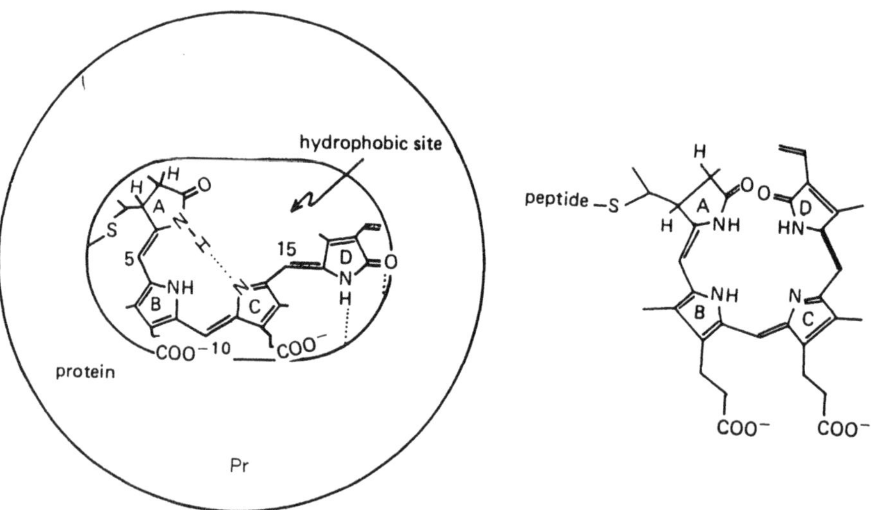

Fig. 4. The presumed semi-cyclic extended and cyclic conformations of the tetrapyrrole chromophore of native and denatured phytochromes (Pr), respectively. The ratio of the integrated intensity of the red to blue absorption bands is markedly different for the cyclic and semi-extended conformations, as predicted from the discussion in Section II (see Fig. 10 for the absorption spectra of native phytochrome).

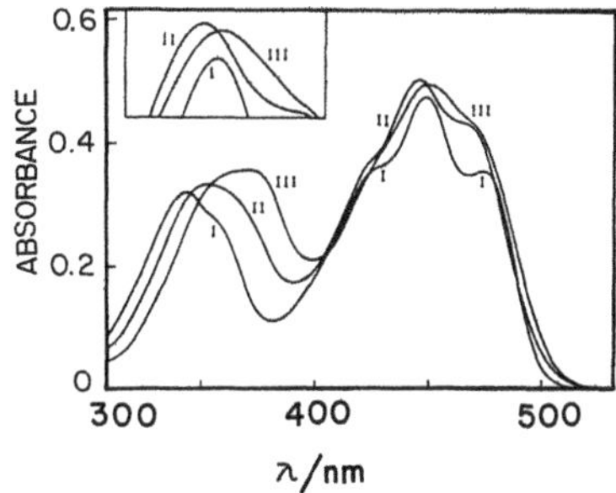

Fig. 5. Effect of hydrogen bonding on the absorption spectrum of riboflavin tetrabutyrate in carbon tetrachloride at room temperature. (a) Free riboflavin tetrabutyrate; (b) hydrogen-bonded riboflavin tetrabutyrate in the presence of 10 mM trifluoroacetic acid (TFA). I, 0 mM TFA (no H-bonding); II, 1 mM TFA; III, 10 mM TFA (nearly fully H-bonded). The inset shows the enlarged figures at the first absorbance band. (Redrawn from Yagi et al., 1980).

ding. Figure 5 shows as an example the effect of hydrogen bonding on the absorption spectrum of a flavin (Yagi et al., 1980).

Figure 5 shows a marked red shift of the near UV absorbance band of the flavin by hydrogen bonding, whereas the blue absorbance band maximum remains nearly unshifted, except for the loss of vibrational structure, by hydrogen bonding. Hydrogen bonding also enhanced the photoreactivity of flavin (Yagi et al., 1980).

Ionic Interactions Including Protonation

Ionic bonding and protonation of the chromophore molecules can strongly modulate the spectral absorbance maximum and its shape. For example, covalent linking of retinal to the amino group via Schiff's base formation only slightly affects the spectral properties of retinal. However, protonation causes a red shift of the main absorbance band by about 80 nm (from 360 nm to ca. 440 nm). Protonation also appears to be important in phytochrome (Fodor et al., 1988). Ionic bonding between the apoprotein and the chromophore, particularly through the two propionate carboxyl groups of the latter, is also likely to play an important role in maintaining the native conformation of the phytochrome chromophore (Parker and Song, 1990). In addition, charge transfer interactions between the chromophore and specific amino acid residues of the apoprotein are likely to contribute to the wavelength modulation phenomenon. However, there are no clearly established examples among the known photoreceptor molecules for a discussion in this paragraph. A strong charge-transfer interaction between the electron donor and the acceptor molecule is usually accompanied by the appearance of a charge transfer absorption band, thus visibly modifying the absorption spectrum of the free chromophore.

Hydrophobic Forces

In addition to more classical van der Waals interactions including dipole-dipole, dipole-induced dipole and London dispersion forces, which contribute to the hydrophobic forces in proteins, spectral properties of the chromophore are often markedly affected by hydrophobic forces. Hydrophobic forces and other forms of interactions enlisted above can modify the overall spectral shape of the photosensor chromophore absorbance bands to varying degrees, and their spectral and photochemical roles *in vivo* and in action spectroscopy must be carefully assessed, particularly in identifying a photoreceptor chromophore from the action spectrum.

Inter-chromophore Interactions

A significant spectral shift can occur as the result of chromophore-chromophore interactions within a photoreceptor protein or complex. Such interactions also affect the fate of excitation energy and thus the photoreactivity of the photoreceptor molecules. The special sandwitched dimer of bacteriochlorophyll *a* in the bacterial photosynthetic reaction center illustrates the functional importance of chromophore-chromophore interactions (Deisenhofer et al., 1986). Bacteriochlorophyll *a* in the reaction center absorbs maximally at 800 nm, whereas the bacteriochlorophyll *a* dimer shows λ_{max} at 865 nm (Norris and Schiffer, 1990).

Peridinin (Fig. 6) is responsible for the color of "Red Tides" which is due to dinoflagellate populations such as *Glenodinium sp., Gonyalaux polyedra, Amphidinium carterae,* and *Gyrodinium dorsum*. This carotenoid pigment harvests blue light (λ_{max}) in the solar spectrum and transfers its excitation energy to chlorophyll *a* within the light-harvesting pigment protein complexes of the dinoflagellate photosynthetic apparatus with essentially 100 % efficiency (Song et al., 1976; Koka and Song, 1977). It is possible that peridinin not only serves as an efficient antenna pigment for photosynthesis in marine dinoflagellates, but may also function as the photosensor for blue light-mediated phototaxis in *Gyrodinium dorsum* (Forward, 1973). Based on circular dichroic and fluorescence polarization data for the peridinin component of dinoflagellate antenna complexes, a special dimer model has been proposed (Song et al., 1976; Koka and Song, 1977). Although other interpretations are possible for the observed spectroscopic data, the dimer model appears to satisfactorily account for the spectral shift and the high energy transfer efficiency of peridinin in the dinoflagellate light-harvesting complexes. X-Ray crystallography remains to be applied to these complexes to determine the precise topography of the carotenoids (four molecules per chlorophyll *a* per protein).

Another interesting case of the potential chromophore-chromophore interactions in aneural photoreceptor systems is implicated in the heterodimer model of phytochrome action as simplified in Figure 7 (vanDerWoode, 1985). Also see Figure 8 for a general scheme of phytochrome action. The calculated absorbance maximum of the Pr-Pfr dimer is at 706 and 416 nm, whereas the Pr-Pr dimer exhibits absorbance maxima at 668 and 372 nm. There are no evidence for the direct/electronic chromophore-chromophore interactions in Pr-Pr and Pfr-Pfr dimers. However, it is possible that an indirect/cooperative interactions between the monomer subunits (and thus between the chromophores) produce a functionally active Pr-Pfr conformation and an absorption spectrum unique to the Pr-Pfr heterodimer. This remains to be studied.

Inter-chromophore interactions also modulate both spectral and redox properties of flavins and flavoproteins. Free FAD in solution exists in both stacked and unstacked forms between the flavin nucleus and adenine moiety.

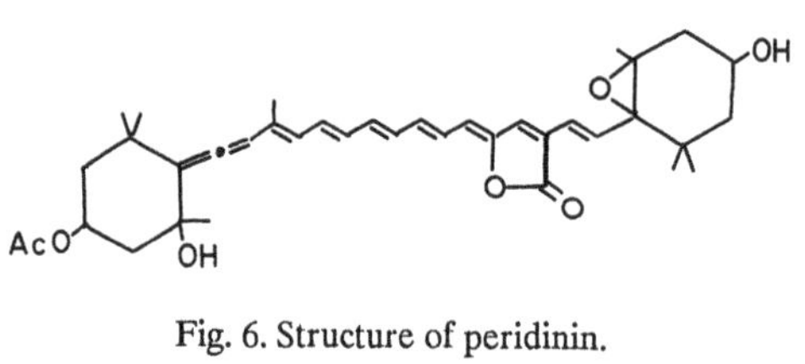

Fig. 6. Structure of peridinin.

$$\begin{array}{c} h\nu \quad\quad h\nu \\ \text{Pr-Pr} \leftrightarrow \text{Pr-Pfr} \leftrightarrow \text{Pfr-Pfr} \\ \downarrow \\ \downarrow \\ \text{VFLR and HIR} \end{array}$$

Fig. 7. Dimeric model of phytochrome action. According to this model, heterodimer Pr-Pfr triggers very low fluence response (VLFR) as well as high irradiance response (HIR). See Fig. 8 for the photochromic scheme of phytochrome action.

Photoreactivity of Photosensor Molecules

The spectral photosensitivity of light responsive organisms (Fig. 1) is determined by the chromophore structure and its interactions with the apoprotein and/or membrane, as discussed in the previous section. The photoreactivity of photosensor molecules can also be modulated by the apoprotein/membrane to which the photosensor molecule is bound. However, with the possible exceptions of bacterial photosynthetic reaction centers and bacteriorhodopsin (and possibly rhodopsin), the role of the apoprotein in modulating the photoreactivity of the photosensor molecule has not been elucidated satisfactorily at a molecular mechanistic level. Nevertheless, it is reasonable that ultrafast primarily photoprocesses required for highly efficient photobiological reactions (Song, 1987) by the light-responsive organisms depend on the intrinsic photoreactivity of the photosensor molecules involved. In this section, we take a bird's-eye view of the photoreactivity of various photoreceptor chromophores some of which are yet to be definitively identified as the chromophore for specific light responses, particularly blue light responses. Since bacteriorhodopsin and stentorin are covered elsewhere in this book and photosynthetic reaction center complexes are beyond the scope of this book and covered elsewhere extensively, we will focus our discussion on phytochrome as a ubiquitous red light sensor and blue light receptors (crytochromes).

Phytochrome

Phototransformation and molecular model. Phytochrome plays a key role as the photosensor for a number of light-regulated gene expression and other photo-responses in plants. The polarotropic rotation of chloroplast disc in *Mougeotia* and other related algae and the polarotropic response in fern protonemata (Wada and Kadota, 1987) are triggered by phytochrome (for review, Furuya, 1987; Song and Poff, 1989). Phytochrome works as the red/far-red sensor according to the following red/far-red reversible scheme (Fig. 8):

$$\text{Pr} \underset{730\text{ nm}}{\overset{660\text{ nm}}{\leftrightarrow}} \text{Pfr} \quad \text{gene regulation and other signal transduction}$$

Fig. 8. Phototransformation of phytochrome. Pr, red-absorbing form; Pfr, far red-absorbing form (physiologically active form). See Fig. 7 for the dimeric model of phytochrome action.

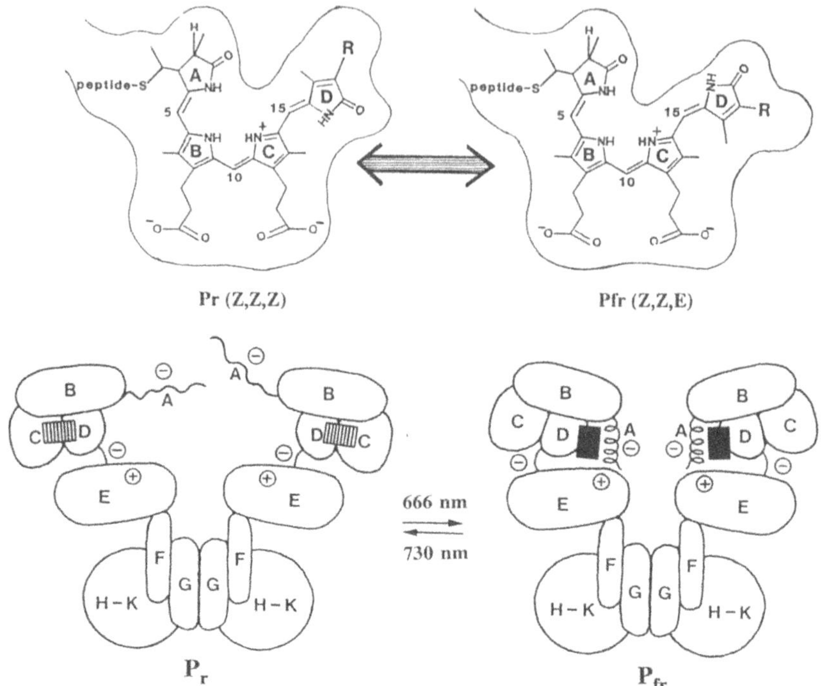

Fig. 9. A two-dimensional schematic model for the phototransformation of dimeric phytochrome. The Pr chromophore is shaded to indicate its relative inaccessibility, whereas the Pfr chromophore is shown as a solid rectangle and relatively exposed. Inset shows the photoisomerization of the chromophore consistent with NMR, resonance Raman and absorption spectroscopic data. Domains A (N-terminal segment) through K (C-terminal segment) are shown in their relative sizes, as calculated by the cross correlation function analysis of the sequence homologies of phytochromes. Domain A is shown to interact with the Pfr chromophore, which induces Á-helical folding of the former. The Pr chromophore linked to a cysteine residue in domain D additionally interacts with domain C, whereas the Pfr chromophore interacts specifically with domains A, B and possibly E.

Thus, phytochrome is a photochromic light sensor. The reversible phototransformation of phytochrome involves observable structural changes in the chromophore and the apoprotein, including a configurational/conformational isomerization (Rüdiger et al., 1983; Farrens et al., 1989; Holt et al., 1989; Rospendowski et al., 1989), secondary and tertiary conformational changes. A specific interaction also occurs between the chromophore and the amino terminal chain in the Pfr form of phytochrome which results in a photoreversible α-helical folding of the amino terminal segment (Chai et al., 1987). Peptide mapping (Vierstra and Quail, 1983; Lagarias and Mercurio, 1985; Grimm et al., 1988; Jones and Quail, 1989; Choi et al., 1990), chromophore accessibility (Hahn et al., 1984), chromophore reorientation (Ekelund et al., 1985), monoclonal antibodies (Cordonnier et al., 1985), phosphorylation (Wong et al., 1986), quaternary structure (Jones and Quail, 1986), electron microscopy (Jones and Erickson, 1989; Tokutomi et al., 1989) and small angle x-ray scattering (Tokutomi et al., 1989) have been used to probe different aspects of the conformation (Pr form) and/or its changes induced by the Pr ↔ Pfr phototransformation. Based on these studies and our recent semi-empirical calculations of the secondary structure and cross correlation analysis (domain structure) of eight phytochrome sequences (Parker and Song, 1990; Romanowski and Song, 1990), a schematic model for the transformation of phytochrome has been proposed (Parker et al., 1990), as shown in Figure 9.

Spectral modulation. Absorption spectrum and cross section/molar extinction coefficient (ϵ) of the chromophore determine its efficiency as a light sensor in photoresponsive organisms. The absorption spectra of phytochrome (Pr and Pfr) are qualitatively similar to those of porphyrins and chlorophylls with characteristic visible and Soret bands, except that the Soret band intensity is significantly weaker than the corresponding Soret band in porphyrins and chlorophylls (Fig. 10). Although the relative intensity of the Soret band in phytochrome "appears" to be considerably lower than the red and far-red bands ("Q_y" band), its integrated intensity (oscillator strength) is comparable to that of the latter, when the spectra are expressed in wavenumber instead of wavelength (Song and Yamazaki, 1987). In fact, the ratio of the Q_y to Soret band intensities is nearly conserved for the Pr and Pfr forms of phytochrome (1.24 and 1.19, respectively), suggesting that the semi-extended π-electron conjugated network of the chromophore (inset in Fig. 9) is retained by the apoprotein in both forms of phytochrome.

As mentioned earlier, absorption spectral properties of the tetrapyrrole skeleton depend strongly on the symmetry/shape of the conjugated system. With the open tetrapyrrole chromophore in phytochrome, spectral shapes (λ_{max} and ϵ) are predominantly determined by the chromophore conformation which is fixed by the apoprotein. Molecular orbital calculations have also shown that rotation about the 14,15-single bond (between ring C and D) greatly affects the intensity ratio.

Fluorescence polarization yields an angle of about 50° between the polarization axes of the Q_y and Soret transition dipoles. The calculated angle ranges from 56-63°, depending on the principal electronic transition chosen for the Soret band (Song and Chae, 1979; Song and Suzuki, 1990). The calculated polarization directions for the two bands greatly depend on the shape/conformation of the tetrapyrrole chromophore. In the photoconversion of Pr to Pfr, the chromophore reorients/rotates by a substantial angle (Ekelund et al., 1985). Though tentative, it is likely that polarotropic responses exhibited by *Mougeotia* and fern protonemata, *vide supra*, entail a reorientation of the chromophore and/or rotation of the protein (Song, 1983b), as the Q_y-transition dipole directions of Pr and Pfr remain essentially unaltered (Song and Yamazaki, 1987).

Phytochrome equally efficient as "on" and "off" light sensor switches. Perhaps the phytochrome molecule has evolved as a light sensor in such a way that the critical role of the apoprotein is to fix a specific chromophore conformation(s), thus modulating the absorbance maximum, which is clearly demonstrated, especially in its photochromic transformation. The conservation of the chromophore conformations in the Pr ↔ Pfr phototransformation after the primary photoisomerization step is responsible for the finely tuned light-sensing efficiency of phytochrome. Molar absorptivities at 668 nm for Pr and at 730 for Pfr are comparable (Fig. 10), although the free chromophores of the two phytochrome forms are markedly different in their molar absorptivity. To maintain the spectral characteristics for the Pr and Pfr forms, the apoprotein ensures that the conformations of both forms are conserved. Thus, phytochrome is not only a highly efficient sensor of red light, but also of far red light, in order to turn on and off the light signal transduction machinery with more or less equal efficiencies. A high efficiency of light sensing by phytochrome is also assured by an ultrafast (Braslavsky et al., 1984,1985; Song et al., 1986,1989) and efficient primary photoisomerization reaction. For example, the quantum yield for the primary step from the excited state of Pr is 0.5 (Heihoff et al., 1987).

Blue Light Sensors/Cryptochromes

Flavin vs. carotenoid, again?. Flavins and carotenoids are the most commonly assumed candidates for a variety of blue light responses in both prokaryotic and eukaryotic light-responsive organisms (for a recent review, Lipson and Horwitz, 1990). Un-

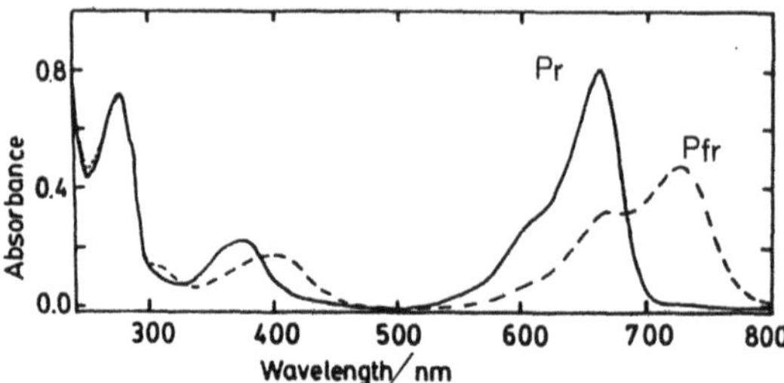

Fig. 10. Absorption spectra of *Avena* phytochrome in its Pr and Pfr forms.

fortunately, neither has been definitively established in terms of molecular characterization and mechanism (photochemical basis) of specific photosensors involved in blue light responses. However, there is good evidence that phototropic responses to blue light in higher plants (for review, Song and Poff, 1989) and in Phycomyces (for review, Lipson and Horwitz, 1990) are based on a flavin as the photosensor chromophore.

At present, only one photobiological reaction has been chemically identified with a flavin photoreceptor. Namely, photoreactivating enzyme/DNA photolyases contain a flavin as the photoreceptive coenzyme, apparently as the reduced form of flavin (e.g., $FADH_2$; Sancar et al., 1987; Payne et al., 1990). 5-Deazaflavins have also been found in some DNA photolyases; they can substitute for the native flavin with retention of enzymatic activity (for review on 1- and 5-deazaflavo-coenzymes, Walsh, 1986). Most interestingly, (5,10-methyltetrahydrofolyl)polyglutamate, a pterin, occurs in *E. coli* DNA photolyase, along with a stable FAD radical (Heelis et al., 1990; Payne et al., 1990). In the next paragraphs, we discuss the intrinsic photoreactivity of flavins, pterins and carotenoids to ascertain their potentiality as blue light sensor chromophores.

Flavins. Flavins appear to be an excellent candidate as a blue light sensor, as they absorb blue wavelength light strongly and are highly photoreactive. Some of the photoreactions mediated by flavins (F) are shown below:

Intramolecular reactions:

$$F + h\nu \rightarrow {}^{1}F \text{ (singlet excited state)}$$

$${}^{1}F \rightarrow F + \text{fluorescence and heat}$$

$${}^{1}F \rightarrow {}^{3}F \text{ (intersystem crossing)}$$

$${}^{1}F \rightarrow \text{lumichrome} \qquad (1)$$

$${}^{3}F \rightarrow \text{2'-oxoflavin} \qquad (2)$$

$${}^{3}F \rightarrow \text{9-formylmethylflavin} \qquad (3)$$

These reactions are self-destructive and it seems unlikely that they play a crucial role in the primary photoprocess of blue light signal transduction. However, reaction (2) may be coupled with a reductase to reduce the 2'-oxoribityl side chain and regenerate the original flavin molecule (see Fig. 11 for the chemical structures of lumichrome, 2'-oxoflavin, and 9-formylmethylflavin).

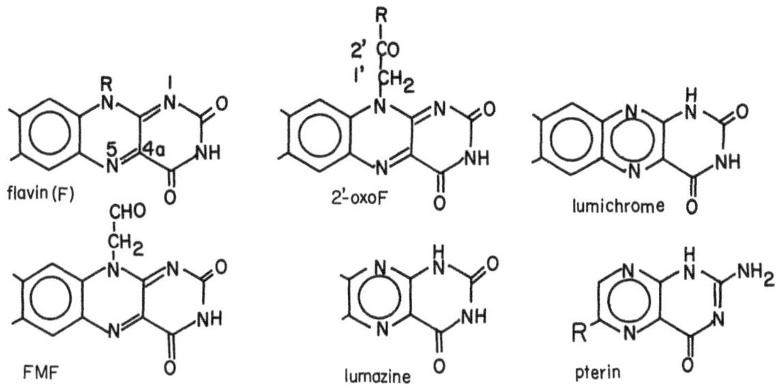

Fig. 11. Chemical structures of flavin, 2'-oxoflavin, lumichrome, 9-formylmethylflavin, lumazine and pterin.

Intermolecular reactions:

Perhaps the most likely type of reaction for the primary photoprocess of a flavin blue light receptor is electron and/or hydrogen atom transfer. For example, flavin and deazaflavin photooxidize NADH with high quantum yields (Sun and Song, 1973):

$$^{3}F + NADPH \rightarrow FH_2 + NAD^+ \quad (4)$$

$$^{3}deazaF + NADH \rightarrow deazaFH_2 + NAD^+ \quad (5)$$

Flavins are excellent photosensitizers. Oxidative reactions of Type I and Type II mechanisms are sensitized by the triplet state flavin. In Type I photosensitization,

$$^{3}F + S \rightarrow F^{\cdot -} + S^{\cdot +} \quad (6)$$

$$S^{\cdot +} + {}^{3}O_2 \rightarrow S_{ox} \quad (7)$$

where S represents an electron/hydrogen-donating substrate.

In Type II photosensitization, reactive oxygen is generated. Singlet oxygen is often the primary oxidant:

$$^{3}F + {}^{3}O_2 \rightarrow F + {}^{1}O_2 \quad (8)$$

$$^{1}O_2 + S \rightarrow S_{ox} \quad (9)$$

Flavins generate singlet oxygen with a high quantum efficiency (quantum yield 0.5; Song and Moore, 1968).

Whether or not these intermolecular reactions, including light-induced absorbance changes (LIAC; see review, Lipson and Horwitz, 1990), are relevant to the primary photochemistry of blue light responses remains to be studied. FH_2 generated in reaction (5) and from disproportionation reactions involving flavin radicals (from reaction (6) may be significant in terms of its ability to release protons, or alternatively, hydrogen peroxide):

$$FH_2 + A \rightarrow F + D + 2\,H^+ \quad (10)$$

$$FH_2 + {}^3O_2 \rightarrow F + H_2O_2 \quad (11)$$

where A and D are electron acceptor and donor, respectively. Protons released in the form of a proton motive force or a transient pH gradient can then be employed as a signal transducing signal in blue light responses (Song, 1987). Reduced flavin ($FADH_2$) also acts as the photoreactive species in DNA photolyase, *vide infra*.

There are several reactions of photobiological relevance mediated by flavins. These are listed below (for review, Song, 1987; Lipson and Horwitz, 1990):

$$^1F + IAA \rightarrow F\text{-}4a\text{-}IA + CO_2 \quad (12)$$

$$^1F + PAA \rightarrow F\text{-}4a\text{-}PA + CO_2 \quad (13)$$

where IAA and PAA are plant auxins, indole 3-acetic acid and phenylacetic acid, respectively. The 4*a* adducts produced are readily cleaved oxidatively, releasing IA (3-methyl possibly 3-methylindole and toluene, respectively, and regenerating flavin. These *in vitro* reactions may or may not be relevant to phototropic responses in higher plants (Song, 1987).

Another highly efficient reaction mediated by flavin is the photoproduction of ethylene, a plant growth regulator, from the oxidative degradation of methionine:

$$^{1\,or\,3}F + Met \rightarrow FH_2 + Methional\ (4a\text{-}adduct?) + NH_3 + CO_2 \quad (14)$$

$$^{1\,or\,3}F + Methional \rightarrow FH_2 + 2\,CH_2{=}CH_2 + 2\,HCOOH + (CH_3\text{-}S)_2 \quad (15)$$

Ethionine, cystathione, homocysteine, and homocystine can also be photooxidized by flavins to produce ethylene (Yang et al., 1967). It is possible that ethylene thus produced by flavin sensitization elicits those blue light phenomena involving morphogenic, developmental and transcriptional responses such as carotenogenesis (Song, 1987).

The role of flavin in photobiology is clearly established at least in one case involving DNA photolyase, as mentioned earlier. $FADH_2$ yields a stable flavin semiquinone radical photochemically(?) within the enzyme. Excitation of the radical to its excited quartet state abstracts a hydrogen atom from a tryptophan residue of the apoenzyme. Electron transfer from the thymine-thymine dimer of UV-damaged DNA leads to the dissociation/repair of the thymine cyclobutane adduct (Heelis et al., 1990). Customarily, the oxidized form of flavins has been considered as a candidate for blue light responses. In view of the fact that $FADH_2$ acts as the photocatalytically active form in

E. coli DNA photolyase (Jorns et al., 1987, 1990), it important to examine the reduced flavin as a primary photosensor chromophore for some of the blue light responses and ascertain the possible role of the apoprotein in modulating their absorbance band maxima and photoreactivities.

Reduced, non-planar flavins lack characteristic flavin absorbance band maxima in the blue wavelength region. Molecular orbital calculations have shown that the planar form of reduced flavin absorbs at wavelengths longer than 450 nm (Song, 1969). The long wavelength absorbance band maximum depends on the planarity of the reduced flavin nucleus (Hemmerich and Haas, 1975). Binding of the reduced flavins to their apoproteins is likely to promote the planarity of the flavin nucleus, thus modulating the blue light wavelength maximum and enhancing its intensity.

The absorption spectrum of the reduced flavin of D-amino acid oxidase extends to above 500 nm (Visser et al., 1979). From fluorescence polarization studies, the angle between the transition dipoles for the first band and the second one at 320 nm are found to be 30°. Theoretically, these two transitions are perpendicularly polarized with each other, with the first band being polarized along the short axis of the reduced flavin molecule. This discrepancy may result from the complete planarity assumed in the calculation (Song, 1969). Polarized single crystal spectroscopy of reduced flavins is needed to ascertain the planarity and orientations of the transition dipoles of the reduced flavin in D-amino acid oxidase. Information about the polarization properties will be helpful in interpreting any polarotropic responses of organisms/organelles to blue light mediated by reduced flavins.

Pterins and folates. Interestingly, a folate cofactor appears to be part of the photoreactive center in *E. coli* DNA photolyase, although its role is secondary, i.e. as antenna/energy transfer to the photoactive flavin (Heelis et al., 1990; Payne et al., 1990). Pterin(s) have also been detected in the presumed photoreceptor organelle, paraflagella body of *Euglena gracilis* (Galland and Senger, 1988; Galland et al., 1990). Pterin/folate appears to occur in association with flavin in other flavoprotein complexes. For example, sulfite reductase, xanthine oxidase, and nitrate reductase, a presumed candidate for cryptochrome in *Neurospora* (Klemm and Ninnemann, 1979), contains pterin (Johnson et al., 1981). It may also be photobiologically relevant that lumazine, a flavin biosynthetic precursor, with its structure tautomerically related to pterin (Fig. 11), occurs as an antenna pigment in association with flavin in bacterial luciferase (Lee et al., 1981).

Pterins seem to exhibit photoreactivities similar to those displayed by flavins. For example, pterins photooxidize NADPH via the triplet pathway (cf. reactions (4) and (5); Ledbetter and Tyner, 1990), and the pterin triplet can sensitize the formation of singlet oxygen (with efficiency comparable to the flavin reaction (8); Chahidi et al., 1981; Aubailly and Santus, 1986). It appears that pterins serve mainly as the antenna in DNA photolyase or as the radical stabilizer in nitrate reductase, sulfite reductase and xanthine oxidase. In fact, *E. coli* DNA photolyase is fully active with $FADH_2$ as the only photocatalytically reactive cofactor in the absence of the folate cofactor (Jorns et al., 1987,1990). Whether or not pterins mediate blue light responses as a primary photosensor remains to be seen. It is conceivable, with no evidence for or against at this time, that pterins and lumazine participate in a near-UV $\leftrightarrow$ blue photochromic manner, analogous to the excited state tautomerism between lumichrome, an alloxazine, and flavin, an isoalloxazine, as shown in Figure 12 (Song et al., 1974). One can also speculate that pterins mediate light-induced electron transfer in certain blue light responses via a LIAC mechanism.

Finally, caution should be exercised in identifying a pterin from cellular extract and organelles by using optical spectroscopic techniques only, because flavin photoproducts such as lumichrome (Reaction 1) with its structure similar to pterins (Fig. 11) can be misidentified as the latter on the basis of spectral similarities. For example, it is con-

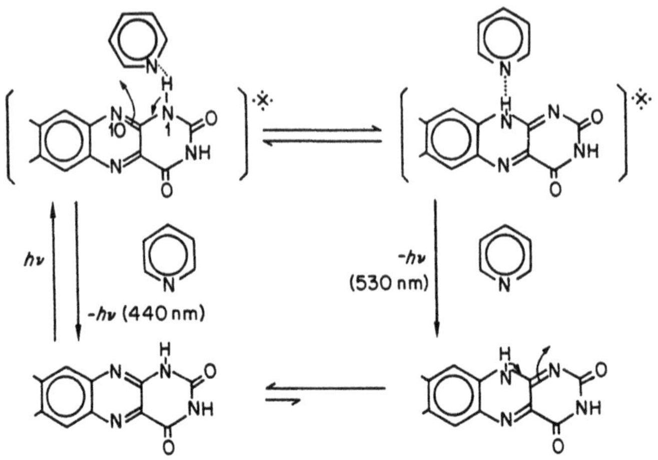

Fig. 12. The base(pyridine)-catalyzed tautomerism between lumichrome (an alloxazine or and flavin (an isoalloxazine) in the excited singlet state (asterisks). Similar phototautomeric reactions are likely with pterins (see Fig. 11).

ceivable that a flavin photoproduct such as lumichrome accumulates during microspectrophotometric and fluorometric detections.

Carotenoids other than retinal. The student is referred to the previous NATO ASI chapter (Song, 1985) and other review articles (Song, 1983a,1987) for a general description of the electronic spectroscopy of carotenoids. In particular, discussion as to the symmetry assignment of the lowest excited singlet state of linear polyenes (B_u vs. A_g state) and its potential implications in photochemistry and photobiology was presented therein. In this lecture, more recent information on the spectroscopy and role of carotenoids as blue light sensors will be reviewed. Hashimoto and Koyama (1989) assigned the lowest excited singlet state of all-*trans* carotenoids to a forbidden $^1A_g^-$ state. The same assignment has been made by others, as mentioned below. However, a well-resolved absorbance band corresponding to this electronic transition has not been observed even in carotenoids without C_{2h}-symmetry, which is expected to substantially remove the forbiddenness of this state for a single-photon absorption. The situation remains a puzzle. Interestingly, some carotenoids containing aldehyde function do exhibit a weak but distinct absorbance band in the wavelength region longer than the main blue absorbance band (Chae et al., 1977). The nature of the long wavelength band has not been elucidated.

Photoreactivity of long conjugated carotenoids is not well understood. However, photoisomerization of shorter chain polyenes including retinal Schiff's base in rhodopsin, bacteriorhodopsin and sensory rhodopsin occurs efficiently, playing the functionally important role in light signal transduction (see other relevant chapters on these retinal-based photoreceptors). Most recently, all-*trans*/13-*cis* isomerization of retinal has been demonstrated for phototaxis-sensory rhodopsin in *Halobacterium halobium* (Yan et al., 1990). This finding contrasts with the report that blue light-mediated phototaxis in *Chlamydomonas* does not involve photoisomerization of retinal (Foster et al., 1989). Unfortunately, *Chlamydomonas* "rhodopsin" has not been isolated and characterized, whereas *Halobacterium* sensory rhodopsin is now well characterized. Under catalytic conditions involving charge transfer interactions between the carotenoid and a catalytic donor (oxygen or iodine), carotenoid polyene can be photoisomerized, though apparently inefficiently and poorly elucidated. With carotenoids in bacterial photosynthetic reaction center complexes, the presence of *cis*-carotenoids and the *cis-trans* photoisomerization has been demonstrated (*vide infra*; Boucher and Gingras, 1984). It remains to be studied if carotenoids containing functional groups, for example, apocarotenals and their Schiff's bases, exhibit more diverse and efficient photoprocesses (for review, Song, 1983a,1985,1987).

The role of carotenoids as light-harvesting antenna pigments is well known, although its mechanism of energy transfer from carotenoids to chlorophylls in photosynthetic reaction centers has just begun to be elucidated (Boucher et al., 1977). An extremely short lifetime of the excited state of carotenoids has been an enigma in understanding the role of carotenoids as potential primary photosensor and as antenna accessory pigments in photosynthesis. Recently, an average excited state lifetime of carotenoids (mainly rhodopsin and spirilloxanthin) in *Chromatium vinosum* has been determined to be 6 ps (Hayashi et al., 1990). In novel synthetic models, energy transfer from a carotenoid to the covalently linked porphyrin base has been demonstrated, with a Forster critical distance of 5 Å (Dirks et al, 1980; Moore et al., 1980). Direct measurement of the lowest excited state (apparently A_g) lifetime of all-*trans* ß-carotene, canthaxanthin and ß-8'-apocarotenal yielded lifetimes of 8.4, 5.2 and 25.4 ps, respectively, suggesting that energy transfer from carotenoids to chlorophylls occurs at a subpicosecond rate via an electron exchange mechanism (as opposed to the Forster dipole-dipole coupling mechanism) (Wasielewski and Kispert, 1986). Femtosecond transient absorption spectroscopy showed that energy transfer occurs in 0.5 and 2.0 ps in the diatom *Phaeodactylum ticornutum* and 0.24 ps in the eustigmatophyte *Nannochloropsis* sp. (Trautman et al., 1990). Energy transfer in the reverse direction, namely from the triplet state chlorophyll *a* to ß-carotene, has also been observed by measuring the chlorophyll-sensitized population/decay of the triplet carotene (Nechushtai et al., 1988). This type of energy transfer may play an important role in protecting the photosynthetic apparatus from photodynamic damages, in addition to the carotenoid's quenching of singlet oxygen spuriously generated by triplet chlorophylls (Davidson and Cogdell, 1981). It appears that all-*trans* carotenoids occur as the preferred pigments for the antenna complex of *Rhodospirillum rubrum*, whereas 15-*cis* carotenoids are present in the reaction center complex. Since the triplet state of the latter is readily isomerized to all-*trans* and other isomers (Hashimoto and Koyama, 1988), it makes sense that *cis*-carotenoids are present in the bacterial reaction center complex to ensure protection from photodynamic damage to the photosynthetic apparatus (Koyama et al., 1990).

Stentorins and blepharismins. These photosensor molecules of the ciliates *Stentor coeruleus* and *Blepharisma japonicum*, respectively, are structurally derived from hypericin. Thus, the hypericin chromophore represents a new class of the photoreceptors for the photoresponsive ciliates. Since these photosensors are treated separately in this volume, they will not be discussed in this lecture.

Concluding Remarks

Molecular biology has been a powerful tool for understanding the properties and evolution of photobiological sensor molecules. For example, the amino acid sequences of several monocot and dicot phytochromes are now known (For review, Tomizawa et al., 1990; Thomas, 1990). The combined approach of spectroscopy/photobiochemistry and molecular biology is likely to deepen our understanding of the properties and mechanisms of photosensor molecules. This is particularly timely and desirable in the area of cryptochromes about which we seem to know very little, in terms of the identity of sensor molecules and their primary photoreaction mechanisms. In the meantime, the question "Flavin vs. Carotenoid, Again?" will continue to persist.

Acknowledgements

This work was supported in part by grants from NIH GM36956 and NS15426 (to PSS) and from the Japan Ministry of Education (to SS). The work on the spectroscopy of phytochrome was in part supported by visiting professorships from the Korea Ministry of Education and Korea Science and Engineering Foundation (to IDK and JHK). We thank William Parker for reviewing an initial draft of the manuscript.

References

Aubailly, M., and Santus, R., 1986, Oxidations photosensitized by pterins and diaminopterins, in: "Chemistry and Biology of Pterines," R. A. Cooper, and V. M. Whitehead, eds., de Gruyter, Berlin, pp. 99-102.

Boucher, F., and Gingras, G., 1984, Spectral evidence for photo-induced isomerization of carotenoids in bacterial photoreaction center, *Photochem. Photobiol.*, 40:277.

Boucher, F., van der Rest, M., and Gingras, G., 1977, Structure and function of carotenoids in the photoreaction center from *Rhodospirillum rubrum, Biochim. Biophys. Acta*, 461:339.

Braslavsky, S. E., Holzwarth, A. R., Wendler, J., Ruzsicska, B. P., and Schaffner, K., 1984, Picosecond time-resolved and stationary fluorescence of oat phytochrome highly enriched in the native 124 kdalton protein, *Biochim. Biophys. Acta*, 791:265.

Braslavsky, S. E., Ruzsicska, B. P., and Schaffner, K., 1985, The kinetics of the early stages of the phytochrome phototransformation Pr Pfr. A comparative study of small (60 kdalton) and native (124 kdalton) phytochromes from oat, *Photochem. Photobiol.*, 41:681.

Chae, Q., Song, P. S., Johansen, J. E., and Liaaen-Jensen, S., 1977, Linear dichroic spectra of cross-conjugated carotenals and configurations of in-chain substituted carotenoids, *J. Am. Chem. Soc.*, 99:5609.

Chai, Y. G., Song, P. S., Cordonnier, M.-M., and Pratt, L. H., 1987, *Biochemistry*, 26:4947.

Chahidi, C., Aubailly, A., Momzikoff, A., Bazin, M., and Santus, R., 1981, Photophysical and photosensitizing properties of 2-amino-4-pteridone: a natural pigment, *Photochem. Photobiol.*, 33:641.

Choi, J. D., Fugate, R. D., and Song, P. S., 1980, Nanosecond time-resolved fluorescence of phototautomeric lumichrome, *J. Am. Chem. Soc.*, 102:5293.

Choi, J. K., Kim, I. S., Kwon, T. I., Parker, W., and Song, P. S., 1990, Spectral perturbations and oligomer/monomer formation in 124-kilodalton *Avena* phytochrome, *Biochemistry*, 29:6883.

Clayton, R. K., 1965, "Molecular Physics in Photosynthesis," Blaisdell, New York, p. 196.

Cordonnier, M.-M., Greppin, H., and Pratt, L. H., 1985, Monoclonal antibodies with differing affinities to the red absorbing and farred absorbing forms of phytochrome, *Biochemistry*, 24:3246.

Davidson, E., and Cogdell, R. J., 1981, Reconstitution of carotenoids into the light-harvesting pigment-protein complex from the carotenoidless mutant of *Rhodopseudomonas spheroides* R26, *Biochim. Biophys. Acta*, 635:295.

Deisenhofer, J., Epp, O., Miki, K., Huber, R., and Michel, H., 1986, Structure of the protein subunits in the photosynthetic reaction center of *Rhodopseudomonas viridis* at 3Å resolution, *Nature*, 318: 618.

Dirks, G., Moore, A. L., Moore, T. A., and Gust, D., 1980, Light absorption and energy transfer in polyene-porphyrin esters, *Photochem. Photobiol.*, 32:277.

Ekelund, N. G. A., Sundqvist, C., Quail, P. H., and Vierstra, R. D., 1985, Chromophore rotation in 124-kilodalton *Avena* phytochrome as measured by light-induced changes in linear dichroism, *Photochem. Photobiol.*, 41:2212.

Farrens, D. L., Holt, R. E., Rospendowski, B. N., Song, P. S., and Cotton, T. M., 1989, Surface-enhanced resonance Raman scattering spectroscopy applied to phytochrome and its model compounds. 2. Phytochrome and phycocyanin chromophores, *J. Am. Chem. Soc.*, 111:9162.

Fodor, S. P. A., Lagarias, J. C., and Mathies, R., 1988, Resonance Raman spectra of the Pr-form of phytochrome, *Photochem. Photobiol.*, 48:129.

Forward, Jr., R. B., 1976, Light and diurnal vertical migration: Photobehavior and photophysiology of plankton, *Photochem. Photobiol. Rev.*, 1:157, and references by the author therein.

Foster, K., Saranak, J., Derguini, F., Zarilli, G., Johnson, R., Okabe, M., and Nakanishi, K., 1989, Activation of *Chlamydomonas* rhodopsin *in vivo* does not require isomerization of retinal, *Biochemistry*, 28:819.

Furuya, M., ed. , 1987, "Phytochrome and Photoregulation in Plants," Academic Press, Tokyo and New York, pp. 354.

Galland, P., Keiner, P., Dornemann, D., Senger, H., Brodhun, B., and Häder, D.-P., 1990, Pterin- and flavin-like fluorescence associated with isolated flagella of *Euglena gracilis, Photochem. Photobiol.*, 51:675.

Galland, P., and Senger, H., 1988, The role of pterins in the photoreception and metabolism of plants, *Photochem. Photobiol.*, 48:811.

Grimm, R., Eckerskorn, Ch., Lottspeich, F., Zenger, C., and Rüdiger, W., 1988, Sequence analysis of proteolytic fragments of 124-kilodalton phytochrome from etiolated *Avena sativa* L.: Conclusions on the conformation of the native protein, *Planta*, 174:396.

Hahn, T. R., Song, P. S., Quail, P. H., and Vierstra, R. D., 1984, Tetranitromethane oxidation of phytochrome chromophore as a function of spectral form and molecular weight, *Plant Physiol.*, 74:755.

Hashimoto, H., and Koyama, Y., 1988, Time-resolved resonance Raman spectroscopy of triplet ß-carotene produced from all-*trans*, 7-*cis*, 9-cis, 13-*cis* and 15-*cis* isomers and high-pressure liquid chromatography analyses of photoisomerization via the triplet state, *J. Phys. Chem.* 92:2101.

Hashimoto, H., and Koyama, Y., 1989, The C=C stretching Raman lines of ß-carotene isomers in the S_1 state as detected by pump-probe resonance Raman spectroscopy, *Chem. Phys. Lett.*, 154:321.

Hayashi, H., Kolaczkowski, S. V., Noguchi, T., Blanchard, D., and Atkinson, G. H., 1990, Picosecond time-resolved resonance Raman scattering and absorbance changes from carotenoids in light-harvesting systems of photosynthetic bacterium *Chromatium vinosum, J. Am. Chem. Soc.*, 112:4664.

Heelis, P. F., Okamura, T., and Sancar, A., 1990, Excited-state properties of *Escherichia coli* DNA photolyase in the picosecond to millisecond time scale, *Biochemistry*, 29:5694.

Heihoff, K., Braslavsky, S. E., and Schaffner, K., 1987, Study of 124-kilodalton oat phytochrome photoconversions *in vitro* with laser-induced optoacoustic spectroscopy, *Biochemistry*, 26:1422.

Hemmerich, P., and Haas, W., 1975, Recent developments in the study of fully reduced flavin, in: "Reactivity of Flavins," K. Yagi, ed., University Park Press, Baltimore, MD. pp. 1.

Holt, R. E., Farrens, D. L., Song, P. S., and Cotton, T. M., 1989, Surface-enhanced resonance Raman scattering spectroscopy applied to phytochrome and its model compounds. 1. Biliverdin photoisomers, *J. Am. Chem. Soc.*, 111:9156.

Johnson, J. C., Hainlive, B. E., and Rajagopalan, K. V., 1981, Characterization of the molybdenum cofactor of sulfite oxidase, xanthine oxidase and nitrate reductase. Identification of a pteridine as a structural component, *J. Biol. Chem.*, 255:1783.

Jones, A. M., and Erickson, H. P., 1989, Domain structure of phytochrome from *Avena sativa* visualized by electron microscopy, *Photochem. Photobiol.*, 49:479.

Jones, A. M., and Quail, P. H., 1986, Quaternary structure of 124-kilodalton phytochrome from *Avena sativa*, *Biochemistry*, 25:2987.

Jones, A. M., and Quail, P. H., 1989, Peptide fragments from the amino-terminal domain involved in protein-chromophore interactions, *Planta*, 178:147.

Jorns, M. S., Wang, B., and Jordan, S. P., 1987, DNA repair catalyzed by *Escherichia coli* DNA photolyase containing only reduced flavin: Elimination of the enzyme's second chromophore by reduction with sodium borohydride, *Biochemistry*, 26:6810.

Jorns, M. S., Wang, B., Jordan, S. P., and Chanderkar, L. P., 1990, Chromophore function and interaction in *Escherichia coli* DNA photolyase: Reconstitution of the apoenzyme with pterin and/or flavin derivatives, *Biochemistry*, 29:552.

Koyama, Y., Takatsuka, I., Kanani, M., Tomimoto, K., Kito, M., Shimamura, T., Yamashita, J., Saiki, K., and Tsukida, K., 1990, Configurations of carotenoids in the reaction center and the light-harvesting complex of *Rhodospirillum rubrum*. Natural selection of carotenoid configurations by pigment protein complexes, *Photochem. Photobiol.*, 51:119.

Koka, P., and Song, P. S., 1977, The chromophore topography and binding environment of peridinin· chlorophyll a· protein complexes from marine dinoflagellate algae, *Biochim. Biophys. Acta*, 495:220.

Lagarias, J. C., and Mercurio, F. M., 1985, Structure function studies on phytochrome. Identification of light-induced conformational changes in 124-Kda Avena phytochrome *in vivo*, *J. Biol. Chem.*, 260:2415.

Lagarias, J. C., and Rapoport, H., 1980, Chromopeptides from phytochrome. The structure and linkage of the Pr form of the phytochrome chromophore, *J. Am. Chem. Soc.*, 102:4821.

Ledbetter, J. W., and Tyner, S., 1990, Photostimulated reaction of pterin, biopterin and folic acid with ß-NADPH, *Photochem. Photobiol.*, 51:7s.

Lee, J., Carreira, L. A., Gast, R., Irwin, R. M., Koka, P., Small, E. D., and Visser, A. J. W. G., 1981, Properties of a lumazine protein from the bioluminescent bacterium *Photobacterium phosphoreum*, in: "Bioluminescence and Chemiluminescence," M. A. DeLuca, and W.D. McElroy, eds., Academic Press, New York. pp. 103.

Lipson, E. D., and Horwitz, B. A., 1990, Photosensory reception and transduction. in: "Sensory Receptors and Signal Transduction," J. Spudich, ed., in: Modern Cell Biology, vol. 7 (Series Editor B. Satir) Academic Press, New York, In press.

Moore, A. L., Dirks, G., Gust, D., and Moore, T. A., 1980, Energy transfer from carotenoid polyenes to porphyrins: a light-harvesting antenna, *Photochem. Photobiol.*, 32:691.

Nechushtai, R., Thornber, J. P., Patterson, L. K., Fessenden, R. W., and Levanon, H., 1988, Photosensitization of triplet carotenoid in photosynthetic light-harvesting complex of photosystem II, *J. Phys. Chem.*, 92:1165.

Norris, J. R., and Schiffer, M., 1990, Photosynthetic reaction centers in bacteria. *Chem. Eng. News*, July 30, p. 22.

Parker, W., Romanowski, M., and Song, P. S., 1990, Conformation and its functional implications in phytochrome, in: "Phytochrome Properties and Biological Action," B. Thomas, ed., Springer, Berlin, in press.

Parker, W., and Song, P. S., 1990, Location of helical regions in tetrapyrrole containing proteins by a helical hydrophobic moment analysis: Applications to phytochrome, *J. Biol. Chem*, 265:17568.

Payne, G., Wills, M., Walsh, C., and Sancar, A., 1990, Reconstitution of *Escherichia coli* photolyase with flavins and flavin analogs, *Biochemistry*, 29:5706.

Romanowski, M., and Song, P. S., 1990, Structural domains of phytochrome deduced from homologies in amino acid sequences, *Biophys. J.*, submitted.

Rospendowski, B. N., Farrens, D. L., Cotton, T. M., and Song, P. S., 1989, Surface-enhanced resonance Raman scattering (SERRS) as a probe of the structural differences between the Pr and Pfr forms of phytochrome, *FEBS Lett.*, 258:1.

Rüdiger, W., Thümmler, F., Cmiel, E., and Schneider, S., 1983, Chromophore structure of the physiologically active form (Pfr) of phytochrome, *Proc. Natl. Acad. Sci. USA*, 80:6244.

Sancar, G. B., Jorns, M. S., Payne, G., Fluke, D. J., Rupert, C. S., and Sancar, A., 1987, Action mechanism of *Escherichia coli* DNA photolyase, *J. Biol. Chem.*, 262:492.

Song, P. S., 1969, Theoretical considerations of the electronic spectra of methyl flavins, *Int. J. Quantum Biol.*, 3:303.

Song, P. S., 1983a, The electronic spectroscopy of photoreceptors (other than rhodopsin), *Photochem. Photobiol. Rev.*, 7:77.

Song, P. S., 1983b, Protozoan and related photoreceptors: Molecular aspects, *Annu. Rev. Biophys. Bioengin.*, 12:35.

Song, P. S., 1985, Primary molecular events in aneural cell photoreceptors. in: "Sensory Perception and Transduction in Aneural Organisms," G. Colombetti, F. Lenci, and P. S. Song, eds., NATO ASI Series A, Life Sci. Vol. 89, Plenum, New York. pp. 47.

Song, P. S., 1987, Possible primary photoreceptors. in: "Blue Light Responses: Phenomena and Occurrence in Plants and Microorganisms," H. Senger, ed., vol. II, CRC Press, Boca Raton, FL. pp. 3.

Song, P. S., 1988, The molecular topography of phytochrome: Chromophore and apoprotein, *J. Photochem. Photobiol. B: Biol.*, 2:43.

Song, P. S., and Chae, Q., 1979, The transformation of phytochrome to its physiologically active form, *Photochem. Photobiol.*, 30:117.

Song, P. S., Koka, P., Prezelin, B. B., and Haxo, F. T., 1976, Molecular topology of the photosynthetic light-harvesting complex, peridinin-chlorophyll a-protein complex from marine dinoflagellates, *Biochemistry*, 15:4422.

Song, P. S., and Moore, T. A., 1968, Mechanism of the photodephosphorylation of menadiol diphosphate. A model for bioquantum conversion, *J. Am. Chem. Soc.*, 90:6507.

Song, P. S., and Poff, K. L., 1989, Photomovement. in: "The Science of Photobiology," 2nd ed., K. C. Smith, ed., Plenum, New York. pp. 305-346.

Song, P. S., Singh, B. R., Tamai, N., Yamazaki, T., Yamazaki, I., Tokutomi, S., and Furuya, M., 1989, Primary photoprocesses of phytochrome. Picosecond fluorescence kinetics of oat and pea phytochromes, *Biochemistry*, 28:3265.

Song, P. S., Sun, M., Koziolowa, A., and Koziol, J., 1974, Phototautomerism of lumichromes and alloxazines, *J. Am. Chem. Soc.*, 96:4319.

Song, P. S., and Suzuki, S., 1990, Properties and evolution of photoreceptor, in: "Photoreceptor Evolution and Function," M. G. Holmes, ed., Academic Press, London, in press.

Song, P. S., Tamai, N., and Yamazaki, I., 1986, Viscosity dependence of primary photoprocesses of 124 kDalton phytochrome, *Biophys. J.*, 49:645.

Sun, M., and Song, P. S., 1973, Excited states and reactivity of 5-deazaflavine. Comparative studies with flavine, *Biochemistry*, 12:4663.

Thomas, B., ed., 1990, "Phytochrome Properties and Biological Action," NATO ASI Cell Biology Series, Springer, Berlin, in press.

Tokutomi, S., Nakasako, M., Sakai, J., Kataoka, M., Yamamoto, K. T., Wada, M., and Furuya, M., 1989, A model for the dimeric molecular structure of phytochrome based on small angle X-ray scattering, *FEBS Lett.*, 247: 139.

Tomizawa, K., Nagatani, A., and Furuya, M., 1990, Phytochrome genes: Studies using the tools of molecular biology and photomorphogenesis mutants, *Photochem. Photobiol.*, 52:265.

Trautman, J. K., Shreve, A. P., Owens, T. G., and Albrecht, A. C., 1990, Femtosecond dynamics of carotenoid-to-chlorophyll energy transfer in thylakoid membrane preparations from *Phaeodactylum tricornutum* and *Nannochloropsis* sp, *Chem. Phys. Lett.*, 166:369.

VanDerWoode, W. J., 1985, A dimeric mechanism for the action of phytochrome: Evidence from photothermal interactions in lettuce seed germination, *Photochem. Photobiol.*, 42:655.

Vierstra, R. D., and Quail, P. H., 1983, Purification and initial characterization of 124-kilodalton phytochrome from *Avena*, *Biochemistry*, 22:2498.

Visser, A. J. W. G., Ghisla, S., Massey, V., Muller, F., and Veeger, C., 1979, Fluorescence properties of reduced flavins and flavoproteins, *Eur. J. Biochem.*, 101:13.

Wada, M., and Kadota, A., 1987, Photo- and polarotropism in fern protonemata, *in*: "Phytochrome and Photoregulation in Plants," M. Furuya, Academic Press, Tokyo and New York. pp. 239-248.

Walsh, C., 1986, Naturally occurring 5-deazaflavin coenzymes: Biological redox roles, *Acc. Chem. Res.*, 19:216.

Wasielewski, M., and Kispert, L. D., 1986, Direct measurement of the lowest excited singlet state lifetime of all-trans-ß-carotene and related carotenoids, *Chem. Phys. Lett.*, 128:238.

Wong, Y. S., Cheng, H. C., Walsch, D. A., and Lagarias, J. C., 1986, Phosphorylation of *Avena* phytochrome *in vitro* as a probe for light-induced conformational changes, *J. Biol. Chem.*, 261:12089.

Yagi, K., Ohishi, N., Nishimoto, K., Choi, J. D., and Song, P. S., 1980, Effects of hydrogen bonding on the electronic spectra and reactivity of flavins, *Biochemistry*, 19:1553.

Yan, B., Takahashi, T., Johnson, R., Derguini, F., and Nakanishi, K., 1990, All-*trans*/13-*cis* isomerization of retinal is required for phototaxis signaling by sensory rhodopsin in *Halobacterium halobium*, *Biophys. J.*, 57:807.

Yang, S. F., Ku, S. H., and Pratt, H. K., 1967, Photochemical production of ethylene from methionine and its analogues in the presence of flavin mononucleotide, *J. Biol. Chem.*, 242:5274.

Zechmeister, L., 1962, "Cis-Trans Isomeric Carotenoids, Vitamins A and Arylpolyenes," Academic Press, Vienna, pp. 80.

Photoresponses in Eubacteria

Judith P. Armitage

Microbiology Unit
Department of Biochemistry
University of Oxford
South Parks Road
Oxford OX1 3QU
England

Introduction

In 1882 Engelmann carried out some of the earliest experiments on bacterial behavior, using a facultative photosynthetic bacterium, probably *Chromatium*, isolated from the Rhine outside his laboratory (Engelmann, 1883). I think it is fair to say that in the last one hundred years our understanding of photosynthetic eubacterial responses to light has not advanced far from those early detailed observations by Engelmann.

There have been perhaps several reasons for this slow progress. The major reason may be that chemotaxis in enteric bacteria and photoresponses in halobacteria have both turned out to be independent of cellular metabolism. The sensory mechanisms depend on dedicated sensory pathways and therefore mutations within the pathways and inhibition of particular sections of the pathways do not cause changes in other systems that could mask or interfere with interpretation of the results (Macnab, 1987; Spudich and Bogomolni, 1988). This is not the case with photoresponses (or responses to any other electron transport dependent effector), where the responses are intimately linked to cellular metabolic processes. A second reason may be that studies on photosynthetic bacteria have used a wide range of species, each worker having his or her own pet species and then attempted to extrapolate to other species. We know so much about halobacterial photoresponses because only one species has been studied in detail, similarly the chemotactic responses of enteric bacteria are understood in great detail, because only *Escherichia coli* and *Salmonella typhimurium* were used initially in the investigations. The basic framework was therefore laid down before the idiosyncrasies of other species became apparent. This has the positive effect of allowing the behavior of one species to be described in minute detail, but the disadvantage that it may blinker the approach to other species.

What I will briefly describe here is a limited range of the accumulated knowledge of a century or so from *Chromatium, Rhodospirillum rubrum and Rhodobacter sphaeroides* and I will attempt to pick out the beginning of a common framework but

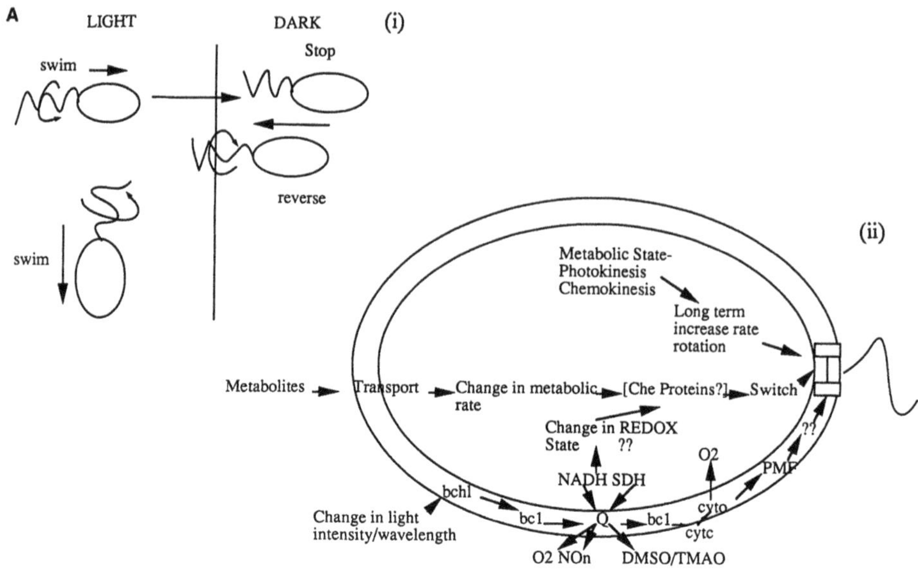

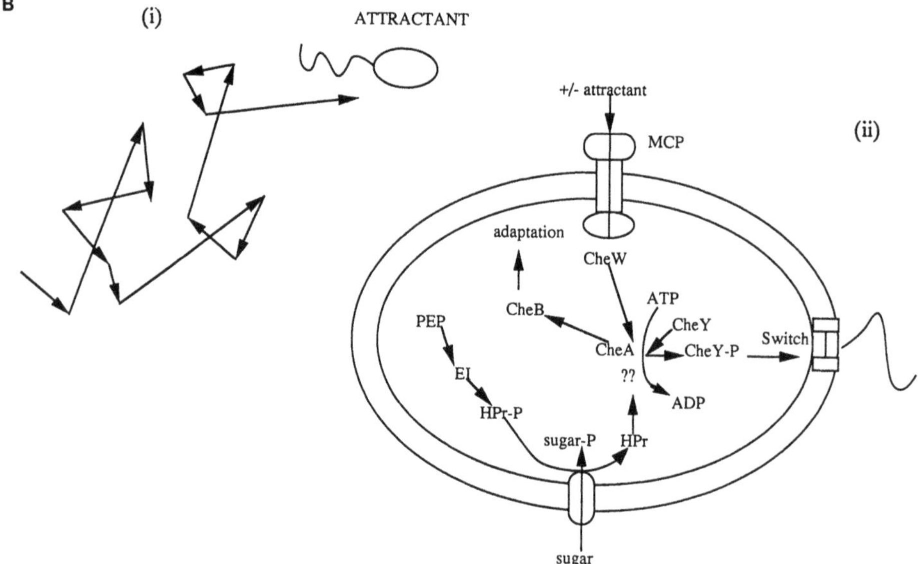

Fig. 1. Possible sensory pathways involved in bacterial responses to changes in light or chemical concentrations.

A. Photophobic responses

(i) The cell (*Chromatium*) swims across a light/dark boundary, stops and reverses the direction of flagella rotation, resuming normal flagella rotation after re-entering the light.

(ii) Possible receptor-independent sensory pathway involved in sensing changes in light and redox compounds. Changes in metabolic state of the cell causes long term changes in the rate of flagella rotation - "kinesis". Changes in the redox state and metabolic rate in response to changes in light, electron acceptors, or metabolites cause a short term signal changing the switching frequency of the motor.

SDH = succinate dehydrogenase; NOn = nitrogen oxides; bchl = bacteriochlorophyll; Q = quinone; PMF = electrochemical proton gradient; DMSO = dimethylsulphoxide; TMAO = trimethylamine-N-oxide.

mainly emphasize the areas that are missing, and the approaches that could be taken to fill in these gaps.

I will attempt to be careful with my language. In common with many bacteriologists I am guilty of misuse of the language of animal behavior. We tend to describe the change in swimming behavior seen in bacteria as phototaxis, whereas strictly this should be restricted to the orientation of cells in response to a light gradient. Bacteria are, in general, too small to show this type of response. They do however change their behavior in response to light. A change in the direction changing frequency is more accurately called a photophobic response (positive or negative dependent on whether the response is to a step-up or step-down in intensity), and a change in speed in a photokinetic response.

Motility in Photosynthetic Bacteria

Flagellate bacteria swim by rotating semi-rigid helical flagella, using the proton gradient across the cytoplasmic membrane as the driving force for that rotation. The mechanism involved in converting the energy of the electrochemical proton gradient into mechanical rotation within the flagellar motor is unknown, although there have been a wide variety of ideas over the past few years (Macnab, 1985). The $\Delta\mu_{H^+}$ is formed as a consequence of either respiratory or photosynthetic electron transport, or the breakdown of ATP via the ATP synthase/hydrolase. It might therefore be expected that anything that affects the rate of electron transport will in turn effect the size of the $\Delta\mu_{H^+}$ which may then also affect the rate of flagellar rotation, making the analysis of electron transport dependent taxis more complex. As will be shown later this is true but it is also an oversimplification, control of flagellar rotation is probably much more complex.

The pattern of flagella on photosynthetic bacteria varies from species to species, and this may have some bearing on the development of different photostrategies. The majority of work on photoresponses has been carried out either in *R. rubrum* or *C. okenii*. *R. rubrum* is a long spiral shaped cell with a tuft of flagella at each end of the cell body. It swims equally well with either end of the cell leading and therefore reverses simply to change direction, the nature of the medium usually meaning that this is not directly down the old path. It should be added that the mechanism involved in coordinating the two ends of the cell so that both bundles of flagella switch simultaneously is not understood, the rapid synchronization shown by large spirilla species suggests that the signal is electrical, but the switching signal in enteric bacteria is known to be chemical. The large rod shaped *C. okenii* has a single large tuft of flagella (large enough to be seen in the light microscope) which tends to push the cell (Pfennig, 1968). To change direction the cell briefly reverses, being pulled by the flagellar bundle, before resuming its forward swimming (Fig. 1A). Again interaction with the medium usually means that this is in a new direction.

Fig. 1. continued

B. Chemotaxis

(i) A biased random pattern of swimming in response to a spatial gradient of a chemoeffector, i.e., swims are longer in the favorable direction.

(ii) Mechanisms involved in chemosensory reception. An attractant binds to a sensory receptor (MCP) and a signal passes through a series of phosphorylated sensory proteins (CheW and CheA) to CheY. CheY-phosphate binds to the switch component of the motor and causes flagella reversal. CheA also interacts with an esterase CheB; CheB is involved in controlling the methylation state of the MCP to allow adaptation. Sugars transported through the phosphotransferase pathway (PEP = phosphoenolpyruvate and E1 = enzyme 1 of the PTS pathway) cause a sensory signal because the phospho-protein (HPr) interacts with CheA (for further detail see Macnab, 1987).

The small cocco-bacillus shaped species *R. sphaeroides* swims very differently, and as will be seen later this requires a different response to changes in light. It has a single flagellum arising from the middle of the long side of the cell. The flagellum is only rotated in a clockwise direction. Direction changing is the result of periodic stops in rotation, during which time the flagellum may relax its structure to a short wavelength, large amplitude form coiled against the cell body (Armitage and Macnab, 1987). These stops in flagellar rotation occur even though there is still a full $\Delta\mu_{H^+}$, suggesting that the flagella motor can be disengaged from the driving force, or the flow of protons stopped by a structural change. The cell is reoriented by Brownian motion during stops, the functional helix reforming when the motor starts to rotate and causes the cell to swim, usually in a new direction.

Historical Overview of Photoresponses

Engelmann saw over a century ago photosynthetic bacteria "take fright" when moving from an illuminated to a dark region. Pfennig working on *C. okenii* and Clayton investigating *R. rubrum* both saw these bacteria rapidly back up when crossing a light/dark boundary (Engelmann, 1883; Pfennig, 1968; Clayton, 1953a). The response was stronger the greater the difference in light intensity between the two regions, i.e., the response was shown more rapidly and by a larger percentage of cell passing over the boundary. The bacteria swam quite normally within the illuminated region and this lead to the straight forward idea that photosynthetic eubacteria become trapped within illuminated regions simply because they reverse when passing over a light/dark boundary back into the light, but show no response when moving from dark to light.

Characteristics of the Reversal Response

Clayton identified many of the characteristics of photoresponses in *R. rubrum*, which are probably similar in other eubacterial species. He showed that a visible response occurred about 90 ms after a reduction in light intensity and that there was a refractory period of 0.25 s before another response could be elicited. The time for complete recovery was 3.25 s (Clayton, 1953b). Interestingly, he noted that if the light level remained low after a step-down in intensity *R. rubrum* would rhythmically reverse for up to 30 s before returning to a normal reversal frequency. The least perceptible change in light intensity that caused a response was proportional to the background level of illumination, therefore as the background level of illumination increased there was a decrease in contrast sensitivity. It is difficult to put absolute figures onto the change in intensity that can cause a response as this will also change with the illumination conditions during growth which controls the concentration of light harvesting pigment. However, data from Clayton suggests that a 2% change in intensity over a 100fold range of background intensities could cause a behavioral response. The strength of the response depends both on the strength and the duration. A change near to the threshold will need about 1.5 s to cause a response but a change of twice the strength will cause a response in 0.5 s. The responses are also additive, the response to two sub-threshold stimuli resulted in a response as long as the time between the stimuli was less than 3 s but greater than 250 ms. The measurements led to the idea that the reversal response to a reduction in light intensity was the result of perturbation in the photosynthetic pathways, therefore it was not the absolute level of photosynthetic activity that mattered but the change, and particularly the rate of change that caused a behavioral response.

Relationship to the Photosynthetic Action Spectrum

Engelmann initially noted that *Chromatium* accumulated in specific regions of a light spectrum, even suggesting that bacteriochlorophyll absorbed in the far red because the bacteria specifically accumulated in that region, at about 850 nm with minor bands around 570 nm and about 510 and 410 nm. He carried out detailed investigations using sharp slits of light of different wavelengths to identify the wavelengths and intensities that caused a reversal response in bacteria swimming over the boundary into either dark or another wavelength. He concluded that the strongest responses were always to the far-red region of the spectrum. This work was carried further by Clayton who again showed a good correspondence between the photosynthetic action spectrum for *R. rubrum* and that of phototaxis, suggesting that bacteriochlorophyll was the major pigment involved in chemotaxis with carotenoids of secondary importance, corresponding to their roles in light harvesting (Clayton, 1953c).

Since then experiments have been carried out using inhibitors of photosynthetic electron transport which support the idea that photosynthesis is involved in phototaxis. Inhibition of photosynthetic electron transport by antimycin A resulted in a loss of the phototactic response. Both electron transport and phototaxis were relieved by by-passing the site of inhibition with phenazine methosulphate (Harayama, 1977). In addition, clamping the membrane potential with a combination of valinomycin and potassium caused a reduction in the phototactic response, although photosynthetic electron transport continued. Mutants of *R. rubrum*, *R. sphaeroides* and *Rhodobacter capsulatus* that have their full complement of photosynthetic pigments but lack a functional reaction centre show no phototactic responses (Armitage and Evans, 1981). These data support the theory that phototaxis is mediated by a change in electron transport causing a change in the pmf, and there is no specific photosensory pigment.

Many motile bacteria, including both photosynthetic and non-photosynthetic species show strong negative responses to high intensity light, particularly blue light (Taylor et al., 1979). In *E. coli* this results in prolonged tumbling, followed by loss of motility and cell death if continued for any length of time. Aerobically grown *R. rubrum* showed an accumulation in a white light spot, but not in far-red light (Harayama and Iino, 1976). The first phenomenon is probably caused by the photosensitive flavoproteins being inactivated by blue-light, eventually resulting in inhibition of respiration. The second response was thought to be caused by the breakdown of ferric ions causing a positive chemotactic response, it could, however, equally be the result of a similar phenomenon to that shown by enteric bacteria. High light could cause accumulation by increasing the reversal frequency again by perturbation of electron transport components. The physiological role of these responses may be limited, but both suggest that changes in electron transport can have an effect on motor function.

All these observations taken together suggest that photosynthetic bacteria accumulate in the light because when they swim over a light/dark boundary there is a change in the rate of photosynthetic electron transport which results in a signal being sent to the flagellar motor causing a reversal response. There is no apparent response when swimming over a dark/light boundary, although there must be a change in electron transport rate. An increase in the electron transport rate does not therefore appear to produce a major behavioral response, whereas a reduction in electron transport rate does. This is in contrast with the aerotactic response mediated through respiratory electron transport in which both a reduction in electron transport rate and an increase cause a response. The data from step-down experiments also suggests that photosynthetic bacteria cannot respond to shallow gradients of light as slow or small changes in intensity would not perturb the photosynthetic apparatus enough to cause a measurable response, again in contrast to the aerotactic behavior of bacteria.

As the flagellar motor is driven by the electrochemical proton gradient, and this is perturbed by the change in photosynthetic electron transport it was reasonable to assume that the signal controlling reversals was a change in the pmf, leading to the idea of the "protometer", a specific receptor that sensed changes in pmf and signalled these changes to the motor (Manson et al., 1977; Glagolev, 1984). This "protometer" could be part of the flagellar motor or a cytochrome type molecule. The idea of the "protometer" was supported by investigations into the aerotactic responses of both enteric bacteria and *R. sphaeroides*. These investigations found that inhibitors of respiratory electron transport inhibited aerotaxis, and aerotaxis only occurred if the respiratory electron transport branch operating was coupled to an increase in pmf (Shioi et al., 1988; Armitage et al., 1985). In addition it was found that a pulse of oxygen given to photosynthetic cells incubated in the light caused both a negative behavioral response and a reduction in the membrane potential, a mechanism that could be involved in preventing bacteria growing photosynthetically entering an aerobic environment. Unfortunately, several recent experiments have suggested that a simple "protometer" sensing changes in pmf and signalling to the flagella motor may not be all the story.

Chemotaxis in Rhodobacter sphaeroides

Investigations into phototaxis cannot be separated from studies into the behavioral responses caused by other environmental stimuli. This is because an understanding of the mechanisms involved in other responses may throw light on those involved in photoresponses, with which they must integrate. Bacteria are also rarely faced with a single stimulus, but they must be able to balance signals from a wide range of very different environmental components, light, oxygen, nutrients, even temperature. The response may vary depending on the overall environment in which the stimulus changes, for example if incubated in the dark *R. sphaeroides* swims towards oxygen, but if the same population of cells is incubated in the light oxygen causes a negative response.

Rhodobacter sphaeroides is the first bacterium identified lacking the specific membrane spanning sensory transducing proteins identified in enteric bacteria and many other species (including *R. rubrum* and *H. halobium*) (Fig. 1.B) (Sockett et al., 1987). Instead *R. sphaeroides* shows two independent behavioral responses dependent on the type of gradient, and the growth state.

When presented with a spatial gradient of a chemoeffector, all of which are metabolites, there is a transient increase in stopping frequency which results in movement up an attractant gradient, the change in stopping frequency biases the overall direction of swimming (Poole and Armitage, 1989). This response is dependent on transport of the chemoeffector and at least limited metabolism as mutants lacking the chemoattractant response tend to be pleomorphic, showing abnormal metabolism as well as behavior (W. A. Havelka and J. P. Armitage, unpublished data).

If, however, there is a step-up in the concentration of a nutrient that is currently growth limiting there is a long term increase in swimming speed. This is the result of individual cells stopping less often and also rotating their flagella faster. This chemokinetic response is only shown in bacteria grown on organic acids, not by sugar grown cells. The speed increase is also only strong to a member of the currently limiting nutrient group, therefore if the cells are nitrogen limited there will be a speed increase in response to an increase in the concentration of nitrogen in the medium whether it is supplied as ammonia or glutamate but not to an increase in a carbon source (Poole et al., 1990a). When the nitrogen limitation is overcome any further increase in nitrogen has no effect on swimming speed, but there will now be a speed response to an increase in concentration of the new limiting compound, e.g., a carbon source. The response will be seen whether the carbon source given is, e.g., pyruvate or acetate. It is therefore only when that nutrient is limiting that chemokinesis is seen. The new swimming speed continues as long as the metabolic state of the cell is high, and obviously when all growth limitations are overcome the population is swimming at maximum speed. Because there is no adaptation shown to this response it serves to spread the population of cells within the environment rather than concentrate them, and dispersal may be its role in the natural environment. If a spatial gradient of a chemical is encountered (sensed temporally because of the size of the cells) the cells may still show chemokinesis but there will also be an increase in stopping frequency, which is independent of whether that chemoattractant is growth limiting or not.

An increase in a currently limiting nutrient can therefore cause an increase in the speed of rotation of the flagellar motor and a decrease in the motor stopping frequency. We were therefore interested to identify whether this chemokinetic response was the result of a change in the proton motive force (pmf). If the idea of a protometer is correct then any change in pmf should result in a behavioral response. Examination of chemokinesis however showed that it was independent of any change in either the rate of respiratory electron transport or any change in the size of pmf measured by the absorption shift of the membrane potential sensitive pigment carotenoids, under conditions where the membrane potential was the major component of the proton gradient (Poole et al., 1990b). The chemokinetic increase occurred under saturating light when there was no measurable change in the membrane potential and also under conditions where the membrane potential was actually reduced. It was also possible to increase the size of the membrane potential without an increase in swimming speed or change in stopping frequency. All this data suggests that the rate of rotation of the flagella motor can alter independently of the size of the pmf, and that a change in pmf does not necessarily lead to a behavioral response suggesting that there is not a "protometer" in its simplest form (Fig. 1A).

If the pmf is artificially reduced by the addition of an uncoupler such as CCCP it was found that bacteria continued swimming normally and at full speed until the membrane potential was as low as 15 mV (less than 10% of the maximum size), stopping and starting normally (Armitage et al., 1990). The pmf threshold for rotation and that for saturation of flagella rotation must be very low compared to other pmf dependent activities. However, contrary to the data presented above there is evidence to suggest that conditions causing transient changes in pmf can effect speed and behavior under certain conditions. It has also been observed both in *R. rubrum* and in *R. sphaeroides* that cells that have been incubated in the dark and are suddenly exposed to light show a brief period of very fast swimming, lasting only a second or so after which time the cells return to normal speed. It is possible that when the light is first turned on photosynthetic electrontransport operates at maximum rate and after a few seconds the build up of the new level of pmf exerts feed back control on the electron transport rate, or activation of the ATP synthase enzymes, returning the pmf to a steady state level. Under all these conditions not only does the pmf change but also the redox state of many of the electron transport components, and probably several intracellular metabolites. Any of these may have a direct effect on the speed of flagella rotation. The chemokinetic data suggests that an intracellular metabolite changed as a result of the change in metabolic state of the cell may interact directly with the flagellar motor altering the interaction of the motor proteins with the proton gradient to change the coupling efficiency, rather than there being a direct effect of a change in pmf (Fig. 1A).

Further experiments on electron transport taxis in *R. sphaeroides* has complicated the picture still further, but also points to the future direction for research. Electron transport taxis is complicated because of the number of interacting pathways in an organism such as *R. sphaeroides* (Fig. 1) respiratory electron transport has at least two branches, DMSO and TMAO can be used as terminal acceptors under anaerobic dark conditions as well as oxides of nitrogen (Ferguson et al., 1987). In addition, the photosynthetic electron transport chain shares cytochromes with these pathways, therefore operation of different branches depends on the concentration of the different acceptors, the affinities of the cytochromes and the redox state of the components. Recently, we carried out experiments looking at the interaction of electron transport to DMSO, oxygen and photosynthetic electron transport, and the effect on the tactic responses. We found that when grown in the dark with DMSO as a terminal acceptor *R. sphaeroides* showed good chemoresponses to DMSO. As expected the response was inhibited by aerobic incubation in both the dark and the light as previous experiments have shown that respiratory and photosynthetic electron transport both inhibit electron transport to DMSO, presumably by competing for electrons and changing the redox poise within the system. Unexpectedly however, chemotaxis to DMSO continued under anaerobic illuminated conditions although DMSO reductase activity was not measurable under these conditions (J. P. Armitage, unpublished) (Jones et al., 1990). The tactic response therefore depends on more than acceptor binding and electron transport activity, but what is unknown. We are currently making "knock out" mutants in different components of the electron transport pathway of *R. sphaeroides* to change the activities of the different branches. We will then investigate individual taxes and their interactions.

Photoresponses in R. sphaeroides

Despite a unidirectional flagella motor and an unusual stop-start motility *R. sphaeroides* does accumulate in regions of light (and in oxygenated environments if incubated in the dark). The simple mechanism of cell reversal described for other photosynthetic species cannot apply to the accumulation of *R. sphaeroides* in the light, the equivalent response of stopping when passing over a light/dark boundary would result in accumulation in the dark rather than the light. If, however, there was "memory" in the system, as with chemotactic responses, an increased stopping frequency could result in accumulation in the light, rather than the dark.

Models show that an increase in speed without adaptation will result in dispersal, and increase in stopping frequency without adaptation will result in accumulation where there is increased stopping or tumbling (Berg, 1983). As stopping or tumbling tend to increase in response to a negative stimulus (reduction of attractant or increase in repellent) this would result in accumulation in a negative environment. For accumulation in a positive environment to occur bacteria must possess both a "memory" and the ability to adapt. For *R. sphaeroides* this would mean that crossing a light/dark boundary would result in a period of stop/start swimming, if a swim resulted in the cell swimming back over the dark/light boundary the increased stopping would cease but if the swim period resulted in continued movement in the dark the increased stopping frequency would continue. This type of response would increase the chances of the bacteria swimming back into the light. After a period of time the bacteria would adapt to the new dark or lower light conditions and resume their normal pattern of swimming. This prolonged response and adaptation are in fact seen in both *R. sphaeroides* and *R. rubrum* to a step-down in light intensity. Clayton showed that after a step-down *R. rubrum* continued rapid reversals for up to 30 s before resuming normal swimming. This response stopped immediately if the original light intensity was restored (Clayton, 1953b). Similarly, we find that there is a period of rapid stop-start swimming if *R. sphaeroides* is subjected to a step-down in light levels or a change in wavelength of light. This behavior lasts for several seconds before normal behavior is resumed and is cancelled if the original conditions are restored. Both of these results suggest that bacteria have a "memory" of the their environment. This memory is unlikely to be the change in photosynthetic electron transport rate or pmf level as experiments show that a new steady state is formed within 100 ms. There are however, many, other cellular consequences of a change in light level that occur on time scales from seconds, e.g., ATP synthase activity, to hours, e.g., change in pigment synthesis. One of the metabolic mechanisms controlling these events may also control the activity of the flagella motor. In enteric bacteria PTS sugars cause a chemotactic response because the process of phosphorylation-dependent transport into the cell causes an indirect change in the phosphorylation state of the cytoplasmic chemotaxis proteins (CheA and CheY), the phosphorylation level of which is responsible for controlling the switching frequency of the flagellar motor (Lengeler and Vogler, 1989) (Fig. 1B). A similar change in the metabolic state of the cell in response to a change in photosynthetic electron transfer may change the state of an equivalent switch protein (although no homology has been found at the DNA or antibody level between any *R. sphaeroides* and *E. coli* sensory component, W. A. Havelka, E. R. Sockett and J. P. Armitage, unpublished data). Adaptation would then occur as a result of the pathway equilibrating.

Conclusions

Photosynthetic eubacteria respond to changes in light levels but the response is probably not a simple reversal of swimming direction caused by a decrease in the rate of photosynthetic electron transport causing a transient reduction in pmf. Although these components are probably involved the longer term responses and adaptation outside the timescale of photosynthetic electron transport suggest a role for an intracellular metabolite controlling the frequency of flagellar switching/stopping. An electron transport dependent cytoplasmic sensory protein might also explain the mechanism of integration with sensory signals from other environmental sensing systems (Fig. 1A). Lastly not all species may have the same mechanism, some may be simpler and others more sophisticated.

Acknowledgements

The work on *R. sphaeroides* described here was funded by the SERC and the Wellcome Trust.

References

Armitage, J. P., and Evans, M. C. W., 1981, The reaction centre in the phototactic and chemotactic responses of *Rhodopseudomonas sphaeroides, FEMS Microbiol. Letts.*, 11:89.

Armitage, J. P., Ingham, C., and Evans, M. C. W., 1985, Role of proton motive force in phototactic and aerotactic responses in *Rhodopseudomonas sphaeroides, J. Bacteriol.*, 161:967.

Armitage, J. P., and Macnab, R. M., 1987, Unidirectional intermittent rotation of the flagellum of *Rhodobacter sphaeroides, J. Bacteriol.*, 169:514.

Armitage, J. P., Poole, P. S., and Brown, S., 1990, Sensory signalling in *Rhodobacter sphaeroides. in*: "Molecular biology of membrane bound complexes in phototrophic bacteria," G. Drews., ed. , Plenum press, New York., in press.

Berg, H. C., 1983. Random Walks in Biology, Princeton, New Jersey.

Clayton, R. K., 1953a, Studies in the phototaxis of *Rhodospirillum rubrum*. I. Action spectrum, growth in green light and Weber-law adherence, *Arch. Microbiol.*, 19:107.

Clayton, R. K., 1953b, Studies in the phototaxis of *Rhodospirillum rubrum*. III. Quantitative relationship between stimulus and response, *Arch. Microbiol.*, 19:141.

Clayton, R. K., 1953c, Studies in the phototaxis of *Rhodospirillum rubrum*. II. The relation between phototaxis and photosynthesis, *Arch. Microbiol.*, 19:125.

Engelmann, T. W., 1883, *Bakterium photometricum*. Ein Beitrag zur vergleichenden Physiologie des Licht- und Farbensinnes, *Pfügers Arch. gesamte Physiol. Menschen Tiere*, 42:183.

Ferguson, S. J., Jackson, J. B., and McEwan, A. G., 1987, Anaerobic respiration in the Rhodospirillaceae: characterisation of pathways and evaluation of roles in redox balancing during photosynthesis, *FEMS Microbiol. Rev.*, 46:117.

Glagolev, A. N., 1984, Bacterial $\Delta\mu_{H^+}$-sensing, *Trends Biochem. Sci.*, 9:397.

Harayama, S., 1977, Phototaxis and membrane potential in the photosynthetic bacterium *Rhodospirillum rubrum*, *J. Bacteriol.*, 131:34.

Harayama, S., and Iino, T., 1976, Phototactic responses of aerobically cultivated *Rhodospirillum rubrum*, *J. Gen. Microbiol.*, 94:173.

Jones, M. R., Richardson, D. J., McEwan, A. G., Ferguson, S. J., and Jackson, J. B., 1990, *In vivo* redox poising of the cyclic electron transport system of *Rhodobacter capsulatus* and the effects of the auxiliary oxidants, nitrate, nitrous oxide andtrimethylamine N-oxide, as revealed by multiple short flash excitation, *Biochim. Biophys. Acta*, 1017:209.

Lengeler, J. W., and Vogler, A. P., 1989, Molecular mechanisms of bacterial chemotaxis towards PTS-carbohydrates, *FEMS Microbiol. Rev.*, 63:81.

Macnab, R. M., 1985, The proton-driven bacterial flagellar motor, *Meth. Enzymol.*, 125:563.

Macnab, R. M., 1987, Motility and chemotaxis. *in: "Escherichia coli* and *Salmonella typhimurium*: Cellular and Molecular Biology," vol. 1, F. C. Neidhardt, J. L. Ingraham, K. B. Low, B. Magasanik, M. Schaechter, and H. E. Umbarger., eds., American Society for Microbiology, Washington, D. C. . p. 732.

Manson, M. D., Tedesco, P., Berg, H. C., Harold, F. M., and van der Drift, C., 1977, A proton motive force drives bacterial flagella, *Proc. Natl. Acad. Sci. USA*, 74:3060-3064.

Pfennig, N., 1968, *Chromatium okenii* (Thiorhodaceae). in: "Encyclopedia cinematographica," G. Wolf., ed., Institut für den Wissenschaftlichen Film, Göttingen. p. 3.

Poole, P. S., and Armitage, J. P., 1989, Role of metabolism in the chemotactic response of *Rhodobacter sphaeroides* to ammonia, *J. Bacteriol.*, 171:2900.

Poole, P. S., Williams, R. L., and Armitage, J. P., 1990a, Chemotactic responses of *Rhodobacter sphaeroides* in the absence of apparent adaptation, *Arch. Microbiol.*, 153:368.

Poole, P. S., Brown, S., and Armitage, J. P., 1990b, Swimming changes and chemotactic responses in *Rhodobacter sphaeroides* do not involve changes in the steady state membrane potential or respiratory electron transport, *Arch. Microbiol.*, 153:614.

Shioi, J., Tribhuwan, R. C., Berg, S. T., and Taylor, B. L., 1988, Signal transduction in chemotaxis to oxygen in *Escherichia coli* and *Salmonella typhimurium*, *J. Bacteriol.*, 170:5507-5511.

Sockett, R. E., Armitage, J. P., and Evans, M. C. W., 1987, Methylation-independent and methylation-dependent chemotaxis in *Rhodobacter sphaeroides* and *Rhodospirillum rubrum*, *J. Bacteriol.*, 169:5808-5814.

Spudich, J. L., and Bogomolni, R. A., 1988, Sensory rhodopsins of halobacteria, *Annu. Rev. Biophys. Biophys. Chem.*, 17:193.

Taylor, B. L., Miller, J. B., Warrick, H. M., and Koshland, D. E. Jr., 1979, Electron acceptor taxis and blue light effect on bacterial chemotaxis, *J. Bacteriol.*, 140:567.

Mechanisms and Strategies of Photomovements in Flagellates

Giuliano Colombetti

Istituto di Biofisica CNR
Via S. Lorenzo 26
I-56127 Pisa
Italy

and

Roberto Marangoni

Universita di Genova
Istituto di Biofisica CNR
Salita Superiore Noce 35
I-16132 Genova
Italy

Foreword

Among the many fascinating aspects of the life of microorganisms there is certainly their capacity to sense signals coming from the outside world and to react by altering their motile behavior (Colombetti and Petracchi, 1988). We will be dealing with a particular type of stimulus, light, and with its effect on the motor response of flagellated algae. There are several reasons that justify the choice of this type of stimulus (light plays a fundamental role for life on Earth, just think of vision and of photosynthesis). There are also practical reasons: light stimuli can be given and removed at will, it is easy to vary their spectral characteristics, intensity, polarization and also their time behavior can be kept relatively easily under control. It is thus possible to give very short light stimuli or to vary light intensity according to a predetermined temporal pattern or to apply different stimuli at the same time and also to different parts of a cell body. The motor responses of different cells may differ considerably, also according to the different structures of their motor apparatuses, and there have been (and there are) debates in the literature about terminology. We will simply and schematically speak of photophobic and phototactic reactions. In photophobic reactions a change in light intensity causes a transient alteration in the activity of the organism's motor apparatus. Where the occurrence of a response depends upon an increase or decrease in stimulus intensity, an increase is indicated by the term "step-up" and a decrease by the term "step-down". A movement oriented with

Biophysics of Photoreceptors and Photomovements in Microorganisms
Edited by F. Lenci *et al.*, Plenum Press, New York, 1991

respect to the direction of the light stimulus is referred to as phototaxis, and a movement oriented with respect to the polarization of the light stimulus is often called a polarotaxis.

Motile Activity

Leewenhoek (17th century) thus describes the motion of a flagellated alga: "..they had two short thin instruments which stuck out a little from the round contour and wherewith they performed the motion of rolling around and going forward".

This is only to remind the reader that cell motile behavior is brought about by specialized structures, cilia or flagella in the case of eukaryotes.

In discussing the problems of photomotile reactions one is often dealing with pigments, sensory transduction chains etc., but much less attention is usually paid to the motor apparatus of the cell which is the only structure that is responsible for the observed behavior. What we want to stress is that very often we look at behavior without considering that what is observed is just the result of molecular interactions taking place in the flagellar or ciliary structure, at least in the case of eucaryotes. Therefore, even if it is important to investigate and characterize the light induced behavior of a photoresponsive cell, it is also very important, and often disregarded, to carefully study its unbiased motion, with particular reference to the beating pattern of cilia and flagella. It should be noted that important contributions to the understanding of the photomotile behavior of *Chlamydomonas* have come from people working more specifically with the structure and function of the motor apparatus of this alga (Witman, 1990, and references therein). We also would like to acknowledge the very interesting work by Prof. Nultsch and coworkers on the flagellar beating of *Chlamydomonas* (Rüffer and Nultsch, 1987 and 1990, Sineschchokov, 1988), and the work of Paolo Gualtieri and coworkers on the ultrastructure of the flagellar apparatus of *Euglena* (Gualtieri, private communication).

Before giving some very general details on the structure and function of the ciliary and flagellar apparatus, we would like to stress once more that photobehavior in these cells is brought about by a coupling between light stimulation and motility; but cells are, of course, capable of sensing light even if there is no motor response; this may happen for instance if the flagellar structure has been paralyzed, or damaged, by the improper use of metabolic poisons, or when a cell has been deciliated or when a flagellum deprived mutant is used. In these cases other techniques can be used to study light reception, say for instance electrophysiology, *in vivo* microspectroscopy etc.

To sum up, a photomotile reaction, as we usually mean it, is the result of a modulation by light of the functioning of the motor apparatus of a cell; the latter does not require light to operate, except indirectly in the cases where light is the source of energy for the organism.

Hydrodynamics of Cell Motion

The motion of microorganisms takes place in a very particular hydrodynamic regime, known as a low Reynolds number regime. It would be very complicated to examine in detail what this means and how the equations of motion can be derived. It suffices to say that the so called Stokes force, acting on a sphere moving through a fluid and given by

$$F = 6\pi\, r\eta\, u \qquad (1)$$

where r is the radius of the sphere (in cm), η the medium viscosity (in g/cm s) and u the speed of motion (in cm/s), is a typical example of a force which is proportional to a

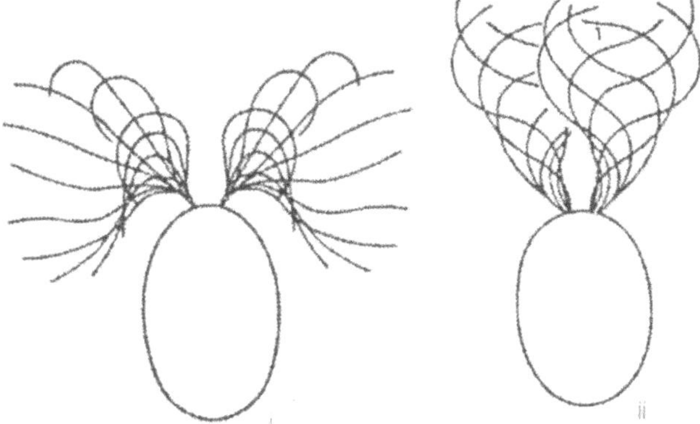

Fig. 1. Single frame analysis of the flagellar beating of *Chlamydomonas reinhardtii* i) breast stroke (ciliary-like; ii) reverse beating (flagellar beating mode). (Redrawn from Nultsch, 1988).

velocity. This is indeed characteristic of the motions at low Reynolds numbers: inertial terms are not important and the motion is only determined by the viscous forces acting the object moving in the fluid. The Reynolds number is defined as:

$$R = \rho \, u \, l / \eta \qquad (2)$$

where ρ is the medium density, l the linear dimension of the object, u its speed and η the medium viscosity.

For microorganisms it is of the order of 10^{-2}, and it is much smaller for cilia and flagella. Hydrodynamics at low Reynolds numbers has three major consequences, to put it in a simplistic way:

i) no reciprocal motion can take place at low R: if a ciliate would beat the recovery stroke just by going through the sequence of the effective stroke in reverse it would simply come back to its starting point, exactly retracing its trajectory;
ii) when something moves, there must be a force acting: inertia does not play any important role;
iii) any change in the force acting on a microorganism, e.g., any change in flagellar or ciliary beating will almost instantaneously affect the motion pattern of the cell.

Motor Apparatus

This discussion will follow the lines of the paper by Witman (1990), to which the reader is referred for more details.

The structure and function of eukaryotic cilia and flagella is practically the same; the difference in name is probably due to historical reasons. Normally we use the term cilia for cells having a large number of propelling units and the term flagellum when few of such structures are present. Cilia are usually shorter than flagella and the length can have some importance in determining the type of beating, flagellar or ciliary like, the ciliary type being less probable in long flagella because of physical constraints. Cilia and flagella have generally a different type of motion: flagella show a more symmetric beat which propagates bending wave from base to tip, moving the fluid parallel to the flagellar axis, as shown in Figure 1.

The fluid reaction thus brings about forward motion of the cell body. Cilia have a particular type of beating, which consists of an effective stroke and a recovery stroke. In this case fluid is moved both in the right and in the wrong direction, but the geometry of the recovery stroke and the hydrodynamics near the cell surface are such that much less fluid is moved in the wrong than in the right direction, so that a net forward movement of the cell body takes place. There are flagella, such as that of *Chlamydomonas*, that show a ciliary type motion during unperturbed swimming and a flagellar-like motion during reactions to external stimuli, as shown in Figure 1.

A typical eukaryotic flagellum is an array of microtubules connected in a characteristic structure, known as the 9+2 axoneme (see Fig. 2).

The nine outer doublets consist of a two components: the so-called A tubule with a circular cross section and the C-shaped B tubule which has part of its wall in common with tubule A. The flagellar microtubules are made up by heterodimers of α- and β-tubulin; tubulin is a protein with a molecular weight of about 110000, the α and ß subunits are similar but not identical to each other. The heterodimers form a protofilament with a diameter of 4 nm that runs along the length of the microtubule; the central microtubule and the A tubules consist of 13 protofilaments, whereas the B tubules have only 10 or 11. The central microtubules are singlet and are more similar to the cytoplasmic microtubules. The axoneme is anchored to the cell body by a special apparatus, called the basal body. In the basal body each external doublet becomes a triplet, but the central pair does not extend to this region of the cell body. Between the distal end of the basal body and the base of the axoneme there is a region known as the transition region, where the outer doublet microtubules lose their dynein arms, radial spokes and nexin links. This region contains very complex structures linking the A to the B tubules, not to be confused with the nexin links, and to the surrounding membrane. There are also intramembraneous particles that have been suggested to be the sites for Ca passage into the cilium or flagellum. Rupture of cilia or flagella occurs at the level of the transition region: it is a Ca-dependent process and may be due to a contractile, Ca-binding protein called centrin. It is a protein similar to calmodulin and has been localized in the transition region of *Chlamydomonas*. It should be noted that during deciliation or deflagellation, the distal portion of the transition region remains with the isolated cilia or flagella, whereas the proximal portion with its associated structures remain within the cell body.

Each subfiber A of the outer doublet has extensions known as arms which protrude towards the B subfiber of the adjacent doublet. These arms are built up by

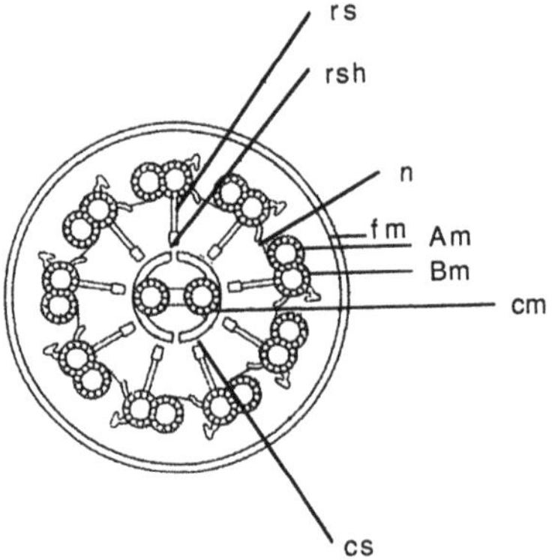

Fig. 2. Schematic representation of the cross section of a eukaryotic flagellum. Am = A-microtubules with inner and outer dynein arms; rs = radial spoke; rsh = radial spoke head; Bm = B-microtubules; fm = flagellar membrane; n = nexin links; cs = central sheat; cm = central microtubule. (Redrawn from Nultsch, 1988).

dynein, a protein that has ATPase activity and that is currently thought to be the molecule responsible for the chemo-mechanical transduction. The transient link between the outer dynein arm of subfiber A and the adjacent subfiber B is the force generating step in the mechanism of flagellar bending. In the absence of ATP the link is irreversible and the flagellum is in a state of permanent rigor. The A tubule of one doublet is connected to the B adjacent tubule also by a thin fiber, termed nexin, whose protein composition is unknown. Another structure, the radial spoke, extends from the A subfiber towards the core terminating in an enlargement known as the head of the radial spoke. Two pairs of projections originate from each central microtubule, once known as the central sheath.

In many flagella there are other types of structures between the flagellar membrane and the axoneme. We will only mention here the paraflagellar rod of *Euglena*, whose structure and function is presently being studied by the group of Dr. Gualtieri at the Biophysics Institute in Pisa.

It is currently accepted that the axoneme possesses the capability of generating flagellar bending by itself and that this process requires at least ATP and Mg. In fact, isolated demembranated flagella can generate normal waveforms when ATP is added to the system. Experiments have shown that the axonemal microtubules do not change their length during flagellar motion, but rather slide relative to one another. This sliding is brought about by the action od the dynein arms of a doublet on the adjacent doublet. The sliding has been shown directly in the classical experiments by Summers and Gibbons (1973). By partially digesting fragments of sea urchin sperm axonemes and adding ATP they observed an active sliding of the microtubules past one another, with the final result of an elongation and a disintegration of the axoneme. An electron microscope analysis showed that the radial spokes and the nexin links were destroyed, whereas the dynein arms were intact. This indicates that the tubulin dynein interaction is sufficient to initiate the sliding process and that the central spokes and the interdoublet links are probably involved in the mechanism that transforms sliding into bending. Sliding occurs in an asynchronous way between the different doublets and there is a phase shift between different doublets or groups of doublets. The speed of sliding and/or the transfer of active sliding to neighboring pairs of doublets determine the frequency of flagellar beating and is regulated by the concentration of ATP, dynein and Mg, but not so much by Ca. Flagellar bending and the general coordination of the beating do not seem to be very sensitive to these factors. The way sliding turns into bending is primarily determined by the fact that the proximally fixed microtubules retain a constant length and spatial arrangement; it depends probably also on the elastic properties of the radial spokes, central projections and nexin links.

It should be clear by now that there may be different sites at which the activity of a flagellum or cilium can be regulated. Regulation may involve symmetry and synchronization of active sliding, flagellar waveform and frequency or the base angle of the cilium or flagellum with respect to the cell body.

From most of the works published in the literature one infers that Ca ions play an important role as regulatory molecules. This will be dealt in more detail in the chapters devoted to sensory transduction. Here we only want to mention an interesting hypothesis that has been put forward very recently by Hans Machemer and his coworkers (Machemer, 1990, personal communication). According to it, Ca would be the universal regulatory messenger for ciliary activity via a competitive mechanism with Mg in the binding to a regulatory axonemal protein.

How can a flagellum or cilium propel a cell? We have already considered this question at the beginnning and we know the answer: cilia or flagella move together with their bodies just because they do not execute a reciprocal motion. This is accomplished in three different ways: a flagellum can propagate planar waves, helicoidal waves or move like a cilium. In most cases, such as *Euglena, Chlamydomonas, Hematococcus,*

Blepharisma (sometimes), *Paramecium*, the cell body moves along a helicoidal path (see Fig. 3).

In a limited number of cases we know reasonably well how the ciliary of flagellar beating is altered by external stimuli. For example, there is consensus that protozoa exposed to mechanical or luminous stimuli stop their ciliary motion, then, after a change in the base angle of the cilium with respect to cell body, reverse for a while their motion during the so-called avoiding response. There is also a general agreement on the overall molecular mechanisms underlying this behavior, even if not all the questions have been definitely answered.

Euglena beats its flagellum, which is bent backwards at the aperture of the reservoir, along the cell body in a helicoidal pattern and this causes the cell to move along a helical path. The frequency of body rotation is of the order of 2 Hz, as measured by high speed cinematography or light scattering techniques. It is thought that light and probably chemo-response are brought about by a change in the position of the flagellum with respect to the cell body; this change may be more or less pronounced according to the strength of the stimulus (phobic and tactic reactions?). There seems to be no variation in flagellar frequency, at least during the step-up photophobic response.

Chlamydomonas (and probably *Hematococcus*) swim showing a steady counterclockwise rotation modulated by an oscillatory component that reflects the forward rotation during each effective phase of flagellar beating and the backward rotation during the recovery stroke. The rotation of the cell body in *Chlamydomonas* and in *Haematococcus* can be visualized under the microscope thanks to the fact that between crossed polarizers the stigma is seen as a bright spot (Boscov and Feinleib, 1979), which appears and disappears while the cell swims. The frequency of body rotation in these algae can also be measured by using light-scattering techniques. In both cases it turns out to be of about 2 Hz. During normal motion the flagella beat in a cilia-like pattern, with a very asymmetric waveform. Upon stimulation with a step-up stimulus (see, however, the most recent results of Prof. Nultsch, 1990) the beating mode changes dramatically to the so-called reverse mode. The effective direction of the flagellar beat pattern is now almost parallel to the long axis of the cell and both flagella show relatively symmetric undulatory waves (flagellum-like beating) going from base to tip. It is a planary beating with an increase in beat frequency. This transition requires about 3 beat cycles (about 50 ms). It is not clear whether there is a significant change in the angular orientation of the flagellar base during this response. Some authors think that this is the case; other think that a simple change in the timing of the sliding machine could explain the observed pattern; this could be ascribed to a change in the modulation property of the radial spoke-central pair interactions.

A differential response of the cis-flagellum (the one closer to the stigma region) and trans-flagellum to unilateral illumination had been suggested and has recently been shown (Nultsch, this volume, Sineschchokov, this volume) to take place. This could also give some hints with respect to the sensory transduction chain, since it has been shown (Kamiya and Witman, 1984) that the two flagella have different sensitiveness to Ca ions.

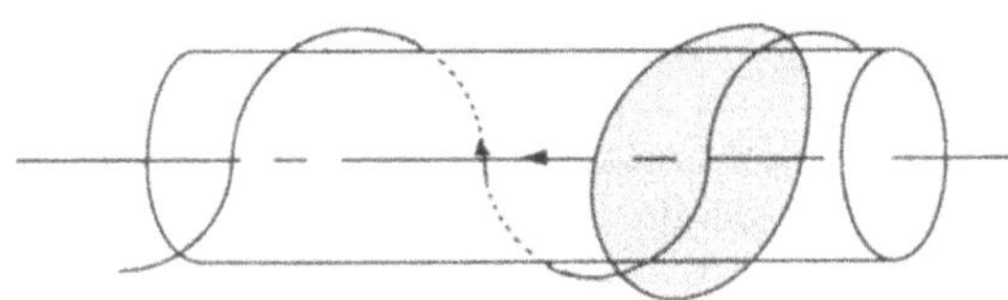

Fig. 3. Helicoidal motion of a ciliate. (Redrawn from Blake and Sleigh, 1974).

To sum up, it looks as though many different parts of the flagellar machinery can influence the type and mode of beating; this is not surprising, given the high degree of structural and functional complexity of the flagellum.

Phototaxis

Experimental Techniques

Other people will deal with this argument in more detail. Here we only want to make some general considerations on the measurement techniques which have been used in the last years and of their value and utility in understanding the cell behavior. There is, we think, general consensus that measuring behavior on single cells is the best choice. In reality, very often single cells measurements do not deal with the behavior of a single cell followed for a certain time (even if there are exceptions to this), but rather with the behavior of many single cells that is then analyzed statistically over the whole population (see e.g. Rayleigh test in phototaxis). This requires measurements on many single cells in order to have a reliable statistics; on the other hand, when the behavior of single cells under a certain light stimulation is precisely known, then it is possible to use the so-called population techniques. These are usually based on the measurement of variations of optical density or of scattered light in a certain cell population (Uhl and Hegemann, 1990). It can be done both for the phobic responses, when one is usually dealing with filling or emptying of lighted regions, and for the phototactic response. In both cases the usual parameter measured is the initial rate of variations of the optical density or number of cells. One should note, however, that this parameter does not depend only on the motor response to the light stimulus, but also on the speed of cells. For instance, if the phototaxis of a population is measured, one should always check very carefully that there are no major variations in the cell average speed during measurements or that different samples of cells have the same average velocity.

Experimental Results

The nature of the mechanisms of cell orientation with respect to light direction is still an open question.

It is not yet clear, by the way, whether cells orient with respect to light direction, or if they follow a light gradient. There are not too many precise and quantitative measurements on this aspect of the problem in the literature, though some experiments have been designed to answer this question. Here we only want to mention the work by Buder (1917), Halldal (1962) and by Song et al. (1980).

It is generally assumed that the driving signal is light direction, but more work is probably necessary to clarify whether this is true and also whether it is true for all the systems that show phototaxis. In the following discussion it is assumed that phototactic cells follow light direction.

There is also not much information available about the nature of the mechanisms of phototaxis. As already seen the phototaxis of cell populations has often been measured by determining the rate of change of cell density under light stimulation - the photoaccumulation rate. This parameter, however, depends not only on cell responsiveness to light, but may also be influenced by other factors, such as initial number of cells and their speed. In addition, it is not possible to use measurements of photoaccumulation rates to determine the manner (and timing) of cell orientation to light; this information is of unique importance in verifying the mechanisms involved in phototaxis. Some measurements on this aspect have been reported in the literature and will be discussed below.

Even when single cells are being tracked, the parameter used as an index of phototaxis is the fraction of cells whose velocity vector lies within a certain angular range with respect to light direction. This parameter does not, in itself, give useful information on the phototactic mechanisms.

Kaneda and Furuya (1986) have investigated the light orientation of individual cells of *Cryptomonas* using infrared videomicroscopy. They have studied in particular the temporal change in swimming direction and the relationship between swimming direction and phototactic orientation of individual cells exposed to unilateral light stimulation. They found out that there is a significant population orientation towards light 2 s after the onset of the light stimulus; in addition to this, only those cells that were swimming within a cone of 120° with respect to light direction showed a course correction significantly different from the random course variation taking place in dark kept cells. Their conclusion is that only cells illuminated in their anterior side can detect light direction.

Meyer and Hildebrand (1988) have described how *Euglena* cells reorient their motion after a 180° change in the direction of the actinic light. Two different behaviors have been observed with more or less the same probability: one that the authors call "arched reversal", and a second, termed "reversal in place"

The former behavior is very similar to the tracking of light in positive phototaxis observed sometimes in *Chlamydomonas* and has been already reported by Jennings (1906).

In cells showing the "reversal in place" type of behavior the average orientation time is of the order of 5 s. No data are reported for the time taken by cells in the "arched reversal", which depends very much on cell speed.

Cantatore et al. (1989) have measured by means of anisotropic laser light scattering the time taken by a cell population of *Haematococcus* to reverse their motion when subjected to stimuli alternatively coming from opposite direction. From their measurements it was concluded that the minimum time required to have a detectable change in the velocity direction was of the order of 800 ms.

The possibility that a modulated light is necessary in order to induce phototaxis has suggested experiments in which the stimulus is modulated externally. The idea is that it should be possible to increase or decrease the cell phototactic response by using appropriate light modulation frequencies with definite relations to the supposed natural modulation frequency. One of the first papers to deal with this problem appeared in the late sixties; recently Watanabe and Furuya (1982) and Kaneda and Furuya (1982) have addressed this question in *Cryptomonas*. They have investigated the phototactic activity of cells as a function of the length of dark and light periods in intermittent light stimulations. They were able to show that there is a critical dark period, independent of the length of the light period and of the fluence rate (250 ms in normal conditions), that strongly decreases the cell orientation to light. This dark period corresponds to about one half of the time taken by the cells to complete a body rotation during their helicoidal motion. This was confirmed by experiments in wich cells were exposed to a higher environment viscosity. Under these conditions the period of helicoidal movement increases and the critical dark period increases in a parallel fashion, so that the number of helical turns per critical dark period is approximately constant.

Orientation Mechanisms

It is generally thought that cells can detect the direction of light in two different ways (Feinleib, 1985):

i) it may have two separate photoreceptors, in addition to optically absorbing structures, such as the stigma, which can create a difference in light intensities between the two photoreceptors. The cell can detect the light direction by comparing the signals perceived by the two photoreceptors at the same instant (one-instant mechanism). Very little attention has been devoted to the one-instant mechanisms, at least in the case of algae, even if there may be experimental evidences that could be explained in these terms, as will be seen later. The one-instant mechanism is, presumably, that used by slowly moving, gliding organisms such as *Anabaena* (Nultsch, 1984), while it is generally accepted that flagellates use a two-instant mechanism.

ii) it may have only one receptor system, coupled with a screening device (the stigma). In this case, the cell must rotate periodically around its axis as it swims, so that the screening device can modulate the light falling on the photoreceptor. In such a case at least one body rotation (≈400 ms) would be necessary for a cell to detect light direction. The two-instant mechanism is essentially that postulated by Jennings (1906) and Mast (1911) and revisited by Foster and Smyth (1980), as discussed below. It requires a comparison in time between light absorbed during the screened and unscreened phases of the cell motion. The results by Kaneda and Furuya (1986), Meyer and Hildebrand (1988) and Cantatore et al. (1989) are compatible with this type of mechanism.

In some formulations the two-instant mechanism is strictly related to step-down or step-up responses. If this is true, phobic and tactic responses are just two different aspects of the same molecular mechanism. Positive phototaxis has been always thought of as associated with step-down responses - both produce accumulation inside a light trap; in this connection, one could, however, imagine a model in which a course correction occurs whenever the cell perceives a step-up signal, the correction being in the opposite direction with respect to the shading body. In any case, a programmed form of behavior is supposed, which enables cells to swim towards the light source.

According to Kaneda and Furuya (1982) there is no detectable correlation between positive phototaxis and photophobic responses in *Cryptomonas*. The idea that negative phototaxis in *Euglena* is brought about by repetitive phobic reactions has recently been questioned by Häder et al. (1986) and Häder (1987). In a first series of experiments this author measured the average cell orientation under two perpendicularly oriented beams of light; from his results, Häder concluded that cell orientation is probably brought about by a combination of the shading hypothesis with a dichroic orientation of the photoreceptor molecules perpendicular to the long axis of the cells - as had been suggested by Creutz and Diehn (1976) (see also below). In a subsequent work he measured the negative phototaxis of negatively gravitactic cells under linearly polarized light (fluence rate of about 100 W/m^2). To explain the results obtained, the author suggests that photoreceptor dichroism is responsible for negative phototaxis but that the dipole transition moments of the photoreceptor are not oriented as was previously thought. He estimates that they are oriented 25° clockwise off the long axis of the cell and 60° clockwise off the flagellar plane as seen from the front end of the cell.

There is a difference between the results of Häder (1987) and those of Creutz and Diehn (1976), who found that the photoreceptor for positive phototaxis under blue light (at a fluence rate of about 2 mW/m^2) is dichroic, but that its transition moment is oriented perpendicularly to the long axis of the cell. Häder (1987) suggests that this might indicate that there are two different sets of photoreceptors oriented at different angles or that the same molecule possesses different transition moments for positive and negative phototaxis.

Recently, Häder has obtained new results, showing that cell orientation is perpendicular to the light polarization plane at low fluence rates and parallel to it at high fluence rates; this can be expected if the transition moment of the receptor pigment is perpendicular to the long axis of the cell body, and if a cell tries to maximize light absorption in positive phototaxis and minimize it in negative phototaxis.

The use of polarized light could also help in clarifying the nature of the photoreceptor pigments. One should, for instance, be able to observe differing orientations under different wavelengths of polarized light if the transition dipole moments in the different absorption bands have different orientations.

It should also be noted that a dichroic photoreceptor can only detect the axis of light propagation, but not its direction; for the detection of light direction, the help of a screening device is again necessary.

A general remark that can be made about both Creutz and Diehn's (1976) and Häder's (1987) papers is that what they describe is not really phototaxis. In fact, the experimental conditions are such as to make impossible a movement toward or away from light and what is analyzed is a sort of transverse taxis, where the cells are forced to move in a plane perpendicular to the light direction.

Even if the interpretation of the periodic-shading mechanism holds true, qualitative differences may be expected between phobic responses and phototaxis; a system can, in fact, react with a different dynamics to a sudden change in its state and to small perturbations around it. One could, for instance, imagine a system where light modulation by a screening device causes a small membrane depolarization followed by a smooth course correction, whereas a sudden stimulus like the one experienced in a typical step-up or step-down photophobic response could bring about an action potential-like response with a dramatic change in trajectory.

There are also systems where phobic and tactic responses have been shown to have different receptor pigments, and systems with the same pigments and differing transduction chains.

Another interpretation of the two-instant mechanism supposes that the signal which determines the course correction is a continuously varying signal (Foster and Smyth, 1980) and that the feedback mechanism acts by small, gradual corrections.

In fact, it is our opinion that there is less difference between the model of Jennings and Mast and that of Foster and Smith than is usually thought: in both cases cells orient to light by a step-wise trajectory correction and their orientation would require a minimum amount of time (at least one body rotation, or about 400 ms).

In *Cryptomonas* the results of Watanabe and Furuya (1982) and Kaneda and Furuya (1982) reported above show that i) the rotation of the cells plays an important role in the detection of light direction, ii) they are consistent with the shading hypothesis and iii) they do not conflict, however, with the hypothesis that cells can detect light only at a certain phase of their rotation (one instant mechanism?).

If the periodic shading due to the stigma is the driving signal in the feedback loop which controls the tracking of the light source, the value of the stigma transmittance is a very important parameter. A statistical investigation on the optical densities of the stigma and their distribution in a cell population could be very useful in evaluating the minimum amount of modulation achievable in a phototactic sample.

A different interpretation of the stigma function has been suggested by Foster and Smyth (1980); they suppose that interference effects between waves reflected by the different layers of the stigma can produce an increase in the light falling on the photoreceptor. They were referring to the *Chlamydomonas* stigma, which has been described as a regularly layered structure. In this hypothesis, the modulation of the light falling on the photoreceptor should be greater than that which can be deduced from measurements of the stigma absorbance. This is a reasonable hypothesis, as the dimensions of the cell structures are of the same order of magnitude as the wavelength of visible light, and diffraction and interference should be expected in such a situation. We must, however, bear in mind that the structure of the stigma is not stable; for instance, in *Haematococcus lacustris* the stigma consists of a single layer in young cells and two or even three layers in old cells (Ristori, personal communication). A similar phenomenon has been observed in an eyeless mutant of *Chlamydomonas*; here the stigma appears in old cells. It should be recalled that the pigment content of the stigma and of the photoreceptor can vary from one cell to another according to cell age and metabolic situation, and, perhaps, the "history" of the single cell; this could be relevant in interpreting the responses of cell populations.

As already mentioned, the two-instant mechanism has been generally accepted for flagellated algae; There are, however, some experiments (Feinleib, 1985; Boscov and Feinleib, 1979) on the response of a population of *Chlamydomonas* to short flashes of light (flash duration of a few μs), which show a partial orientation in the population for fluence rates within a range of 0.1 - 2 W/m^2, and for responses going from positive to negative phototaxis. Since the duration of the flash is very short compared with the time for a full body rotation, it is tempting to explain these results in terms of a one-instant mechanism. A possible alternative explanation - proposed by the authors - is that only those cells whose photoreceptor is not screened by the stigma change their motion direction. This explanation does not conflict with the data obtained by recording the response of single cells - which were undergoing negative phototaxis under a laser beam - to flashes of 1 W/m^2. In these experiments only half of the cells responded to flashes, probably because of the high intensity of the orienting laser beam.

One hypothesis suggested by this series of experiments is that phobic reactions are programmed with respect to the morphology of the microorganisms (only if this hypothesis holds can the orientation resulting from single short flashes be accounted for without invoking the one-instant mechanism).

Models

A model for phototaxis has been proposed by Colombetti and Lenci (1983), by analogy with the chemotaxis of bacteria. According to this model, phototaxis can be thought of as a biased random walk. In unperturbed conditions the cells swim with periods of straight swimming followed by sudden jerks, or tumbles, after which they start again moving in a new direction. The new direction may or may not be correlated with the previous motion direction. If an external stimulus, such as unilateral illumination is present, the cells continue in their random motion, but the probability of a longer run or of a preferential turn along the direction of light is slightly increased, depending on cell adaptation, light intensity etc. (see Fig. 4). As a final result the cells move towards the light source, but there is no active orientation towards it.

A very simple model of a biased random walk is offered by the so-called random walk with drift. In this case the cells move randomly but are subjected to an external force acting in a given direction. The random walk takes place as in the absence of the external force (in our case this simulated force will depend on wavelength and light intensity) but there will be a net displacement in the direction of the force, with an average velocity proportional to the force itself. Another possible way of looking at this model is that of a random walk in which the probability of going towards the light is slightly higher than that of going opposite to it.

From a macroscopic viewpoint it is expected that single cells do not show a clear orientation of their paths towards light; this is true at least under not too strong illuminations. In fact, if the bias becomes very strong, as may happen under strong light, then the paths become almost oriented. It should be kept in mind that such a model allows one to understand the importance of the magnification used to observe the cells behavior. In fact, a biased random walk is a random walk, and if the magnification used to observe the cell trajectories (true or simulated) is too high, it will be very difficult to detect any motion toward a given direction.

The mathematical form of the bias is very important in determining the overall behavior of a simulated population and there are no hints as to this form. One should start from a measure of the main parameters of motion, such as average cell speed, number of jerks per unit time, mean free path etc.; then these parameters should be used to simulate the unperturbed motion and possibly refined in order to fit in the best possible way the real motion. This should be followed by a thorough analysis of the modifications of these parameters brought about by a non-directional light stimulation; eventually one should introduce a light dependent bias along light direction. This should be fitted to the experimental data, taking into account, for instance, the time taken by a cell population to accumulate in a lighted region, at different wavelengths and different light intensities, given a certain average speed.

The Fluence Rate-Response Curves

An interesting feature of some intensity-effect curves reported in the literature (Foster and Smyth, 1980) is their bell-shaped form. At low light intensities there is no orientation with respect to light direction; this orientation increases with increasing light intensities until a maximum response is reached; a further increase in light intensity produces a decrease in the light response, which eventually falls to zero, or becomes negative.

It should be recalled that the interpretation of the intensity-effect curves is based on hypotheses about how the system functions; the experimental determination of the curves is not enough to allow conclusions to be drawn on the underlying mechanisms.

The statement "the response is proportional to" implies that the response is an internal variable of a single cell - the signal produced by its photoreceptor - while the experimental data are the response as measured on a cell population. The latter kind of response is proportional to the percentage of cells moving towards the light source. The

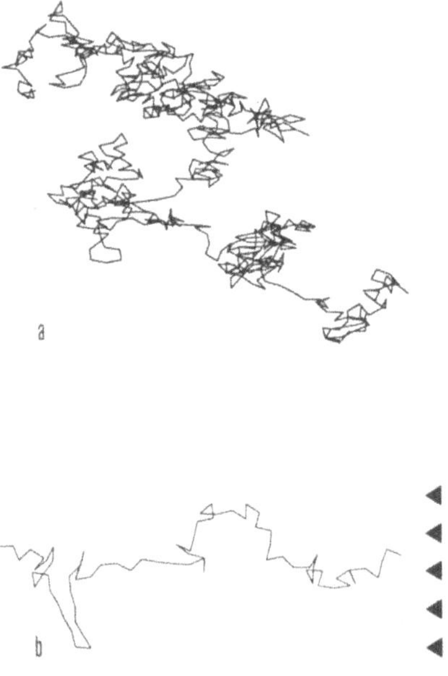

Fig. 4. Random walk of a cell. a) Pure random walk; b) biased random walk: the cell motion is biased with respect to light direction (indicated by the arrows).

identification between these two kind of responses can certainly be made if the signal inside a cell affects the probability that a cell will be oriented towards the light source; if it is supposed, on the other hand, that a cell is oriented when its internal signal is greater than the signal quantum noise, the response of a cell population could be explained by taking into account the differences between cells. So the model of the response determines the interpretation of experimental data.

We will now discuss the stimulating formulation of this problem put forward by Foster and Smyth (1980). As we have seen, the basic idea of these authors is that microorganisms can track the direction of light by exploring the environment through a directional antenna; this directional scanning produces a modulation of the light perceived by the system. The antenna consists of the stigma and the adjacent photoreceptor. In the traditional view, the stigma acts as a screening device; if the stigma is considered to be a quarter-wave plate - an interference mirror - it acts as an intensifier of the light falling on the photoreceptor, whenever this is suitably oriented. The quarter-wave plate model of the stigma has been considered the highlight of Foster and Smyth's paper. In reality, beside the considerations reported above on the reliability and generality of such a model, we believe that the most important concept proposed by these authors is that of a system which is capable of using the information carried by a directional antenna - a time-modulated light signal - to align the direction of cell motion with that of incoming light.

The modulated signal can then be "translated" by the cell machinery into an electric signal, into the concentration of some biochemical species or into a reaction rate; what counts is that the initial steps of the transduction chains will be determined by the number of molecules that are excited by light.

Foster and Smyth suggest that the bell-shaped form of the intensity-effect curves in positive phototaxis and the decrease in phototactic efficiency at relatively high light intensities can be explained in terms of a decrease in the availability of excitable pigment with increasing light intensity; but this conflicts with the fact that at higher light intensities phototaxis becomes negative.

A more quantitative analysis of this point requires an estimate of the fraction of pigment which is brought to an excited level when the system is illuminated at a certain light intensity I.

A general remark about the kinetics of photocycles is that the efficiency of a photoenergetic receptor is directly related to the speed of the photocycle, whereas this is not the case for sensory receptors. If it is assumed that signal amplification occurs after a photon has been absorbed, and that this amplification is achieved using metabolic energy, then a long-lived active form of the photoreceptor must exist, which should be able to catalyze many biochemical cycles. This is what happens in the visual system; there is no evidence that this is the case in algae or bacteria, but it is certainly a reasonable working hypothesis.

The optimum value for the lifetime of the active form of the pigment is the longest compatible with the time resolution needed for sensory response; this time resolution is of the order of some tenths of a second in the transduction of the visual signal and it could well be of the same order of magnitude in algae, as the frequency of body rotation is about 0.5 s.

An estimate of the amount of pigment in its active form can be given by simply modelling the photocycle according to Foster and Smyth. Let N_r be the number of photoreceptor molecules, σ_r its absorption cross-section, Φ_r the quantum efficiency for absorbed photon, Γ the time constant of the dark reaction that converts the pigment back to its native form, and N the number of pigment molecules in their native form. We can write:

$$dN/dt = - \Phi_r \sigma_r NI + (N_r - N)/\Gamma \qquad (3)$$

and at equilibrium

$$N = N_r/(1 + \Phi_r \sigma_r I \Gamma) \tag{4}$$

The factor $\Phi_r \sigma_r$ has been estimated to be of the order of 10^{-16} cm^2 for rhodopsin, and the values of I at which phototaxis occurs are of the order of 10^{14}-10^{16} quanta/cm^2. This implies that, for a rhodopsin-like pigment, N differs appreciably from N_r only for values of Γ of the order of a few seconds; on the other hand, if the model considered is true, the Γ values must be much shorter than the rotation period of the cell, which is about 0.5 s. It follows that the amount of pigment in its active form remains constant as long as phototaxis is positive. This is not in agreement with the conclusions of Foster and Smyth.

Foster and Smyth have also proposed an interesting interpretation of the intensity-effect curves, which is based on an analysis of the threshold values. We will now discuss this in some detail, as we believe that such an analysis ought to be applied whenever threshold values are determined; it should, however, be borne in mind that the measurements of threshold values in photomotile responses depends on what is called a response and on the sensitivity of the experimental apparatus.

This kind of approach makes it possible i) to estimate the number of photoreceptor molecules in a cell, starting from phototaxis threshold measurements; ii) to deduce some spectral properties of the antenna pigment. Let N_r be the number of photoreceptor molecules, Φ the photoreceptor quantum efficiency, and σ the photoreceptor cross section. Then: (Foster and Smyth, 1980, Colombetti and Lenci, 1983)

$$E_0 = N_r \Phi \sigma I_T \tau \tag{5}$$

where I_T indicates the threshold intensity and τ the duration of the non-screened phase. From this we obtain:

$$N_r = (1 + T) / [I_T \Phi \sigma \tau (1 - T)^2] \tag{6}$$

If T, τ and σ are known, a measurement of I_T allows us to estimate N_r.

The authors have used equation (12) to estimate the minimum number of molecules of receptor pigment per photoreceptor. For all algae except *Euglena gracilis* they chose the quantum efficiency Φ equal to 0.67 and the absorption cross section of the photoreceptor pigment equal to 1.5×10^{-16}. For *Euglena gracilis* the chosen values were 0.3 and 3.8×10^{-17}, respectively. The screen transmission T was assumed to be equal to 0.3 in all cases. By substituting these values, together with the threshold intensity measured by other authors (see Table 1 in their paper) in equation (11), they were able to calculate the number of photoreceptor molecules in several flagellated algae. They found, for example, that this number is within a range of 20000-200000 for *Chlamydomonas reinhardtii*, and of the order of 10^6 for *Euglena gracilis*. Colombetti and Lenci (1981) have used the same model to calculate the phototactic threshold intensity in *Euglena gracilis*. The calculated value is about 10^{12} photons/cm^2s, which corresponds to an average absorption of about 20 photons/s. The calculated threshold intensity is rather similar to that for positive phototaxis measured by Häder et al. (1986). The close agreement between calculated and measured values might indicate that Foster and Smyth's model is reliable for the system. The fact that the behavior of such a system depends on the absorption of a very few photons implies that there must be a great amplification to transduce the signal to the motor apparatus.

The formula given in (6) can be rearranged to give:

$$\Phi \sigma \approx [(1 + T)/(1 - T)^2] \, 1/I_T \tau \tag{7}$$

This allows us to determine the form of $\Phi\sigma$, if the threshold value and the transmittance of the screening organelles are known.

It also possible to determine action spectra at equal response. This will be dealt with in detail in another chapter devoted to action spectra determination. Here we only want to mention that under appropriate conditions (Colombetti and Lenci, 1980) it is possible to obtain the shape of the absorption spectrum of the photopigment also by plotting $1/N(\lambda)$ versus λ for a given value of the response. If the mechanism of orientation by light modulation is operating, it is possible to obtain information on the transmittance of the screening organelle through a comparison of the action spectra determined as threshold values and at equal response.

Phobic Responses

Typically, photophobic responses take place whenever microorganisms are subjected to a sudden variation in light intensity. In nature, this usually happens when the cells experience steep spatial light gradients, which the cell motion transforms into temporal gradients. As an example, consider the motion of *Euglena gracilis* in crossing the border of a light trap. A typical spatial gradient is of the order of 10 mW/m^2 in about 100 μm, which, for an average cell speed of about 100 $\mu m/s$, corresponds to a temporal gradient of about 10 mW/m^2s.

As we have seen, photophobic responses are usually brought about by changes in the physical parameters that characterize flagellar beating - such as flagellar base angle, beating frequency and waveform. The behavioral expression of these changes differs from case to case.

In *Euglena* a step-down of the stimulating light induces tumbling behavior; the response pattern consists of two different phases, an initial prolonged tumbling followed by a period of shorter tumblings. The time durations of the different phases have been used to characterize this photoresponse.

A typical phobic response in *Chlamydomonas* consists of a stop reaction followed by backward motion; the cell then stops again and eventually resumes its forward motion in a new direction. In *Chlamydomonas* and in *Haematococcus* it is sometimes possible to observe stop responses which are not followed by reverse motion. This usually occurs at lower light intensity.

Despite the different behavioral patterns of their photoinduced movements, all flagellated algae can detect temporal changes in the stimulus and adapt to steady stimuli. There are no detailed studies in the literature on cell responses to temporal light gradients of different forms. The information deriving from such measurements could greatly improve our understanding of the kinetics of the processes involved. It should be recalled that the overall behavior of cells is determined both by excitation and adaptation, which act concurrently in the process of light response. The adaptation process presumably determines the time duration of the phobic response, while the excitation process can be thought of as the trigger of a programmed behavioral pattern.

What is usually measured to quantify photophobic reactions - the percentage of responding cells in a population - is a continuous function of the stimulus parameters, and this has been used to plot intensity or dose-effect curves.

A possible model of how a photophobic system could work is that the stimulus affects an internal variable of the cell (the concentration of a molecular species and the rate of a chemical reaction are two of the alternatives to have been considered); the measured response would then depend upon such a variable. In such a model, the internal variable might continuously control response probabilities.

An alternative model, accepted more or less explicitly by some authors (Angelini et al., 1986) is that phobic responses occur whenever the amount of light absorbed by the photoreceptor in a given time - which is a characteristic of the system - is enough to induce the accumulation of some molecular species over a certain threshold value. This is a deterministic model.

In the deterministic model the fact that the form of the dose-effect curves is independent of wavelength is interpreted in a different way. Let us consider responses to stimuli of short and constant duration; the response of a single cell as a function of light intensity I should be a stepwise function and the same should be true of a population of identical cells. The difference between cells must then be invoked to explain the graded dose-effect curves.

It should be noted that for both these models the fact that the form of the intensity-effect curves is independent of wavelength does not prove that there is only one photoreceptor pigment. If the difference between cells is considerable, it may dominate any other effect and might even completely mask the dependence on wavelength of the intensity-effect curves.

If more than one pigment contributes to the response, the interpretation of the data is more complicated in both models; more refined models are then necessary, together with an experimental approach which makes it possible to reduce the contribution of screening or auxiliary pigments.

The difference between these two models may become considerable when the behavior of single cells is investigated. In a deterministic model, for instance, the lag time between stimulus and response should be constant for a single cell, whereas it should be randomly distributed in a probabilistic model. Repeated measurements are possible on single cells by using low-motility mutants (such as the *Chlamydomonas reinhardtii* mutant used by Berg to measure flagellar beating after light flashes), by immobilizing the cell with micropipettes, or by tracking the cells during their motion, as will be reported in this NATO-ASI.

The measurement of photomotile responses can give information not only on the photoreceptor pigments but also on the kinetics of the receptor system, in particular by using spatial gradients, as in a light trap, or temporal gradients, as in experiments with light flashes. These two cases allow different stimulus parameters to be controlled, and information about different parts of a photosensory transduction chain to be obtained.

An interesting experiment was performed by Creutz and al. (1978) on the photophobic behavioral responses of *Euglena gracilis* in a light intensity gradient. The idea was that of inducing step-down photophobic reactions in a light trap formed by an illuminated region (light intensity I) surrounded by a region with a lower light intensity, $I - \delta I$. This experimental set-up, obtained quite simply through a combination of neutral density filters and the optical system of a microscope, made it possible to vary I and $\delta I/I$ independently and, therefore, to study the dependence of the step-down photophobic response upon the intensity within the light trap and the step increase of intensity at the boundary of the trap. The experiments were carried out by measuring the transmittance of the sample, using the same beam as stimulating and as analyzing light; the parameter used to measure the strength of the response was the initial rate of variation of the sample transmittance. This parameter should be proportional to the percentage of cells which undergo photophobic responses when crossing the border of the light trap. The results show that the strength of the behavioral response depends on I as well as on δI, and is moreover directly proportional to the ratio $\delta I/I$. The authors interpreted these results in terms of a simple kinetic model. According to this, the strength of the step-down photophobic response should be proportional to the rate of change of the concentration of a photochemically active form of the photoreceptor pigment, which is converted into a photochemically quiescent form at a rate directly proportional to the light intensity.

Useful information on the dynamics of photoreceptor systems can be obtained by varying flash intensity I and flash duration δT. (Angelini et al., 1986). The step-up reactions of *Haematococcus pluvialis* in response to flashes of white light have been investigated by analyzing the anisotropically scattered light, as mentioned in the chapter on light scattering techniques. The percentage of cells that show variations in flagellar beating frequency after a flash stimulation has been chosen as an index of the step-up photophobic response. The findings of these authors show that the response depends on the dose of light (flash duration times flash intensity); reciprocity between light intensity and flash duration holds for durations not exceeding 60-80 ms, which could indicate that the rate-limiting step of the overall process is an accumulation process with a time constant of 20-30 ms.

It would be very interesting to combine both these approaches in a single experiment on the same microorganism; in reality, the first steps towards an understanding of the functioning of the system require a knowledge of the combination of the stimulus parameters controlling the response; a measurement of the incremental sensitivity (the measurement of the response as a function of $\delta I/I$) would also offer interesting information about the mechanisms of phototactic orientation.

A knowledge of the minimum number of photons - the threshold - which has to be absorbed to elicit a response, can give important clues as to the photosensory transduction chain of the step-up photophobic responses. Even if step-up photophobic responses occur at relatively high light intensities, they can be induced by short flashes; the integrated number of photons absorbed is therefore small.

It is possible to estimate this minimum number of photons from an evaluation of the number of photoreceptor molecules, but such an estimate is affected by a high error deriving from our poor knowledge of this last quantity.

Another possibility is to set up experiments based on the classical paper on visual perception by Hecht and Pirenne (1942). In this approach, the frequency of a system response - which is in our case the frequency of the occurrence of a motor response in a single cell - is measured at several flash intensities near the perception threshold, and is compared with the theoretical expression of the probability that n photons will be absorbed. Since the process of light absorption is Poissonian, the probability P(n) that n photons are absorbed can be expressed as

$$P(n) = ((p\, I\, \delta t)n/n!)\, e^{-p I \delta t} \tag{8}$$

where p is the probability that one photon will be absorbed per unit time, I is the light intensity, and δt the flash duration. In this way, the minimum number n^* of photons which give rise to a response can be estimated by fitting expression (8) to the experimental data points (Hegemann and Marwan, 1988).

It should be noted that in such an experiment the response is determined by a probabilistic relationship (where the probability is that enough photons will be absorbed), but the lag time between stimulus and response does not depend on stimulus intensity; this is not true in the probabilistic model discussed above, where it is the probability of response that is controlled by the stimulus.

The remarks just made indicate the importance of a thorough analysis of the behavior of single cells for an understanding of the functioning of the system; the current use of the measurement of single cell responses to evaluate average population behavior probably means that insufficient use is being made of the information which this kind of measurement makes available.

References

Angelini, F., Ascoli, C., Frediani, C., and Petracchi, D., 1986, Transient photoresponses of a phototactic microorganism, *Haematococcus pluvialis*, revealed by light scattering, *Biophys. J.*, 50:929.

Berg, H. C., 1985, Physics of bacterial chemotaxis, *in*: "Sensory Perception and Transduction in Aneural Organisms", Colombetti, G., Lenci, F., and Song, P.-S., eds., Plenum Press, New York, London.

Blake, J. R., and Sleigh, M. A., 1974, Mechanics of ciliary locomotion, *Biol. Rev.*, 49:85.

Boscov, J.S ., and Feinleib, M. E., ,1979, Phototactic response of *Chlamydomonas* to flashes of light. II. Response of individual cells, *Photochem. Photobiol.*, 30:499.

Buder, J., 1917, Zur Kenntnis der phototaktischen Richtungsbewegungen, *Jahrb. Wiss. Bot.*, 58:105.

Cantatore , G., Ascoli, C., Colombetti, G. and Frediani, C., 1989, Doppler velocimetry measurements of phototactic response in flagelated algae, *Bioscience Reports*, 9:475.

Colombetti, G., and Lenci, F., 1980, Identification and spectroscopic characterization of photoreceptor pigments, *in* "Photoreception and Sensory Transduction in Aneural Organism", Lenci, F., and Colombetti, G., Eds., Plenum Press, New York, London.

Colombetti, G., and Lenci, F., 1983, Photoreception and photomovement in microorganisms, in "The Biology of Photoreceptor", Cosens, D. and Vince-Prue, D., Eds, Cambridge University Press, Cambridge.

Colombetti, G., and Petracchi, D., 1988, Photoresponse mechanisms in flagellated algae, *Crit. Rev. Plant. Sci.*, 8:309.

Creutz, C., and Diehn, B., 1976, Motor responses to polarized light and gravity sensing in *Euglena gracilis, J. Protozool.*, 23:552.

Creutz, C., Colombetti, G., and Diehn, B., 1978, Photophobic behavioral responses of *Euglena* in a light intensity gradient and the kinetics of photoreceptor pigment interconversions, *Photochem. Photobiol.*, 27:611.

Feinleib, M. E., 1985, Behavioral studies of free-swimming photoresponsive organisms, *in* "Sensory Perception and Transduction in Aneural Organisms", Colombetti, G., Lenci, F. and Song, P.-S., Eds., Plenum Press, New York and London.

Foster, K. W., and Smyth, R. D., 1980, Light antennas in phototactic algae, *Microbiol. Rev.*, 44:572.

Häder, D.-P., 1986, Signal perception and amplification in photomovement of prokaryotes, *Biochim. Biophys. Acta*, 864:107.

Häder, D.-P., 1987, Polarotaxis, gravitaxis and vertical phototaxis in the green flagellate *Euglena gracilis, Arch. Microbiol.*, 147:179.

Halldal, P., 1962, Phototaxis in Protozoa, *in* "Physiology and Biochemistry of algae", Lewin, R. A., Ed., Acad. Press, New York.

Hecht, S. S., and Pirenne, M. H., 1942, Energy, quanta and vision, *J. Gen. Physiol.*, 25:819.

Hegemann, P. and Marwan, W., 1988, Single photons are sufficient to trigger movement responses in *Chlamydomonas reinhardtii, Photochem. Photobiol.*, 48 :99.

Kaneda, H., and Furuya, M., 1982, Effects of calcium and potassium ions on phototaxis in *Cryptomonas, Plant Cell Physiol.*, 23:1377.

Kaneda, H., and Furuya, M., 1986, Temporal changes in swimming direction during the phototactic orientation of individual cells in *Cryptomonas* sp., *Plant Cell Physiol.*, 27:265.

Jennings, H.S., 1906, "Behavior of the Lower Organisms", Columbia University Press, New York.

Mast, S.O., 1911, "Light and the Behavior of Lower Organisms", Wiley, New York.

Meyer, R., and Hildebrand, E., 1988, Phototaxis of *Euglena* at low external calcium concentration, *J. Photochem. Photobiol., B:Biology*, 2:443.

Nultsch, W., 1988, Untersuchungen zum Bewegungs- und Reaktionsverhalten des Flagellaten *Chlamydomonas reinhardtii*, Sitzungsberichte der wissenshafftlichen Gesellschaft an der J. W. Goethe-Universität Frankfurt am Main, XXIV (4).

Nultsch, W., 1990, this volume.

Rüffer, U., and Nultsch, W., 1987, Comparison of the beating of *cis*- and *trans*-flagella of *Chlamydomonas* cells held on micropipettes, *Cell Motil. Cytosk.*, 7:87.

Rüffer, U., and Nultsch. W, 1990, Flagellar photoresponse of *Chlamydomonas* cells held on micropipettes: 1. Change in flagellar beat frequency, *Cell Motil. Cytosk.*, 15:162.

Sineschchokov, O. A., and Litvin, F. F., The mechanisms of phototaxis in microorganisms, *in* "Molecular mechanisms of bioloigical action of optical radiation", A. B. Rubin, ed., Nauka, Moscow (in russian).

Song, P.-S., Häder, D.-P., Poff, K. L., Phototactic orientation by the ciliate *Stentor coeruleus*, *Photochem. Photobiol.*, 32:781.

Summers, K.E., and Gibbons, I.R., 1973, Effects of trypsin digestion on flagellar structures and their relationship to motility, *J. Cell Biology*, 58:618.

Uhl, R., and Hegemann, P., 1990, Adaptation of *Chlamydomonas* phototaxis: 1. A light scattering apparatus for measuring the phototactic rate of microorganisms with high time resolution, *Cell motil. Cytosk.*, 15:230.

Watanabe, M., and Furuya, M. 1982, Phototactic behavior of individual cells of *Cryptomonas* sp. in response to continuous and intermittent light stimuli, *Plant Cell Physiol*, 15:413.

Witman, G. B., 1990, Introduction to cilia and flagella, *in*: "Ciliary and Flagellar Membranes," R. A. Bloodgood, ed., Plenum, New York.

Mechanism and Strategies of Photomovement in Protozoa

Michael J. Doughty

University of Waterloo
School of Optometry
Waterloo
Ontario
Canada N2L 3G1

Overview and Perspectives

The subkingdom of the protozoa contains many different types of single celled organisms and some colonial (aggregate) forms that have been of interest to those interested in photomovement. In an older review by Bendix (1960), an evaluation of the photomovements of these different organisms, as reported by various researchers, was conducted by making reference to the orders to which each of the protozoa belonged. Revisions have been made both in the terminology (or naming of the protozoa) and also in the systematic grouping for classification (Lee et al., 1985). A re-evaluation of photomovements in the protozoa based on a systematics approach will hopefully prove useful if for no other reason than being considered as an update. Emphasis will be placed on the grouping of the protozoa by their subphyla, classes, (sub)orders. Within such a scheme, the current status of knowledge within each group will be briefly reviewed by highlighting aspects of the photochemical basis for the responses, the information available on possible transduction schemes and to compare the characteristics of the overall photomovements. Any order or genus where photobehavior has not obviously been reported has been omitted for convenience but noted in summary at the end. In this systematics approach, all of the subclass and suborder details are not included since there does not seem to be enough information at this time to warrant a systematics evaluation at a level beyond the individual (sub)orders. Full details of the classifications can be found in the recent text published by Lee, Hutner and Bovee (1985).

Photomovement, especially in some protozoan orders (e.g., *Euglena* sp.) is very easily observed (Wager, 1911) and equally well, easy to demonstrate in the laboratory (Cohn, 1866). For example, observations of a preference of *Euglena* to accumulate in a blue versus a red beam of light date back over 100 years (Cohn, 1866). Over the ensuing 30 to 40 years, considerable numbers of studies on a relatively large number of protozoan orders allowed reviews to be written comparing the different spectral sensitivity of these cells (Jennings, 1904; Holt and Lee 1901, Oltmans, 1917). In addition, even at this relatively early date, the ecological significance of the observed photomovements was already being discussed (Jennings, 1904). Such considerations of the pos-

Biophysics of Photoreceptors and Photomovements in Microorganisms
Edited by F. Lenci *et al.*, Plenum Press, New York, 1991

sible ecological significance of the observable photobehaviors did not receive significant further attention in the subsequent 25 to 30 years (Mast, 1941). In the initial reviews of what can be considered as a period of renewed interest in the photomovement of the protozoa (starting some 30 years ago), there can be found little consideration of the ecological consequences of the numerous types of photomovement that had, by then, been reported (Diehn, 1979). This period of research might be considered as a time when the mechanistic aspects of photomovement received a great deal of attention. These aspects covered both the experimental means to allow automatic recording of photo-behaviors (Diehn, 1979) as well as the actual details of the cellular motions that could underly the photomovements (Häder, 1979). This period of photomovement research also saw the proposal of numerous by sometimes diverse terminologies to describe the options (strategies) for photomovement that eventually prompted the introduction of a recommended set of terms (Diehn et al., 1977).

Now, at a time that is nearly 100 years since the start of the major studies on protozoan behaviors (including photobehaviors) in the latter years of last century (Holt and Lee, 1901; Jennings, 1904), the issue of ecological significance has been raised again in ernest (Fenchel, 1987; Häder, 1988). With a wide interest in ecological issues, a renewed concern over solar radiation and especially its ultraviolet components has started to receive attention (Worrest and Häder, 1989). In line with such ecologically-oriented perspectives, this seems an appropriate time to also carefully re-consider the significance of the numerous photobehaviors that have been reported for the different protozoan orders. Particular attention should be given to the fluence rates (light intensities) used to elicit photomovement in the laboratory.

The systematics approach is chosen here merely as a means to logically divide the protozoa into groups. From such an approach, constructive questions can hopefully be asked concerning any commonality of responses or chromophore types and/or physicochemical mechanisms that might underly these photomovements. Furthermore, this approach will serve to highlight the disproportionate attention that has been given to certain classes (or orders). As indeed concluded by others (Kivic and Walne, 1983), no one protozoan (e.g., *Euglena*) can be used as a representative "model" for the photobehavior of single-celled organisms.

Some of these organisms have become well known by investigators of photobehavior and some have yet to be studied in detail. This review will consider the photomovement of the protozoa at three levels. These will be the nature of the chromophore mediating photobehavior(s), the transduction mechanism linking chromophore excitation with changes in the motility-conferring apparatus and finally a consideration of the types and kinetics (time base) of these photomovements. The latter category can be linked to potential strategies since a consideration of the photomovement options should provide a basis for considering the ecological options for the organisms when the illumination characteristics of their habitat transiently or temporarily change or even show sustained (albeit cyclical) changes.

The systematics classification is given below (Table 1).

For the chromophores, it is relatively easy to summarize them (Table 2) even though there have been few complete attempts to both isolate and fully characterize a chromophore. The transduction mechanisms overall still remain largely speculative although it is hopefully useful to try to summarize them (Table 3). A full transduction mechanism has not been elucidated for any of these organisms. With the continued interest in photomovements of the protozoa, the list of photomovement options is now quite substantial (Table 4) even though the relative detail available on each of the photomovements is disproportionate.

As just indicated, two aspects of photomovement in the protozoa are necessarily speculative at this time: these relate to the mechanisms and the strategies for photomovement. It appears, at this time, that the mechanisms underlying photobehavior in

Table 1 Subkingdom Protozoa

Phylum	- sarcomastigophora
Subphylum	- mastigophorea
Class	- phytomastigophorea
Order	- cryptomonadida, e. g., *Cryptomonas*
	- dinoflagellida, e. g., *Gyrodinium, Gymnodinium, Gonyaulax, Peridinium*
	- euglenida, e. g., *Euglena, Astasia, Peranema*
	- chrysomonadida, e. g., *Ochromonas*
	- volvocida, e. g., *Dunalliella, Chlamydomonas, Haematococcus, Volvox*
Subphylum	- sarcodina
Order	- amoebida [suborder tubulina], e. g., *Amoeba*
	- dictyosteliida, e. g., *Dictyostelium, Polyphondylium*
	- physarida, e. g., *Physarum*
Phylum	- ciliophora
Subphylum	- postciliodesmatophora
Class	- spirotricea
Order	- hetertrichida, e. g., *Blepharisma, Spirostomum, Stentor*
Subphylum	- cryptophora
Class	- oligohymenophorea
Order	- loxodida, e. g., *Loxodes*
	- peniculida, e. g., *Paramecium*
	- hymenstomatida, e. g., *Tetrahymena*

the protozoa are diverse. In attempting to overview these possible mechanisms, it is necessary to define what might be included under the heading of "mechanism". This can be viewed at three levels:

(a) the photochemical mechanism or the actual photochemical process(s) that first need to be activated (or supressed) by incident light. The first step usually towards defining the photochemical process can either involve the measurement of action spectra or the use of exogeneous chemical reagents that will hopefully specifically affect the photochemical process (Table 2).

(b) the transduction mechanism(s) or those processes that connect the activation of a (specific) chromophore and produce changes in the cellular membranes (internal or external) of the cells. Such changes are now generally accepted to involve discrete alterations in ionic permeability such that ion flows across cell membranes can be directly or indirectly induced. Examples of both options have been reported (Table 3).

(c) the mechano(chemical) mechanism or the actual changes in the locomotory apparatus of the cells that eventually results in photomovement. These mechanical effects can involve alterations (increase or decreases) in the activity (e.g., frequency of beating) of organelles such as cilia or flagella, involve changes in the cellular cytoskeleton or internal cell contractile mechanisms that mediate gliding-substratum behavior.

In considering the actual characteristics of the reported photomovements, a consideration of the mechanical aspects cannot be easily distinguished from what one can term the strategy of photomovement. In other words, the options available to a cell for directed or non-directed photomovement are determined by the motility options that the cell has in its reperoire. While some organisms clearly can exibit increases or decreases in motion (speed), the added option of stop responses or even time-dependent

Table 2 Chromophores Mediating Photomovement in Protozoa

order	spectral sensitivity[a]	chromophore[b]
cryptomonadida	420 & 560 nm	phycoerythrin (carotenoid)
dinoflagellida	470-490 nm	unknown
euglenida	450-470 nm	flavin
chrysomonadidia	?	?
volvocida[c]	ca. 500 nm	rhodopsin
tubulina	450-470 nm	flavins + cytochromes
dictyosteliida	405-410 nm	cyotochrome + flavin
	580, 640 nm	
physaridia	blue	?
heterotrichida	475, 550	hypericins & 590 nm
loxodida	360 & 435 nm	flavin ?
peniculida	480-560 nm	various[d]
hymenostomatida	?	protoporphyrin ?

[a]Spectral sensitivity indicated by major peak(s) in action spectrum or spectral sensitivity profiles.
[b]Chromophore - the chromophore(s) mediating one or more types of photomovement whether assessed by action spectrum or absorption spectroscopy or on cells or isolated pigments.
[c]Some of these cells may have chlorophyll-containing algae.
[d]Some of these cells have chlorophyll-containing algae so the chromophores are carotenoid and/or chlorophyll linked. However, others are "colorless" yet contain a heme-like pigment (protoporphyrin).

Table 3 Ionic Mechanisms Related to Transduction for Photomovement in Protozoa

order	example	transduction[a]
cryptomonadida	*Cryptomonas*	Ca^{2+} (K^+)
dinoflagellida	*Gyrodinium*	?
euglenida	*Euglena*	Ca^{2+} (K^+, Na^+)
chrysomonadidia	*Ochromonas*	?
volvocida	*Chlamydomonas*	Ca^{2+}
tubulina	*Amoeba*	Ca^{2+} K^+ pH
dictyosteliida	*Dictyostelium*	?
physaridia	*Physarum*	?
heterotrichida	*Stentor*	Ca^{2+} pH (K^+)
loxodida	*Loxodes*	pH O_2 (Ca^{2+}?)
peniculida	*Paramecium*	(Ca^{2+}?)
hymenostomatida	*Tetrahymena*	?

[a]Indications are given of the ionic species that have been primarily proposed or considered to be involved in transduction. Secondary species (regulatory species ?) are given in brackets.

alterations in their motile organelles (phobic responses; step-up or step-down) has not been reported for others or vice versa. It remains uncertain at this time whether all cells can display all of these photomovement options or whether it is simply that someone has not, as yet, found the time to investigate them. For each type of photomotile response, we can at least guess as to what the manifestation of the response actually means to the cell in its natural habitat and these ideas will be indicated where appropriate. However, in attempting to define such strategies, it must be recognized that in many cases, the light stimuli to elicit a photomovement may not have been scrutinized in terms of their physiological relevance.

Table 4 Options for Photomovement in Protozoa (Mechanisms and Strategies)

order	motility system	mechanics (options)	strategy
cryptomonadida	flagella	flagella reorientation phototaxis	avoid light
dinoflagellida	flagella	flagella arrest stop response, phototaxis	water column positioning
euglenida	flagella	flagella reorientation photokinesis phototaxis photophobic responses	selection of optimum light for photosynthesis avoid bright light
chrysomonadidia	flagella	flagella reorientation photophobic responses phototaxis	?
volvocida	flagella	flagella arrest/reversal stop response phototaxis	selection of optimum light for photosynthesis
tubulina	gliding motion (amoeboid)	stop response photokinesis phototaxis	avoid light
dictyosteliida	gliding motion (amoeboid) photokinesis	amoeboid motion arrest phototaxis	light-regulated cell aggregation (development)
physaridia	gliding motion (amoeboid)		light-regulated cell aggregation
heterotrichida	cilia	cilia (somatic or membranelle) reversal/arrest photophobic responses phototaxis photokinesis	avoid light
loxodida	cilia	cilia reversal photokinesis	habitat positioning
peniculida	cilia	cilia reversal, photokinesis photophobic responses	symbiosis; avoid light
hymenostomatida	cilia	cilia reversal? photophobic responses?	photoregulation of growth

A systematics approach to each of the protozoan groups follows.

Subkingdom: Protozoa, Phylum: Sarcomastigophorea, Subphylum: Mastigophorea, Order: Cryptomonadida, e.g., Cryptomonas

The cryptoflagellates (cryptomonads) of the genus Cryptomonas are small oval shaped cells (10-20 μm in length) that contain chloroplasts and possess two, mastigoneme-bearing locomotory flagella (Lee et al., 1985), (Fig. 1). There are free-swimming freshwater and marine species. The cells swim as a result of a base-to-tip undulatory wave of the flagella that results in forward motion with rotation of the cell along its long axis.

Mechanisms

The Photochemical Mechanisms. These cells have been observed to exhibit positive and negative phototaxis and at least some species may show perpendicular orientation (diaphototaxis) to a light source (Rhiel et al., 1988). Neither a photophobic response (step-up or step-down) nor a stop response to light have been reported. There appear, however, to be marked differences in the relative photic sensitivity of freshwater versus marine species. Cell suspensions show peaks of carotenoid absorption at 420 nm and a secondary peak at 560 nm in addition to chlorphyll (Rhiel et al., 1988). The action spectra for positive photoaxis to low fluence rates (e.g., 15 W/m^{-2}) in one species show a clear peak at 550 nm (Kaneda and Furuya, 1987) - a result consistent with a carotenoid or carotenoprotein or a phycobilin in phycoerythrin (Watanabe and Furuya, 1974).

The Transduction Mechanisms. The orientation of single *Cryptomonas* to light (as a mediator of positive phototaxis) has been reported to be influenced by the levels of free calcium in the medium (Kaneda and Furuya, 1987). The addition of high levels of the chelating agent EDTA supressed the photorientation but this effect could be competitively overcome by addition of large excesses calcium salts. The addition of 20 mM KCl decreased the turning angle and eventually reduced it to zero [i.e. the cells, thought still motile, failed to orientate to light (Kaneda and Furuya, 1987)]. Therefore, while actual observations of flagella motion changes have not been reported, it can be concluded that the action of the actinic stimulus alters calcium levels at the cell (or flagellar surface) or even the entry of calcium into the cells. These changes result in a change in flagella beating.

The Mechanochemical Mechanisms. Since the cell rotates on its long axis while swimming and because the photoreorientation can be induced by very short light flashes (less than the period of rotation), a periodic shading of an unknown photoreceptor structure has been proposed (Watanabe and Furuya, 1974). Indeed, based on studies of the angle of cell reorientation with respect to the swimming direction in the dark, it has been suggested that the photoreceptor is on the anterior side of the cell and cannot absorb light through the chlorplasts (Kaneda and Furuya, 1986). The period of rotation can be increased (with reduction in cell swimming speed) by the placement of cells in a

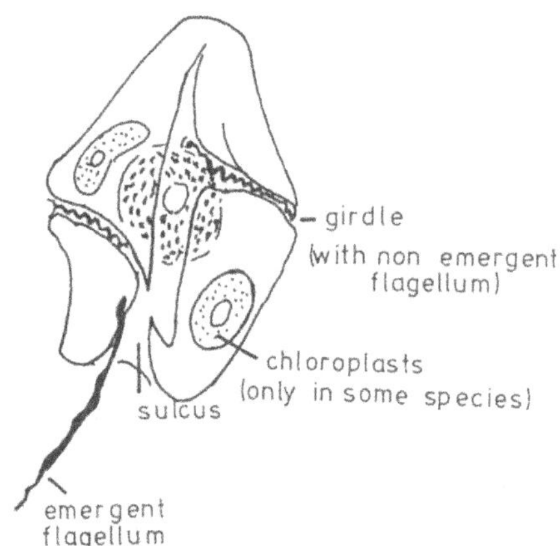

Fig. 1. Basic structural organization of cryptoflagellates (after Lee et al., 1985).

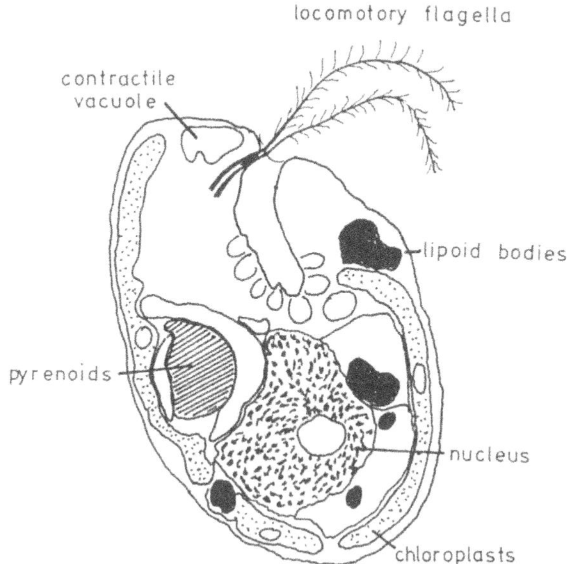

Fig. 2. Basic structural organization of dinoflagellates (after Lee et al., 1985).

viscous methylcellulose medium (Umetsu-Kaneda and Furuya, 1982) with a concommittant increase in the response time. In all cases, the photoorientation of the cells has been observed to be a smooth progressive process with a defined angle (Kaneda and Furuya, 1986) to the direction of the light source.

Since, at least in some species, the percentage responding organisms increases as a function of fluence rate (regardless of whether the phototaxis was positive or negative), the dominant strategy, in terms of the cell population, seems to be to avoid light. It has been proposed that negative phototaxis is a means to avoid photobleaching in "susceptible" species (Rhiel et al., 1988). However, while photodestruction (photobleaching of the cell pigments) has been reported (Rhiel et al., 1988), the ecological significance of this is uncertain since the fluence rates eliciting negative phototaxis appear to be within natural fluence rates, i.e., the cell has the means to avoid high levels of sustained solar irradiation despite the observation that they can show perpendicular orientation to sunlight exposure initially (Häder et al., 1987a). As noted above, some species may show this perpendicular orientation to light or an apparent bias towards positive phototaxis (Rhiel et al., 1988) while other species have been reported to show a negative phototaxis at either moderate fluence rates (100 W/m^2) (Häder and Häder, 1989) or sustained solar radiation exposure (Häder et al., 1987a). The dynamics of these processes need further investigation before definitive conclusions about the ecological significance can be made.

Subkingdom: Protozoa, Phylum: Sarcomastigophorea, Subphylum: Mastigophorea, Class: Phytomastigophorea, Order: Dinoflagellida, e.g., Gyrodinium, Gymnodinium, Gonyaulax and Peridinium

Four examples of a rather large order of free-swimming dinoflagellates (dinophyceae) of widely variant sizes up to 500 μm that possess a locomotory flagellum (sited in a specialized groove along the edges of the cell, the cingulum) anchored, as in other flagellates, in a relatively straight sulcus (Lee et al., 1985) (Fig. 2). At least in some *Gyrodinium*, the sulcus extends to what has been proposed to be a light channel -

at the junction of which (with the sulcus) the photoreceptor region is thought to exist (Hand and Schmidt, 1975). The other flagellum can be emergent or more usually aligned along an equatoral groove, called the girdle. In its simplest case, it is thought that while the emergent (or longitudinal) flagellum exerts a simple planar beat (from proximal to distal end to drive the cell forward), the activity of the equatorial (or lateral) flagellum exerts a complex rippling beat (called helicoidal) that promotes cell rotation. The overall motion has thus been described as "the cell moves forward in the water screwing itself along" (Piccinni and Omodeo, 1985), i.e., exibits a complex spiral motion for which the velocity and pitch have been measured (Hand and Schmidt, 1975).

Mechanisms in Dinoflagellida

The Photochemical Mechanism. Cells in this order have been reported to exibit a marked stop response to illumination as well as positive phototaxis. Several photochemical systems can apparently mediate photobehavior in this order. Action spectra for a photo-stop response (sometimes referred to as a step-up photophobic response) of *Gyrodinium* shows a maximum at 470 - 490 nm (Forward, 1973). The relative sensitivity to blue light appears, however, to be dependent on the past light history of the cell (particularly UV-B and far-red) (Forward, 1974). Equivalent studies have not, however, been carried out on other flagellates such as those of the orders euglenida, chrysomonadidia or volvocida and so the significance of these observations on *Gyrodinium* is uncertain. The nature of the "blue" pigment remains unknown. Similar broad action spectra peaking in the blue have been reported for *Gymnodinium* (Forward, 1974) and *Gonyaulax* (Halldal, 1958). In contrast, an action spectrum for positive phototaxis has been reported for *Peridinium* that shows a well resolved set of peaks clustered around 640 nm (Liu et al., 1990) despite the fact that it has been classified as in the same family as *Gonyaulax*.

The Transduction Mechanism. No studies have been performed on the photosensory transduction mechanisms of this order of protozoa although a variety of bioamines (natural and synthetic) have reported to alter the photic sensitivity of *Gymnodinium* (Foward, 1977) and ions and ionophores have been successfully applied to *Gonyaulax* to investigate membrane events associated with photoperiods of growth (Mergenhagen, 1980) (indicating that its membranes are at least susceptible to these drugs). It is relevant too to note that spheroplast membrane potential changes have been reported to be associated with a marked circadian rhythm (of growth) in these cells (Adamich et al., 1976). The coupling of these membrane potentials to any that might be postulated to be associated with photomovement is unknown but needs to be considered especially because such circadian rhythms also occur in *Gyrodinium* (Hand and Schmidt, 1975; Forward, 1975).

The Mechanochemical Mechanism. The best known response of this order of protozoa is their stop response. *Gyrodinium*, in response to a sudden increase in illumination, stops swimming within 0.5 to 2.5 s (Hand and Schmidt, 1975). The cells then rotate (i.e., the girdle flagellum is still undulating or describing its helical beating profile driving the cell rotation) until that period in time when the emergent, longitudinal flagellum starts beating in a planar fashion: the initiation of this latter motion causes the cell to reorientate to the light. *Gyrodinium* has been reported to swim towards the light (i.e., shows positive phototaxis after a transient stop response) as the emergent flagellum adopts its normal helical (or multiplanar) beating profile (Hand and Schmidt, 1975). The orientation is thought to be achieved as a result of the unique orientation of the sulcus to that the cell stops "turning" as it leaves the stop response when the sulcus is aligned to the light direction. The positive phototaxis of other *Gyrodinium* species has been confirmed by others (Ekelund and Häder, 1988) and also observed in *Peridinium* (Liu et al., 1990; Häder et al., 1990).

The photic sensitivity of *Gymnodinium* is thought to interact with chemical and gravitational sensitivity of these cells to determine their vertical position within the water column (Häder, 1988). The latter components have been proposed as being necessary balances for the positive phototaxis (since an obvious negative phototaxis has not been observed) (Ekelund and Häder, 1988). It has been proposed that *Gyrodinium* fail to show negative phototaxis because "they stand rather high fluence rates without being noticeably photobleached" (Ekelund and Häder, 1988). This proposal can alternatively be viewed as evidence of an indirect regulator of the degree of positive orientation and movement in addition to the cells simply exhibiting lesser degrees of surface-directed motion when the fluence rate at the surface decreases towards dusk (Forward, 1975). Based on laboratory determinations of alterations in positive phototaxis following periodic solor radiation exposure, it has been suggested that freshwater *Peridinium* will position themselves at a water column level equivalent to 20% sunlight (Häder et al., 1990). In this context, it needs to be pointed out that the criteria of resistance to photobleaching in *Gyrodinium* was assigned from laboratory studies at solar fluence rates that were about 25% of those measurable on a sunny day (Ekelund and Häder, 1988).

Subkingdom: Protozoa, Phylum: Sarcomastigophorea, Subphylum: Mastigophorea, Class: Phytomastigophorea, Order: Euglenida, e.g., Euglena, Astasia and Peranema

Members of the *Euglena* genus are often colored (i.e., contain chloroplasts); the others are generally colorless (Bendix, 1960; Lee et al., 1985). The cells, especially for *Euglena* sp. also have a characteristic "eyespot" (the stigma) sited at one side of the sulcus (Fig. 3). The cells are characteristically elongate with a prominently large anterior reservoir (sulcus) from which emerges a locomotory flagellum. This flagellum executes a base-to-tip helicoid motion in *Euglena* and *Astasia* but is more likely to be planar in *Peranema*. In the "euglenids", the 3-D profile and hydrodynamics of the locomotory flagellum are such that it bends backwards to beat alongside the body in most situations. The flagellum length is often similar to that of the body (30-40 μm). The 3-D motion of the locomotory flagellum produces a helical swimming path for which the velocity has been measured under a variety of conditions (Holwill, 1966; Doughty and Diehn, 1979).

Mechanisms in Euglenida

The Photochemical Mechanisms. In a recent review (Colombetti, 1990) it was stated that the "photobehavior of protozoa began when Engelmann reported (in 1882) that the protozoan, *Paramecium bursaria* ... accumulated in a light trap". A correction must be made that this referred to ciliated protozoa. Studies on *Euglena* date back to at least 15 years earlier (Cohn, 1866) when photoaccumulation in a blue light trap was reported! The repertoire of photobehaviors is the most extensive reported because of the extended time period for which *Euglena* has been studied. This list of reported photobehaviors include photoaccumulation in a blue (or even red) light trap, a step-up and a step-down response and positive and negative phototaxis. Data relating to possible light regulation of cell swimming velocity is ambiguous (Doughty and Diehn, 1980). Action spectra for photoaccumulation/photodispersal (Diehn, 1969) as well as step-down photophobic responses of individual cells (Barghigiani et al., 1979) show a clear peak in the blue (450-470 nm). The photopigment is accepted to be a flavin especially because fluorescence microscopy studies show a clear fluorescent green signal eminating from the region of the accepted photoreceptor organelle, the paraflagellar body (Benedetti and Checcucci, 1975). Similar, but less well resolved, spectral sensitivity profiles for step-up photophobic responses have been reported for two chloroplast free (colorless) species of *Astasia* (Suzaki and Williamson, 1983; Mikolajczyk and Walne, 1990). A spectral sensitivity profile for a step-up photophobic response in a *Peranema* species however showed a peak at 500 nm (Shettles, 1937). *Euglena* have also long been known to respond to ultraviolet radiation (or simulated solar radiation) (Jirovec, 1934; Häder, 1986; Doughty and Diehn, 1980) although the photoreceptor is undefined. Such sensitivity could be a consequence of UV-B absorption by the flavin chromophore of the PFB since significant responses have been reported for polarized UV-B (Diehn, 1969). Some explanations still need to be found, however, for as to different action spectra have been observed in the various studies (Doughty and Diehn, 1980). The mechanism(s) by which excitation of a flavin chromophore can produce intracellular signals that eventually result in a changed motion of the flagellar apparatus remain undefined.

The Transduction Mechanism(s). The transduction mechanism has been mainly studied for the flagella response mediating step-down photobehaviors. The duration of the flagellar-reorientation response that can be induced by a reduction in light intensity has been shown in numerous studies to be extraordinarily sensitive to the ionic composition of the medium in which the cells are suspended (Doughty and Diehn, 1979, 1982, 1983, 1984; Doughty et al., 1980). The same flagella reorientation response can be specifically induced by ionophore (A23187) in the presence of suitably high extracellular levels of calcium ions (Doughty and Diehn, 1979). The mechanism of calcium elevation and the membrane(s) involved (for both the step-down response and for phototaxis) are not well defined (Doughty and Diehn, 1983) but a scheme has been proposed by which light alters sodium and potassium permeability of cell membranes and this in turn promotes calcium influx (Lenci et al., 1984). A similar rationale, unfortunately with different extracellular solutions, has been utilized to probe the ionic mechanisms of the response of *Euglena* to directional illumination (a positive or negative phototaxis) (Meyer and Hildebrand, 1988; Häder et al., 1987). The conclusions are rather different but two studies (Doughty and Diehn, 1979; Häder et al., 1987) indicate that drugs targeted at voltage-dependent ionic channels (verapamil or D600) have no effect on either the step-down photobehavior or negative phototaxis. Changes in extracellular calcium levels *per se* have been reported to only have relatively small effects on the duration of the step-down photophobic response (Doughty and Diehn, 1979) or the degree of photoorientation in positive phototaxis (Meyer and Hildebrand, 1988). It has proven to be difficult to separate photoresponses from the effects of extracellular calcium ions or motility *per se* (Doughty and Diehn, 1979; Meyer and Hildebrand, 1988). Further studies are clearly required. For *Peranema*, similar studies (Mikolajczyk, 1986)

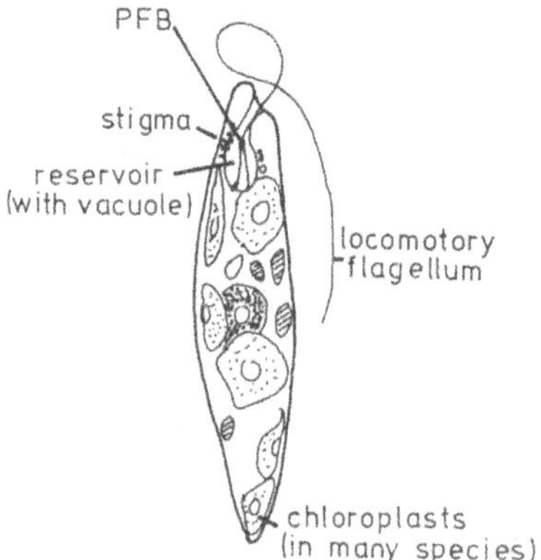

Fig. 3. Basic structural organization of euglenida (after Lee et al., 1985).

with drugs considered active on membrane sodium and potassium permeability can be presented as evidence for similar transduction mechanisms in *Euglena* and *Peranema*.

The Mechanochemical Mechanism. For the step-down photobehavior, and end-over-end cell tumbling is clearly mediated by a flagellar reorientation such that the proximal to distal helical travelling wave takes on a larger amplitude and is essentially perpendicular to the cell axis rather than being parallel in its overall vector. The cell typically shows such a response on crossing a light to dark boundary and thus accumulates in the lightened area (Doughty and Diehn, 1979, 1980). The same type of flagellar reorientation appears to mediate a similar behavioral response seen in response to (high intensity) step-up stimuli (Diehn et al., 1975; and unpublished observations). In response to step-up stimuli, the cells exibit a series of cell tumbling events before adapting to the higher fluence rates and resuming normal swimming (Doughty, 1990a). This response would be expected to prevent the cells from moving into a high intensity (> 100 W/m^2 white light) light trap. The flagellar response during orientated movement has not been clearly defined although the angle of orientation of the cell trajectory with respect to the light source has been investigated in some detail (Häder et al., 1986, 1987). Both *Astasia* (Suzaki and Williamson, 1983, Mikolajczyk and Walne, 1990) and *Peranema* (Shettles, 1937; Mikolajczyk, 1986) have been reported to exibit step-up photophobic responses following increases in light intensity although both are "colorless".

Insufficient information is known about the ecology of the motile *Euglena* species. The cell avoids sunlight since it is photolethal (Doughty and Diehn, 1980; Häder, 1986; Häder and Häder, 1988) and so its avoidance of high light intensities is not surprising. The reason for its positive phototaxis to low light intensities (balanced perhaps by its ability to accumulate in low light intensity regions as a result of being trapped at the light-dark boundary by step-down photophobic responses) is presumably related to preservation of the optimum light intensity for light-regulation of photosynthesis (Kaufman and Lyman, 1982). These mechanisms probably all contribute to determining the position of the cell in the water column (Häder and Griebenow, 1988).

Subkingdom: Protozoa, Phylum: Sarcomastigophorea, Subphylum: Mastigophorea, Class: Phytomastigophorea, Order: Chrysomonadida, e.g., Ochromonas

The genus *Ochromonas* is one of several in the order of chrysomonadidia (Chrysophaecae) flagellates that are not dissimilar in gross form from the euglenida. They can be of a similar size to the euglenida and some also have an eyespot like the euglenida (Fig. 4). This genus, however, possesses two unequal length locomotory flagella that are inserted at the apical surface rather than emerging from an apical reservoir. As with euglenida, a paraflagellar body exists (which is presumably the photoreceptor) but this is now outside of the cell (rather than being encircled by a reservoir) (Lee et al., 1985). The cell is capable of sustained locomotion. Although quantitative details do not appear to have been reported, light micrographs show the cell with a forward-directed locomotory flagellum with a standing wave type beating pattern (Omodeo, 1980; Hochberg et al., 1972). The flagellar wave travels from base to tip but, due to the presence of a well developed mastigoneme system, the net effect is that fluid is displaced towards the organism (rather than the other way). This produces forward progressive motion. It should be noted that the eyespot appears to be directly under the beating plane of the flagellum and aligned along the axis of the cell (Bouck, 1971). The cell normally also rotates around its long axis as it swims forward.

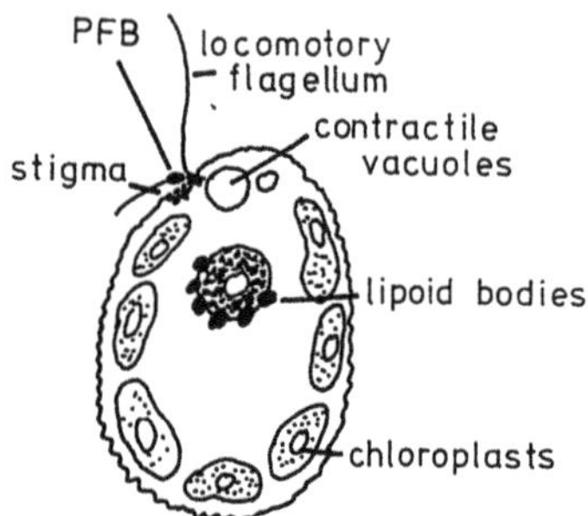

Fig. 4. Basic structural organization of chrysomonadidia (after Lee et al., 1985).

Mechanisms in Chrysomonadida

The Photochemical Mechanism. In common perhaps with the euglenida, chrysomonadida have been reported to exibit both step-up and step-down photoresponses as well as positive and negative phototaxis. However, only white light seems to have been used to study the photobehavior of this alga.

The Transduction Mechanism(s). Nothing is known about the transduction mechanism and, to date, *Ochromonas danica* has been studied in a very heterogeneous, organic culture medium (Häder et al., 1981).

The cell has been reported to show a step-down and step-up photophobic response and both positive and negative phototaxis (Häder et al., 1981). The details of the flagellar acctivities during each of these responses has not been reported. In view of the somewhat unique locomotion of *Ochromonas*, it would seem particularly important to establish the direction and orientation of the main flagellum during positive phototaxis for example and, equally importantly, to know if the secondary (shorter) flagellum plays any role in photophobic responses. It has been proposed that the short flagellum is normally "arched over the anterior eyespot" (Bouck, 1971).

Similar strategies for photobehaviors can be forwarded for *Ochromonas* as for *Astasia* and *Peranema* but without details, no firm proposals can be forwarded.

Subkingdom: Protozoa, Phylum: Sarcomastigophorea, Subphylum: Mastigophorea, Class: Phytomastigophorea, Order: Volvocida, e.g., Dunalliella, Chlamydomonas, Chlorogonium, Haematococcus, Volvox

Many of these biflagellates are characterized by being more oval in shape than most of the euglenid and chrysomonadida flagellates although *Chlorogonium* are slender in profile (Fig. 5). In addition, in the volvocida, both flagella contribute substantially to the locomotion of the cell. A breast-stroke like beating profile is common in this order. Most contain some form of specialized eyespot (Foster and Smyth, 1980) that confers photosensitivity on such cells by acting as a special form of reflecting and light-directing array. As a result of the symmetrical (or asymmetrical) flagellar beating, the cells are swept forward as like oars propelling a boat (or a swimmer performing the breast stroke). Under certain circumstances, the cells can be induced to swim backwards. In such cases the flagella assume a completely different configuration which, in the extreme, results in the flagella adopting a short amplitude, base-to-tip travelling wave: the flagella are then often pointing straight out from the apical end of the cell thus forcing the cell backwards. The flagellar motion has been evaluated in detail (e.g., Rüffer and Nultsch, 1985; Brokaw and Luck, 1983).

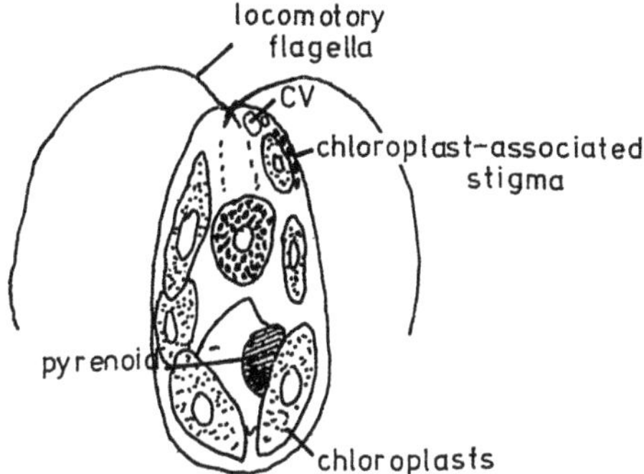

Fig. 5. Basic structural organization of volvocida (after Lee et al., 1985).

Mechanisms in Volvocida

These are diverse and few objective comparisons have been made.

The Photochemical Mechanism(s). These flagellates have been reported to show phototaxis as well as a special form of photoreversal of the locomotory flagella such that the cell swims backwards upon stimulation with high intensity light. A photo-stop-response may occur prior to the induction of backward swimming. Action spectra for phototaxis (photoaccumulation) indicate a general sensitivity to blue-green light. The reported action spectra/spectral sensitivity curves thus all show peaks close to 500 nm, e.g., 490 in *Dunalliella* (Tollin, 1969), 510 nm in *Chlamydomonas* (Nultsch, 1983), 495 nm in *Haematococcus* (Litvin et al., 1978) and 482 nm in *Volvox* (Laurens and Hooker, 1920). The action photoreceptor pigment in *Chlamydomonas* seems to have been unambiguously identified as a rhodopsin-type chromophore (Foster et al., 1984) since a number of retinal analogues have been found to alter phototaxis. It is presumably located in or close to the eye-spot like arrays near the apical end of these organisms.

The Transduction Mechanism(s). While *Dunaliella* (and related species) were the subject of very early research on the actions of extracellular ions on photobehavior (Tollin, 1969; Halldal, 1959), the further definition of the ionic mechanisms has been sporadic. However, in both *Chlamydomonas* (Schmidt and Eckert, 1976) and *Chlorogonium* (Hoops and Witman, 1985) (like *Euglena*) a calcium-dependent change in flagellar beating (from the breast-stroke to the straight, reversed beat profile has been reported. At least in *Chlamydomonas*, the switch of the beating profile occurs via a transitory flagella arrest (Hegemann and Bruck, 1989). The kinetics of the frequency of flagellar beating during the transitions has been recorded using high speed video recording (Hegemann and Bruck, 1989). Extracellular calcium ion levels appear to be able to determine the magnitude of the photoresponses of this alga (Dolle et al., 1987; Nultsch et al., 1986). High speed video recording has also indicated that as the extracellular ionized calcium levels are lowered to less than 0.002 mM, the flagella of *Chlamydomonas* can only respond to illumination and not the removal of the illumination (which occurs at 0.001 mM calcium) (Hegeman and Bruck, 1989). This phenomena has been presented as evidence to support a concept that illumination produces a transient elevation in intraflagellar calcium that is dependent upon the gradient of calcium across the external membranes of *Chlamydomonas*. The membranes involved are not fully defined but it seems plausible that light can, at least indirectly, alter the flagellar membrane permeability. However, it has been pointed out (Hegemann and Bruck, 1989) that since *Chlamydonomas* can show a photo-stop response and reversal when suspended in media containing exceedingly low ionized calcium levels (e.g., < 0.001 mM), calcium cannot be the (major) ion carrying a postulated inward current. Studies on *Haematococcus*

(Litvin et al., 1978) have attracted attention because it seems fairly easy to record a substantial change in membrane potential associated with photic stimulation. The occurrence of such phenomena has prompted considerations that the same might occur in other members of this order. Regardless of whether photic stimulation does activate inward calcium currents in *Chlamydomonas* for example, it needs to be noted that the exercise of suspending large numbers of cells [e.g., 40,000/ml (Hegemann and Bruck, 1989)] in media containing calculated (or even measured) levels of ionized calcium in the micromolar range is unlikely to be a good indicator of the available calcium at the surface of the cells (Dolle et al., 1987). Similarly, the photobehavior of *Chlamydomonas* (in media containing significant calcium levels) has been probed with a number of drugs (eg. D600) characterized as being selectively active on voltage-dependent Ca^{2+} channels in other eukaryotic (mammalian) cell membranes: an inhibitory action has been observed (Schmidt and Eckert, 1976; Dolle et al., 1987; Nultsch et al., 1986) leading to proposals that voltage-dependent calcium permeability mechanisms are operative. This aspect of the transduction would appear therefore to be markedly different from that in the euglenida. It would be extremely useful to know if the inward calcium currents reported for *Hematococcus* are sensitive to the same "calcium channel blocker" drugs used on both *Chlamydomonas* and *Euglena*. Similar transduction mechansisms may operate in other members of this order but studies have been limited to the gross investigation of the addition of cations to the cell medium (e.g., Tollin, 1969; Sakaguchi, 1979).

The Mechanochemcial Mechanism(s). These algae have, to date, been reported to exibit three types of photoresponses: a stop response to a sudden increase in light intensities, a backward swimming response (sometimes) following the stop response and, linked to both of these, an ability to respond to directional light. Both positive and negative phototaxis have been reported (Tollin, 1969; Schmidt and Eckert, 1976; Dolle et al., 1987; Nultsch et al., 1986; Sakaguchi, 1979; Smyth and Berg, 1980) depending on both the light intensity and the culture conditions imposed upon the organisms. For most the volvocida, the cell skews on its axis as it responds to directional stimulation. This appears to be the result of a slightly asymmetrical beating of the two flagella (referred to as the cis-and trans-flagella with respect to the position of the eyespot, at least in *Chlamydomonas* and *Chlorogonium*; (Foster and Smyth, 1980; Nultsch, 1983; Hoops and Witman, 1985). The actual character of flagella activity changes and the resultant cell motions have been studied in detail (Hegemann and Bruck, 1989; Smyth and Berg, 1980; Morel-Laurens and Feinleib, 1983; Hegemann and Marwan, 1988). All these cells will accumulate in a light trap although the actual details of the cell movements on crossing the illumination boundary have not been investigated in any detail. An accummulation response has been reported to be fluence-rate dependent in *Volvox* (Sakaguchi and Iwasa, 1979) and *Chlamydomonas* (Feinleib and Curry, 1971) yet could be by phototaxis or light-induced cessation of cell locomotion.

The function of photobehavior, from an ecological perspective, remains unknown for these flagellates. Perhaps like *Euglena*, the positive phototaxis (photoaccumulation) balanced by negative phototaxis and/or the high-light intensity-induced backward swimming determines the positioning of the cell in an optimum light level for photosynthesis. In such a context, however, the stop response would have to be considered as the response to gross (photic) insult.

Subkingdom: Protozoa, Phylum: Sarcomastigophorea, Subphylum: Sarcodina, Order: Amoebida; Suborder: Tubulina, e.g., Amoeba

Of the numerous free-living amoebal forms in this (sub)order, perhaps the best known is *A. proteus*. Like other amoebae, the cells can adopt a wide range of sizes and

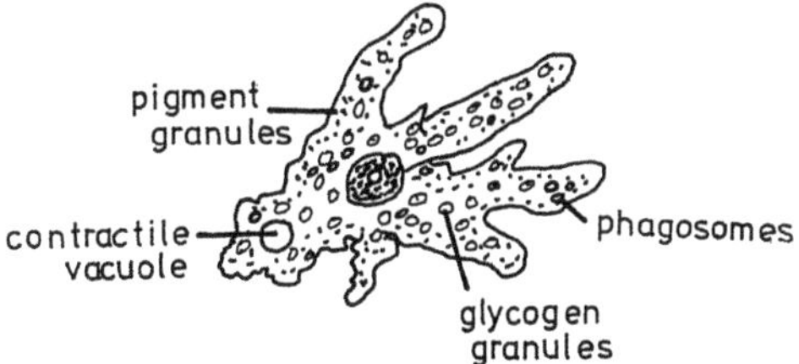

Fig. 6. Basic structural organization of amoebida (after Lee et al., 1985).

shapes ranging from a condensed rounded form some 150 μm across to very extended forms some 500 μm across (Lee et al., 1985). The cytoplasm contains numerous dispersed minute pigment granules that generally do not impart any obvious coloration to the organism (Fig. 6). There are fresh water and sea water species. The freshwater-dwelling, *A. proteus* has long been known for its photosensitivity (Harrington and Leaming, 1900). It deserves special comment in that numerous characteristics of its photomovement have, as yet, not been studied on many other unicells capable of photomovement! Motion is by protoplasmic streaming with the cell essentially continuously in contact with the substratum over which it glides. The responses to light are thus slow.

Mechanisms in Amoebida

The Photochemical Mechanism. The repertoire of photomovement includes phototaxis, photokinesis and a photo-stop response. Action spectra for positive phototaxis indicate a sensitivity to the shorter blue wavelengths (440-470 nm) and even to ultraviolet (Mast, 1910; James, 1987). Similarly, action spectra for a photo-stop response show two peaks at 450 nm and 470 nm (Mast and Hupieu, 1930). It has been suggested that "some complex of flavoproteins and/or cytochromes" compose the chromophore (James, 1987).

The Transduction Mechanisms. Essentially nothing is known from specific studies on photomovement although early studies indicate that monovalent and divalent cation salts (at high concentrations) can effect the time required for light to stop cell motion (Opas, 1975). The stop response was also affected by medium pH (Opas, 1975) as is the overall motility (Schaeffer, 1917).

The Mechanochemical Mechanisms. Early studies reported very distinct pseudopodial changes (direction of cytoplasmic streaming) in response to photic stimulation (Schaeffer, 1917, 1920) and, as a result the amoebae are negatively phototactic. Despite the reported action spectra (Mast, 1910; James, 1987), the reproducibility of positive phototaxis is uncertain (Harrington and Leaming, 1990; Schaeffer, 1920). At lower fluence rates, cell locomotion (rate) has been reported to be unaffected by light (Schaeffer, 1920). However, in later studies (Mast and Stahler, 1937) where the cells were carefully adapted prior to testing, a marked photokinesis was observed. Higher light intensities arrest the cytoplasmic streaming and gliding motion ceases (Harrington and Leaming, 1900; Mast and Hupieu, 1930).

The reason for a bottom-dwelling cell to be responsive to light can only be guessed at. The negative phototaxis could clearly play a role in water column positioning or even determining the level below the detritus and soil that the cell might migrate to. Direct solar radiation may be harmful to the cells.

Subkingdom: Protozoa, Phylum: Sarcomastigophorea, Subphylum: Sarcodina, Class: Mycetozoea, Order: Dictyosteliida, e.g., Dictyostelium, Polysphondylium

These two orders of "slime moulds" exist in single cell amoebal life stages and also develop into an aggregated cell mass (the "slug") (Lee et al., 1985). Both forms are light sensitive but only the amoebal forms will be discussed here other than acknowledging that, as part of this differentiation into slugs, the relative spectral sensitivity changes. The amoebal forms are not dissimilar to those in the suborder tubulina (e.g., *Amoeba proteus*) and the cytoplasm contains numerous dispersed pigment granules (Fig. 7).

Mechanisms in Dictyosteliida

The Photochemical Mechanism. The perception of light can affect both photodispersal and photoaccumulation which are both presumably achieved via a combination of phototaxis and photokinesis. Action spectra for photodispersal of amoebae of *Dictyostelium* from light traps show a complex of peaks that include a primary maximum at 420 nm and a secondary maximum at about 550 nm (Häder and Poff, 1979a). Action spectra for photoaccumulation are similar showing peaks at 405-410 nm and secondary peaks at 580 nm and 640 nm (Häder and Poff, 1979b) while action spectra for true phototaxis (at low and high fluence rates) seem blue shifted by 10-20 nm compared to those for photoaccumulation (Häder et al., 1988). In both orders, light affects (determines ?) the occurrence and rate of aggregation of the cells into multicellular colonial forms. For *Polysphondylium violaceum*, blue and green light were reported more effective than red wavelengths to stimulate the aggregation (Arnal et al., 1984). A crude pigment-containing extract from *Dictyostelium* amoebae, when analyzed by high resolution absorption spectroscopy, showed a predominant peak at 430 nm and a broad doublet of peaks at 550-590 nm (Poff et al., 1974). It has been proposed that this pigment, "phototaxin", is a mixture of cytochrome b and possibly a flavin (Schmidt et al., 1977).

The Transduction Mechanism. These are unknown especially because the localization of the pigment is uncertain and because, at the light-stimulated aggregation phenomena, is complicated by the presence of factors released by the cell on continuous illumination (Teta et al., 1983). A number of membrane-active chemicals (ionophores and lipophillic cations) have been assessed for their effects of photoaccumulation of *Dictyostelium* amoebae (Häder and Poff, 1980) but these agents affect both motility and photoresponses in a similar fashion.

The Mechanochemical Mechanism. Single cells can show photoaccumulation and photodispersal as well as light-induced cellular aggregation (into slugs). For the photodispersal, studies indicate a localized effect of high intensity white light on the peripheral cell cytoplasm in *Dictyostelium* thus causing that part of the cell to retract (Häder et al., 1983). The progressive phases of photomovement have been qualitatively studied (Häder and Poff, 1979b) and later cell track analyses indicated that, at least

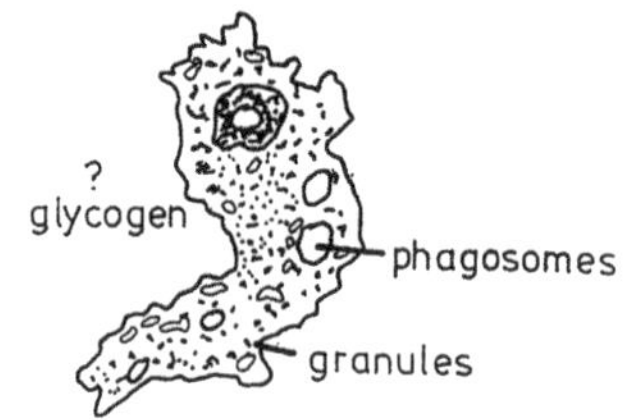

Fig. 7. Basic structural organization of dictyosteliida (after Lee et al., 1985).

under certain laboratory conditions, the amoebal cells will orient themselves parallel to a light source (Häder et al., 1988). Very high light intensities can inhibit cellular aggregation in *Dictyostelium* (Häder and Poff, 1979).

Strategies

The photic sensitivity of *Dictyostelium* and *Polysphondylium* at least plays a role in promoting cellular aggregation into the reproductive slug forms. Presumably the relative fluence rates at which this stimulation occurs reflects a light transition in their natural habitat although systematic studies seem lacking. The photodisperal and photoaccumulation phenomena of the individual cells perhaps reflects some form of fine tuning so that only aggregation competent amoebae persist in a lighted area (since continuous illumination for several hours is required for aggregation). However, excessive illumination (analogous to full solar irradiation) is perhaps damaging and so the cells will disperse from areas so illuminated.

Subkingdom: Protozoa, Phylum: Sarcomastigophora, Subphylum: Sarcodina, Order: Physarida, e.g., Physarum

Organisms in the order physaridia are also often referred to as "slime moulds" but only bear some characteristics in common with the order dictyosteliida. They are best referred to as the myxomeycete slime moulds (as opposed to the acrasiale slime moulds). Both orders however have a unicellular amoebal stage (Dee, 1975) with the myxomycetes also being able to transform in a biflagellated unicell (Clayton et al., 1983). While the photomovement of the giant plasmodial form of *Physarum* has been extensively studied and considered to show "blue light" sensitivity (even though the pigments change as a culture ages) (Rokoczy et al., 1986), the (myx) amoebae have not been studied in any detail and, surprisingly, when motion has been studied, light spectral character or intensity does not seem to have been noted (Jacobson, 1979; Quinlan et al., 1981). The amoebal forms of *Physarum* are mentioned here for completeness since there seems to be no *a priori* reason why they should not show photomovement.

Subkingdom: Protozoa, Phylum: Ciliophora, Subphylum: Postciliodesmatophora, Class: Spirotricea, Order: Heterotrichida, e.g., Blepharisma, Stentor

Organisms in these orders are generally the larger ciliates and are easily visible to the naked eye - especially if pigmented (e.g., coerulean blue in *Stentor coeruleus*). The coloration results from the presence of organized rows of pigment granules sited near the ciliary bases (Fig. 8). These cells are motile as a result of the coordinated beating activity of hundreds to thousands of (somatic) cilia that cover the entire body surface. The motion usually takes the form of a helical path with the cells rotating along their long axis while swimming. The cause of the rotation is partly the result of the orientation of the cilia, partly the result of oral cilia (if a side gullet exists) and, in some cases, due to the presence of an anterior array of specialized cilia (the membranellar band in *Stentor*) (Lee et al., 1985).

Mechanisms in Heterotrichida

The Photochemical Mechanism. The spectrum of photomovements include photo-stop type responses, photophobic responses and forms photoaccumulation

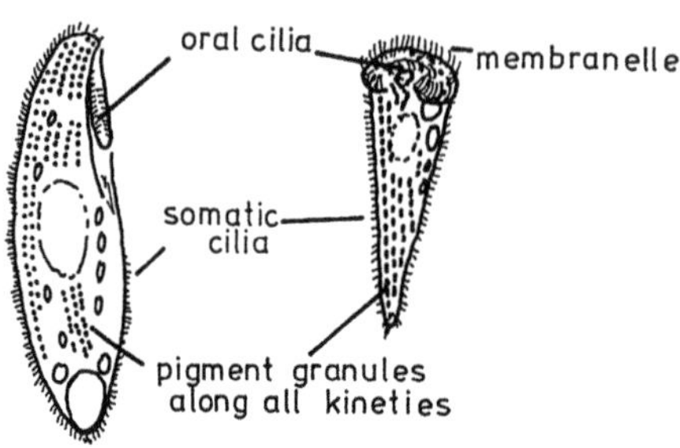

Fig. 8. Basic structural organization of heterotrichida (after Lee et al., 1985).

(photodispersal) that, in part, may result from phototaxis. *Blepharisma* species have a pinkish to red color under direct light microscopy (Inaba et al., 1979). Action spectra for a step-up photophobic (backward swimming-inducing) response has been found to have a single and largely featureless broad peak centered at 400 nm when assessed as that fluence rate required for 50 % of the cells in a sample to respond (Kraml and Marwan, 1983) yet an additional series of peaks at 475 nm, 550 nm and 590 nm when assessed as the slope of a fluence rate-response curve (Scevoli et al., 1987). The reason for this marked difference in spectra is uncertain but may be, in part explained, by the reaction of two pigments. *In vitro* spectroscopy (Gualteri et al., 1989) and absorption/fluorescence spectroscopy of isolated pigments (Kraml and Marwan, 1983; Gualteri et al., 1989) reveal a pigment with peaks at 490 nm, 575 and 595 nm for a "pink" pigment. A yellow pigment (with an absorption maximum at 410 nm and no red peaks) can also be isolated from the cells (Kraml and Marwan, 1983). The pink pigment, blepharismin, is a hypericin-like pigment (Sevenants, 1965). Its photochemistry is presumably similar to that of stentorin. Pigmented *Stentor* species also have pigment granules arranged in rows (Seshachar and Rao, 1959; Weisz, 1950). The uniqueness of the blue pigment in *S. coeruleus* was first recognized by Lankester (1873) and promoted extensive studies by Moller (1962) and Song and colleagues (Walker et al., 1979). Action spectra for step-up photophobic responses (cell backward swimming mediated by somatic ciliary reversal) show peaks at 475 nm, 570 nm and 615 nm (Wood, 1976). Similar action spectra have been measured for photodispersal from a light trap (Song et al., 1983). Studies on the photochemistry of isolated stentorin have led to the proposal that photic stimulation results in rapid deprotonation (and thus proton release from the granules) of the stentorin molecule (Song et al., 1983; Song, 1981). Different *Stentor* strains, however, have been reported to contain fluorescent or non-fluorescent forms of this pink pigment (Moller, 1962) and the significance of this character, with respect to the photochemical cycle, is unknown. The yellow pigment and its mechanisms remains to be investigated in comparative detail.

The Transduction Mechanisms. Recent studies (Scevoli et al., 1987; Song, 1981) have resulted in the proposal of a general scheme by which membrane calcium permeability is indirectly linked to photic stimulation in these ciliates. Such schemes however still require verification and further elaboration especially of what seems to be a unique, but complex, compartmentalization of sensory systems especially in *Stentor* (Doughty, 1990b). At least for *Stentor*, the cells can show vigorous ciliary responses to chemical stimulation (including to pH and monovalent cation salts) (Merton, 1935; Pietroiwica-Kosmynka, 1971; Colombetti et al., 1982). Several investigators have reported effects of cations, pH and membrane active chemicals (ionophores, protonophores) on the photobehavior of *Stentor* (Walker et al., 1979; Song et al., 1983; Colombetti et al., 1982; Iwatsuki and Song, 1989) and *Blepharisma* (Scevoli et al., 1987). However, some of these effects may be on the phototransduction system per se and others arise as a result of the direct action of the chemicals on the (chemo)sensory transduction/ciliary motility regulation systems (Doughty, 1990b).

The Mechanochemical Mechanisms. For *Blepharisma*, being "slipper-like" or "sickle shaped", the body cilia clearly dominate the cells locomotory behavior. Light-induced reversal of the body cilia causes the cell to back away from a lighted area for a finite period of time before resuming forward swimming in a new direction (Kraml and Marwan, 1983). The oral cilia do not apparently reverse upon step-up photic stimulation (Kraml and Marwan, 1983). The cell has also been reported to show an overall swimming velocity that is dependent upon light intensity (i.e., a photokinesis) (Kraml and Marwan, 1983; Matsuoka, 1983). *Stentor* are more complex since the cells have both body cilia and an anterior membranelle band (in addition to anterior oral gullet cilia): both sets of cilia play an important role in determining cell motion. The body cilia can show a beating activity and ciliary reversal while the membranelle can show beating, reversal and arrest (Wood, 1976; Doughty, 1990b, Merton, 1935; Pietroiwica-Kosmynka, 1971). The compartmentalization of transduction mechanisms that allow these to be regulated in turn or together remain to be objectively studied. For example, the forward swimming of *Stentor* can be interupted by the activity of the membranelle cilia to cause the cells to rotate on their long axis extremely rapidly under certain conditions (Doughty, 1990b). On arrival of the cell at the edge of a light trap (of very high intensity) causes a ciliary arrest response (at least of the membranellar band such that these cilia stick out from the anterior cell surface) (Song, 1981); the cell can then show ciliary reversal of the body cilia and swim briefly backwards before resuming forward swimming as in *Blepharisma*. However, whole cell illumination can induce membranellar reversal (Doughty, 1990b) with the behavioral result that the cell stop translational motion and spin on their posterior end (like a top) for several seconds before resuming forward swimming. Observations and measurements of photodispersal from light traps (mediated by the stop and reverse response) and positive and negative phototaxis have also been made for *Stentor* (Song et al., 1983; Song, 1981). Photokinesis has not been reported (Song, 1981) although a photoklinokinesis (light-induced, interupted swimming) can occur (Doughty, 1990b).

One can only surmise that the photoresponse of both organisms to increase in light intensity serve to protect the cell against excessive exposure to sunlight (Kraml and Marwan, 1983). The uniqueness of the pigment, in terms of the unique spectral position compared to other cells showing photomovement, presumably allows the cells to sense the radiant energy of the sun without interference from other material or cells in their aquatic environment. The positive phototaxis has an unknown physiological role and the photoklinokinesis would be expected to completely inhibit feeding behavior (since the cell is busy spinning!). The balances (or selection) of the photoresponse, at least in *Stentor*, is presumably determined by the actual direction of the incident light (Jennings, 1904).

Subkingdom: Protozoa, Phylum: Ciliophora, Subphylum: Cryptophora, Class: Oligohymenophorea, Order: Loxodida, e.g., Loxodes

The (karyoreltid) *Loxodes* is one of a larger group of slender ciliates that also have an side oral gullet (close to the front of the cell) (Fig. 9). The organism has pigment granules arranged in rows close to the ciliary bases (Finlay and Fenchel, 1986; Puytorac and Njine, 1970) and has a faint yellow-brown color. Locomotion is by the beating action of thousands of cilia covering the body and arranged in rows. It swims forward on a spiral path and its motion has been studied photographically in some detail (Finlay and Fenchel, 1986; Fenchel and Findlay, 1984, 1986).

Mechanisms in Loxida

The Photochemical Mechanisms. The cells exibit a complex repertoire of positive photokinesis associated with negative (inverse) photo-klinokinesis. Action spectra with broad peaks in the ultraviolet (360 nm) and in the blue (435 nm) have been reported (Finlay and Fenchel, 1986). A yellow-brown pigment extract from the cells showed a similar absorption spectra. Preliminary experiments have been presented as evidence that this extract could generate superoxide radicals when illuminated (Finlay and Fenchel, 1986). The pigment was tentatively proposed to be flavin-containing.

The Transduction Mechanisms. No systematic studies have been carried out, and it is unknown how, for example, the proposed superoxide species could alter ciliary activity.

The Mechanochemical Mechanisms. In the dark, *Loxodes* generally swims forward continuously with few interuptions (spontaneous tumbling of the cell associated with transient ciliary reversal). Illumination can have two effects. After exposure to white light, the cell swimming velocity can increase by some 50 % and, at the same time, the frequency of the tumbling responses decreases (Fenchel and Finlay, 1986). In contrast, *Loxodes* has been reported to avoid light in the sense that if presented with a light trap (light-dark border), most of the cells slowly migrate to the dark (Fenchel and Finlay, 1986). Details of the actual cellular responses associated with this photodispersal have not been reported.

The photobehavior of *Loxodes* has been proposed to be part of a complex sensory system that allows the cell to position itself relatively deep in the water table. The cell is a microaerophilic ciliate (i.e., prefers water at very low oxygen tensions but short of anoxia). The cells also show a pronounced positive gravitaxis (i.e., tend to swim downwards; Fenchel and Finlay, 1984) that is linked to the cells' perception of oxygen. As the light level increases, the oxygen tension at which the positive gravitaxis is induced is lower, i.e., the higher up the water table the ciliate swims (to increasing light intensity ?), the more sensitive the cells becomes to oxygen and thus swims back downwards again. The increased swimming velocity associated with photic stimulation is thus presumably a reflection of the association of increased gravitaxis with increased swimming velocity.

Subkingdom: Protozoa, Phylum: Ciliphora, Subphylum: Cryptophora, Class: Oligohymenophorea, Order: Peniculida, e.g., Paramecium

These ciliates range in size from small (less than 30 μm in length) to large (300 μm in length). They all have a characteristic side oral gullet and usually the "slipper animalicule" (German: Pantoffeltierchen) shape (Fig. 10). Some of these ciliates exibit photobehavior as a result of containing endosymbiotic algae. However, other colorless species have been sporadically reported, over the last 40 years, to show photomo-

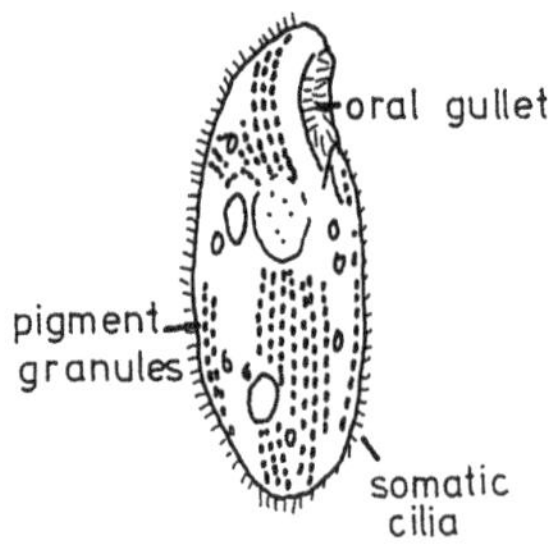

Fig. 9. Basic structural organization of loxida (after Lee et al., 1985).

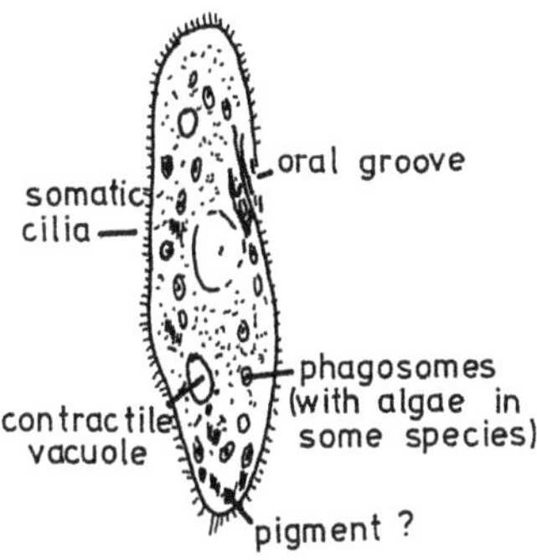

Fig. 10. Basic structural organization of peniculida (after Lee et al., 1985).

vement as well. Motion is achieved as a result of the coordinated beating of thousands of cilia that cover the body surface. The swimming path is usually a helix of the left-handed type although, in response to noxious stimuli, the spiral can shift to a right-handed one. In 2D, the motion appears as a spiral so the swimming form has often been termed forward left hand spiral (FLS) (Doughty and Dryl, 1981). The body cilia can fully reverse their beating (giving rise to periodic or continuous ciliary reversal) or, with only some of the cilia reversed, adopt partial ciliary reversal. The latter activity results in cell spinning or looping responses (Doughty and Dryl, 1981).

Mechanisms in Peniculida

The Photochemical Mechanisms. The cells exibit photokinesis and photoklinokinesis and, as a result, can also show photoaccumulation and photodispersal. The photoreceptor pigments in the algae-containing species are unknown since they have not been studied in any detail. However, they have traditionally been thought to be related to the carotenoid pigment system of the symbiotic green algae of the cell. The cells will accumulate in light trap (Cronkite and Van Den Brink, 1981), and an action spectra for the photoaccumulation response shows a broad peak between 480 and 560 nm (Iwatsuki and Naitoh, 1981) although a trap of 470 nm has been used (Cronkite and Van Den Brink, 1981). The cells may also respond to a red light trap (> 700 nm) (Iwatsuki and Naitoh, 1981) and can also apparently show motile responses to (lethal) ultraviolet irradiation (Jensen, 1959). An alternative photoreceptor system may be linked to the presence of hemoglobin (photoporphyrin-like) molecules in many species of "colorless" *Paramecium* (Usuki and Hino, 1987); the porphyrins could act as sensitizers.

The Phototransduction Mechansisms. Traditionally, the photobehavioral (photoaccummulation) responses of the endosymbiote-containing *Paramecium bursaria* have been considered to result from alterations of light-induced metabolite uptake or release by the endosymbionts (Reisser and Häder, 1984). Alternative metabolites suggested have been carbon dioxide uptake or release by the endosymbiont algae (Neiss et al., 1982) or oxygen from the algae (Cronkite and Van Den Brink, 1981). In short, it seems that the photosynthetic system of the algae needs to be active for photoaccumulation to occur but the mechanism of transduction remains uncertain. However, cell membrane potential changes have been reported to be associated with increased light intensity (Matsuoka and Nakaoka, 1988), and so an ionic transduction step (direct or indirect) could be considered as plausible. Indeed, it has been suggested that the algae-released carbon dixide may alter the pH of the *Paramecium* cytoplasm (Matsuoka and Nakaoka, 1988). Such pH changes could indirectly change ciliary membrane ionic permeability to alter ciliary activity of the host *Paramecium*. Mechanisms of ciliary motion control are reviewed in Doughty and Dryl (1981).

The Mechanical Mechanisms. The photoaccummulation of *Paramecium bursaria* seems to be brought about by a reduction in cell swimming velocity as cells cross a dark-to-light boundary, i.e., a step-up response (Matsuok and Nakaoka, 1981). The cells can even stop swimming in response to such a stimulus (Matsuoka and Nakaoka, 1988). The photosensitivity of the cells to a light trap displays a circadian rhythm (Nakajima and Nakaoka, 1989) indicating that the photomotile response can be observed as both "positive" and "negative" according to the experimental time chosen. The algae-containing cells have also been reported to show a step-down response that has been described as a "spinning of the front end about a stationary posterior" (Cronkite and Van Den Brink, 1981). This response thus resembles the second phase of the sensory response of what are usually termed "colorless" ciliates (e.g., *P. caudatum*) to stimulation with KCl: the gyration response (Doughty and Dryl, 1981). Our understanding of these photobehaviors has been complicated by the recent re-discovery of photobehavior (step-up and step-down and photoaccumulation behavior) by so-called colorless ciliates of the *Paramecium* genus (Dembowski, 1950; Okumura, 1963; Iwatsuki and Naitoh, 1982). It has been reported that *P. caudatum* can be induced to show ciliary reversal (backward swimming in response to a increase in light intensity (Dembowski, 1950) and thus can avoid a light trap rather than accumulating in it. This reaction of *P. caudatum* is characteristic of cells from stationary phase cultures at high density (unpublished observations). The existence of hemoglobin type molecules in such colorless *Paramecium* species (Usuki and Hino, 1987) could conceivably serve as endogeneous photosensitizers.

One might argue that algae-containing *P. bursaria* (or related ciliates) accumulate in light to faciliate photosynthesis by their symbiotic algae and that the sometime-observed photo-avoidance of a lighted area reflects the fact that the light intensity is too high for the algae. The step-up photophobic response (or ciliary reversal induction) of the colorless *Paramecium* sp. might similarly be viewed as an avoidance of light intensities that could cause photooxidative damage to the cell mitochondria for example. The sensitivity to UV-A has been considered as a means of determining the habitat positioning of *Paramecium* (Jensen, 1959; Barcello and Calkins, 1979).

Subkingdom: Protozoa, Phylum: Ciliophora, Subphylum: Cryptophora, Class: Oligohymenophorea, Order: Hymenostomatida, e.g., Tetrahymena

This small ciliate is pear shaped, has a different form of oral gullet to *Paramecium* that is more anteriorly located (Fig. 11). Cilia are, however, still arranged in rows along the body. While relatively sparsely covered with cilia, locomotion is the result of their coordinated beating (although details of the ciliary stroke are sparse). *Tetrahymena* is not well known for its light sensitivity and what studies have been reported lack significant detail on photomovement *per se*. However, the photic sensitivity of colorless *Paramecium* prompts consideration of the same in *Tetrahymena*.

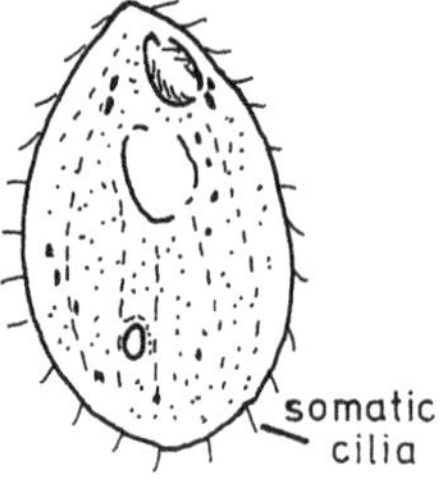

Fig. 11. Basic structural organization of hymenostomatida (after Lee et al., 1985).

Mechanisms in Hymenostomatida

The Photochemical Mechanisms. The photic sensitivity is related more to growth and its regulation rather than a photomovement *per se*. However, while photomovement does not appear to have been systematically studied, the cells can show a photoklinokinesis to high fluence rates (unpublished results) and high levels of UV-B induce arrest of cell motion. Dark grown *Tetrahymena* show an accumulation of a red pigment that has been identified as a protoporphyrin (Willie and Ehret, 1968; Ruben et al., 1982). The cell cycle has been reported to be influenced by switches from far red to far red plus white light (Willie and Ehert, 1968). More recently, light has been reported to affect the ability of the cells to respond to exogenous insulin (Kohidal et al., 1987) although no spectral sensitivity studies were done. The cells are also sensitive to ultraviolet radiation, and action spectra (Calkins et al., 1987) verify a sensitivity to UV-B.

The Transduction Mechanisms. Nothing is known.

The Mechanochemical Mechanisms. Nothing is known. However, as a result of UV-B irradiation, the cells stop swimming (a phototoxic effect ?).

The sensitivity of *Tetrahymena* to light is presumably somehow linked to the regulation of its growth such that it can also avoid intense solar irradiation.

Summary

Nine protozoan orders have been observed to exibit marked photobehaviors. Whether or not there are any behaviors that are unique to a particular order remains to be resolved through further study since, as outlined, the degree of attention paid to each order (or genus or species) has been sporadic and inconsistent. It is hoped that this systematics review will assist in providing a logical basis for provision of data in those clearly deficient areas. The overview provided should not be taken as completely comprehensive. Where only single resports of a photobehavior seem to have been published, space does not permit their being listed. For example, the ultraviolet sensitivity of a marine-dwelling ciliate, *Euplotes vannus* has been reported as a shock reaction (Hildebrand, 1972, 1975). This genus can be considered as belonging to an order between that includes *Paramecium* and that in which *Tetrahymena* is included (Lee et al., 1985). Similarly, UV-B irradiation has been reported to arrest ciliary activity in *Paramecium* (Geise and Leighton, 1935). However, as with the *Euplotes* ultraviolet studies the significance of such observations in relation to photobehavior *per se* requires further study.

Acknowledgements

This work was supported in part by funds from the University of Waterloo and a general operating grant from the Natural Sciences and Engineering Research Council (NSERC) of Canada.

References

Adamich, M., Lais, P. C., and Sweeney, B. M., 1976, *In vivo* evidence for a circadian rhythm in membranes of *Gonyaulax*, *Nature, Lond.*, 261:583.

Arnal, F., Recer, G., and Hanna., 1984, Photostimulation of aggregation in the slime mould *Polysphondylium violaceum*, *Photochem. Photobiol.*, 40:519.

Barcello, J. A., and Calkins, J., 1979, Positioning of aquatic microorganisms in response to visible light and simulated solar UV-B irradiation, *Photochem. Photobiol.*, 29:75.

Barghigiani, C., Colombetti, G., Franchini, B., and Lenci, F., 1979, Photobehavior of *Euglena gracilis*: action spectrum for the step-down photophobic responses of individual cells, *Photochem. Photobiol.*, 29:1015.

Bendix, S. W., 1960, Phototaxis, *Botan. Rev.*, 26:145.

Beneditti, P. A., and Checcucci, A., 1975, Paraflagellar body (PFB) pigments studied by fluorescence microscopy in *Euglena gracilis*, *Plant Sci. Lett.*, 4:47.

Bouck, G. B., 1971, The structure, origin, isolation and composition of the tubular mastigonemes of the *Ochromonas* flagellum, *J. Cell Biol.*, 50:362.

Brokaw, C. J., and Luck, D. J. L., 1983, Bending patterns of *Chlamydomonas* flagella. I. Wild-type bending patterns, *Cell Motil.*, 3:131.

Calkins, J., Colley, E., and Wheeler, J., 1987, Spectral dependence of some UV-B and UV-C responses of *Tetrahymena pyriformis* irradiated with dye laser generated UV, *Photochem. Photobiol.*, 45:389.

Clayton, L., Pogson, C. I., and Gull, K., 1983, Ultrastructural and biochemcial characterization of *Physarum polycephalum* myxamoebae, *Protoplasma* 118:181.

Cohn, F., 1866, Über die Gesetze der Bewegung der Mikroskopischen Pflanzen und Thiere unter Einfluß des Lichtes, *Hedwigia*, 5:161.

Colombetti, G., 1990, New trends in photobiology - photomotile responses in ciliated protozoa, *J. Photochem. Photobiol. B.*, 4:243.

Colombetti, G., Lenci, F., and Song, P.-S., 1982, Effects of K^+ and Ca^{2+} ions on motility and photosensory responses of *Stentor coeruleus*, *Photochem. Photobiol.*, 36:609.

Cronkite, D., and Van Den Brink, S., 1981, The role of oxygen and light in guiding photoaccumulation in the *Paramecium bursaria-Chlorella* symbiosis, *J. Exp. Zool.*, 217:171.

Dee, J., 1975, Slime moulds in biological research, *Sci. Prog. Oxford*, 62:523.

Dembowski, J., 1950, On the conditioned reactions of *Paramecium caudatum* towards light, *Acta Biol. Exp.* 15:17.

Diehn, B., 1969, Action spectra of the phototactic responses in *Euglena*, *Biochim. Biophys. Acta* 177:136.

Diehn, B., 1979, Photic responses and sensory transduction in motile protists, *in*: "Handbook of Sensory Physiology," VII/6a, Autrum, H., ed., Springer-Verlag, Berlin, pp. 23.

Diehn, B., Feinleib, M., Haupt, W., Hildebrand, E., Lenci, F., and Nultsch, W., 1977, Terminology of behavioral responses in microorganisms, *Photochem. Photobiol.*, 26:559.

Diehn, B., Fonseca, J. R., and Jahn, T. L., 1975, High speed cinematography of the direct photophobic response of *Euglena* and the mechanism of negative phototaxis, *J. Protozool.*, 22:492.

Dolle, R., Pfau, J., and Nultsch, W., 1987, Role of calcium ions in motility and phototaxis of *Chlamydomonas reinhardtii*, *J. Plant Physiol.*, 126:467.

Doughty, M. J. 1990a, A kinetic analysis of a step-up photosensory response of the ciliate, *Stentor coeruleus*, *Can. J. Microbiol.*, 36:414.

Doughty, M. J. 1990b, A kinetic analysis of a step-up photophobic response of the flagellate *Euglena gracilis* in culture medium, *J. Photochem. Photobiol. B.*, in press.

Doughty, M. J., and Diehn, B., 1979, Photosensory transduction in the flagellated alga, *Euglena gracilis*. I. Action of divalent cations, calcium antagonists and calcium ionophore on motility and photobehavior, *Biochim. Biophys. Acta* 588:148.

Doughty, M. J., and Diehn, B., 1980, Flavins as photoreceptor pigments for behavioral responses in motile microorganisms, especially in the flagellated alga, *Euglena* sp., *in*: "Structure and Bonding," Dunitz, J. D., Goodenough, J. B., Hemmerich, P., Ibers, J. A., Jorgensen, C. K., Neilands, J. B., Reinen, D., Williams, R. J. P., eds., Springer-Verlag, Berlin, Heidelberg, New York, 41:45.

Doughty, M. J., and B. Diehn, 1982, Photosensory transduction in the flagellated alga, *Euglena gracilis*. III. Induction of calcium-dependent responses by monovalent cation ionophores, *Biochim. Biophys. Acta* 682:32.

Doughty, M. J., and Diehn, B., 1983, Photosensory transduction in the flagellated alga, *Euglena gracilis*. IV. Long-term effects of ions and pH on the expression of step-down photobehavior, *Arch. Microbiol.*, 134:204.

Doughty, M. J., and Diehn, B., 1984, Anion sensitivity of motility and step-down photophobic responses of *Euglena gracilis*, *Arch. Microbiol.*, 138:329.

Doughty, M. J., and Dryl, S., 1981, Control of ciliary activity in *Paramecium*. An analysis of chemosensory transduction in a eukaryotic unicellular organism, *Progr. Neurobiol.*, 16:1.

Doughty, M. J., Grieser, R., and Diehn, B., 1980, Photosensory transduction in the flagellated alga, *Euglena gracilis*. II. Evidence that blue-light effects alternation in Na^+/K^+ permeability of the photoreceptor membrane, *Biochim. Biophys. Acta* 602:10.

Ekelund, N., and Häder, D.-P., 1988, Photomovement and photobleaching in two *Gyrodinium* species. *Plant Cell Physiol.*, 29:1109.

Feinleib, M. E. H., and Curry, G. M., 1971, The relationship between stimulus intensity and oriented phototactic response (topotaxis) in *Chlamydomonas*, *Physiol. Plant.*, 25:346.

Fenchel, T., 1987, "Ecology of Protozoa," Science Tech Publ., Madison, WI.

Fenchel, T., and Finlay, B. J., 1984, Geotaxis in the ciliated protozoan, *Loxodes*, *J. Exp. Biol.*, 110:17.

Fenchel, T., and Finlay, B. J., 1986, Photobehavior of the ciliated protozoan *Loxodes*: tactic, transient and kinetic responses in the presence and absence of oxygen, *J. Protozool.*, 33:139.

Finlay, T., and Fenchel, B. J., 1986, Photosensitivity in the ciliated protozoan, *Loxodes*: pigment granules, absorption and action spectra, blue light perception and ecological significance, *J. Protozool.*, 33:534.

Forward, R. B., 1973, Photoaxis in a dinoflagellate: action spectra as evidence for a two-pigment system, *Planta* 111:167.

Forward, R. B., 1974, Phototaxis by the dinoflagellate, *Gymnodinium splendens* Lebour, *J. Protozool.*, 21:312.

Forward, R. B., 1975, Dinoflagellate phototaxis: pigment systems and circadian rhythm as related to diurnal migration, *in*: "Physiological Ecology of Esturarine Organisms," Vernberg, F., ed., S. Carolina Press, Columbia, SC, pp. 367.

Forward, R. B., 1977, Effects of neurochemicals upon a dinoflagellate photoresponse, *J. Protozool.*, 24:401.

Foster, K. W., and Smyth, R. D., 1980, Light antennae in phototactic algae, *Microbiol. Rev.*, 44:572.

Foster, K. W., Saranak, J., Patel, N., Zarilli, G., Okabe, M., Kline, T., and Nakanishi, K., 1984, A rhodopsin is the functional photoreceptor for phototaxis in the unicellular eukaryote, *Chlamydomonas*, *Nature, Lond.*, 311:756.

Giese, A. C., and Leighton, P. A., 1935, Quantitative studies on the photolethal effects of quartz ultraviolet radiation upon *Paramecium*, *J. Gen. Physiol.*, 18:557.

Gualteri, P., Passarelli, V., and Barsanti, L., 1989, *In vivo* microscpectrophotometric investigation of *Blepharisma japonicum*, *J. Photochem. Photobiol. B.*, 3, 379.

Häder, D.-P., 1979, Photomovement, *in*: "Encyclopedia of Plant Physiology," New Series, vol. 7, Physiology of Movements, Haupt, W., and Feinleib, M. E., eds., Springer-Verlag, Berlin, pp. 267.

Häder, D.-P., 1986, Effects of solar and artificial UV irradiation on motility and phototaxis of the flagellate, *Euglena gracilis*, *Photochem. Photobiol.*, 44:651.

Häder, D.-P., 1988, Ecological consequences of photomovement in microorganisms. *J. Photochem. Photobiol. B.* 1:385.

Häder, D.-P., Claviez, M., Merkel, R., and Gerisch, G., 1983, Responses of *Dictyostelium discoideum* to local stimulation by light, *Cell Biol. Int. Rep.*, 7:611.

Häder, D.-P., Colombetti, G., Lenci, F., and Quaglia, M., 1981, Phototaxis in the flagellated, *Euglena gracilis* and *Ochromonas danica*. *Arch. Microbiol.*, 130:78 (and ref. cit.).

Häder, D.-P., and Griebenow, K., Orientation of the green flagellate, *Euglena gracilis*, in a vertical column of water, *FEMS Microbiol. Ecol.*, 53:159.

Häder, D.-P, and Häder, M. A., 1988, Ultraviolet-B inhibition of motility in green and dark-bleached *Euglena gracilis*, *Curr. Microbiol.*, 17:215.

Häder, D.-P., and Häder, M., 1989, Effects of solar radiation on photoorientation, motility and pigmentation in a freshwater *Cryptomonas*, *Botanica Acta* 102:236.

Häder, D.-P., Häder, M., Liu, S-M., and Ullrich, W., 1990, Effects of solar radiation on photoorientation, motility and pigmentation in a freshwater *Peridinium*, *Biosystems* 23:335.

Häder, D.-P., Lebert, M., and DiLena, M. R., 1986, New evidence for the mechanism for phototactic orientation of *Euglena gracilis*, *Curr. Microbiol.*, 14:157.

Häder, D.-P., Lebert, M., and DiLena, M. R., 1987, Effects of culture age and drugs on phototaxis in the green flagellate, *Euglena gracilis*, *Plant Physiol.*, 6:169.

Häder, D.-P., and Poff, K. L., 1979a, Photodispersal from light traps by amoebae of *Dictyostelium discoideum*, *Exptl. Mycol.*, 3:121.
Häder, D. P., and Poff, K. L., 1979b, Light-induced accumulations of *Dictyostelium* amoebae, *Photochem. Photobiol.*, 29:1157.
Häder, D.-P., and Poff, K. L., 1979c, Inhibition of aggregation by light in the cellular slime mould, *Dictyostelium discoideum*, *Arch. Microbiol.*, 123:281.
Häder, D.-P., and Poff, K. L., 1980, Effects of ionophores and $TPMP^+$ on light-induced responses in *Dictyostelium discoideum*, *Arch. Microbiol.*, 126:97.
Häder, D.-P., Rhiel, E., and Wehrmeyer, W., 1987, Phototaxis in the marine flagellate *Cryptomonas maculata*, *J. Photochem. Photobiol. B.*, 1:115.
Häder, D.-P., Watanabe, M., and Furuya, M., 1988, Multiple photoreceptors in phototaxis of *Dictyostelium* amoebae, *Protoplasma Suppl.* 1: 155.
Halldal, P., 1958, Action spectra of phototaxis and related problems in Volvocales, *Ulva* gametes and Dinophyceae, *Physiol. Plant.* 11:118.
Halldal, P., 1959, Factors affecting light response in phototactic algae, *Physiol. Plant.* 12:742.
Hand, W. G., and Schmidt, J., 1975, Phototactic orientation by the marine dinoflagellate, *Gyrodinium dorsum* Kofoid. II. Flagellar activity and overall response mechanism, *J. Protozool.*, 22:494.
Harrington, H. R., and Leaming, E., 1990, The reactions of *Amoeba* to light of different colors, *Am. J. Physiol.*, 3:9.
Hegemann, P., and Bruck, B., 1989, Light-induced stop response in *Chlamydomonas reinhardtii*: occurrence and adaptation phenomena, *Cell Motil. Cytoskel.*, 14:501.
Hegemann, P., and Marwan, W., 1988, Single photons are sufficient to trigger movement response in *Chlamydomonas reinhardtii*, *Photochem. Photobiol.*, 48:99.
Hildebrand, E., 1972, Avoiding reaction and receptor mechanism in protozoa, *Acta Protozool.*, 11:361.
Hildebrand, E., 1975, Bedeutung der Konkurrenz zwischen Calcium und anderen Kationen für die Steuerung der Leitfähigkeit sensorischer Membranen, *Verh. Dtsch. Zool. Ges.*, 24:62.
Hochberg, A., Pimstein, R., and Rahat, M., 1972, Properties of an ethionine-resistant (ER) mutant of *Ochromonas danica*, *J. Protozool.*, 19:66.
Holt, E. B., and Lee, F. S., 1901, The theory of the phototactic response, *Am. J. Physiol.*, 4:460.
Holwill, M. E. J., 1966, The motion of *Euglena viridis*: the role of flagella, *J. Exp. Biol.*, 44:579.
Hoops, J. H., and Witman, G. B., 1985, Basal bodies and associated structures are not requires for normal flagellar motion or phototaxis in the green alga, *Chlorogonium elongatum*, *J. Cell Biol.*, 100:297.
Inaba, F., Nakamura, R., and Yamaguchi, S., 1979, An electron-microscopic study on the pigment granules of *Blepharisma*, *Cytologia (Tokyo)* 23:72.
Iwatsuki, K., and Naitoh, Y., 1981, The role of symbiotic *Chlorella* in photoresponses of *Paramecium bursaria*, *Proc. Jpn. Acad. Ser. B.*, 57:318.
Iwatsuki, K., and Naitoh, Y., 1982, Photoresponses in colorless *Paramecium*, *Experentia*, 38:1453.
Iwatsuki, K., and Song, P-S., 1989, The ratio of extracellular Ca^{2+} to K^+ ions affects the photoresponses in *Stentor coeruleus*, *Comp. Biochem. Physiol.*, 92A:101.
Jacobson, D. N., 1979, The role of regulation of cell speed in the behavior of *Physarum polycephalum* amoebae, *Exp. Cell Res.*, 122:219.
James, T. W., 1987, Photomechanical transduction in *Amoeba proteus*: an action spectrum, *J. Photochem. Photobiol. B.*, 1:203.
Jennings, H. S., 1904, Reactions to light in ciliates and flagellates, *in*: "Contributions to the Study of the Behavior of Lower Organisms," Carnegie Institute, Washington, DC, pp. 31.
Jensen, D. D., 1959, A theory of the behavior of *Paramecium aurelia* and behavioral effects of feeding, fission and UV microbeam irradiation, *Behavior*, 15:82.
Jirovec, O., 1934, Der Einfluß von ultravioletten Strahlen auf grüne und farblose Stämme von *Euglena gracilis*, *Protoplasma*, 21:577.
Kaneda, H., and Furuya, M., 1986, Temporal changes in swimming direction during the phototactic orientation in cells of *Cryptomonas* sp., *Plant Cell Physiol.*, 27:265.
Kaneda, H., and Furuya, M., 1987, Effect of calcium ions on phototactic orientation of individual *Cryptomonas* cells, *Plant Sci.*, 48:31.

Kaufman, L. S., and Lyman, H., 1982, A 600 nm receptor in *Euglena gracilis*: its role in chlorophyll accumulation, *Plant. Sci. Lett.*, 26:293.

Kivik, P. A., and Walne, P. L., 1983, Algal photosensory apparatus probably represent multiple parallel evolutions. *BioSystems* 16:31.

Kohidal, L., Darvas, Z., and Csaba, G., 1987, The effect of varying illumination on imprinting of *Tetrahymena* by insulin, *Acta Microbiol. Hung.*, 34:179.

Kraml, M., and Marwan, W., 1983, Photomovement respones of the heterotrichous ciliate, *Blepharisma japonicum*, *Photochem. Photobiol.*, 37:313.

Lankester, E. R., 1873, Blue stentorin - the coloring matter of *Stentor coeruleus*, *Quart. J. Microscop. Sci.*, 13:139.

Laurens, H., and Hooker, H. D., 1920, Studies on the relative physiological value of spectral lights. II. The sensibility of *Volvox* to wavelengths of equal energy content, *J. Exp. Zool.*, 30:345.

Lee, J. J., Hutner, S. H., and Bovee, E. C., 1985, "An Illustrated Guide to Protozoa," Society Protozoologists, Lawrence, KS.

Lenci, F., Häder, D.-P., and Colombetti, G., 1984, Photosensory responses in freely motile microorganisms, *in*: "Membranes and Sensory Transduction," Colombetti, G., and Lenci, F., eds., Plenum Press, New York, pp. 199.

Litvin, F. F., Sineshchekov, O. A., and Sineshchekov, V. A., 1978, Photoreceptor electrical potential in the phototaxis of the alga, *Haematococcus pluvialis, Nature, Lond.*, 271:476.

Liu, S-M., Häder, D.-P, and Ulrich, W., 1990. Photoorientation in the dinoflagellate, *Peridinium gatunense* Nygaard, *FEMS Microbiol. Lett.*, 73:91.

Mast, S. O., 1910, Reactions in *Amoeba* to light, *J. Exp. Zool.*, 9:265.

Mast, S. O., 1941, Motor responses in unicellular organisms, *in*: "Protozoa in Biological Research," Calkins, G. N., and Summers, F. M., Eds., Columbia Univ. Press, New York, pp. 271.

Mast, S. O., and Hulpieu, H. R., 1930, Variation in responses to light in *Amoeba proteus* with special reference to the effects of salts and hydrogen ion concentration, *Protoplasma*, 11:412.

Mast, S. O., and Stahler, N., 1937, The relation between luminous intensity, adaptation to light and the rate of locomotion in *Amoeba proteus* (Leidy), *Biol. Bull.*, 73:126.

Matsuoka, T., 1983, Distribution of photoreceptors inducing ciliary reversal and swimming acceleration in *Blepharisma japonicum, J. Exp. Zool.*, 225:337.

Matsuoka, K., and Nakaoka, Y., 1988, Photoreceptor potential causing phototaxis of *Paramecium bursaria, J. Exp. Biol.*, 137:477.

Mergenhagen, D., 1980, Circadian rhythms in unicellular organisms, *Curr. Topics Microbiol. Immunol.*, 90:123.

Merton, H., 1935, Zwangsreaktionen bei *Stentor* als Folge bestimmter Salzwirkungen, *Biol. Z.*, 55:268.

Meyer, R., and Hildebrand, E., 1988, Phototaxis of *Euglena gracilis* at low external calcium concentrations, *J. Photochem. Photobiol. B.*, 2:443.

Mikolajczyk, E., 1986, Na^+/K^+ transport and photosensitivity of the colorless flagellate, *Peranema trichophorum* (Euglenida), *Photochem. Photobiol.*, 43:455.

Mikolajczyk, E., and Walne, P. L., 1990, Photomotile response and ultrasturcture of the euglenoid flagellate, *Astasia fritschii, J. Photochem. Photobiol. B.*, 6:275.

Moller, K. M., 1962, On the nature of stentorin, *C. R. Trav. Lab. Karlsberg*, 32:471.

Morel-Laurens, N. M., and Feinleib, M. E., 1983, Photomovement in an "eyeless" mutant of *Chlamydomonas*, *Photochem. Photobiol.*, 37:189.

Nakajima, K., and Nakaoka, Y., 1989, Circadian change of photosensitvity of *Paramecium bursaria, J. Exp. Biol.*, 144:43.

Niess, D., Reisser, W., and Wiessner, W., 1982, Photobehavior of *Paramecium bursaria* infected with different symbiotic and aposymbiotic species of *Chlorella, Planta*, 156:475.

Nultsch, W., 1983, The photocontrol of movement in *Chlamydomonas*, *in*: "The Biology of Photoreception," Cosens, D. J., and Vincent-Price, D., eds., Soc. Exptl. Biol., Cambridge, UK, pp. 521.

Nultsch, W., Pfau, J., and Dolle, R., 1986, Effects of calcium channel blockers on phototaxis and motility of *Chlamydomonas reinhardtii*, *Arch. Microbiol.*, 144:393.

Okumura, H., 1963, Response to light in *Paramecium*, *J. Fac. Sci. Hokkaido Univ. Ser. VI. Zool.*, 15:225.

Oltmans, F., 1917, Über Phototaxis, *Z. Botanik*, 9:257.

Omedo, P., 1980, The photoreceptive apparatus of flagellated algal cells: comparative morphology and some hypotheses on functioning, *in*: "Photoreception and Sensory Transduction in Aneural Organisms," Lenci, F., and Colombetti, G., eds., Plennun Press, New York, pp. 127.

Opas, M., 1975, Studies on the locomotion of *Amoeba proteus*. I. The response to hydrogen ion concentration of the medium, *Acta Protozool*., 13:285.

Piccinni, E., and Omodeo, P., 1975, Photoreceptors and phototactic programs in protista, *Boll. Zool*., 42:57.

Pietrowica-Kosmynka, D., 1971, Chemotactic effects of cations and pH on *Stentor coeruleus*, *Acta Protozool*., 9:235.

Poff, K. L., Loomis, W. F., and Butler, W. L., 1974, Isolation and purification of the photoreceptor pigment associated with phototaxis in *Dictyosteliuum discoideum, Proc. Natl. Acad. Sci. USA*, 249:2164.

Puytorac, P., and Njine, T., 1970, Sur l'ultrastructure des *Loxodes* (cilies holotriches), *Protistologica*, 6:427.

Quinlan, R. A., Roobol, A., Pogson, C. I., and Gull, K., 1981, A correlation between *in vivo* and *in vitro* effects of the microtubule inhibitors colchicine, parbendazole and nocodazole on the myamoebae of *Physarum polycephalum, J. Gen. Microbiol*., 122:1.

Reisser, W., and Häder, D.-P., 1984, Role of endosymbiotic algae in photokinesis and photophobic responses of ciliates, *Photochem. Photobiol*., 39:673.

Rhiel, E., Häder, D.-P., and Wehrmeyer, W., 1988, Photoorientation in a freshwater *Cryptomonas* species, *J. Photochem. Photobiol*. B., 2:123.

Rokoczy, L., Majcherczyk, A., and Huttermann, A., 1986, Changes in plasmodial pigments of *Physarum polycephalum* in relation to the age of the culture medium, *Can. J. Microbiol*., 33:217.

Ruben, L., Lageson, J., Hyzy, B., and Hooper, A. B., 1982, Growth cycle-dependent overproduction and accumulation of protoporphyrin IX in *Tetrahymena*: effect of heavy metals. *J. Protozool*., 29:233.

Rüffer, U., and Nultsch, W., 1985, High-speed cinematographic analysis of the movement of *Chlamydomonas, Cell Motil*., 5:251.

Sakaguchi, H., 1979, Effect of external ionic environments on phototaxis of *Volvox carteri, Plant Cell. Physiol*., 20:1643.

Sakaguchi, H., and Iwasa, K., 1979, Two photophobic responses in *Volvox carteri, Plant Cell Physiol*., 20:909.

Scevoli, P., Brisi, F., Colombetti, G., Ghetti, F., Lenci, F., and Passarelli, V., 1987, Photomotile responses of *Blepharisma japonicum*. I. Action spectra determination and time-resolves fluorescence of photoreceptor pigments. *J. Photochem. Photobiol*. B., 1:75.

Schaeffer, A. A., 1917, Reactions of *Amoeba* to light and the effect of light on feeding, *Biol. Bull*., 32:45.

Schaeffer, A. A. 1920, "Amoeboid Movement," Princeton University Press, New Haven, CT.

Schmidt, J. A., and Eckert, R., 1976, Calcium couples flagellar reversal to photostimulation in *Chlamydomonas reinhardtii, Nature, Lond*., 262:713.

Schmidt, W., Thomson, K., and Butler, W. L., 1977, Cytochrome *b* in plasma membrane-enriched fractions from several photoresponsive organisms, *Photochem. Photobiol*. 26:407.

Seshachar, B. R., and Rao, A. V. S. P., 1959, Observations on the pigment from an Indian species of *Blepharisma* (ciliata; protozoa), *J. Sci. Industri. Res*., 18C:76.

Sevenants, M. R., 1965, Pigments of *Blepharisma undulans* compared with hypericin, *J. Protozool*., 12:240.

Shettles, L. B., 1937, Response to light in *Peranema trichophorum* with special reference to dark-adaptation and light-adaptation, *J. Exp. Zool*., 77:215.

Smyth, R. D., and Berg, H. C., 1982, Change in flagellar beat frequency of *Chlamydomonas* in response to light, *Cell. Motil. (suppl.)* 1:211.

Song, P-S., 1982, Photosensory transduction in *Stentor coeruleus* and related organisms, *Biochim. Biophys. Acta*. 639:1.

Song, P-S., Tapley, K. J., and Berlin, J. D., The photoreceptor in *Stentor coeruleus*, *in*: "The Biology of Photoreception," Cosens, D. J., and Vince-Price, D., eds., Soc. Exptl. Biol., Cambridge, UK.

Suzaki, T., and Williamson, R. E., 1983, Photoresponse of a colorless euglenoid flagellate, *Astasia longa, Plant Sci. Lett*., 32:101.

Teta, L. A., Ellsaesser, C. F., and Hanna, M. H., 1983, The role of light and an aggregation-stimulating factor suring aggregation of *Polysphondylium violaceum, J. Gen. Microbiol*., 129:167.

Tollin, G., 1969, Energy transduction in algal phototaxis, *Curr. Topics Bioenerget.*, 3:417.

Uemetsu-Kaneda, H., and Furuya, M., 1982, Effects of viscosity on phototactic movement and period of cell rotation in *Cryptomonas* sp., *Physiol. Plant*, 56:194.

Usuki, I., and Hino, A., 1987, Hemoglobin content in various stocks of different species of the *Paramecium aurelia* group, *Cell. Molec. Biol.*, 33:601.

Wager, H., 1911, On the effect of gravity upon the movements and aggregations of *Euglena viridis* Ehrb. and other microorganisms, *Phil. Trans. Roy. Soc. Lond., Ser. B.* 201:333.

Walker, E. B., Lee, T. Y., and Song, P-S., 1979, Spectroscopic characterization of the *Stentor* photoreceptor, *Biochim. Biophys. Acta*, 587:129.

Watanabe, M., and Furuya, M., 1974, Action spectrum of phototaxis in a cryptomonad alga, *Cryptomonas* sp., *Plant Cell Physiol.*, 15:413.

Weisz, P. B., 1950, On the mitochondrial nature of the pigmented granules in *Stentor* and *Blepharisma*, *J. Morphol.*, 86:177.

Willie, J. W., and Ehret, C. F., 1968, Light synchronization of an endogenous circadian rhythm of cell division in *Tetrahymena*, *J. Protozool.*, 15:785.

Wood, D. C., 1976, Action spectrum and electrophysiological responses correlated with the photophobic response of *Stentor coeruleus*, *Photochem. Photobiol.*, 24:261.

Worrest, R. C., and Häder, D.-P., 1989, Effects of stratospheric ozone depletion on marine microorganisms. *Environm. Conservat.*, 16:261.

Image Analysis Techniques for Studying Photomovements

Paolo Gualtieri

CNR Istituto di Biofisica
via S. Lorenzo 26
56100 Pisa
Italy

Introduction

Motility of swimming microorganisms is among some of the well documented activities that have been observed by light microscopy. This activity and other physological responses are becoming increasingly understood at the subcellular and molecular level; however, for a thorough understanding of the broad range of behavioral and physiological aspects of motility a complete, automatic analysis of microorganism motion under environmental stimuli could represent a very useful tool for the the cell biologist, by providing him with the exact determination of fundamental parameters as speed, direction, and acceleration.

Many cellular activities can be investigated by analyzing videotapes of microorganisms moving on a microscope slide. A good review of the problems connected with this type of analysis is provided by the paper of Yashida et al. (1981). The first step of this procedure allows a coarse location of the moving cells; these cells are then exactly detected, and their quantitative parameters measured. This procedure ends with a control analysis for objects which could have been misinterpreted during the tracking process. There are other systems which use a videotape for the tracking of moving microorganisms (Gualtieri et al., 1985, 1988). The cells are detected in single frames by means of a labelling procedure, and the trajectories are reconstructed by joining cell baricenters in the successive frames. All these systems, however satisfactory their results may be, are time consuming because of the time necessary to drive the video recorder.

In order to achieve accurate statistical results, the number of reconstructed tracks must be higher than 500; therefore, the analysis time represents a fundamental constraint. Häder and Lebert (1985) set up a system which shortens the time of analysis by detecting and following one microorganism at a time in the microscope field, thus avoiding the use of the videotape. Kondo et al. (1988) achieve this same result by storing successive video frames on a floppy disk and by thinning the recorded tracks.

In our system, a fast analysis is linked with an exact track reconstruction. It detects moving microorganisms, reduces the size of the acquired frames, stores them in a RAM board in real time (40 ms/frame, using CCIR standard), and then traces the trajectories of the moving objects by means of a very fast algorithm. The flagellate *Euglena gracilis* has been used as an experimental subject.

Hardware Configuration

We describe the experimental set-up we use for motion analysis of microscope images; this configuration applies for whatever method of analysis of temporal images is required (Fig. 1). Our system consists of a Zeiss Axioplan microscope (Zeiss, FRG), equipped with a black and white TV camera (saticon target, NICAL, Italy). The microorganisms can freely swim within a narrow layer of growth medium placed between a slide and a cover slip. The microscope field is acquired by the TV camera, and its video signal is acquired and digitized by an image system board (Imaging technology, U.S.A.) which is plugged into the bus of a personal computer (IBM, U.S.A.). The board consists of a frame memory, Look-up-tables (LUT) and an A/D converter. In the 1024x1024x12 bit board, the operator can select a window of 640x512x8 bits. This size corresponds to a digitized image with equal vertical and horizontal spacing. This image can be reduced, moved to the left or to the right, scrolled up and down, and stored in a selected position of the frame memory by means of the Zoom, Pan, and Scroll hardware capabilities of the board. Up to 192 images can be stored in the frame memory.

Using an image board which cannot operate in real time, a video recorder should be used, in order to store the frames acquired by the TV camera. The tape can be analyzed successively, by connecting the video recorder to the personal computer through a parallel interface.

The menu-driven software, whose commands are selected by means of a mouse, is written in the C language (Microsoft, USA). A black and white monitor (Nical, Italy) is used to display the signal output of the TV camera, while an RGB high persistence monitor (Mitsubishi Model C-3479, Japan) is used to display the digital image.

In order to study the behavior of microorganisms under chemical stimuli, the chemical compound should be added to the medium, or placed onto the slide. In order to study the photobehavior we have to monochromatically illuminate the microscope field of view. The light stimuli can be controlled by the computer.

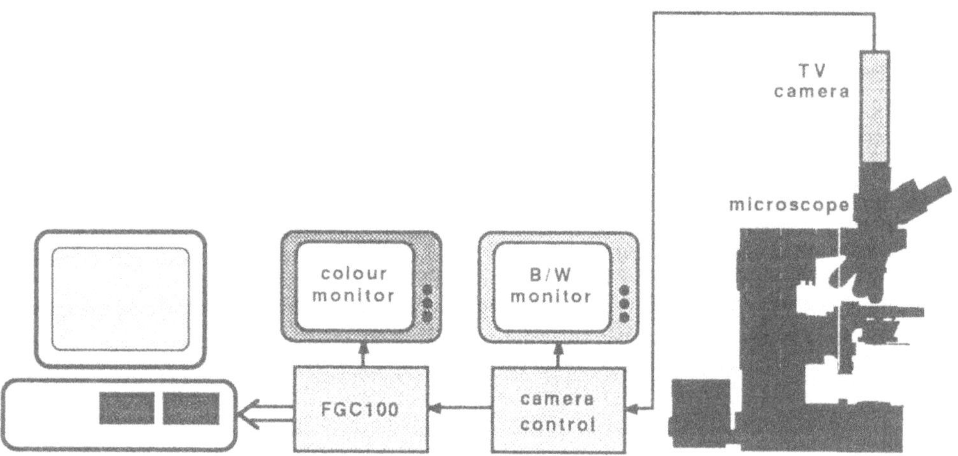

Fig. 1. The hardware set-up of the tracking microscope.

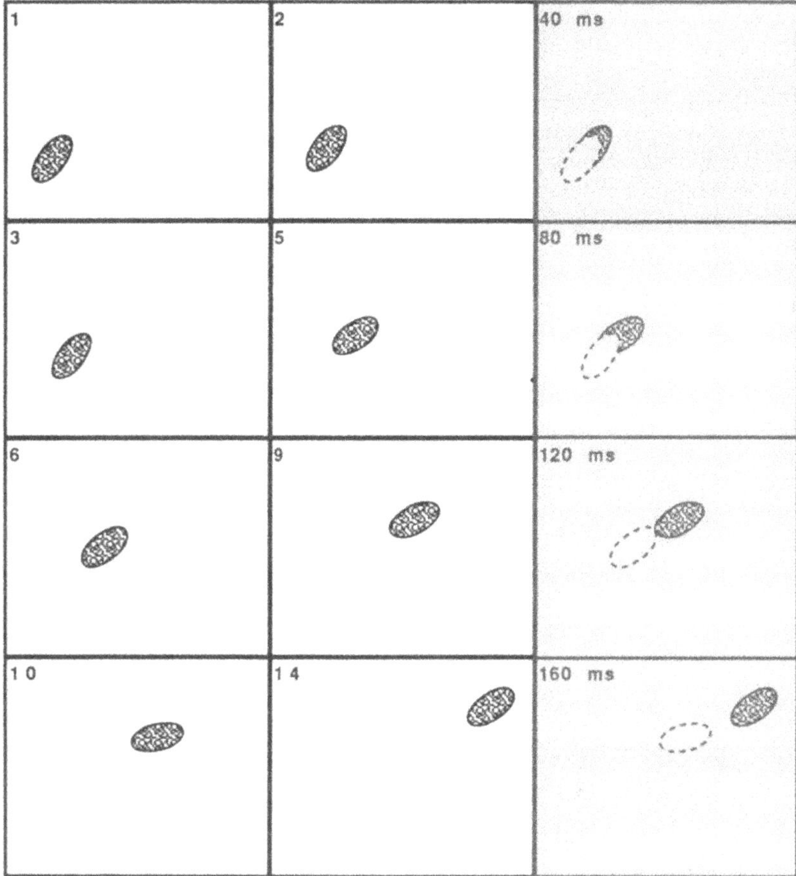

Fig. 2. An example of the storing procedure: each difference image is represented as framed.

Methods

In the following we will briefly describe our methods; for their exhaustive treatment, we refer the reader to more technical papers (Gualtieri and Coltelli, 1989; Gualtieri and Coltelli, 1990a,b).

Real time detection of moving objects is performed by subtracting continuously each frame of the video image from a previous frame, with an experimentally determined delay. Thanks to a feed-back circuit between the frame memory and the input LUT, operations can be made on combinations of stored and newly acquired data.

In order to store several difference images in the frame memory (in our examples the difference images are 12), we reduce the spatial resolution of the image being acquired by means of the hardware Zoom. The result of this operation is a 320x256 pixel image, whose size corresponds to 1/4 of the original image size. In order to store the reduced image into its proper position of the frame memory, the X and Y coordinates of its origin are shifted by means of hardware Pan and Scroll operations. In this way we can store 12 images, by moving the coordinates of their origin toward the right and downward. At the end of this procedure, the frame memory is displayed as a patchwork of reduced images. Each difference image is represented as a framed image (Fig. 2). Due to our hardware facilities, the zooming and storing operations are performed during the frame acquisition. Using this method, more images could be stored, up to 192; a further reduction would be meaningless, because of the very low space resolution achievable.

In order to identify moving cells and to extract their features such as baricenter coordinates and areas, a labelling procedure is applied to each difference image (Fig. 3). After cell identification, an algorithm of track reconstruction selects a cell baricenter in the first difference image, and starting from this element scans the successive difference images in order to find the cell whose baricenter lies in a pre-established distance and direction with respect to the previously detected baricenter. When a track is completed, the procedure selects a new starting element, i.e., the baricenter of a new cell, and another cell track will be reconstructed.

The reconstruction of the tracks of microorganisms freely swimming on a microscope slide can present the problem of overlapping cells, i.e., intersecting tracks. However, the peculiar characteristics of our experimental apparatus allow to minimize this problem.

The microscope has a low depth of focus, and the microorganisms swim within a narrow layer of medium between a slide and a cover slip. In this situation the cells, which swim turning along their major axis in the focused layer, usually are not occluded by cells that swim in out-of-focus layers.

However, it can happen that two cells hide one another, but these rare events are left unsolved in the routine work, since for this kind of analysis it is important to reconstruct more than five hundred tracks. In these cases the labelling algorithm detects

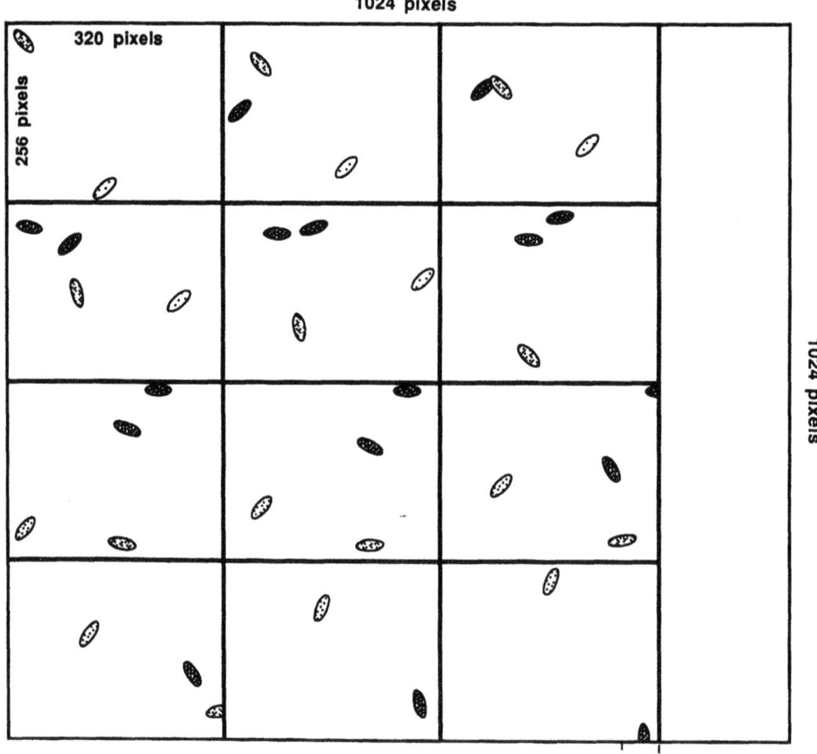

Fig. 3. An example of the labelling procedure: the identified cells are labelled with the corresponding frame number.

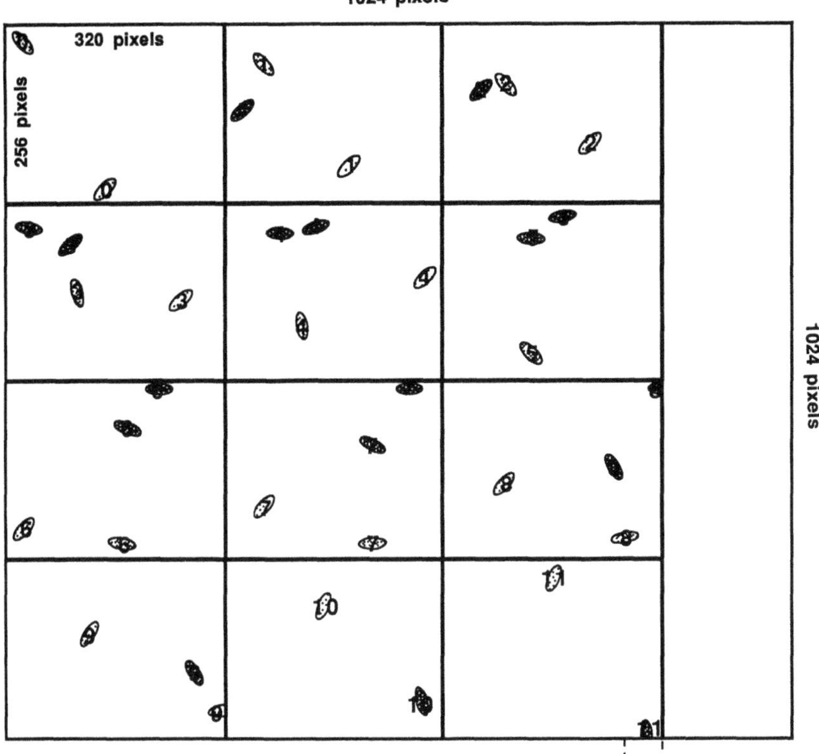

Fig. 4. The differential track reconstruction menu.

a cell with an area value about twice that of the average area value, but only one baricenter. Therefore, this baricenter is inserted in the track of one cell, while the other cell will have two tracks: the first track ends just before the intersection, the second track starts just after the intersection.

For the statistical analysis our samples demand, it has no importance to link the two sections of the incomplete track. Therefore, in order to save time, in the routine work, we do not implement any procedure for a second-step analysis of incomplete tracks. However, for a more precise analysis, we apply a procedure to solve these misinterpreted tracks. Because the data structure contains the positions of the cells, we can calculate the motion vectors in each difference image, and on the basis of this calculus we can forecast the behavior of the cells. As the choice of the delay avoids the occurrence of the occlusion phenomena in more than one difference image, the problem has an easy solution.

Operation Procedures

Automatic systems need specific procedures in order to obtain a reliable analysis of data. First of all, the microscope slide must be very clean, and the microscope well aligned. The microorganisms can freely swim within a very narrow layer, in order to reduce the presence of out-of-focus cells as much as possible.

The system has a user friendly interface. Nested menus, presented on the computer monitor and selected by a mouse, help the operator. From the root the operator selects the tracking menu (Fig. 4). In order to select its options, the operator should have an *a-priori* knowledge of the threshold between cells and background, the minimum and the maximum cell area values, the minimum and the maximum distance travelled by the cell, the minumum and the maximum angle of deviation from its trajectory, the speed of the microorganisms, the number of difference images per second, and the reduction factor of the difference images.

Some of these parameters can be easily determined using the automatic labelling procedure menu (i.e., maximum and minimum area values). Other parameters need longer processing for their determination. We describe the automatic procedure used for the speed value determination. The system acquires a couple of frames utilizing for each frame six bits of the frame memory and performs the subtraction operation as previously explained. Usually, we choose a delay that varies linearly, but the delay can progress in a different way as well. In our case, because of the CCIR standard, the delay between two successive frames is 40 ms, or a multiple of 40 ms. After the subtraction, every difference image is zoomed and placed in real time in its proper position of the frame memory by means of pan and scroll operations.

The first image (40 ms) represents the real time difference between the first acquired frame and the second acquired frame; the second image (80 ms) is the real time difference between the third acquired frame and the fourth acquired frame (Fig. 5). For the determination of the speed value of the cell we have to measure the distance covered by the cell and the time lapse; if no stimuli are applied to the environment, the swimming speed of the cells can be considered constant. Therefore, the time a cell takes to cover a distance equal to its long axis can be used for the determination of its speed. As the difference procedure suggests, until the cell does not cover a distance equal to its size, the area value will be lower than the real one. The greater the delay between the frames, the less two successive images of the cell are superimposed; there is a delay for which the subtraction operation gives two separate images of the same cell, i.e., the area value of the cell is the real value. A higher delay between the two frames still separates the two images of the cell, but the area value of the cell will remain constant. In the top corner of the whole frame memory, the cells can be hardly recognized because the difference between two successive images of the same cell consists of a small agglomerate of pixels. In the bottom corner of the whole frame memory, the cells are instead easily recognizable, because in this case the difference between the two images of the same cell is the whole cell area.

In order to determine quantitatively the cell area variation every reduced image is labelled. Because of the limited depth of the medium, the swimming path of the cells is planar, i.e., the cells are always in focus. The average area of the cells moving in the

```
            PAPRI MOVEMENT ANALYSIS SYSTEM - ISTITUTO DI BIOFISICA

                    Differential Track Reconstruction Menu

                       Set Movement Analysis Parameters
                  Continous Acquisition - reduction factor 1
                               Film Acquisition
                    Default Threshold 5 - select to change
                     Labelization & Track reconstruction
                            Or of images - dark field
                               Save track data

                            EXIT to previous Menu

angle=60  reduct=2  fr_del=1  cp_del=10  last_th=5  magn=20
min_radius=10  max_radius=30      av_file=  last_file=
                                                          AOI: Whole
```

Fig. 5. An example of the operation procedure for the microorganism speed determination. The rows show the frame couple on which the difference operation has been performed and the resulting difference image.

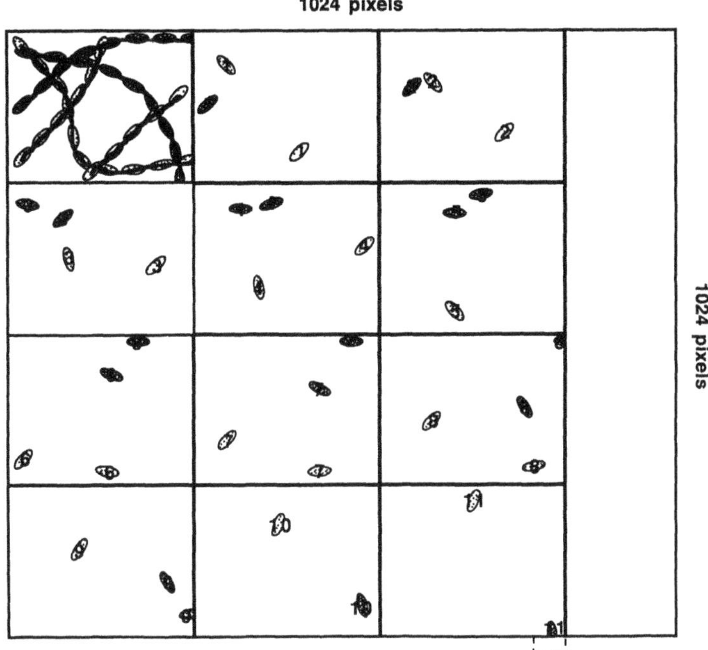

Fig. 6. An example of the reconstructed tracks: the tracks are shown superimposed on the first difference image.

field is calculated and the detected cells are labelled with a dot superimposed on their baricenters.

For each reduced image a preestablished standard deviation value determines the selection of an area value range. Therefore, touching cells are automatically rejected because their area is too big; similarly, small agglomerate of pixels, produced by the subtraction operation in the case of moving cells which intrudes onto the area formerly occupied by a different cell, or when a cell enters the field of view between the two frames on which the subtraction operation is performed, are rejected.

The variation of cell number in the microscope field during the acquisition, is not critical for the analysis, because we calculate the average area of the labelled cells present in each difference image. By interpolating the data of the plots of the average cell area versus the delay progression, we obtain two intersecting straight lines. The first line shows that the average area value increases with the increasing of the delay because the overlapping of two successive images of the same cell decreases. The second lines shows that the average area value becomes steady because there is no more overlapping between the two successive images of the same cell. The intersection of these two lines identifies the time delay which has to be used for the determination of the exact swimming speed of the microorganisms. In our case 640 ms is the time *Euglena* cells need to cover a distance equal to its long axis. When all the values are defined, the operator can start the experimental section of the photobehavior phenomena analysis, using these values as input. The sequence of the procedure is the following: acquisition, automatic segmentation, labelling, and track reconstruction.

Figure 6 shows the whole frame memory, for a 12-image time sequence. The reconstructed trajectories are superimposed on the first difference image. In order to verify the accuracy of the reconstructed tracks, we apply the logical OR operation to the first stored frame and to all the successive frames, as if we were using a long term exposition with a photographic camera. The cells are represented by dark areas. The reconstructed tracks are the heavy black lines. The computation time of the whole procedure depends on the number of the cells swimming in the microscope field. For the case shown in here, it was 0.3 s/frame. Cell parameters such as average speed, acceleration, travelled distance, and directions are stored in a file.

The wrong selection of an operation parameter can produce unpredictable results. In the case of too low a threshold value between cells and background, fuzzy agglomerates of cells will appear; on the other hand, if too high a threshold value is selected, cells may be not found, during the labelling procedure. If the reduction factor and/or the number of difference images is too low, the microorganism tracks may be incomplete; if the reduction factor is too high, cells can be suppressed. A correct speed value allows the selection of the proper delay between two frames in order to obtain a difference image with an entire cell as difference. The right range of area values allows to label true cells and to reject agglomerates or debris. The right distance range and the right direction range choices allow to connect the baricenter of the same cell in successive difference images.

Acknowledgements

This work has been partially supported by "Progetto Finalizzato Sistemi Informatici e Calcolo Paralello" of CNR.

References

Coltelli, P., and Gualtieri, P., 1990a, A procedure for the extraction of object features in microscope images, *J. Biomed. Comput.*, 25:169.

Coltelli, P., and Gualtieri, P., 1990b, A real time, automatic velocity determination of moving objects in microscope images, *J. Comput. Assist. Microsc.*, in press.

Gualtieri, P., Colombetti, G., and Lenci, F., 1985, Automatic analysis of the motion of microorganisms, *J. Microsc.*, 139:57.

Gualtieri, P., Ghetti, F., Passarelli, V., and Barsanti, L., 1988, Microorganism track reconstruction: an image processing approach, *Comput. Biol. Med.*, 18:57.

Gualtieri, P., and Coltelli, P., 1989, A digital microscope for real time detection of moving microorganism, *Micron Microsc. Acta*, 20:99.

Häder, D.-P., and Lebert, M., 1985, Real time computer-controlled tracking of motile microorganisms, *Photochem. Photobiol.*, 42:509.

Kondo, T., Kubota, M., Aono, Y., and Watanabe, M., 1988, A computerized video system to automatically analyze movements of individual cells and its application to the study of circadian rhythms in phototaxis and motility in *Chamydomonas reinhardtii*, *Protoplasma Suppl* 1:185.

Yachida, M., Asada, M., and Tsuji, S., 1981, Automatic analysis of moving images, *IEEE Trans. Pattern Anal. Machine Intell*, *PAMI*-3, 1:12.

Light Scattering Techniques in Studying Photoresponses

Cesare Ascoli and Donatella Petracchi

Istituto di Biofisica del CNR
Via S.Lorenzo 26
56127 Pisa
Italy

Introduction

Photoresponses of microorganisms, especially of flagellated algae, to light level variations are called phobic responses. Typical phobic responses are the stop and the change of direction of *Haematococcus* or *Chlamydomonas* after a flash.

A much more complex response is given by phototactic microorganisms (such as *Haematococcus* and *Chlamydomonas*) when there is a directional illumination. These cells, when illuminated from a definite direction, are able to track the light direction and move towards the light source.

In the last ten years the study of these phenomena has developed mainly through the use of image analysis. Nevertheless some motion parameters can also be measured by analyzing the light scattered from swimming microorganisms. Light scattering based methods have the advantage of being quick, and, in the case of phobic reactions, they provide more direct data on the transduction process, as they measure the frequency of flagellar beating.

In the first part of this paper, measuring methods will be described for determining the distribution of the velocity and of the flagellar beating frequency. The second part will discuss which kind of experiments provide significant information on the biological process. In fact, it is the way the experiment is devised which determines whether the results we take from it provide clear information on the biological system we are studying or not.

Experimental Methods

The light scattered from swimming microorganisms can be used for measuring both velocities as vectors and flagellar beating frequencies (Ascoli et al., 1978a,b; Racey et al., 1981; Chen and Hallett, 1982; Ascoli and Frediani, 1983; Angelini et al., 1986; Pfau et al., 1983).

To measure velocities, the Doppler shift of the scattered light can be revealed, and this entails the use of coherent laser light and of interferometric methods. Laser light can be schematized as a monochromatic plane wave, which undergoes a Doppler frequency

Biophysics of Photoreceptors and Photomovements in Microorganisms
Edited by F. Lenci *et al.*, Plenum Press, New York, 1991

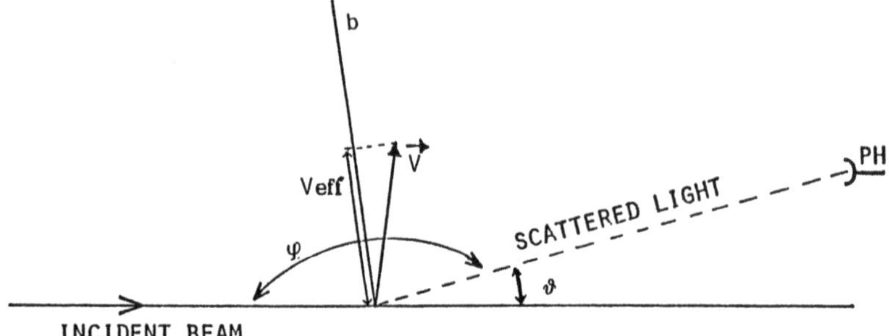

Fig. 1. **b** is the bisector of the angle Φ, *V* is the scatterer velocity, its component V_{eff} in the b direction is the effective component for the Doppler shift. *V* is not necessarily lying on the scattering plane.

shift when scattered by a moving particle. The Doppler shift of the scattered light is given by:

$$\delta f = (V/\lambda) \cos\alpha \sin\Theta/2 = (V_{eff}/\lambda) \sin\Theta/2 \quad (1)$$

where λ is the wavelenght of the incident light, Θ is the observation angle, i.e., the angle between the directions of the incident and scattered light, α is the angle between the velocity *V* and the bisector **b** of the angle Φ supplementary to Θ and $V_{eff} = V \cos \alpha$ is the effective component of the velocity in the **b** direction (see Fig. 1). For a given observation angle the maximum Doppler shift is produced by cells moving along the bisector **b.** Varying Θ for a given velocity of the scatterer the maximum Doppler shift is obtained for $\alpha = 180°$, i.e., for the back scattered light.

The most direct detection method to measure the Doppler shift is the heterodyne detection, schematically shown in Fig. 2. The light scattered from the sample and a part of the unperturbed beam, often called the local oscillator, fall on the photodetector and their beating makes it possible to measure the Doppler shift. Calling E_o the amplitude of the electric field of the unperturbed wave and E_j the contribution from the jth scatterer, and assuming that E_o and E_j are parallel vectors, the electric field on the photodetector is given by:

$$E(t) = E_o \sin 2\pi\, f_o\, t + \sum_{j=1}^{N} E_j \sin (2\pi\, f_j\, t + \psi_j) \quad (2)$$

where ψ_j is a phase term depending on the actual position of the jth scatterer inside the scattering volume. By computing from (2) the mean value of $E^2(t)$ and neglecting terms at optical frequency the photocurrent is:

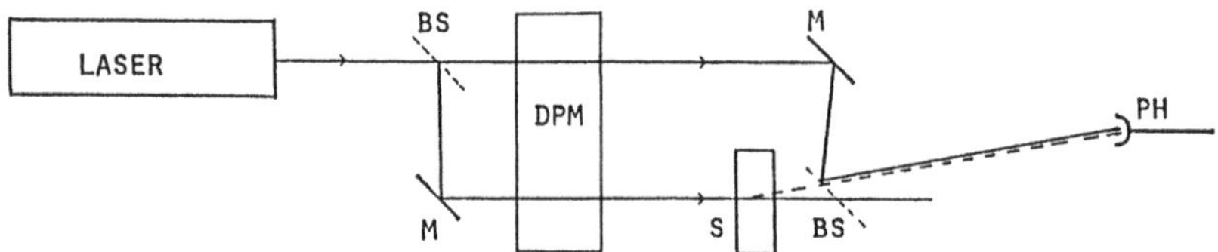

Fig. 2. Scheme of the heterodyne spectrometer: BS beam splitter, M mirror, DPM differential phase modulator, Ph photodiode. The differential phase modulator is used to obtain a frequency shift between the local oscillator and the beam which falls on the sample S.

Fig. 3. Scattering patterns from *Euglena gracilis*. At left a single cell (the arrow shows the cell orientation), in the medium several cells with random orientation, at right a sample oriented by the application of an electromagnetic field (Ascoli et al., 1978).

$$i(t) \approx E_o^2 + \sum_{1}^{N}{}_j E_o E_j \sin(2\pi\, \delta f_j\, t + \psi_j) \qquad (3)$$

A good alignment of the scattered and heterodyning lights is required to have a good signal to noise ratio; in fact, if wave surfaces are not parallel, several interference fringes arise on the photodetector, thereby reducing the useful signal (Cummins and Swinney, 1970).

If there is only one scatterer in the measuring volume, the power spectrum of i(t) is a line at a frequency equal to absolute value of δf, which is proportional to V_{eff}. With many scatterers in the measuring volume, the photocurrent is a sum of many independent contributions, and its power spectrum provides information on the statistical distribution of V_{eff}.

Let us assume that all the cells scatter the same light intensity on the photodetector. In this hypothesis and if all the cells move in the same direction, the power spectrum P(f) of the signal obtained by heterodyne detection gives the speed distribution directly. In general, for isotropically scattering cells, P(f) is homologous to the distribution of the absolute value of the effective component V_{eff}. For instance, if the scatterers are moving at the same speed in all directions in the space, we will have a rectangular distribution, since the distribution of V_{eff} presents a rectangular distribution.

All this holds for scatterers moving on straight lines and if the scattering diagram is isotropic around the unperturbed beam. But this is far from true. Figure 3 shows the pattern of the light scattered from a single *Euglena*. The picture was taken with the film placed perpendicularly to a laser beam, 20 cm from the sample. The scattering pattern of a single *Euglena* may easily be observed by the naked eye: there is a sharp intensity peak at right angles to the major axis of the cell and a large region with a weak, almost uniform intensity. Looking at the screen for a few seconds we can observe that the sharp peak oscillates by some degrees around its mean position; this is because the cell axis oscillates during the motion. All this generally holds for other flagellated algae, which are greater than the wavelength of the light and which move in a complex roto-translatory motion.

Therefore, for the anisotropy of the light scattering pattern, the scattered light intensity is modulated at the frequencies of the cell body rotation and of the flagellar beating, and these frequencies appear in the heterodyne spectrum as lines around the Doppler line. On the other hand the frequency of flagellar beating and of body rotation can be detected directly, by measuring the power spectrum of the scattered light.

Usually heterodyne detection of Doppler shift enables the distribution of a component of the velocity in the sample to be measured, regardless of its sign. To study phototaxis, the sign of the measured component of the velocity must be detected, so

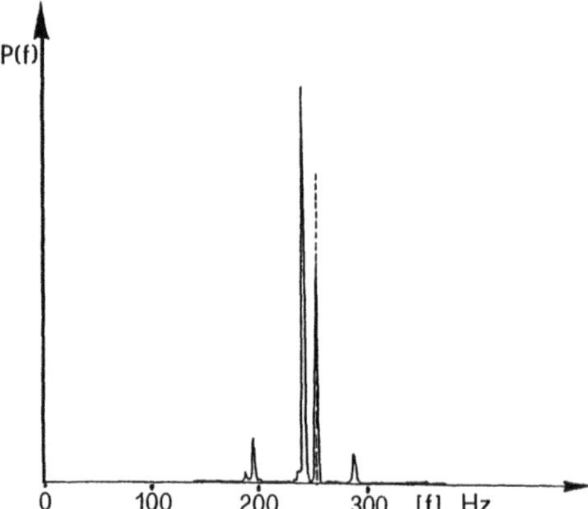

Fig. 4. Heterodyne spectrum obtained from a single cell of *Haematococcus pluvialis*; dashed line shows the frequency of the local oscillator. The frequency shifted Doppler line and two lateral bands can be observed. Redrawn from Ascoli and Frediani (1983).

that an improved apparatus is required. The differential phase modulator (DPM) block in the heterodyne spectrometer of Figure 2 makes it possible to measure the sign of velocity component. Applying a suitable sawtooth voltage to the DPM, the local oscillator beam is frequency shifted with respect to the beam which falls on the sample. As a result, the spectrum of the photocurrent gives the distribution of V_{eff} directly, without averaging left and right bands around the line of the local oscillator.

The following figures are reported to illustrate the heterodyne detection method. Figure 4 shows the frequency shifted heterodyne spectrum obtained from a single cell of *Haematococcus pluvialis*. There is a very narrow Doppler line, without a meaningful structure, and two low symmetrical lines 30 Hz far from the Doppler line, which are due to the amplitude modulation produced by the flagellar beating.

Figure 5 shows two heterodyne spectra. The lower is obtained from a randomly swimming population of *Haematococcus pluvialis*. It appears to be symmetrical around the frequency of the local oscillator. The two lateral bands are due to the flagellar beating. The upper is still obtained from a sample of the phototactic *Haematococcus pluvialis*, but laterally stimulating the cells with green light. A large asymmetry clearly appears in the Doppler spectrum and the peak on the left of the local oscillator frequency indicates that the preferred direction of motion is towards the light source. Nevertheless the orientation is not complete as contributions, which cannot be ascribed to lateral bands, appear on the right of the local oscillator. For comparison, Figure 6 reports the speed distribution of a sample of *Euglena* oriented by an electromagnetic field (Ascoli et al., 1978b); left and right swimming speeds are separately detected and both are clearly separated from the local oscillator.

To measure flagellar beating frequencies a different method can be used which does not require a coherent source and is easier to use than the heterodyne detection. The fluctuations in the intensity of the scattered light due to the flagellar beating and to the non-isotropic scattering properties of the microorganisms can be measured simply by using a dark field microscope and collecting on the photodetector the light scattered into the microscope lens. The dark field condenser prevent the incident light falling on the sample from reaching the photodetector and the small modulation at the flagellar beating frequency can be measured. Figure 7 shows the set up used for these measure-

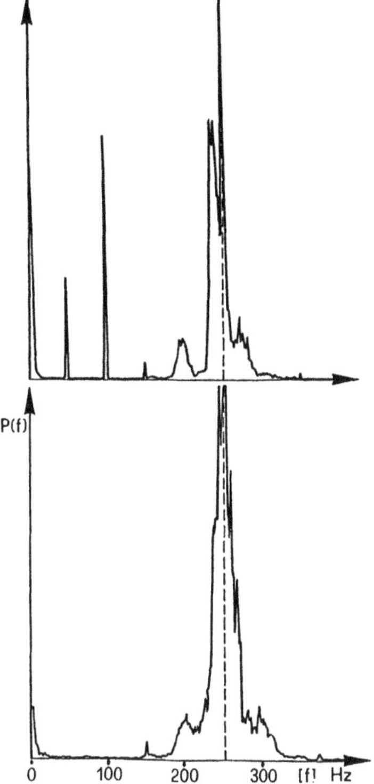

Fig. 5. Heterodyne spectrum obtained from a random swimming sample of *Haematococcus pluvialis* (lower curve) and from a sample oriented by lateral illumination (upper curve). Dashed line shows the frequency shift of the local oscillator. Redrawn from Ascoli and Frediani (1983).

ments. By this method, photocurrents due to light scattered from single swimming cells have been recorded (Angelini et al., 1986; Petracchi, 1988). As shown in Figure 8, the photocurrent from a single microorganism varies about sinusoidally in time at two different frequencies: the higher (at about 30-50 Hz) corresponds to the flagellar beating frequency and the lower, at about 2 Hz, to the cell body rotation.

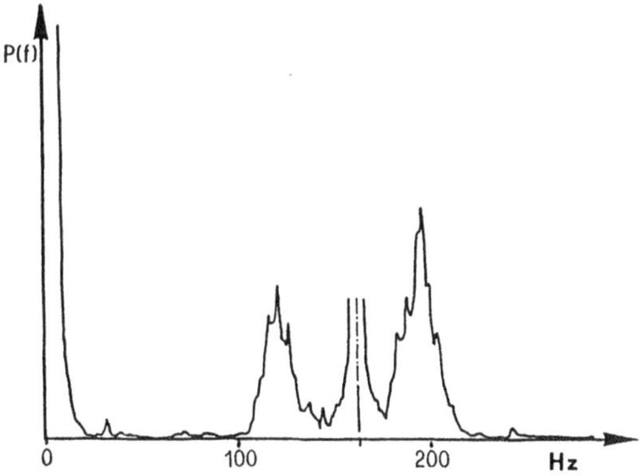

Fig. 6. Heterodyne spectrum obtained from a population of *Euglena gracilis* oriented by application of an electromagnetic field; dashed line shows the frequency shift of the local oscillator. Redrawn from Ascoli and Frediani (1983).

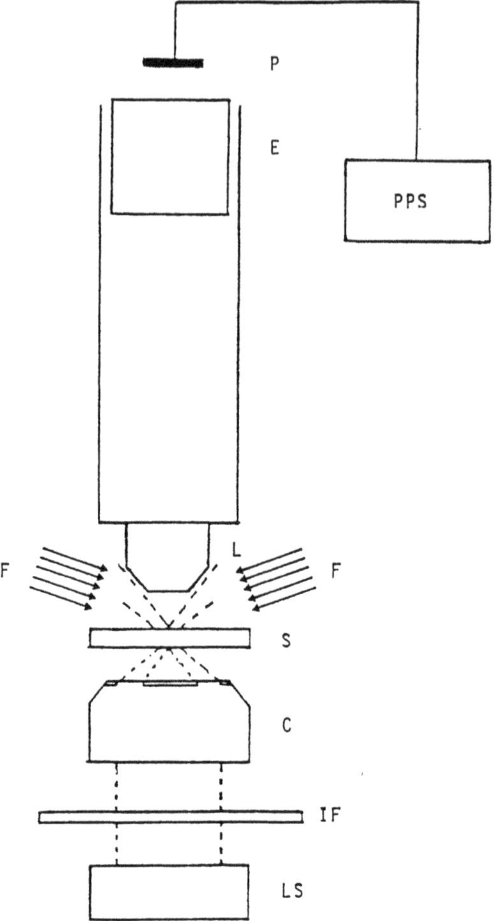

Fig. 7. Set-up to measure flagellar beating frequency of swimming algae in dark field. LS light source, IF infrared filter, C dark field condenser, S sample, L microscope lens, E eyepiece, P photodiode, PPS photocurrent power spectrum analyzer, F flash units.

When a sample with many swimming cells is in the microscope field the photocurrent power spectrum gives the distribution of flagellar beating and body rotation frequencies (Figs. 9, 10).

Experiments and Results

Typically, to use light scattering techniques in studying photoresponses, the analyzing light must be in a spectral range where the microorganisms are not sensitive (red or infrared light is suitable for many flagellated algae) and the photodetector must be screened from the light stimulus. Moreover, to have stable results, a number of equivalent spectra must be averaged, i.e., spectra with the same delay from the delivery of the light stimulus. To do this several data acquisitions are stored in a computer, and spectra at the same distance from the light stimulus are averaged. All this can be performed on line.

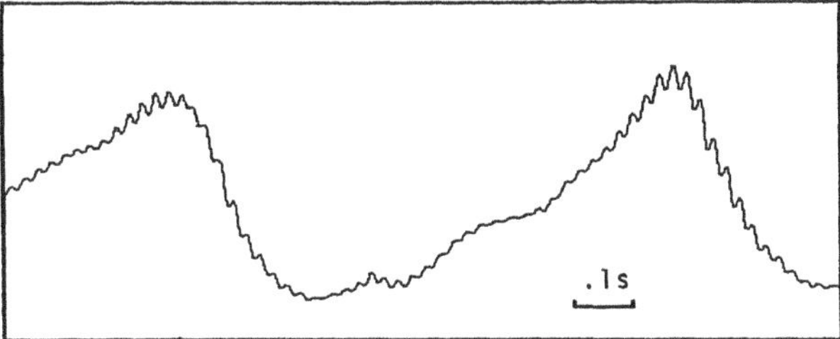

Fig. 8. Photocurrent detected with a single *Euglena* cell swimming in the microscope field. Redrawn from Angelini et al. (1986).

Both heterodyne detection of the Doppler shift and flagellar beating measurements in dark field have been used to study photobehavior in *Haematococcus pluvialis* (Angelini et al., 1986; Cantatore et al., 1989; Petracchi, 1988). In *Haematococcus pluvialis*, which is very similar to *Chlamydomonas reinhardii*, the photorecptor is placed very near to the stigma, as shown by loose patch clamp recording (Ristori et al., 1981); during the motion the stigma shades, more or less periodically, the photoreceptor, yielding a signal which probably allows the cell to detect light direction, with a mechanism similar to the one used by tracking radars (for a discussion see the review of Foster and Smith, 1980). This is a reasonable model, but there is no direct evidence for it and the nature of the mechanisms responsible for cell orientation with respect to light direction is still an open question. If a periodic shading is responsible for the detection of light direction, the time spent to orient a sample should be of the order of, or greater than,

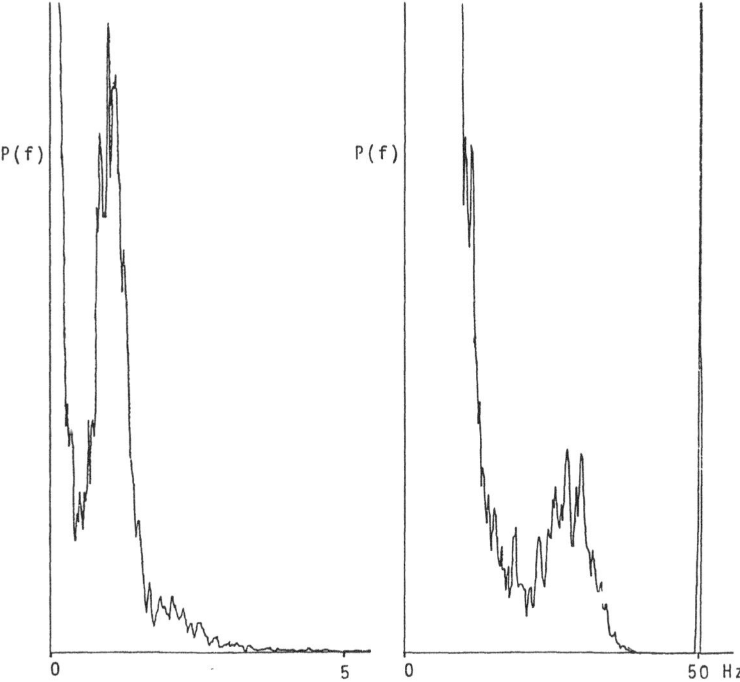

Fig. 9. Power spectrum of the photocurrent detected from a sample of *Euglena gracilis* with the set-up of Figure 7, showing the flagellar beating frequency distribution (on the right) and the body rotation frequency distribution (on the left).

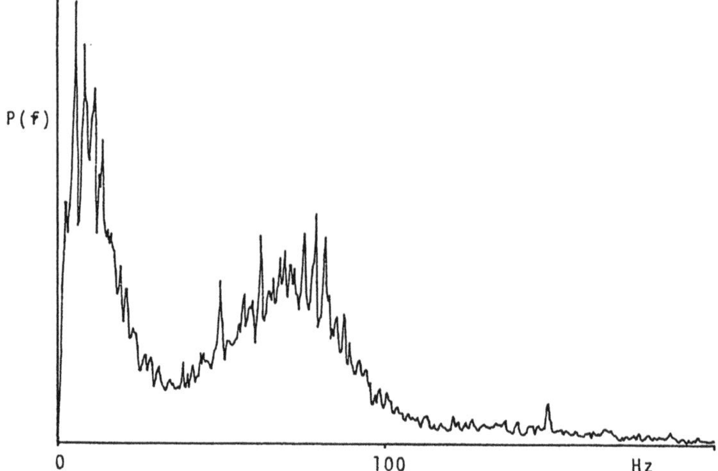

Fig. 10. Power spectrum of the photocurrent detected from a sample of *Chlamydomonas reinhardii* with the set-up of Figure 7, showing the flagellar beating frequency distribution.

the shading period (about 0.5 s). Then the measurement of the time needed by the cells to reorient themselves towards the light source can provide some useful information.

This has been done (Cantatore et al., 1989) by measuring the average of the component along the stimulus axis of the swimming velocity of a population of *Haematococcus pluvialis* under positive phototaxis conditions. The stimulus was given by two beams of actinic light, directed opposite each another and shone alternately for about 5 s. Thus, phobic reactions were avoided and the time spent on reaching reorientation could be measured. Using the heterodyne spectrometer with shifted local oscillator (Ascoli and Frediani, 1983) the distribution of the effective velocities (with their sign) was measured. Figure 11 which reports the mean velocity after the stimulus switchting, shows the time course of the reorientation in a population of *Haematococcus pluvialis* (Cantatore et al., 1989). The results show that the technique is suitable for studying phototaxis, but further work is necessary. In particular, smaller changes of stimulating beam direction should be used and the dependence of the reorientation time on the stimulus intensity should be studied.

As far as the measurements of step-up responses in *Haematococcccus pluvialis* are concerned, the data (Angelini et al., 1986) show a transient increase in the flagellar beating frequency after a flash (Figs. 12, 13). This increase does not depend on the stimulus strength, the change of the flagellar beating can only occur fully, or not at all. The percentage of reacting cells can be measured by the ratio between the areas of shifted and unshifted bands (Fig. 13). This percentage, on the other hand, depends on the light dose (light dose = flash intensity X flash duration) only for flash durations not exceeding 60-80 ms. In fact, for flashes of low intensity the increase in the duration over

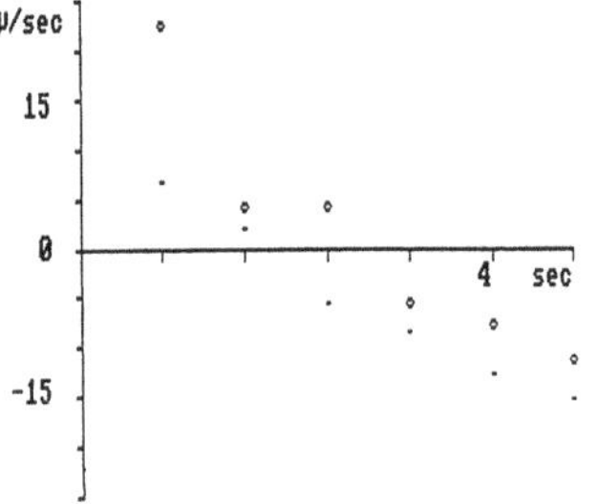

Fig. 11. Trend of the mean velocity of a phototactic sample of *Haematococcus pluvialis* against the time after switching the light direction. Redrawn from Cantatore et al. (1989). The mean velocity obtained with left stimulation (small dots) are changed of sign and superimposed with that obtained with right stimulation.

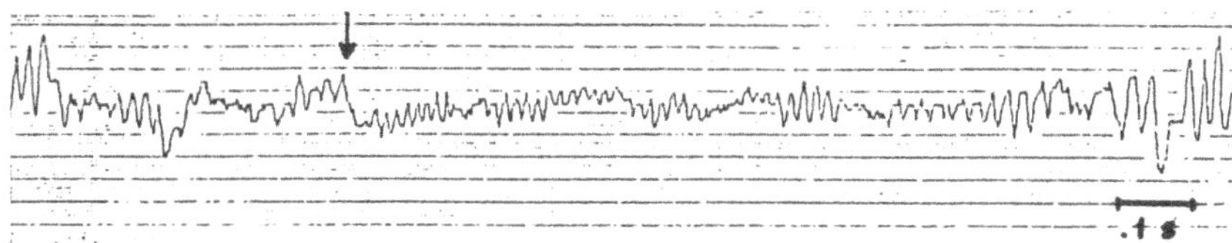

Fig. 12. Changes of flagellar beating frequency in a single cell of *Haematococcus pluvialis* after a flash.

60-80 ms does not produce an increase in the percentage of cells which react to the stimulus. Thus an accumulation process of an unknown molecular species with a decay time constant of 20-30 ms has been assumed to drive the step-up response in *Haematococccus pluvialis*, a response occuring each time the accumulation exceeds a given threshold value S (Angelini et al., 1986). The graded shape of the dose-effect curves can be explained assuming that the threshold and/or the absorbed light is not identical in different cells. In fact, during the motion each cell can be oriented in a more or less favorable way with respect to the light source. As shown in Figure 14 the light which falls on the photoreceptor depends on the relative orientation between photoreceptor, stigma and light source.

This idea has been the basis for experiments on the pigments involved in photoreception (photoreceptor and stigma) (Petracchi, 1988). Dose-effect curves for step-up reactions were determined by measuring the percentage of reacting cells for different light intensities and for two wavelengths of the stimulating light (450 and 600 nm). Quite surprising results were obtained (Fig. 15), especially in old samples (8-10 days). The shape of the dose-effect curves depends strongly on the wavelength. The slope at 450 nm is allways lower than at 600 nm, but the difference depends on the age of the culture and, in two day samples, slopes at different wavelengths do not differ very much.

Figure 16, obtained in a two days old sample, illustrates a different comparison: the two curves are both obtained at 450 nm (±25 nm), but they have been obtained by using a different geometry of illumination. In the first curve the source was a single flash unit, while in the other curve two flash units facing each other were used simultaneously to illuminate the sample. Figure 16 shows that the shape of the dose-effect curve very much depends on the geometry of illumination.

Now let us look in some detail at the computer simulation of the model for phobic responses suggested above: we assume that photoreceptors are isotropical structures and that the number of light quanta which are absorbed depends on the absorption cross section and number of the pigment molecules, and on the intensity of the light falling on the photoreceptor, but it does not depend on the incidence angle of the light beam falling on the photoreceptor. The intensity of light on the photoreceptor in turn depends not only on the light intensity of the stimulus but also on the cell structures interposed between the photoreceptor and the light source. The position of each cell in the sample can be considered constant during the flash and therefore the response in a population is due to a combination of responses coming from cells in a favourable situation and from those which are screened or partially screened (Fig. 14).

We can imagine the stigma as an absorbing disk with a tickness L (other hypotheses for the role of the stigma could be assumed (Foster and Smyth, 1980), but the results only differ in the details). We will call α the angle between the normal to the stigma plane and the light beam, I the light intensity which falls on the stigma, D its optical density, σ the cross section of the photoreceptor pigment, and M the number of molecules in the photoreceptor.

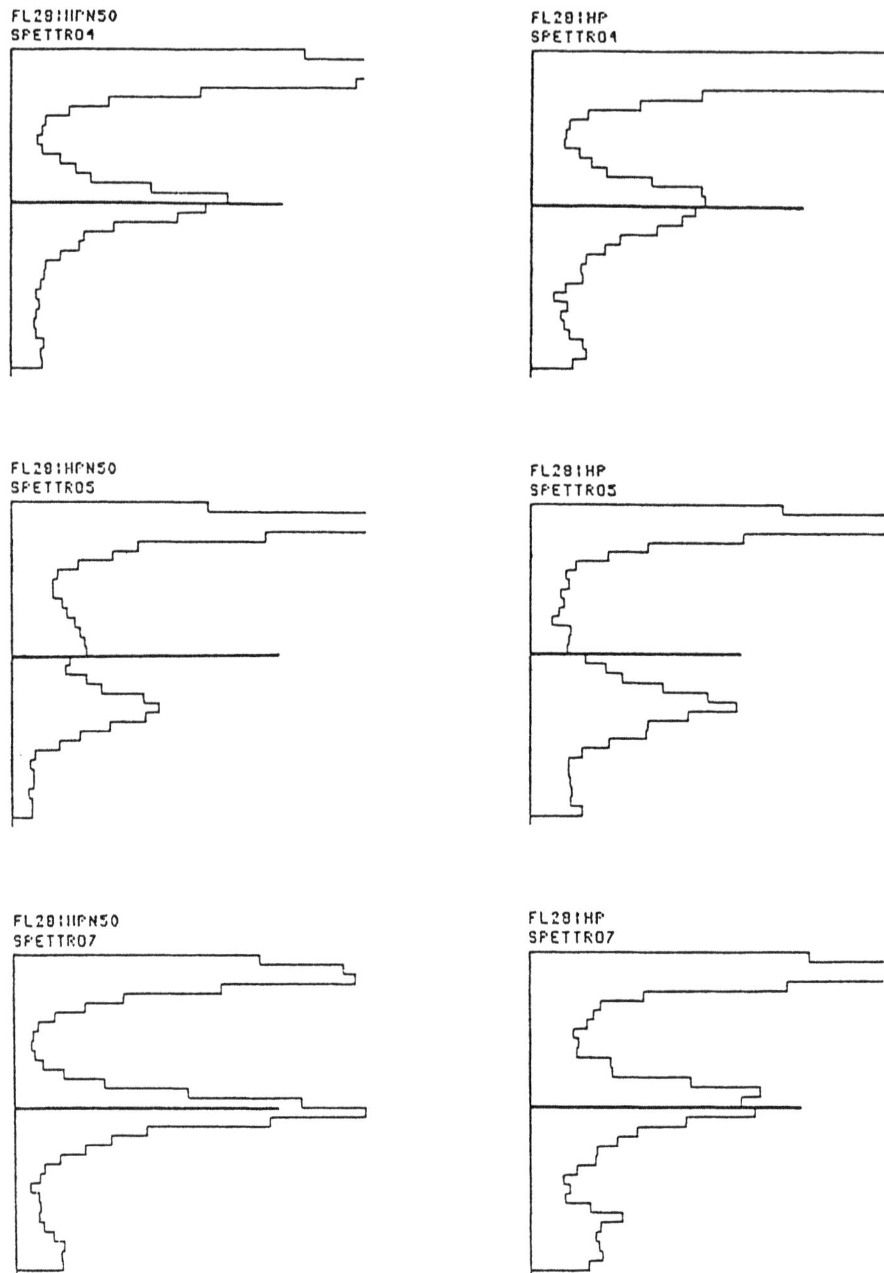

Fig. 13. Changes of flagellar beating frequency in a population of *Haematococcus pluvialis* after a flash. Spectra before the flash, just after the flash and about 1 s after the flash are shown. The black line shows the center of the distribution before the flash. In the upper row there is a complete shift of the flagellar beating; in the second row, obtained with a lower energy flash, only a part of the population is frequency shifted.

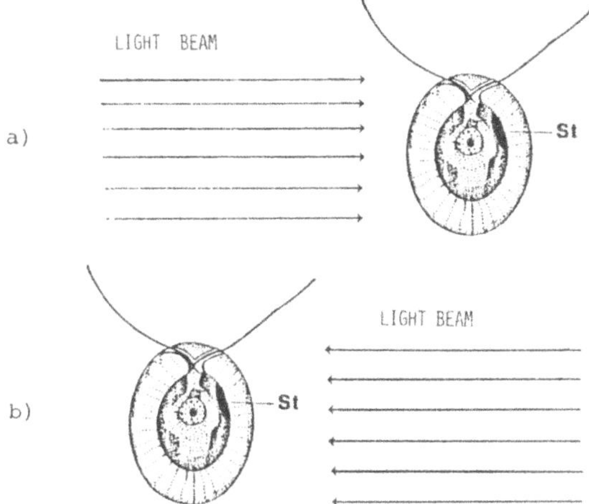

Fig. 14. Relative orientation of cell body, stigma and light source.

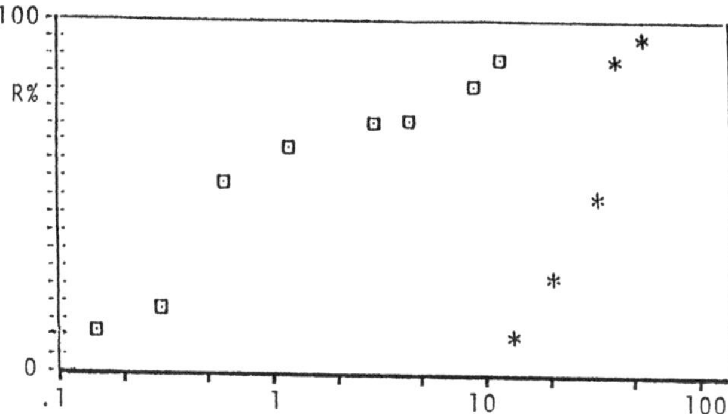

Fig. 15. Dose-effect curves for phobic responses in *Haematococcus pluvialis* at 450 nm (squares) and at 600 nm (stars). Note the different slopes of the two curves. Redrawn from Petracchi (1988).

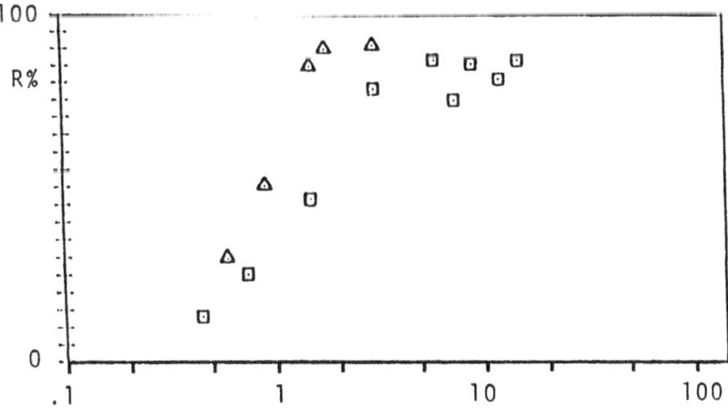

Fig. 16. Dose-effect curves for phobic responses in *Haematococcus pluvialis* with unilateral illumination (squares) and when using two flash units facing each other (triangles). Note the strong effect of the geometry of the illumination on the shape of dose-effect curves. Redrawn from Petracchi (1988).

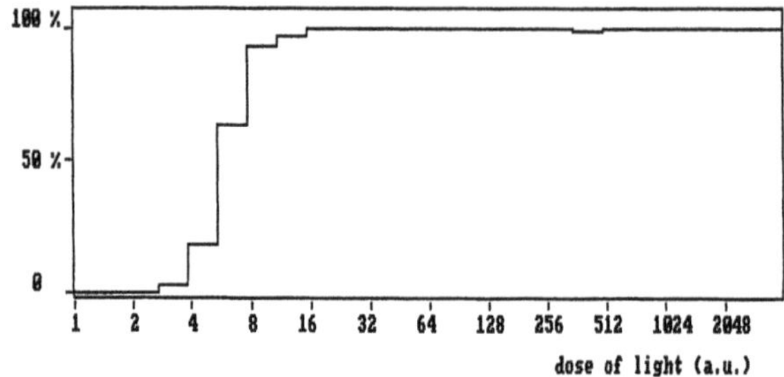

Fig. 17. Simulated dose-effect curve assuming the optical density of the stigma is zero and the variation coefficient of the cell threshold for phobic responses is 0.3.

When a light beam with intensity I falls on the stigma the light intensity I_T emerging from the other side is:

$$I_T = I \cdot 10^{-DL/\cos\alpha} \tag{4}$$

The mean number N of photons which are absorbed in the unit time is given by two different expressions depending on the value of angle α:

$$N = I\sigma\, M\, 10^{-DL/\cos\,\alpha} \qquad \alpha < \pi/2 \tag{5}$$

and

$$N = I\sigma\, M \qquad \alpha > \pi/2 \tag{6}$$

In our model the response to short flashes (short compared to the time constant of the accumulation) should occur only if $N\,\delta t > S$. In a population, the threshold S can be assumed as a characteristic parameter of each cell, with values in a range which depends on the sample age, on its methabolic situation and so on.

Figures 17 and 18 were obtained by a computer simulation of this model: for each cell the angle α and the threshold value S are randomly selected from a uniform and a Gaussian distribution, respectively. Then, by comparing expression (5) or (6) with the threshold S, the program determines whether a response occurs or not. Finally, the percentage of reacting cells is computed and plotted against I.

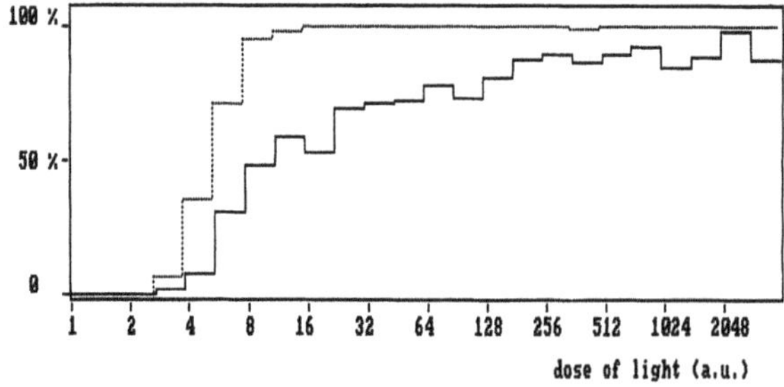

Fig. 18. Simulated dose-effect curves assuming the optical density of the stigma is 0.5 and the variation coefficient of the cell threshold is 0.3; the continuous curve corresponds to unilateral illumination while the dashed one is obtained by simulating two opposite flash units.

Obviously the model predicts a stepwise dose-response curve for a population of identical cells without screening structures (D=0). A graded dose-effect curve can be obtained for D=0 if one assumes that the threshold S is different in different cells. Figure 17 was obtained by assuming a Gaussian distribution for S with a variation coefficient of 0.3. Figure 18 was obtained assuming the same distribution for the threshold S as in Figure 17 and a value of 0.5 for the optical density D. The continuous curve shows the predicted dose-effect curve for unilateral illumination while the dashed line is obtained simulating two light sources, simultaneously illuminating the sample from opposite sides. The use of two opposite lamps greatly reduces the effect of the screening structures and the dashed line in Figure 18 looks similar to the one in Figure 17. Thus, the computer simulation shows that the dose-effect curve for phobic responses, which in order to erase the effect of the screening should be made using omnidirectional illumination, can be measured quite well using only two opposite lamps.

There is a qualitative agreement between experimental results and computer simulated experiments, which confirms the relevance of the stigma, and thereby of the geometry of the illumination, in determining dose-effect curves.

References

Angelini, F., Ascoli, C., Frediani, C., and Petracchi, D., 1986, Transient photoresponses of a phototactic microorganism, *Haematococcus pluvialis*, revealed by light scattering, *Biophys. J.*, 50:929.

Ascoli, C., Barbi, M., Frediani, C., and Muré, A., 1978a, Measurements of *Euglena gracilis* motion parameters by laser light scattering technique, *Biophys. J.*, 24:585.

Ascoli, C., Barbi, M., Frediani, C., Petracchi, D., 1978b, Effects of electromagnetic fields on the motion of *Euglena gracilis*, *Biophys. J.*, 24:601.

Ascoli, C., and Frediani, C., 1983, The application of laser light scattering to the study of photoresponses of unicellular motile algae, *In* "The Application of Laser Light Scattering to the Study of iological Motion," Earnshaw, J. C. and Steer, M. V., eds., Plenum, New York, pp. 667.

Cantatore, G., Ascoli, C., Colombetti, G., and Frediani, C., 1989, Doppler velocimetry measurement of phototactic response in flagellated algae, *Bioscience Reports*, 9:475.

Chen, S. H., and Hallett, F. R., 1982, Determination of motile behaviour of prokaryotic and eukaryotic cells by quasi-elastic light scattering, *Q. Rev. Biophys.*, 15:131.

Cummins, H. Z., and Swinney, P. N., 1970, Light beating spectroscopy, *in* "Progress in Optics," vol. 8, Wolf, E., ed., North-Holland Publishing Co., Amsterdam, p. 135.

Foster, K. W., and Smyth, R. D., 1980, Light antennas in phototactic algae, *Microbiol. Rev.*, 441:572.

Petracchi, D., 1988, Advanced techniques in photobehavioural studies: light scattering techniques, *In* "Light in Biology and Medicine," vol 1, Douglas, R. H., Moan, J., and Dall'Acqua, F., eds., Plenum Press, New York, p. 373.

Pfau, J., Nultsch, W., and Rüffer, U., 1983, A fully automated and computerized system for simultaneous measurements of motility and phototaxis in *Chlamydomonas*. Arch. Microbiol., 135:259.

Racey, T. J., Hallett R., and Nickel, B., 1981, A quasi-elastic light scattering and cinemetographic investigation of motile *Chlamydomonas reinhardii*, *Biophys. J.*, 35:557.

Ristori, T., Ascoli, C., Banchetti, R., Parrini, P., Petracchi, D., 1981, Localization of photoreceptor and active membrane in the green alga *Haematococcus pluvialis*, "VI International Congress of Protozoology," Warsawa, July 5-11, p. 314

Optical Absorption and Emission Spectroscopy of Photoreceptor Pigments

Francesco Lenci

CNR Istituto di Biofisica
Via San Lorenzo 26
56127 Pisa
Italy

Introduction

The common purpose of all sensory photopigments is the detection of light signals from the environment and the conversion of these external inputs into triggers for the transduction process, which, in our case, ends in the motile response of microorganisms to photic stimuli. To be efficient light detectors and signal converters, photoreceptor molecules usually have to be arranged in relatively ordered structures, such as, e. g., membranes, which they are embedded in or associated with, or paracrystalline-type structures which they are fixed in.

In nature a variety of physiological molecular matrices may exist, and in each of them a single pigment molecule can have quite diverse spectroscopic, photophysical and photochemical properties. For example, energy gaps between ground and excited states, vibrational width and structure of the electronic states, lifetimes of the excited states, ground and excited states interactions of a photoreceptor molecule can be markedly affected by the nature of the molecular microenvironment.

In the case of photomotile responses of microorganisms various molecules are, at present, thought to play the role of photosensory pigments for light signals (e.g., flavins, carotenoids, rhodopsins, stentorins and blepharismins), but in some cases even their chemical structure has not yet been unambiguously ascertained. The spectroscopic characterization of these photopigments is undoubtedly a prerequisite for their unambiguous identification and an important step towards the understanding of the mechanisms by which they collect and transduce photic stimuli into biochemical/biophysical signals the cell can amplify and utilize (Lenci and Colombetti, 1980; Song, 1983; Song, 1987; Lenci and Ghetti, 1989).

In the following paragraphs some of the properties of electronic states and transitions of polyatomic molecules and the basic principles of steady state optical absorption and emission spectroscopy will be outlined, of necessity neither exhaustively nor with fine details. More complex and/or advanced spectroscopies, like, e. g., circular dichroism, light induced absorbance changes (LIAC), time-resolved fluorescence, will not be discussed, nor will be the different experimental techniques. A few case-examples of

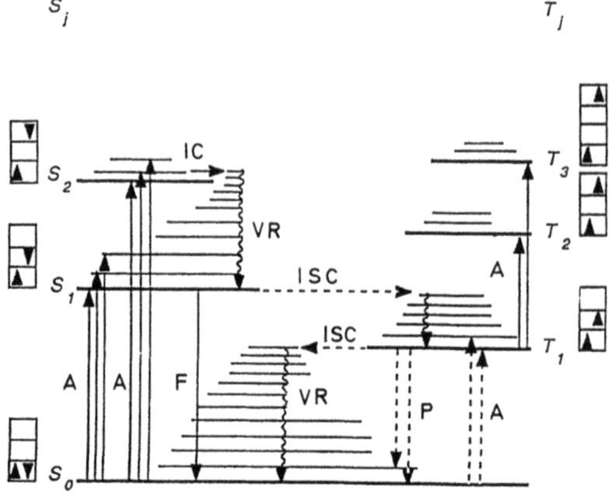

Fig. 1: Schematic diagram of electronic singlet, S, and triplet, T, states and of intramolecular transitions (Jablonski diagram) for a polyatomic molecule. VR: vibrational relaxations; IC: internal conversion; ISC: intersystem crossing; A: absorption; F: fluorescence; P: phosphorescence.

spectroscopic studies on photoreceptor pigments for photomotile responses of microorganisms will also be briefly discussed.

Electronic States and Transitions in Polyatomic Molecules

The familiar Jablonski diagram, reported in Figure 1, gives a schematic view of the energy levels of a typical polyatomic organic molecule containing an even number of electrons (as most molecules actually contain) and of the different possible transitions among the corresponding states.

Electronic States

In any molecule with an even number of electrons, the lowest energy electronic state (ground state), in accordance with Pauli's principle, will have a total spin angular momentum, S, equal either to zero (singlet state, with antiparallel spins and multiplicity $M = 1$ ($M = 2S + 1$)) or to one (triplet state, with parallel spins and $M = 3$). Since the electronic repulsion is lower in the first excited triplet state, T_1, than in the first excited singlet state, S_1, T_1 energy is lower than S_1 energy (Fig. 1). Because for only very few molecules the ground state is a triplet state ($^3O_2$ is one of the most interesting of these rare exceptions), unless otherwise stated, by ground state of a molecule we will mean its ground singlet state, S_0.

In a polyatomic molecule, the positively charged nuclei tend to repel each other, whereas bonding electrons tend to keep them together; due to the fact that in any molecule the electronic distribution is, in principle, different for its various electronic states, each electronic state is subdivided in a peculiar set of quantized vibrational substates; as also rotational motions of nuclei are possible, each vibrational state is, in its turn, subdivided in a set of rotational substates. According to the Franck-Condon principle, both absorption and emission electronic transition occur in a time ($\sim 10^{-15}$-10^{-16} s) which is much shorter than the time of a nuclear vibration ($\sim 10^{-13}$) and *a fortiori* of a rotation, so that both the momentum of the nuclei and their relative position can be considered as fixed during the electronic transition itself.

For the sake of simplicity, in Figure 1 only the vibrational sublevels of the electronic states are shown.

Radiationless Transitions

Once a molecule has been excited by the absorption (A) of a photon, it returns to the ground state releasing the absorbed energy through radiative and non-radiative transitions.

The main intramolecular non-radiative de-excitation pathways are (see Fig. 1):
VR: vibrational relaxations (within 10^{-14}-10^{-12} s) to the lowest vibrational state of the electronic excited state with emission of infrared quanta or through collisional transfer of kinetic energy.
IC: internal conversions (within about 10^{-12} s) between electronic states of the same multiplicity by means of vibrational coupling (the lower vibrational level of the higher electronic state overlaps the higher vibrational level of the lower electronic state, from which the molecule vibrationally relaxes).
ISC: intersystem crossing between electronic states of different multiplicity (for example from S_1 to T_1, in Fig. 1). Actually spin selection rules for electronic transitions require that the initial and final state have the same multiplicity, M; in principle, therefore, $S_i \rightarrow T_j$ and $T_j \rightarrow S_i$ transitions are spin forbidden. However, in a molecule the interaction between spin and orbital angular momenta can be strong enough to make triplet and singlet state no longer pure states and to give the triplet state some singlet state character and *vice versa*; as a result of this mixture, transitions between singlet and triplet states can occur, even though they are usually very weak ($S_0 \rightarrow T_1$ transitions, for example, are approximately 10^6 times less probable than $S_0 \rightarrow S_1$ transitions). The stronger the spin-orbit coupling and the weaker the spin-spin coupling is, the higher is the intersystem crossing probability; as an excited electron occupies an orbital different from that of its partner, spin-orbit and spin-spin couplings are respectively stronger and weaker in an excited state than in the ground state. The triplet character of S_1 is, therefore, higher than that of S_0 and the intersystem crossing to T_1 is much more probable from S_1 than from S_0 . Moreover the presence of heavy atoms within the molecule itself, as well as in the surrounding solvent or in the medium in which the molecule is embedded, can strengthen the spin-orbit coupling and, consequently, enhance the intersystem crossing probability.

It is worthwhile to stress that actual photosensing events do not take place through radiative deactivation pathways and that transitions and reactions which can have low yields in *in vitro* systems may well be highly efficient processes in physiological molecular environments.

Radiative Transitions

Fluorescence and phosphorescence radiative transitions, F and P in Figure 1, are discussed in the paragraph dedicated to fluorescence and phosphorescence spectroscopy.

Simplifying and schematizing, molecular optical absorption spectroscopy provides data on the probabilities of the electronic transitions from the ground state to the excited states and on the energy associated with each of these transitions as well as on the effect of the molecular framework on these parameters.

Similarly, fluorescence and phosphorescence measurements can give information on the fate of electronic excitation energy, i.e., on the probabilities of the different radiative and radiationless transitions from the excited levels to the ground level of a molecule and on the lifetime of the different excited states, as well as on the influence the molecular environment has on these processes.

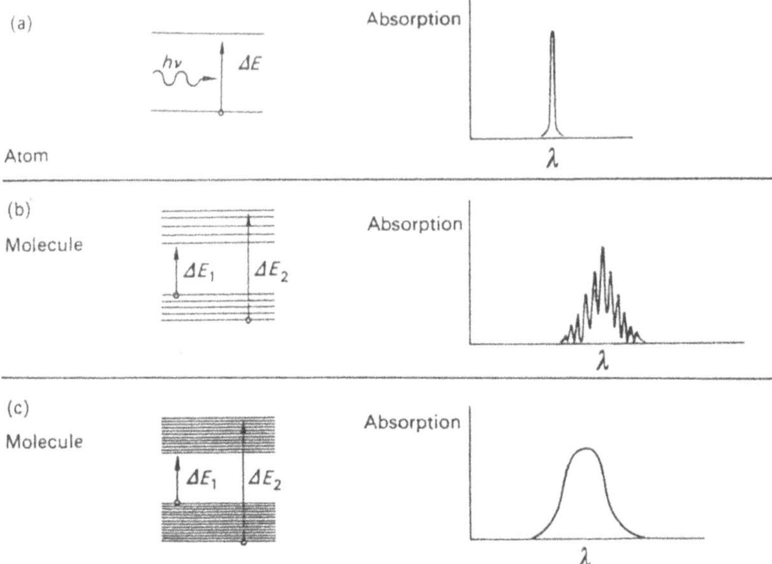

Fig. 2 : Energy levels for an atom and for two typical molecules and resulting absorption spectra (redrawn from Clayton, 1970).

Absorption Spectroscopy

Each wavelength at which absorption occurs, corresponds to the energy gap between a substate of the ground electronic state and a substate of the excited electronic states, E_{tot}. In the Born-Oppenheimer approximation (adiabatic approximation) the relatively slow motion of nuclei can be separated from the relatively fast motion of electrons and E_{tot} is given by:

$$E_{tot} = E_{el} + E_{vib} + E_{rot}$$

where E_{el}, E_{vib} and E_{rot} represent the electronic, vibrational and rotational energies.

Whereas electronic energy gaps are of the order of some eV (for example, from ~ 4.4 eV for absorption at 280 nm to ~ 1.9 eV for absorption at 650 nm) and vibrational ones of the order of some tenth of eV, the separation of the rotational energy levels is of the order of a few hundredths of eV, corresponding to absorption in the microwave region. From a comparison of these orders of magnitude, it follows that, to a fair approximation, the total energy of a molecule is given by the sum of its electronic and vibrational energies. The wavelength at which absorption is maximum corresponds to the most probable of these transitions, at room temperature all absorption transitions originating from the lowest vibrational substate of the ground electronic state.

For our present purposes, it is sufficient to state that the experimental measure of the transition probability at a certain wavelength, λ, is given by the molar extinction coefficient, $\epsilon(\lambda)$, at that wavelength, defined by the Beer-Lambert law:

$$A(\lambda) = \log_{10}(I_0/I) = \epsilon(\lambda)\, C\, l$$

where $A(\lambda)$ is the absorbance at that λ, I_0 and I are the incident and transmitted light intensities, C is the molar concentration (mol/l) of the absorbing species and l is the optical path length (in cm).

As shown in Figure 2, the fine structure of an optical absorption spectrum depends on the energy separation between substates: whereas for an atom the absorption spectrum consists in a narrow band centered at the wavelength corresponding to the well defined energy required for the transition from the ground to the excited state, for a molecule the absorption spectrum is made up of a set of more or less closely spaced lines, each corresponding to a transition from a particular ground substate to a particular excited substate. If these lines are broadened, they cannot be resolved and give rise to a broad structureless absorption band.

Effects of the Molecular Environment

For a molecule, all these spectral characteristics can be markedly influenced by a large variety of factors: from the chemical and physical properties of solvent, to the occurrence of weak ground state interactions or of covalent bonds between the molecule and its microenvironment.

Non-specific solute-medium interactions depend only on multipole and polarizability properties of the solute and the solvent and are responsible for extensively investigated displacements of electronic spectra (solvatochromic shifts) (Suppan, 1990).

In the case of flavins, for example, the near UV band, at 375 nm in aqueous solutions, is shifted to shorter wavelengths with decreasing solvent polarity, whereas the 445 nm band, whose position is hardly affected by the solvent, splits in several inflections in non polar environments (see, for instance, Koziol and Szafran, 1990).

Among specific solute-solvent interactions, the most important is the formation of H bonds, which, contrary to the dielectric interactions, imply a fixed stoichiometry of the interacting molecules (usually 1:1) and fixed bond lengths and angles (Suppan, 1990).

Again in flavins, the presence of H-bonding agents causes a characteristic red shift of the near UV maximum; for instance, for flavin-derivatives like riboflavin tetrabutyrate (RFTB) in 1,2-dichloroethane in the presence of 0.8 M trifluoroacetic acid as H-bonding agent, the absorption spectrum substantially consist in a band centered at about 400 nm, with a weak shoulder around 450 nm, as shown in Figure 3 (Koziol and Szafran, 1990).

From these examples it should result quite clear that *in vivo*, even in the absence of other chromophores absorbing in the same spectral region, the unambiguous identification of a flavin-type photoreceptor pigment can require a conspicuous amount of data on the chemical nature and behavior of the physiological molecular environment.

Basically, the same considerations hold for carotenoids and other candidate photoreceptor pigments. In the case of all-*trans*-lycopene, for example, the main absorption band in the blue decreases and a new band at 350-360 nm appears upon addition of water to an ethanol solution. The resulting altered absorption spectrum artifactually resembles more a flavin-type spectrum (Song and Moore, 1974).

Also solvent viscosity, and therefore temperature, can affect the structure of absorption spectra, usually enhancing the vibronic structure of the electronic transitions bands at high viscosities and/or low temperatures. This spectral behavior is also observed, e.g., in flavins upon binding to some apoproteins.

To reveal the occurrence, in ternary systems, of ground state intermolecular interactions which can be responsible for tiny changes in the absorption spectrum, the simple system shown in Figure 4 can easily be used, just to measure small changes in a large total absorption due to interactions between the two molecular species, M_1 and M_2, dissolved in the solvent L.

In every molecule light absorption is maximal when the electric field vector, $\vec{E}$, of the incident light and the transition dipole moment, $\vec{\mu}$, of the molecule are parallel and zero when they are perpendicular; When $\vec{E}$ and $\vec{\mu}$ form an angle Θ, the efficiency of

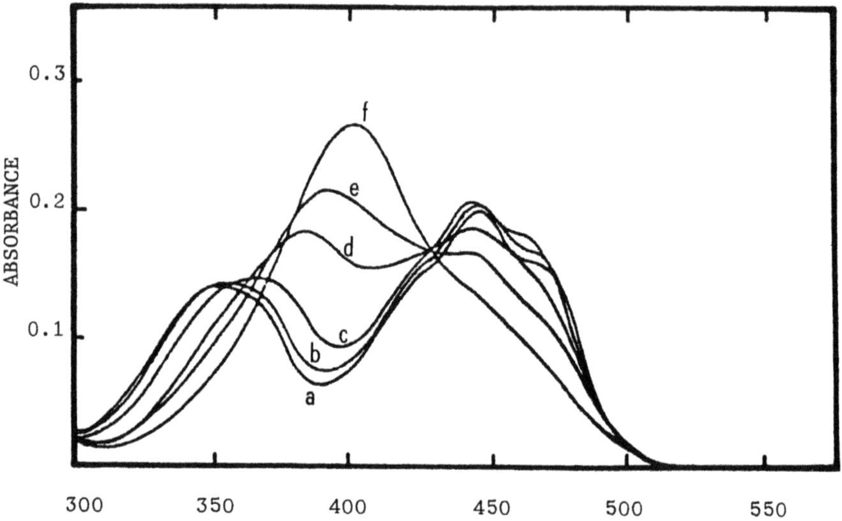

Fig. 3 : Optical absorption spectra of RFTB in dichloroethane in the presence of different concentrations of the hydrogen-bonding agent trifluoroacetic acid (TFA). a: no TFA; b: TFA 0.001 M; c: TFA 0.01 M; d: TFA 0.1 M; e: TFA 0.25 M; f: TFA 0.8 M (redrawn after Koziol and Szafran, 1990).

absorption is proportional to $\cos^2\theta$. A dichroic photoreceptor is a relatively rigid structure in which the chromophores, and hence their electric dipoles, are oriented with respect to a fixed axis and which, consequently, shows a preferential direction for light absorption, both for polarized and non-polarized light, as drafted in Figure 5.

If the measuring light is linearly polarized, the linear dichroic ratio, d, can be defined:

$$d = \frac{A_{paral} - A_{perp}}{A_{paral} + A_{perp}}$$

where A_{paral} and A_{perp} are the absorbances for light polarized parallel and perpendicular to the axis along which the dipole moments are oriented. Under appropriate conditions, measurements of the linear dichroic ratio, d, allow to find out if an absorption band corresponds to a single electronic transition or to have information on the structural organization of the system or, finally, to measure the direction of the transition electric dipoles.

In a phototactic microorganism, a dichroic photoreceptor would act as a directional detector and allows the cell to sense the position of the light source, as has been suggested for the green flagellate *Euglena* (Häder, 1987; Häder, this volume).

Since photoreceptor pigments are often localized in particular subcellular structures, it is possible to investigate the optical properties of the chromophores in their physiological environment by means of sophisticated microspectroscopic techniques. Advantage of this approach includes the possibility of performing measurements in the intact cell on the photoreceptor structure in its physiological state.

Here we will mention only a couple of works, referring to Gualtieri (1990) for a comprehensive review of the subject. Benedetti et al. (1976) confirmed that carotenoids are indeed the pigments contained in the stigma of *Euglena* and suggested that the chromophores neither are closely packed nor have an ordered arrangement in this organelle.

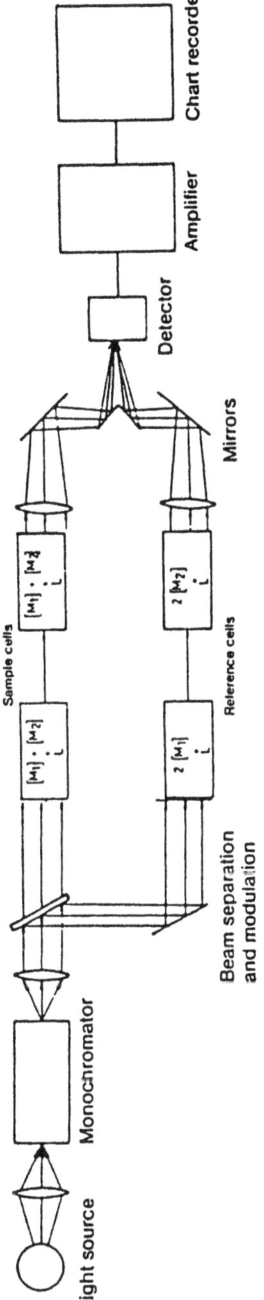

Fig. 4. Scheme of a simple system to detect intermolecular interactions in ternary systems (redrawn from Hoppe et al., 1983).

Recently, the optical absorption spectrum of a single isolated photoreceptor organelle (paraflagellar body, PFB) of *Euglena* has been measured by means of a computer-assisted microspectrophotometer (Gualtieri et al., 1989). The absorption maximum, centered around 500 nm and with an optical density at this wavelength of about 1.4, indicates that rhodopsin-like molecules, rather than flavins (Benedetti and Lenci, 1977; Ghetti et al., 1985), are present in this organelle and therefore suggests that rhodopsin may play the role of photoreceptor pigments in *Euglena* (Gualtieri, et al., 1989).

Fluorescence and Phosphorescence Spectroscopy

Fluorescence

In Figure 6 the simplified potential energy curves of the ground, S_0, and first excited, S_1, singlet states of a hypothetical diatomic molecule are sketched.

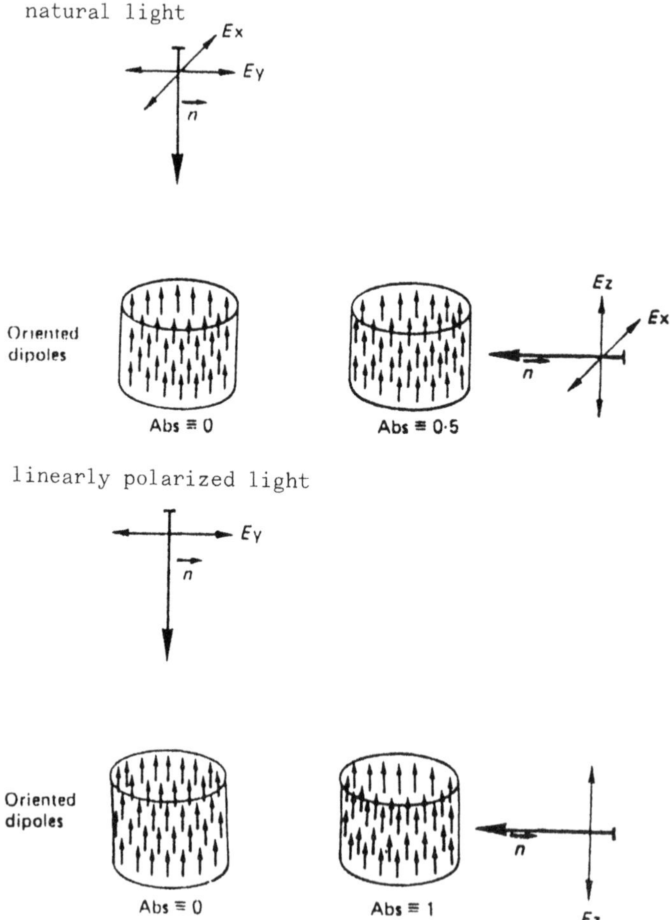

Fig. 5. Schematic representation of a dichroic photoreceptor and its absorption properties of polarized and non-polarized light n: direction of propagation of the light beam (redrawn from Colombetti and Lenci, 1983).

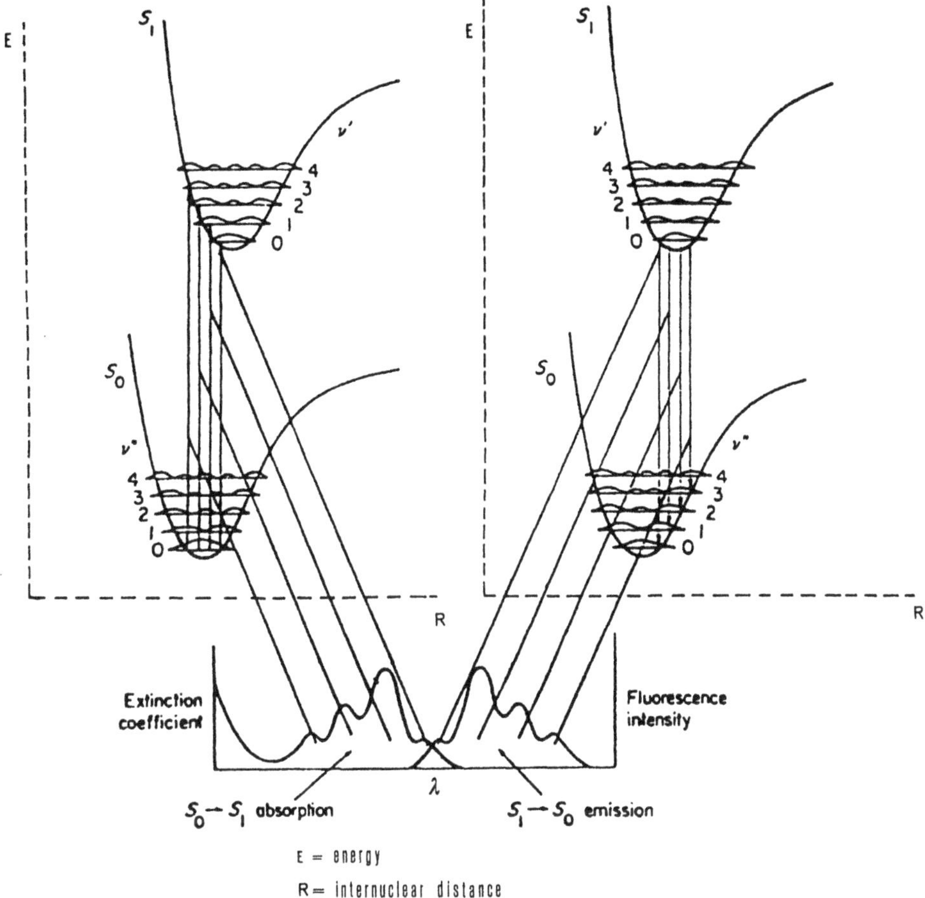

Fig. 6. Schematic representation of the potential energy curves for S_0 and S_1 and of a fluorescence spectrum mirror image of the $S_0 \to S_1$ absorption spectrum (redrawn from Burnett and North, 1969).

As already mentioned, the excited molecule releases part of its energy through very fast, within 10^{-14}-10^{-12} s, transitions among vibrational substates (vibrational relaxation) down to the lowest vibrational state of S_1. From this state the molecule can decay, within about 10^{-9}-10^{-8} s, to the various vibrational levels of S_0 emitting fluorescence quanta, the spectral distribution of which is red shifted with respect to the absorption spectrum (Stokes shift).

When the electron distributions of S_0 and S_1 are similar enough not to significantly alter the vibrational energy levels of the two states, the fluorescence emission spectrum ($S_1 \to S_0$) of an "isolated" molecule is often the mirror image of the $S_0 \to S_1$ absorption band, whereas its fluorescence excitation spectrum coincides with its optical absorption spectrum. In practice both fluorescence emission and excitation spectra have to be corrected for instrumental distortions due, for example, to the spectral sensitivity of the photomultiplier tube and the emission spectrum of the exciting light source.

Fluorescence Quantum Yield and Lifetime

The ratio of the number of photons emitted to the number of photons absorbed, the fluorescence quantum yield (Φ_f), depends on the rates of non-radiative processes, such as intersystem crossing and internal conversion, and that of fluorescence. Φ_f is given by:

$$\Phi_f = \frac{k_f}{k_f + k_{nr}}$$

where k_f is the rate constant for fluorescence and k_{nr} is the sum of all the rate constants for non-radiative deactivation of S_1.

In practice Φ_f values can be determined by measuring fluorescence intensities and absorbances and applying the Beer- Lambert law: if I_0 is the light intensity incident on the sample (measured in number of photons/cm^{2}s), I_a the light intensity absorbed by the chromophore with molar extinction coefficient ϵ at the excitation wavelength used, C the chromophore concentration, l the optical path, I the transmitted light intensity and I_f the fluorescence intensity, then:

$$\Phi_f = \frac{I_f}{I_a}$$

Since, from the Beer-Lambert law,

$$I_a = I_0(1 - 10^{-C\epsilon l}) = I_0(1 - 10^{-A})$$

Φ_f is given by:

$$\Phi_f = \frac{I_f}{I_0(1 - 10^{-C\epsilon l})} = \frac{I_f}{I_0(1 - 10^{-A})}$$

This formula allows to determine the fluorescence quantum yield of an unknown sample, Φ_{fx}, with respect to the fluorescence quantum yield of standard samples, Φ_{fs}, emitting in the same spectral region as the sample under investigation when excited at the same wavelength, according to the expression:

$$\Phi_{fx} = \Phi_{fs} \frac{I_{fx}}{I_{fs}} \frac{(1 - 10^{-A})_s}{(1 - 10^{-A})_x}$$

The reciprocal of the rate constant for fluorescence, k_f, is the radiative lifetime τ°_f, which represents S_1 mean lifetime if fluorescence were the only deactivation pathway of S_1. The actual S_1 mean lifetime, τ_f, depends not only on k_f, but also on the sum of the rate constants for non-radiative deactivation of S_1, k_{nr}, and is given by:

$$\tau_f = \frac{1}{k_f + k_{nr}}$$

or, in terms of Φ_f,

$$\tau_f = \frac{\Phi_f}{k_f}$$

Phosphorescence

As already mentioned, the $T_1 \rightarrow S_0$ phosphorescence radiative transition is forbidden by spin selection rules and can take place only because of the spin-orbit coupling. This spin-forbidness makes the lifetime of T_1 much longer than that of S_1 (from 10^{-5} to several seconds); as a result of its long lifetime, T_1 is a most reasonable candidate as the reactive state for several molecular reactions and can play a role in a number of photobiological processes (see, e.g., photosensitized reactions, (Spikes, 1989)).

Since T_1 is populated via intersystem crossing from S_1, the phosphorescence quantum yield, Φ_p, is given by an expression quite similar to that of Φ_f multiplied for the intersystem crossing quantum yield, Φ_{isc}:

$$\Phi_p = \Phi_{isc} \frac{k_p}{k_p + k_{nr}}$$

Similarly to τ_f, the phosphorescence lifetime, τ_p, is given by the ratio between the phosphorescence quantum yield and the phosphorescence rate constant:

$$\tau_p = \frac{\Phi_p}{k_p}$$

Because of T_1 long lifetime, collisional deactivations, photochemical reactions and energy transfer processes can severely quench the phosphorescence in fluid media; that is why phosphorescence is usually studied in low temperature glasses or in very viscous solvents.

Fluorescence and Phosphorescence Polarization

We already mentioned that, in any molecule, optical absorption is proportional to $\cos^2\theta$, being θ the angle formed by the electric field vector, $\vec{E}$, and the dipole moment of the molecule $\vec{\mu}$. If linearly polarized light is used for exciting a molecule, the fluorescent light too may be polarized, and the fluorescence polarization degree, p_f, can be defined as:

$$p_f = \frac{I_{fparal} - I_{fperp}}{I_{fparal} + I_{fperp}}$$

where I_{fparal} and I_{fperp} are the components of the fluorescence emission, respectively, parallel and perpendicular to the $\vec{E}$ vector of the exciting light. The theoretical limits for p_f are +1 for a system with $\vec{\mu}$ perfectly parallel to $\vec{E}$ and -1 for a system with $\vec{\mu}$ perfectly perpendicular to $\vec{E}$. In non-rigid systems randomly oriented, i.e. isotropic systems, if rotational relaxation times of the molecule are shorter than the molecule fluorescence lifetime, any degree of polarization is lost.

Measurements of p_f can provide information about, for example, size, shape, degree of immobilization of the fluorophore, mutual orientations of the dipoles of the different electronic transitions (Schmidt, 1979) as well as about the occurrence of intermolecular interactions, like electronic energy transfer from a "donor" to an "acceptor" molecule.

The case of phosphorescence polarization is complicated by the fact that light absorption raises the molecule to a state, S_1, different from the state from which emission occurs, T_1. Here we only want to remind that in phosphorescence studies, usually performed in rigid glasses, phosphorescence depolarization is due only to non-collisional energy transfer (Förster type dipole-dipole interactions) since diffusional and rotational motions are hindered.

Emission Spectroscopy of Photoreceptor Pigments

From the few features of fluorescence and phosphorescence phenomena described above, it is clear that fluorescence and phosphorescence spectroscopy can provide information about the interactions of a pigment with its environment and on the effect these interactions have on the various de-excitation processes.

It would be impossible to thoroughly report the extremely numerous cases in which even steady-state emission spectroscopy of photoreceptor pigments has been successful in solving at least part of the problems and in prompting new experimental and theoretical approaches.

The lateral mobility of a photoreceptor pigment embedded in a membrane, for example, can be measured by monitoring the recovery of fluorescence intensity following bleaching of a small membrane area. The emission increase in time is, in fact, due to the diffusion of fluorophores from the unbleached region of the membrane into the bleached spot.

Diffusion-enhanced energy transfer by dipole-dipole interactions between properly mutually oriented donor and acceptor molecules, during the lifetime of the donor excited state, can be used to monitor translational motions and to determine the location of chromophores inside proteins and membranes (Stryer et al., 1982).

Specific quenchers of fluorescence and phosphorescence can help in identifying the primary reactive excited electronic state and in determining whether a protein-bound chromophore is accessible to the quencher itself or if it is buried into the protein framework. The heavy atom gas Xe, for instance, quenches flavin triplet states, whereas flavin excited singlet states are efficiently quenched by potassium iodide and sodium azide.

In the case of *Stentor* and *Blepharisma* the hypothesis of a photoinduced release of protons from the excited state of the photoreceptor molecule and the spectroscopic approaches to check this assumption (Song, 1987; Lenci et al., 1989; Cubeddu et al., 1990) is discussed in detail by Song, by Ghetti and by Cubeddu et al. in this volume, and we will not insist on this point.

Microspectroscopy of Photoreceptor Pigments

Microspectrofluorometry in *Euglena* intact cells has been used to determine both emission and excitation fluorescence spectra of the pigments contained in the photoreceptor organelle, the PFB (Benedetti and Lenci, 1977; Ghetti et al., 1985). Both spectra suggested that the photoreceptor pigments of *Euglena* are flavin-type chromophores, most likely embedded in a rigid and/or hydrophobic molecular matrix. In this physiological molecular environment the fluorescence quantum yield is of the order of 0.005, a value which has to be compared with the value of 0.25 for free riboflavin solution. This low value of Φ_f indicates that the first excited singlet of PFB flavins can efficiently undergo deexcitation pathways other than radiative decay. These findings are fully in agreement with the hypothesized photoreceptor function of PFB flavins, since, as already mentioned, no photoreceptor pigment is likely to decay from its first excited singlet state mainly through a radiative transition, but rather through reactions which can serve to trigger the first molecular steps of the photosensory transduction chain.

Our previous identification of flavins as the photoreceptor pigments contained in the PFB is now challenged by absorption microspectroscopy measurements (Gualtieri et al., 1989; Gualtieri, 1990) which indicate that a rhodopsin-like pigment is contained in the PFB.

Interestingly, flavins and pterins have recently been detected fluorometrically in isolated flagella of *Euglena* and reclaimed as photoreceptors for photobehavioral responses of this alga (Galland et al., 1990).

The rhodopsin hypothesis for *Euglena* is relevant to the suggestion that also for the flagellated alga *Chlamydomonas* a rhodopsin would serve as the photoreceptor pigment for photomotile responses (Foster et al., 1984; Hegemann et al., 1988; Foster et al., 1989).

Acknowledgements

The author is sincerely grateful to Pill-Soon Song for critical reading of the manuscript and for helpful suggestions.

References

General Reference Textbooks

Burnett, G. M., and North, A. M. (eds.), 1969, "Transfer and Storage of Energy by Molecules," Wiley, New York.

Christophorou, L. G., 1971, "Atomic and Molecular Radiation Physics," Wiley, New York.

Clayton, R. K., 1970, "Light and Living Matter," McGraw-Hill, New York.

Di Bartolo, B., Pacheco., and Goldberg, V. (eds.), 1975, "Spectroscopy of the Excited State," Plenum, New York.

Grell, E. (ed.), 1981, "Membrane Spectroscopy," Springer, Berlin.

Guilbault, G. G. (ed.), 1973, "Practical Fluorescence," Dekker, New York.

Hoppe, W., Lohmann, W., Markl, H., and Ziegler, H. (eds.), 1983, "Biophysics," Springer, Berlin.

Lakowicz, J. R., 1983, "Principles of Fluorescence Spectroscopy," Plenum, New York.

Schulman, S. G., 1977, "Fluorescence and Phosphorescence Spectroscopy: Physicochemical Principles and Practice," Pergamon, Oxford.

Specialized References

Benedetti, P. A., Bianchini, G., Checcucci, A., Ferrara, R., Grassi, S., and Percival, D., 1976, Spectroscopic properties and related functions of the stigma measured in living cells of *Euglena gracilis, Arch. Microbiol.*, 111:73.

Benedetti, P. A., and Lenci, F., 1977, *In vivo* microspectrofluorometry of photoreceptor pigments in *Euglena gracilis, Photochem. Photobiol.*, 26:315.

Colombetti, G., and Lenci, F., 1983, Photoreception and photomovements in microorganisms, *in* "The Biology of Photoreception," D. J. Cosens and D. Vince-Prue, eds., Oxford, pp. 399.

Cubeddu, R., Ghetti, F., Lenci, F., Ramponi, R., and Taroni, P., 1990, Timegated fluorescence of blepharismin, the photoreceptor pigment for photomovement of *Blepharisma, Photochem. Photobiol.*, 52:567.

Foster, K. W., Saranak, J., Derguini, F., Zarilli, G., Johnson, R., Okabe, M., and Nakanishi, K., 1989, Activation of *Chlamydomonas* rhodopsin *in vivo* does not require isomerization of retinal, *Biochemistry*, 28:819.

Foster, K. W., Saranak, J., Patel, N., Zarilli, G., Okabe, M., Kline, T., and Nakanishi, K., 1984, A rhodopsin is the functional photoreceptor for phototaxis in the unicellular eukaryote *Chlamydomonas, Nature*, 311:756.

Galland, P., Keiner, P., Doernemann, D., Senger, H., Brodhun, B., and Häder, D.-P., 1990, Pterin- and flavin-like fluorescence associated with isolated flagella of *Euglena gracilis*, *Photochem. Photobiol.*, 51: 675.

Ghetti, F., Colombetti, G., Lenci, F., Campani, E., Polacco, E., and Quaglia, M., 1985, Fluorescence of *Euglena gracilis* photoreceptor pigment: an *in vivo* microspectrofluorometric study, *Photochem. Photobiol.*, 42:29.

Gualtieri, P., 1990, Microspectroscopy of photoreceptor pigments in flagellated algae, *Critical Reviews in Plant Sciences*, in press

Gualtieri, P., Barsanti, L., and Passarelli, V., 1989, Absorption spectrum of a single isolated paraflagellar swelling of *Euglena gracilis*, *Biochim. Biophys. Acta*, 993:293.

Häder, D. P., 1987, Polarotaxis, gravitaxis and vertical phototaxis in the green flagellate *Euglena gracilis*, *Arch. Microbiol.*, 147:179.

Koziol, J., and Szafran, M. M., 1990, Spectral properties of riboflavin tetrabutyrate in the presence of hydrogen-bonding agents, *J. Photochem. Photobiol.* B, 5:429.

Lenci, F., and Colombetti, G., eds., 1980, "Photoreception and Sensory Transduction in Aneural Organisms," Plenum, London.

Lenci, F., and Ghetti, F., 1989, Photoreceptor pigments for photomovement of microorganisms: some spectroscopic and related studies, *J. Photochem. Photobiol.* B, 3:1.

Lenci, F., Ghetti, F., Gioffré, D., Passarelli, V., Heelis, P. F., Thomas, B., Phillips, G. O., and Song, P.-S., 1989, Effect of molecular environment on some spectroscopic properties of *Blepharisma* photoreceptor pigment, *J. Photochem. Photobiol.* B, 3:449.

Schmidt, W., 1979, On the environment and the rotational motion of amphiphilic flavins in artificial vesicles as studied by fluorescence, *J. Membrane Biol.*, 47:1.

Song, P. S., 1987, Possible primary photoreceptors, in: "Blue Light Responses," H. Senger, ed., CRC Press, Boca Raton, pp. 3-17.

Song, P.-S., 1983, Protozoan and related photoreceptors: molecular aspects, *Ann. Rev. Biophys. Bioeng.*, 12:35.

Song, P.-S., and Moore, T. A., 1974, On the photoreceptor pigment for phototropism and phototaxis: Is a carotenoid the most likely candidate?, *Photochem. Photobiol.*, 19:435.

Spikes, J. D., 1989, Photosensitization,. in: "The Science of Photobiology," K. C. Smith, ed., Plenum, New York, pp. 79-110.

Stryer, L., Thomas, D. D., and Meares, C. F., 1982, Diffusion enhanced fluorescence energy transfer, *Ann. Rev. Biophys. Bioeng.*, 11:203.

Suppan, P., 1990, Solvatochromic shifts: the influence of the medium on the energy of electronic states, *J. Photochem. Photobiol.* A, 50: 293.

Application of Laser Flash Photolysis to Study Photoreceptor Pigments

Suppiah Navaratnam and Glyn O. Phillips

The North East Wales Institute
Deeside, Clwyd CH5 4BR
Wales, UK

Introduction

In any photobiological process, the initial step is the absorption of a photon by the receptor pigment, which can transform it into the excited singlet state. This state can undergo a number of processes such as internal conversion to the ground state, photochemical reactions (e.g. isomerisation in rhodopsin, charge separation in chlorophyll), intersystem crossing to the triplet state, etc. The triplet, in turn, can also undergo comparable reactions. Thus a knowledge of the reactivities of excited states is necessary to understand the primary processes of pigments on exposure to light. Such processes can conveniently be studied using flash photolysis, a technique devised and developed by Norrish and Porter (1949), Porter (1950, 1963) and Norrish (1965) to study the absorption spectra and reaction kinetics of triplet states and other transient species.

Boag (1968) described a typical microsecond flash photolysis apparatus and the factors limiting the time resolution and the sensitivity. Flash photolysis in the microsecond range is limited by the intensity and the duration of the light flash used for excitation. Various types of lamps have been devised to shorten the flash duration and increase its intensity. Unfortunately these two characteristics are mutually exclusive. The advent of pulsed lasers has transformed the situation. Flash photolysis equipments based on a pulsed laser can operate even in the femtosecond time scale and at the same time produce sufficient light intensity for monitoring the nature and kinetics of the transient species produced.

Flash Sources

Lasers which are currently being used in flash photolysis systems are pulsed solid state lasers (both Q-switched and mode locked), dye lasers, gas lasers and excimer lasers. They are capable of delivering pulses in the femtosecond (Fork et al., 1981; Shank et al., 1982) and nanosecond range and energies up to several joules. Rodgers (1985) has described the operating characteristics of various pulsed lasers which are commercially available.

Biophysics of Photoreceptors and Photomovements in Microorganisms
Edited by F. Lenci *et al.*, Plenum Press, New York, 1991

Detection Methods

The following techniques can be used to detect the transient species produced after flash photolysis:

Optical Absorption Spectroscopy

This was the first method employed in flash photolysis and remains the most popular. It is capable of monitoring a wide range of wavelengths from the UV to near infrared and can be used for both spectral and kinetic studies.

Optical Emission Spectroscopy

The range and capability is comparable to the absorption spectroscopy method. It is more extensively used in the picosecond time scale because there is no need for an analysis light source. Direct detection of excited singlet oxygen was based upon this method.

Diffuse Reflectance Spectroscopy

The scope is comparable to kinetic absorption spectrophotometry. This monitoring system was devised and developed by Wilkinson (1986), and is useful for opaque, heterogenous solutions and suspensions where there is a great deal of light scattering. This method has potential for biological systems and solid surfaces.

Resonance Raman Spectroscopy

This technique reveals vibrational features in the spectra of the transient species which can lead to an understanding of their structure and chemical bonding (Phillips et al., 1986 and references cited therein).

Conductivity

This technique can be applied to follow the decay and mobility of any ions formed. Changes in H^+ concentration can also be readily detected and quantified in this way (Beck, 1969).

Photoacoustic Spectroscopy

Such spectroscopy is useful for the study of non-radiative processes. Braslavsky and co-workers at Mülheim (Nitsch et al., 1988) have used this technique to obtain detailed information about the energy partitioning between fluorescence, photochemistry and heat dissipation for photosystems I and II particles isolated from the cyanobacterium *Synechococcus* sp.

ESR Spectroscopy

Free radicals can be detected using this method, leading to a better understanding of the nature and structure of the transient species.

Nanosecond Flash Photolysis

The nanosecond flash photolysis system (Navaratnam et al., 1985) such as is used in our Institute is shown in Figure 1. The excitation source is a Q-switched Nd.YAG

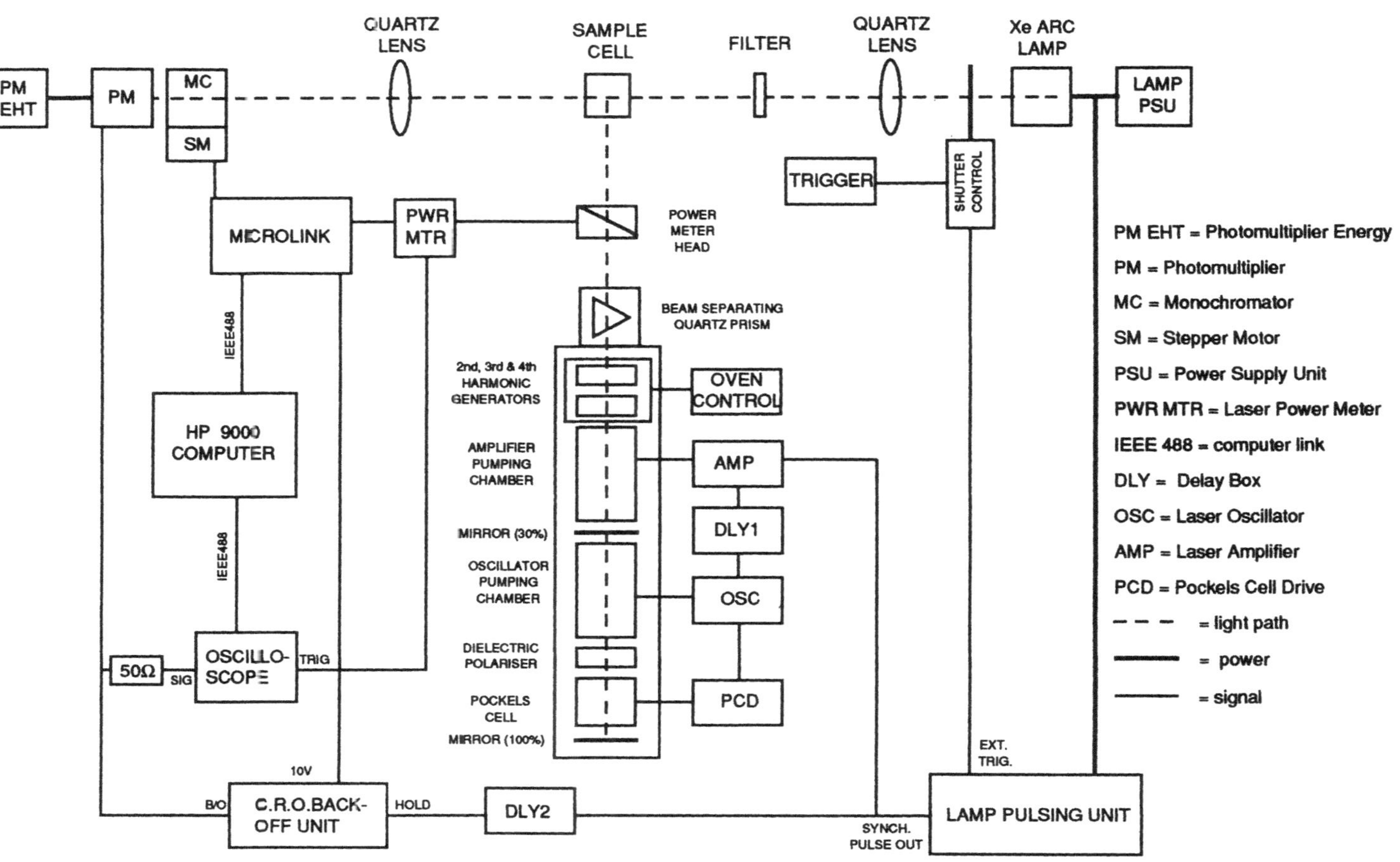

Fig. 1. Laser flash photolysis system at N. E. W. I. Deeside.

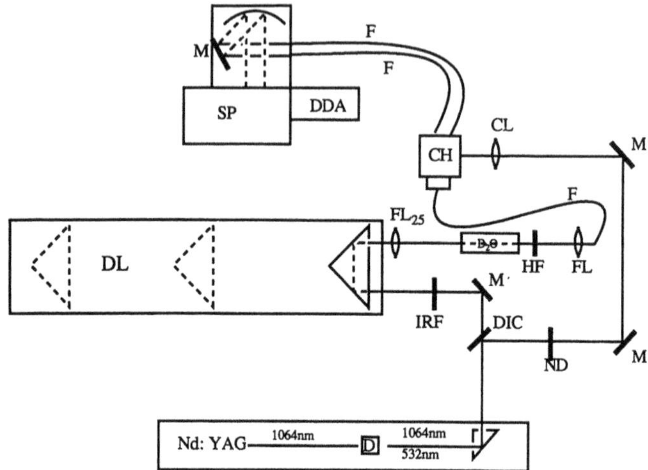

Fig. 2. Experimental apparatus: D, frequency doubler; DDA, double diode array; DIC, dichroic mirror; DL, delay line; IRF, infrared pass filter; FL, focussing lens; H, heat filter; CL, cylindrical lens; M, mirror, F, fiber optics; CH, cell holder; SP, spectrograph; ND, neutral density filter.

(JK Lasers 2000 series) laser capable of delivering 1 J of energy at 1064 nm in pulses of 12 ns duration. The fundamental beam is frequency doubled, tripled and quadrupled to give maximum pulse energies of 500, 100 and 100 mJ at 532, 355, and 266 nm respectively. A prism is used to separate the harmonics. A laser pulse of the required wavelength traverses the reaction cell. The sample under investigation is contained within a quartz cell through which a monitoring beam of light passes at right angles to the direction of the laser pulse. The monitoring beam is generated by a xenon lamp, which can be pulsed to increase the intensity of the monitoring beam particularly at short wavelengths. To minimize photolysis of the sample by the monitoring light, a shutter and a filter are placed in between the sample cell and the analyzing lamp. The shutter opens a few hundred microseconds before the laser pulse and closes after the event under examination is completed. The monitoring beam, after passage through the sample, is dispersed in wavelength by a monochromator and then on to a photodetector. The transmittance of light at the particular wavelength is detected by this photodetector before, during and after the laser flash. The detector is also connected to an automatic back-off box which enables changes in transmittance to be observed by feeding back a signal equal and opposite to the detector anode current prior to the laser pulse, thus maintaining a the anode current close to zero.

The signal from the detector is captured and stored using a programmable digital oscilloscope. Back-off and energy meter readings are digitized using a 8-digit Analog to Digital Convertor (ADC) attached to a Microlink data acquisition system. The Microlink also controls the stepper motor which drives the monochromator. Data acquisition and processing are carried out using a Hewlett Packard 9000 Series 300 computer. Time-resolved spectra are obtained by repeating the above procedure at successive wavelengths.

An alternative method is to use a spectrograph with a diode array detector at its focal plane. In this way a complete spectrum, at a fixed time, can be obtained with one laser pulse. The time resolutions in the nanosecond time scale is obtained by an electrical gating .

Picosecond flash photolysis is necessary to directly study the primary photochemical processes in certain photoreceptor pigments. The picosecond laser flash photolysis system at CFKR, Texas (Atherton et al., 1987) is schematically illustrated in Figure 2. The excitation source is a mode locked Nd.YAG laser. The fundamental and frequency doubled beams fall on a dichroic mirror (DIC) which reflects the 532 nm light through right angles, whereas the fundamental pulse continues undeviated. The 532 nm pulse is attenuated using a neutral density filter and focussed on to the cell. The fundamental pulse traverses a variable delay line (DL) before being focused into a 10 cm cuvette containing a mixture of D_2O and H_2O to produce a white light continuum which is passed through a heat filter and focussed on to the cell. The analyzing pulse enters the cell at right angles to the excitation pulse and travels through the cuvette as two beams one of which traverse the excited sample and the other the unexcited sample. The two beams fall onto a spectrograph fitted with a cathode image intensifier at its focal plane and a double diode array at the exit slit. The output from the detector is passed via a Multichannel analyser (MCA) to a computer for storage, analysis and display.

Methods for Measuring Extinction Coefficients of Triplet-triplet Absorption

Energy Transfer Method

Here the donor (D) and the acceptor (A) compounds are mixed and the donor is initially excited by the laser pulse to produce the triplet state, which can transfer energy to the acceptor by collisional quenching. A sufficient concentration of the acceptor compound is used to quench all the donor triplet. Under these conditions the relationship

$$\epsilon_A = \epsilon_D (OD_A/OD_D) \qquad (1)$$

where ϵ_D and ϵ_A are extinction coefficients of donor and acceptor triplets ; OD_D the maximum optical density of the donor triplet in the absence of the acceptor; OD_A the maximum optical density of the acceptor triplet when both the donor and the acceptor were present. Although under ideal conditions equation (1) is valid, in practice a number of corrections must be applied. The following reactions must be considered in the kinetic scheme.

$$^3D^* \rightarrow {}^1D \qquad (2)$$

$$^3D^* + {}^1D \rightarrow {}^3A^* + {}^1D \qquad (3)$$

$$^3A^* \rightarrow {}^1A \qquad (4)$$

If k_2 cannot be neglected compared with $k_3[^1A]$ then the acceptor triplet yield will be reduced by a factor P given by

$$P = k_3[^1A] \,/\, (k_2 + k_3[^1A])$$

where k_2 and k_3 are the first order decay rate constants of the donor triplet in the absence and presence of the acceptor. If the acceptor triplet is long lived then equation (1) must be modified to

$$\epsilon_A = \epsilon_D (OD_A/OD_D)/P \quad (5)$$

However, if k_4 is not negligible compared to $(k_2 + k_3[^1A])$ equation (5) must be modified (Capellos and Bielski, 1980) to

$$\epsilon_A = \epsilon_D (OD^*_A/OD_D)/P$$

where

$$OD^*_A = OD_A \exp(k_4 \ln((k_2+k_3[^1A])/(k_4-k_3[^1A]-k_2))$$

Additional corrections are necesary if there is direct excitation of the acceptor molecule. (Amouyal et al., 1974).

Singlet Depletion Method

Here the assumption is made that all excited singlet molecules either return to the ground state or intersystem crosses to the lowest triplet state within the time scale of the observations. The change in absorption at any given wavelength is given by

$$OD = (\epsilon_t - \epsilon_s)[T]l \quad (6)$$

where ϵ_t and ϵ_s are extinction coefficients of the excited triplet and singlet states; [T] the concentration of the triplet state and l the path length.

If at any wavelength $\epsilon_s >> \epsilon_t$ then equation (6) reduces to

$$OD = -\epsilon_s[T]l \quad (7)$$

From this equation the concentration of triplet state can be measured. This allows the extinction coefficient at any other wavelength to be calculated. Only an upper limit to the extinction coefficient is given by this method. Hadley and Keller (1969) and Pavlopoulos (1973) have described extended versions of this method.

Total Depletion Method

This method is simple, provided the laser intensity is sufficient to completely convert all the ground state into triplet and that no other processes occur. Carmichael and Hug (1985) have shown that complete conversion can be attained only when there is sufficiently fast singlet decay or long pulse duration. Generally, a lower estimate for the extinction coefficient is given by this method.

Determination of Quantum Yields

The principle of the comparative method is to successively irradiate optically thin solutions of a standard compound whose quantum yield is accurately known and of the test compound, with the optical densities of their ground states at the excitation wavelength are equal. Since both solutions absorb the same number of photons quantum yield (Φt) of the test compound is given by:

$$\Phi t = \Phi s \, OD_t \, \epsilon_s/(OD_s \, \epsilon_t) \quad (8)$$

where Φs is the quantum yield of the standard; OD_t and OD_s observed optical densities of the triplet state of the test compound and the standard at their respective absorption maxima; ϵ_t and ϵ_s are the respective extinction coefficients. Equation (8) is valid only if the ground state depletion is negligible. Bensasson and co-workers (1978) have modified equation (8) to accommodate large laser intensities.

Applications

The technique of flash photolysis has ben used to study the primary photoprocesses of photoreceptor pigments in photosynthetic reaction centers, bacteria and other organisms; we will discuss only a few examples in order to illustarate the applications of the technique. Readers are referred to reviews by Hoff (1979) and Birge (1990) for more examples. Photochemistry of the photoreceptor pigment flavin and related compounds have been studied in our laboratories using the system described above (Heelis, 1982 and references cuted therein; Heelis et al., 1985; Heelis and Phillips, 1985). Initial laser flash photolysis experiments were also undertaken in our laboratories using isolated stigma of *Euglena gracilis* (Heelis et al., 1981) and *Blepharisma* (Lenci et al., 1989). For *Euglena* the transient absorption spectrum obtained by laser flash photolysis of an hexane extract could be assigned to a C_{40} carotenoid pigment triplet. Moreover, it could be demonstrated from the rate of growth of the absorption over 500 ns that the triplet was formed by sensitization and not by direct excitation. The quantum yield for the formation of this species was determined to be 0.01 ± 0.005 using the method described in section 7 and anthracene as standard. The triplet-triplet absorption spectrum of blepharismin was similar to that of hypericin but with more structured bands. Moreover, the triplet life time in deareated ethanol was found to be 62.5 μs, whereas in air-equlibrated solutions it decreased to 0.34 μs.

Mathis and Setif (1981), using a similar technique, observed that in chlorophyll *a* extracted from spinach the triplet-triplet absorption spectrum consisted of two bands centered at 760 nm and 1100 nm. Formation of triplet states of carotenoids and photooxidation of the primary donor of photosystem II, P-680, in chloroplasts in which oxygen evolution was inhibited, was observed by Kramer and Mathis (1980). The rate of triplet-triplet energy transfer from chlorophyll *a* to carotenoids in chloroplasts and in several other light harvesting pigment-protein complexes was also determined by them and found to be in the order of $8 \times 10^7\ s^{-1}$. Durrant et al. (1990), using a flash photolysis technique, studied the isolated D1/D2 cytochrome b-559 reaction center complex at 4° C. They found that the P-680 triplet state was quenched by oxygen and not by carotenoids. Moreover, they estimated the quantum yield for this state to be about 30%.

Acknowledgements

Much of the applications of laser flash photolysis in our laboratory with pigments and model compounds related to biological photomovement has been undertaken in collaboration with our colleagues in the CNR Laboratory for Biophysics at Pisa, particularly Professors Sandro Checcuci, Francesco Lenci, Giuliano Colombetti and Francesco Ghetti. We are grateful for support from NATO for promoting the collaboration, led from our side by Paul Heelis and Diana Heelis.

References

Amouyal, E., Bensasson, R., and Land, E. J., 1974, Triplet states of ubiquinone analogs studied by ultra violet and electron nanosecond irradiation, *Photochem. Photobiol.*, 20:415.

Atherton, S. J., Hubig, S. M., Callan, T. J., Duncanson, J. A., Snowden, P. T., and Rodgers, M. A. J, 1987, Photoinduced charge separation in a micelle-induced charge-transfer complex between methylviologen and ethidium ions. A picosecond absorption spectroscopy study, *J. Phys. Chem.*, 91:3137.

Beck, G., 1969, Elektrische Leitfähigkeitsmessungen zum Nachweis geladener Zwischenprodukte der Pulsradiolyse, *Int. J. Radiat. Phys. Chem.*, 1:361.

Bensasson, R., Goldschmidt, C. R., Land, E. J., and Truscott, T. G., 1978, Laser intensity and the comparative method for determination of triplet quantum yields, *Photochem. Photobiol.*, 28:277.

Birge, R. R., 1990, Nature of the primary photochemical events in rhodopsin and bacteriorhodopsin, *Biochim. Biophys. Acta*, 1016:293.

Buchert, J., Stefancic, V., Doukas, A. G., Alfano, R. R., Callender, R. H., Pande, J., Akita, H., Balogh-Nair, V. and Nkanishi, K., 1983, Picosecond kinetic absorption and fluorescence study of bovine rhodopsin with a fixed 11-ene, *Biophys. J.*, 43:279.

Boag, J. W., 1968 Techniques of flash photolysis, *Photochem. Photobiol.*, 8:565.

Capellos, C., and Bielski, B. H. J., 1980, "Kinetic Systems", Krieger, Huntinton, N.Y.

Carmichael, I., and Hug, G. L., 1983, A note on the total depletion method of measuring extinction coefficients of triplet-triplet transitions, *J. Phys. Chem.*, 89:4036.

Durrant, J. R., Giorgi, L. B., Barber, J., Klug, D. R., and Porter, G., 1990, Characterisation of triplet states in isolated Photosystem II reaction centres: oxygen quenching as a mechanism for photodamage, *Biochim. Biophys. Acta*, 1017:167.

Fork, R. L., Greene, B. I., and Shank, C. V., 1981, Generation of optical pulses shorter than 0.1 psec by colliding pulse mode locking, *Appl. Phys. Lett.*, 38:671.

Hadley, S. G., and Keller, R. A., 1969, Direct determination of the extinction coefficient for triplet-triplet transitions in naphthalene, phenanthrene and triphenylene, *J. Phys. Chem.*, 73:4351.

Heelis, D. V., Heelis, P. F., Bradshaw, F., and Phillips, G. O., 1981, Does the stigma of *Euglena gracilis* play an active role in the photoreception processes of this organism? A photochemical investigation of isolated stigma, *Photobiochem. Photobiophys.*, 3:77.

Heelis, P. F., 1982, The photophysical and photochemical properties of flavins (isoalloxazines), *Chem. Soc. Rev.*, 11:15.

Heelis, P. F., Parsons, B. J., Thomas, B., and Phillips, G. O, 1985, One electron oxidation of the flavin triplet state as studied by laser flash photolysis, *J. Chem. Soc. Chem. Commun.*, 954.

Heelis, P. F., and Phillips, G. O., 1985, A laser flash photolysis study of the triplet states of lumichromes, *J. Phys. Chem.*, 89:770.

Hoff, A. J., 1979, Applications of ESR in photosynthesis, *Phys. Reports* 54:75.

Kramer, H., and Mathis, P, Quantum yield and rate of formation of the carotenoid triplet state in photosyntheic structures, *Biochim. Biophys. Acta*, 593:319.

Lenci, F., Ghetti, F., Gioffre, D., Passarelli, V., Heelis, P. F., Thomas, B., Phillips, G. O., and Song, P-S., 1989, Effects of the molecular environment on some spectroscopic properties of *Blepharisma* photoreceptor pigment, *J. Photochem. Photobiol. B: Biology*, 3:449.

Mathis, P., and Setif, P., 1981, Near infra-red absorption spectra of the chlorophyll *a* cations and the triplet state *in vitro* and *in vivo*, *Israel J. Chem.*, 21:316.

Navaratnam, S., Hughes, J. L., Parsons, B. J. and Phillips, G. O., 1985, Laser flash photolysis and steady-state photolysis of benoxaprofen in aqueous solution, *Photochem. Photobiol*, 41:375.

Nitsch, C., Braslavsky, S. E., and Schatz, G. H., 1988, Laser induced optoacoustic calorimetry of primary processes in isolated Photosystem I and Photosystem II particles, *Biochim. Biophys. Acta*, 934:201.

Norrish, R. W. G., and Porter, G., 1949, Chemical reactions produced by very high light intensities, *Nature*, 164:658.

Norrish, R. W. G., 1965, The kinetics and analysis of very fast chemical reactions, *Chem. Britain*, 1:289.

Pavlopoulos, T. G., 1973, Measurement of molar extinction coefficients of organic molecules by means of cw laser excitation, *J. Opt. Soc. Am.*, 63:180.

Phillips, D., Moore, J. N. and Hester, R. E., 1986, Time-resolved resonance Raman spectroscopy applied to anthraquinone photochemistry, *J. Chem. Soc. Faraday Trans.* II 82:2093.

Porter, G., 1950, Flash photolysis and spectroscopy: A new method for the study of free radical reactions, *Proc. Roy. Soc., A*, 200:284.

Porter, G., 1963, *in* "Technique of Organic Chemistry," Wessberger, A., ed., chapter 19, Wiley Interscience, New York, pp. 1055.

Rodgers, M. A. J., 1985, Instrumentation for the generation and detection of transient species, in "Primary Photo-Processes in Biology and Medicine," Bensasson, R. V., Jori, G., Land, E. J., and Truscott, T. G., eds., Plenum Press, New York.

Shank, C. V., Fork, R. L., Yen, R., Stolen, R. H., and Tomlinson, W. J., 1982, Compression of femtosecond optical pulses, *Appl. Phys. Lett.*, 40:761.

Wilkinson, F., 1986, Diffuse reflectance flash photolysis, *J. Chem. Soc. Faraday Trans.*, 82:2073.

Time-gated Fluorescence Spectroscopy of Photoreceptor Pigments

Rinaldo Cubeddu, Roberta Ramponi, Paola Taroni

C.E.Q.S.E., CNR
Politecnico of Milan
Milan
Italy

and

Francesco Ghetti and Francesco Lenci

Istituto Biofisica
CNR
Pisa
Italy

Introduction

Time-resolved fluorescence spectroscopy using ultrashort pulsed lasers as the excitation source has found a wide number of applications both in photochemistry and photobiology (Schneckenburger et al., 1988). In fact, the most common techniques based on continuous wave (cw) measurements are inadequate whenever the fluorescent signal is very low and/or cannot be separated from the background by spectral discrimination (i.e., by a suitable choice of the excitation or observation wavelength). Different fluorophores, however, even when overlapped in spectra, usually appreciably differ in their fluorescence decays and their presence can be detected by measurements in the time domain. Moreover, the decay time of the fluorescence emission is a parameter very sensitive to the molecular environment and its variations can provide useful information on the photophysical properties of the chromophore. If the time-resolved fluorescence analysis is performed at different wavelengths, it is also possible to obtain time-resolved spectra, by measuring the fluorescence intensity at given time intervals with respect to the peak of the waveform. By choosing suitable time gates, the spectral characteristics of the molecular species with slow decay time constants can be discriminated with respect to the fast ones and this information allows an accurate characterization of the emitting molecular species that are present in the sample.

In the last few years, the availability of picosecond excitation sources with high pulse repetition rate and of photomultipliers with low jitter (less than 40 ps) (Kinoshita et al., 1981) allowed the development of time-correlated single photon counting systems with a temporal resolution down to 45 ps. With these systems, deconvolution or iterative convolution and procedures to correct the experimental curves for the temporal response of the detection apparatus are not strictly required, except in the case of very fast transients. Therefore, in most practical situations of biological interest, direct acquisition of both fluorescence decays and time-resolved spectra can be straigthforwardly obtained.

This paper describes a computer-controlled system for time-resolved fluorescence spectroscopy able to directly measure gated fluorescence spectra with variable gate width and its application to the study of the photophysical properties of blepharismin, which is the photoreceptor pigment for photomovement of *Blepharisma japonicum*.

Experimental Apparatus

Time-correlated single photon counting is a technique which consists in detecting the time delay between the excitation laser pulse and the first fluorescence photon emitted by the sample. The measurement is repeated over a large number of excitation pulses and the emission probability of a single photon is recorded as a function of time, the laser pulse corresponding to time zero. The temporal statistics of the photon emission is representative of the intensity versus time profile of the fluorescence decay waveform (O'Connor and Phillips, 1984). By fitting the curve obtained with a multi-exponential function, it is possible to evaluate the fluorescence decay times and the relative peak amplitudes of the different exponential components for the sample under study. These parameters give an evaluation of both the temporal behavior of the fluorescence and the relative proportion of the excited chromophores. Since a large number of pulses is required to obtain a good statistics, lasers with a high pulse repetition rate (typically mode-locked lasers) are most suitable for this technique in order to keep the measuring time within reasonable values.

By performing the single photon counting measurements at a large number of different emission wavelengths, it is possible to obtain the time-gated spectra (Meech et al., 1981). This can be done by counting only the fluorescence photons in the decay waveform detected within preselected time intervals as a function of the emission wavelength and by repeating the acquisition over the whole range of wavelengths for a fixed counting time. Of course, if all the photons in the fluorescence decay are counted, a time-integrated spectrum is obtained, which closely resembles a cw spectrum. In fact, as compared to the cw spectrum, the photons falling outside the time scale of the measurement are not acquired, but, if the scale is properly chosen, their contribution is negligible. Assuming, for example, that the fluorescence decay under study is the sum of two exponential components given by two independent fluorescing molecular species, the time-gated spectrum, obtained with a sufficiently long delay, will be that of the long-living component, the contribution of the short-living one being negligible. This can be subtracted (after normalization for the gate (Cubeddu et al., 1988) from the time-integrated one and the spectrum of the short-living component is obtained. In general, to evaluate the single contribution to the spectrum of n exponential components, it is necessary to acquire the time-integrated spectrum and n-1 time-gated spectra. Interesting information can be obtain also by measuring an undelayed gated-spectrum with a gate width significantly shorter than the fastest lifetime. This gives an emission spectrum in which the contribution of each molecular species is proportional to the number of its molecules in the excited state, almost independently of their fluorescence decay.

The scheme of the experimental apparatus developed in our laboratory for time-gated fluorescence spectroscopy is shown in Figure 1 (Cubeddu et al., 1988). The excitation source is a mode-locked laser, either an ion (Argon or Krypton, both with pulses

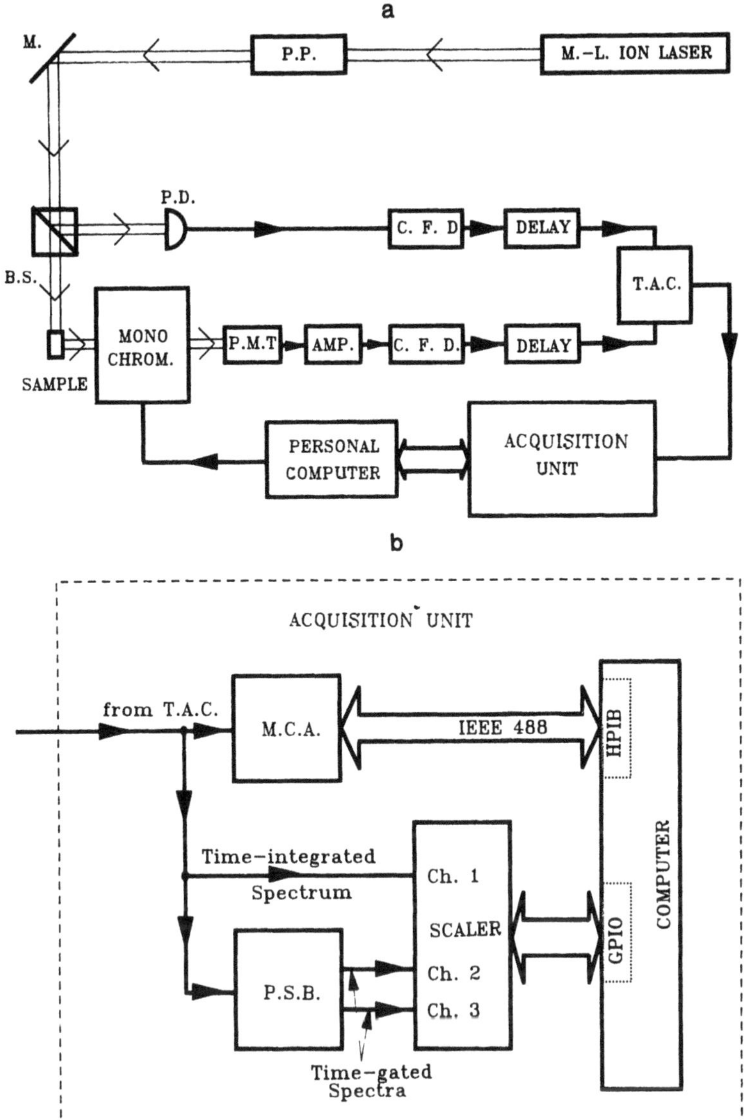

Fig. 1.a: Block diagram of the system for time-resolved fluorescence spectroscopy with picosecond gating. P.P.: pulse picker; M.: mirror; B.S.: beam splitter; P.D.: p-i-n photodiode; AMP.: signal preamplifier and amplifier circuitry; C.F.D.: constant fraction discriminator; T.A.C.: time to amplitude converter. 1.b: Details of the acquisition unit for decay waveforms, and for time-integrated and time-gated spectra. M.C.A.: multichannel analyzer; P.S.B.: pulse selecting board.

of 70-120 ps depending on the wavelength), or a synchronously-pumped dye laser (with pulses of ≈ 5 ps), depending on the wavelength and pulse duration requirements. Thus excitation wavelengths ranging from the UV to the near infrared can be achieved. To prevent fluorescence pile up, the repetition rate of the laser pulses of 76 MHz is reduced of a factor of 100 by an acousto-optic pulse picker (for the ion lasers) or a cavity dumper (for the dye laser). The fluorescence photons are detected through a monochromator by a microchannel-plate photomultiplier (Hamamatsu 1564U-01) operating in single photon counting regime. Depending on the laser pulse, the time response of the system ranges from 45 ps to 150 ps. As shown in the scheme, part of the laser beam is sent to a fast p-i-n photodiode to provide a reference signal. Both pulses from the photodiode and from the photomultiplier are processed by a constant fraction discriminator (to obtain a timing independent of the pulse height variations) and by a pulse-delay generator. The output pulses are sent to the start and stop inputs of a time-to-amplitude converter (TAC). The start pulse initiates the time sweep of the TAC and, if a pulse is received by the stop input during the sweep, it blocks the voltage ramp. The TAC then generates an output signal whose voltage is proportional to the time difference between start and stop. To profit of the high repetition rate of the laser, in our apparatus the TAC is working in the inverted arrangement, i.e., with the photomultiplier providing the start pulse (O'Connor and Phillips, 1984). The output pulses from the TAC are sent to a multichannel analyzer operating in the pulse-height-acquisition mode, to measure the fluorescence decay, and to a home-made computer-controlled acquisition unit. At each wavelength, this unit collects in one channel all the TAC output pulses and in the other two channels only those pulses falling within preselected voltage windows. After a fixed time, the computer increments by a step the monochromator and the measurement is repeated at the new wavelength. Thus, taking into account the correspondence between time and voltage, scanning a selected wavelength range, a time-integrated spectrum and two time-gated spectra are obtained. The experimental data are collected by the computer for storage and analysis. In particular, fluorescence decay waveforms are analyzed by means of a non-linear least square fitting method and the quality of the fitting is judged by the weighted residuals and their autocorrelation function. It must be noted that all the data are collected simultaneously within a single measurement. This can be relevant especially for measurements On photodegradable samples that can safely be exposed to the light only for a very short time.

Time-gated Spectroscopy of Blepharismin

The experimental apparatus described above has been used for several applications in biological problems (Cubeddu et al., 1989, 1990a,b, in press). In this paper it will be reported that on the photoreceptor pigment blepharismin (BP). The analysis will be given to elucidate the potentiality of the technique and the type of information that can be obtained. A more accurate discussion on the biological aspects is given in Cubeddu et al. (1990b) and Lenci and Ghetti (1989).

Blepharismin is localized in deeply colored granules just beneath the cell membrane of the ciliate protozoan *Blepharisma japonicum*. To investigate the primary steps of the photomovement, fluorescence studies were performed on the pigment crude extract in different solution environments. The BP concentration was adjusted to provide an optical density of 0.1-0.3 at 364 nm. The cw fluorescence spectrum of BP in ethanol exhibits a single band centered around 600 nm. By adding NaOH to the solution, the appearance of a second band centered at about 650 nm was observed. At high NaOH concentrations (100 mM), this band becomes predominant while that at 600 nm is present as a small shoulder. The effect of a polar solvent as water (pH 7) resulted in a spectrum similar to that obtained in pure ethanol. However, by increasing the pH, the 650 nm emission grows and is dominant at the pH value of 12.4. Since only minor changes were noticed in the absorption spectra, fluorescence lifetimes and time-gated spectra were measured under excitation at 364 nm. The decay waveforms were obtained both from an average over the whole emission spectrum and for observation at 600 nm and 660 nm. In the ethanol solutions the fluorescence decay curves were characterized by three exponential components with decay times of 0.2-0.5 ns, ≈ 1 ns and 4-6 ns,

Table 1. Fluorescence decay-time constants and relative amplitudes of BP in different solvents. Full spectrum emission

Solvent	τ_1 (ns)	A_1 (%)	τ_2 (ns)	A_2 (%)	τ_3 (ns)	A_3 (%)
EtOH	4.06	3.52	1.03	78.61	0.50	17.87
EtOH + 20 mM NaOH	5.69	22.93	1.16	27.39	0.20	49.68
EtOH + 100 mM NaOH	6.22	50.00	1.40	6.68	0.19	43.32
H_2O (pH 7)			0.89	9.28	0.23	90.72
H_2O (pH 12.4)	4.90	41.73	1.01	20.65	0.27	37.62

Note: In Table 1, the numerical values result from a non-linear interpolation of the exponential decays. Different measurements provided variations in the time decays from 50 to 300 ps, depending on the decay value, and the fluctuations in the relative amplitudes were a few percents.

respectively. In water at pH 7 the slow component was not observable. The relative amplitudes strongly depended on the solvent characteristics as shown in Table 1.

From the reported data, it appears that the presence of NaOH alters the equilibrium among the molecular species. Taking into account the value of the decay times, the long-living one is dominant in the cw spectrum for high NaOH concentrations and is responsible for the emission peak at 650 nm. The other two molecular species are related to the emission at 600 nm. For a better evaluation of the spectral distribution among the molecular species, the time-integrated spectrum together with two gated spectra were taken in all samples, being the gates set with no delay, 100 ps width and 8 ns delay, 2 ns width respectively.

Figure 2 shows the spectra obtained for BP in ethanol. The time-integrated and the undelayed spectrum both peak at 600 nm according with the attribution of this band to the fast and intermediate components. The 8 ns-delayed spectrum, related to the long-living component, presents a sharp band at 605 nm. By comparing this gated spectrum with the corresponding one observed for BP in aqueous solution at pH 12.4 (see Fig. 3c), it is evident that the long-living component observed in ethanol is related to a different molecular species. It is worth mentioning that this result can be obtained only by time-gated spectroscopy; in fact, due to the similarity among the values of the decay times, an analysis based only on fluorescence decays could have led to wrong attributions. The shapes of the other two spectra in Figure 3 are consistent with the above attribution. The time-integrated one is peaked at 650 nm with a small shoulder around 600 nm due to the contribution of the fast and intermediate components. This shoulder becomes a secondary peak in the undelayed spectrum according to the relative amplitudes of the exponential components reported in Table 1. In the case of BP in water at pH 7 only the fast and intermediate components emitting around 600 nm were detected. Accordingly, the delayed gate is not reported in Figure 4, since no significant signal could be measured. However, some differences are observed between the two spectra shown in the figure. The undelayed spectrum is shifted to the blue and is shar-

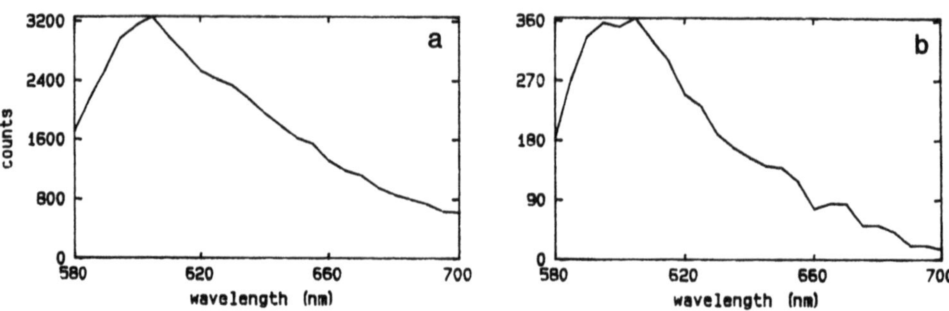

Fig. 2. Emission spectra of blepharismin in ethanol: time-integrated spectrum (a), gated spectrum, 0 delay, 100 ps width (b), and gated spectrum 8 ns delay, 2 ns width (c).

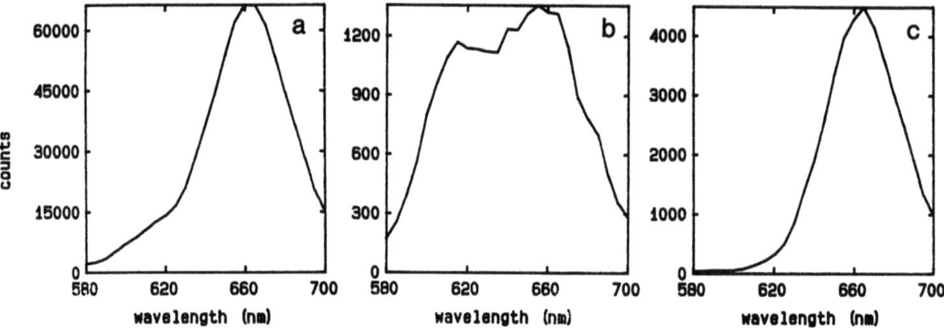

Fig. 3. Emission spectra of blepharismin in aqueous solution at pH 12.4: time-integrated spectrum (a), gated spectrum, 0 delay, 100 ps width (b), and gated spectrum 8 ns delay, 2 ns width (c).

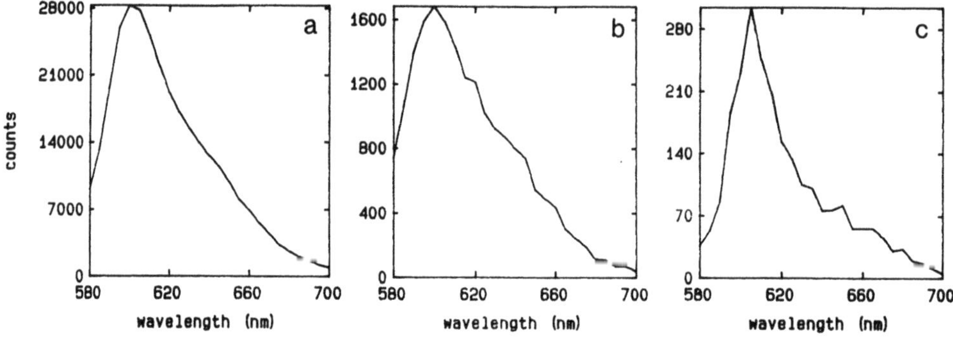

Fig. 4. Emission spectra of blepharismin in aqueous solution at pH 7: time-integrated spectrum (a), and gated spectrum, 0 delay, 100 ps width (b).

per than the time-integrated one. Taking into account that the product $A_i * \tau_i / \Sigma(A_i * \tau_i)$ is roughly proportional to the contribution of each component to the overall cw spectrum, the intermediate component weights for ≈ 30% in the time-integrated spectrum and only for ≈ 10% in the gated one. Thus, the fluorescence spectrum of this molecular species seems to be broader and shifted to the red with respect to that of the fast emitting one.

From the reported data, it appears that two molecular species are present, whose relative abundance depends on the nature of the solvent. In particular the aqueous solution seems to favor the short-living species, while the intermediate-living one is dominant in hydrophobic environment. This assumption seems to be confirmed by measurements performed in micellar systems (Triton X-100), which provided data similar to those described for ethanol. These results can possibly be interpreted in terms of an equilibrium between phenolic (short-living species) and quinonic forms of blepharismin. The presence of OH^- alters the above equilibria both in water and in ethanol leading to the formation of the long-living species. Since this emission is present only at high OH^- concentrations, it can be identified with the deprotonated form of BP.

In conclusion, on the basis of the discussed study, it appears that the time-gated fluorescence spectroscopy provides a very useful tool to evaluate the photophysical properties of a complex molecular system. In fact, it allows to obtain data simultaneously in the spectral and the temporal domains of the fluorescence emission, providing a large set of complementary information. Moreover, due to the high sensitivity of the technique, the instrumentation can be coupled to a microscope to perform measurements on single cells or tissue samples.

References

Cubeddu, R., Docchio, F., and Boulton, M. Time-resolved fluorescence spectroscopy of the retinal pigment epithelium: age related studies. *IEEE J. of Quantum Electronics*, Special Issue on Biomedical Applications of Lasers, in press a.

Cubeddu, R., Docchio, F., Liu, W.-Q., Ramponi, R., and Taroni, P., 1988, A system for time-resolved laser fluorescence spectroscopy with multiple picosecond gating. *Rev. Sci. Instrum.* 59:2254.

Cubeddu, R., Ghetti, F., Lenci, F., Ramponi, R., and Taroni, P., 1990b, Time-gated fluorescence spectroscopy of blepharismin, the photoreceptor pigment for photomovement of *Blepharisma*. *Photochem. Photobiol.*, 52:567.

Cubeddu, R., Ramponi, R., Liu, W.-Q., and Docchio, F., 1989, Time-gated fluorescence spectroscopy of the tumor localizing fraction of HpD in the presence of cationic surfactant. *Photochem. Photobiol.* 50:157.

Cubeddu, R., Ramponi, R., Taroni, P., Canti, G., Ricci, L., and Supino, R., 1990a, Time-gated fluorescence spectroscopy of porphyrin derivatives incorporated into cells. *J. Photochem. Photobiol. B* 6:39.

Kinoshita, S., Ohta, H., and Kushida, T., 1981, Subnanosecond fluorescence lifetime measuring system using single photon counting method with mode-locked laser excitation. *Rev. Sci. Instrum.* 52:572.

Lenci, F., and Ghetti, F., 1989, Photoreceptor pigments for photomovement of microorganisms: some spectroscopic and related studies. *J. Photochem. Photobiol. B* 3:1.

Meech, S.R., O'Connor, D.V., Roberts, A.J., and Phillips, D., 1981, On the construction of nanosecond time-resolved emission spectra. *Photochem. Photobiol.* 33:159.

O'Connor, D.V., and Phillips, D., 1984, *Time-correlated Single Photon Counting*, Academic Press, London.

Schneckenburger, H., Seidlitz, H.K., and Eberz, J., 1988, New trends in photobiology: time-resolved fluorescence in photobiology. *J. Photochem. Photobiol.* B 2:1.

Effects of Enhanced Solar Ultraviolet Radiation on Aquatic Ecosystems

Donat-P. Häder

Institut für Botanik und
Pharmazeutische Biologie
Friedrich-Alexander-Universität
Staudtstr. 5
D-8520 Erlangen
Germany

Introduction

The photosynthetic production of organic biomass using solar energy is the almost exclusive source of energy for life on our planet. The amount of carbon in the form of its dioxide incorporated annually into organic molecules exceeds 100 gigatons which can be visualized by the load filling 10 coal trains spanning the distance from the earth to the moon (Häder et al., 1989). However, only about one third of this enormous production is accounted for by terrestrial plants - forests, savannas, crop plants etc. - while the majority is produced by the phytoplankton organisms (primary producers) in aquatic habitats, especially in the world oceans. The marine phytoplankton communities represent by far the largest ecosystem on earth (Schneider, 1989); therefore even a small percentage decrease in the populations would result in enormous losses in the biomass productivity of these organisms, which could have dramatic effects both for the intricate ecosystem itself and for humans, who depend on this system in many ways (Häder et al., 1989).

Stratospheric Ozone Destruction and Solar UV-B Radiaton

One potential threat to the phytoplanktonic microorganisms is the possible decrease in the stratospheric ozone layer caused by manmade gaseous pollutants such as the chlorinated fluorocarbons (CFC), of which more than 1 million tons are produced and emitted annually, e.g., as coolants in refrigerators and air conditioning devices, as propellents in the production of foams and as cleansing agents in the production of electronic circuitry. Since this chapter is not intended to review the complex stratospheric chemistry involved in the ozone generation and destruction (see for review Caldwell et al., 1989), only a few remarks may suffice: Ozone is generated from atmospheric molecular oxygen which combines with an oxygen atom, which had previously been produced by splitting O_2 by solar short wavelength ultraviolet radiation. O_3 in turn is broken down into O_2 and O by longer wavelength solar ultraviolet radiation, so that there is a constant cycle. While the stratospheric ozone layer extents from about 15 to

Biophysics of Photoreceptors and Photomovements in Microorganisms
Edited by F. Lenci *et al.*, Plenum Press, New York, 1991

45 km above our heads, the concentration is extremely low so that the total layer, when compressed under atmospheric pressure would rarely exceed 3 or 4 mm. The CFCs operate catalytically and break down O_3 by abstracting one oxygen atom from O_3 to produce ClO, which in turn splits so that the chlorine atome can react with thousands of O_3 before it is washed down to the earth (Stolarski, 1988). In addition to operating catalytically, CFCs have a high potential of O_3 destruction since they have a mean life-time in the stratosphere of about 100 years and they take about 10 years to get there from the lower atmospheric layers where they are emitted.

Ozone is an effective filter for short wavelength ultraviolet solar radiation in the range between 280 and 320 nm (UV-B radiation) which, because of its high energy, easily destroys proteins, DNA and other biologically relevant molecules. The ozone loss has been most obvious in the ozone hole which opens up over the Antarctic continent each year in the Southern spring, where the O_3 concentration has been found to decrease progressively ever since 1979 over an area as large as the continental United States; this decrease has reached a maximum of about 50% in the last few years (Stolarski, 1988). Less pronounced, but also potentially harmful, is a smaller but global ozone loss, predicted to amount to as much as 10% over the next decades (Madronich et al., 1989).

The biological hazards of the resulting increased UV-B radiation include enhanced skin cancer fatalities, increases in cataract formation and reduced immune responses (van der Leun, 1989). Higher plants are expected to be negatively affected which may result in lower harvest yields in crop plants (Tevini et al., 1989). The third group of UV-B related damages to the biota concerns the phytoplankton communities (Häder et al., 1989; Smith, 1989) which is the topic of this chapter.

Phytoplankton Microorganisms

For energetic reasons phytoplankton organisms are restricted to the upper layers in their environment. Mutual shading, absorption and light scattering by other particles within the water column prevent light to penetrate below the photic zone which may extend from a few decimeters in turbid costal waters to up to 200 m in clear oceanic waters. The penetration of radiation into a body of water strongly depends on the wavelength: Short wavelength violet and long wavelength red light is more strongly absorbed than blue-green light (Jerlov, 1970). However, even though partially attenuated, solar UV-B radiation penetrates well into the photic zone and thus affects the phytoplankton organisms.

The members of the phytoplankton populations are not equally distributed in the water column but rather optimize their position in the habitat using very precise orientation strategies (Nultsch and Häder, 1988). The clues for this orientation are based on external signals such as light and gravity, chemical and temperature gradients as well as the magnetic field of the earth (see chapter on photo- and graviorientation in *Euglena*, elsewhere in this volume).

While for energetic reasons it is obvious that photosynthetic microorganisms need to move toward the surface in order to absorb sufficient solar radiation, in contrast to terrestrial plants, which are adapted to the high fluence rates of unfiltered sunlight (exceeding 1000 W m^{-2}), most photosynthetic (and also non photosynthetic) microorganisms cannot tolerate high intensity radiation which photobleaches and kills the cells within short exposure times (Nultsch and Agel, 1986; Häder, 1988).

In order to solve the dilemma between the energetic need and the risk of radiation-induced damage, many motile microorganisms utilize two or more antagonistical reactions such as negative gravitaxis, which takes the cells to the surface, and negative phototaxis which guides the cells away from the surface (Häder, 1988). In order to cope with the constantly changing conditions in their environment the organisms have to permanently adjust their vertical position. Any inhibition of these responses due to reduced motility or a decreased precision of orientation would consequently impair the ability of the cells to respond to the changes in their environment and this would negatively affect the chances for growth and survival of the populations. And indeed, there is growing evidence that both freshwater and marine phytoplankton organisms are already under UV-B stress at ambient levels.

In addition to the impaired motility and orientation strategies, UV-B radiation has also been shown to affect the general metabolism, the photosynthetic energy production and nitrogen fixation in most species studied so far. Any decrease in the production of organic material will be relayed through all rugs in the intricate biological food ladder affecting all feeders such as larvae, crabs, fish, birds and mammals including men.

Initial quantitative assays of the UV-B sensitivity in phytoplankton organisms were performed on laboratory "test systems", which had the advantage that their biochemistry and behavior had been established beforehand; therefore UV-B stress-induced changes could be detected experimentally in comparison with untreated controls. More recently, ecologically important algal groups, such as dinoflagellates, cryptophyceae and diatoms have been included in the studies; however, the nanoplankton, defined by its small size compared to other phytoplankton groups, has not yet been studied in any detail; though it is known to be responsible for a large amount of biomass production. Initially, artificial radiation sources were used frequently because of their stability and ease of quantification in contrast to solar radiation, the spectral energetic distribution of which is both highly variable and difficult to measure. Recently, the effects of natural sunlight experienced by the organisms in their habitat has been investigated. In the following some of the UV-B related effects will be described in more detail.

Inhibition of Orientation and Motility

Effects of UV-B Radiation on Motility and Velocity

In a healthy population most individuals are motile at all times. In *Euglena* the percentage of motile flagellates has been shown to be impaired by solar radiation (Häder, 1985, 1986; Häder and Häder, 1988a). When exposed to unfiltered sunlight the percentage of motile organisms decreases within a few hours (Fig. 1). This effect is not due to excess energy absorbed by the chlorophylls, since also the colorless relative of *Euglena, Astasia longa*, and even dark-bleached *Euglena* have been found to respond similarly (Häder and Häder, 1988b, 1989a). This effect is also obvious in other freshwater algae such as *Peridinium gatunense* (Häder et al., 1990) and a freshwater *Cryptomonas* species (Häder and Häder, 1989b) as well as in marine algae such as *Cryptomonas maculata* (Häder and Häder, 1990a) and *Gyrodinium dorsum* (Ekelund and Björn, 1990).

In accordance with the decrease in the percentage of motile cells, the speed of movement of the remaining motile cells is drastically decreased (Fig. 2) in all forms indicated above. Inhibition of motility was also found in gliding green algae (Häder, 1987b), cyanobacteria (Häder et al., 1986; Häder and Häder, 1990b) and also in slime molds (Häder, 1983a).

The ecological consequences for gliding cyanobacteria are similar as in flagellated organisms: the organisms are deprived of their means to escape dark areas or too bright irradiation in exposed areas which may be even more detrimental than in eukaryotic cells since most cyanobacteria are known to be adapted to rather low fluence rates on the order of only a few percent of the unfiltered solar radiation and are photo-

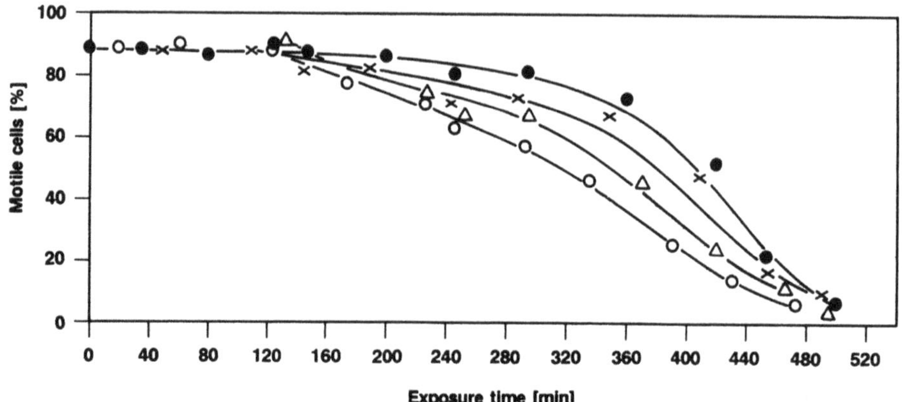

Fig. 1. Percentage of motile cells (ordinate) in samples of green *Euglena gracilis* taken from populations exposed to solar radiation (abscissa, exposure time, min) and covered with UV-B cut-off filters (WG 280, open circles; WG 295, triangles; WG 305, crosses; WG 320, closed circles). At noon the total fluence rate was 1040 W m^{-2} and the total UV-B fluence rate was 0. 919 W m^{-2}.

killed by intensities exceeding a few thousand lx (Walsby, 1968). Both effects on motility are definitely not due to thermal stress by the infrared component of solar radiation since the experiments were carried out under temperature-controlled conditions. In solar irradiation experiments growth chambers with double layered Plexiglas tops were employed which allowed the solar radiation to penetrate. The cavity between the Plexiglas panes could be flooded with air enriched with ozone produced on site, so that the total ozon column was increased by about 5% as compared to the control (Tevini et al., 1989). Under the ozone cuvette motility in the organisms was less impaired than in unfiltered sunlight. Likewise specific UV-B cutoff filters (WG series, Schott & Gen., Mainz, FRG) which removed part or all of the UV-B radiation were found to prolong the tolerated exposure time of the organisms. However, it cannot be denied that also both UV-A and visible radiation exert an effect on the organisms.

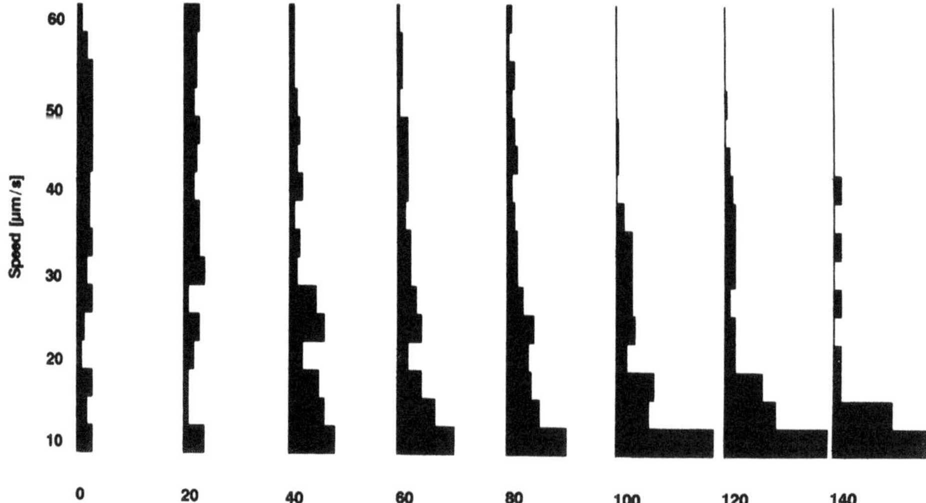

Fig. 2. Effect of the exposure time (abscissa, min) to unfiltered solar radiation on the speed of movement (abscissa, μ m s^{-1}) in *Cyanophora paradoxa*.

Almost all studies performed up to now have proven that most motile microorganisms do not possess a sensor for UV-B radiation - a characteristic similar to human beings. Thus, they are not capable of escaping a possibly hazardous radiation. An action spectrum of the inhibition of motility (percentage of motile cells) by ultraviolet monochromatic radiation was measured for *Euglena gracilis*. The action spectrum is totally different from the ones determined for the photoresponses found in this organism. It shows a major peak at about 270 nm, a smaller one at 305 nm and a shoulder at 290 nm; in addition to covering the UV-B range it extends well into the UV-C range (wavelengths shorter than 280 nm). This is not a typical DNA spectrum (Jagger, 1983). For comparision, the emission spectrum of solar radiation (local noon, 50° north, June 27) has been included in the diagram calculated using a computer simulation developed by Björn and Murphy (1985), which indicates that only part of the UV sensitivity in this organism coincides with solar radiation. The inhibition potential can be determined by integrating over the product of the two curves. The effect of increasing solar UV radiation in a scenario of further ozone depletion can be visualized by including the solar emission spectrum for the equator calculated for the highest solar zenith angle. Under these circumstances the biologically effective UV-B mediated inhibition increases by about 22% as compared to a location in Southern Germany.

Inhibition of Orientation by Ultraviolet Radiation

Inhibition of orientation has the same deleterious effects on the population as impaired motility since the organisms fail to respond to the changing parameters in their habitat. As described above, most phytoplankton organisms orient in the water column using various stimuli, the most important of which may be light (Diehn, 1973; Nultsch, 1974; Haupt, 1959) and gravity (Bean, 1984; Briegleb and Block, 1986; Block et al., 1986; Wolke et al., 1987).

Action spectra for the photoorientation responses (phototaxis, photokinesis and photophobic responses) have revealed that the photoreceptor pigments employed for this purpose differ among the various algal groups (Halldal, 1963, 1964; Haupt, 1965). However, most species use various bands in the visible and long-UV range to orient with respect to light (Nultsch and Häder, 1979; Foster and Smyth, 1980).

Solar radiation has been found to impair these vital responses for optimization and thus survival of phytoplankton populations. Exposure to both artificial and solar radiation has been found to inhibit phototactic orientation within rather short exposure times in freshwater and marine species (Häder, 1984, 1985, 1986). It was found that the degree of orientation, quantified using the Rayleigh test (Batschlet, 1965, 1981; Mardia, 1972), decreases even after exposure times of 15 min to unfiltered solar radiation (Häder et al., 1990). The histograms show that after about 130 min the organisms moved random in the light field (Fig. 3).

The second stimulus to which many microorganisms have been found to respond to is the gravitational field of the earth (Jensen, 1893; Kessler, 1985, 1986; Häder, 1987a). The gravireceptor has not been identified in phytoplankton organisms and it has even been proposed that the orientation is brought about by a passive physical process: this hypothesis assumes that the center of gravity is located in the rear end of the cell so that its front end points upwards to where the propulsion mechanism pulls the organism (Brinkmann, 1968; Kuroda et al., 1986; Taneda et al., 1987). However, gravitaxis - like phototaxis - is impaired by solar and artificial ultraviolet radiation within a few minutes of exposure in *Euglena* and *Peridinium* (Häder and Liu, 1990a,b). After 130 min the cells were totally disoriented (Fig. 4). This result may indicate that an active physiological response based on the action of a gravireceptor organelle may be responsible for this orientation mechanism.

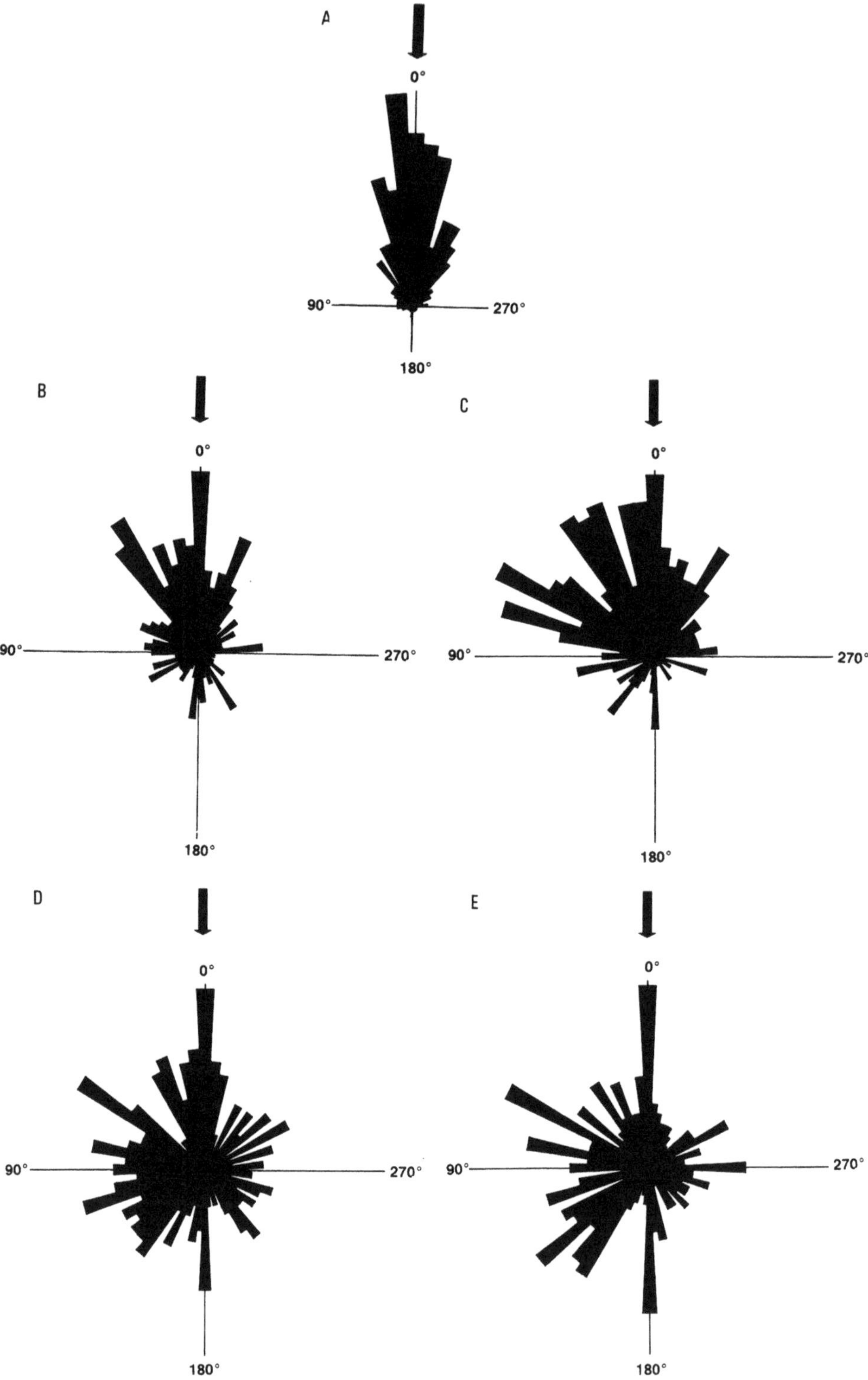

Fig. 3. Circular histograms of the negative phototactic orientation of *Peridinium gatunense* to a test stimulus of 10 klx white light after (a) 0 min, (b) 15 min, (c) 30 min, (d) 75 min and (e) 130 min of solar radiation. The UV-B fluence rate was 1. 24 W m^{-2} and the total solar radiation amounted to 1145 W m^{-2}.

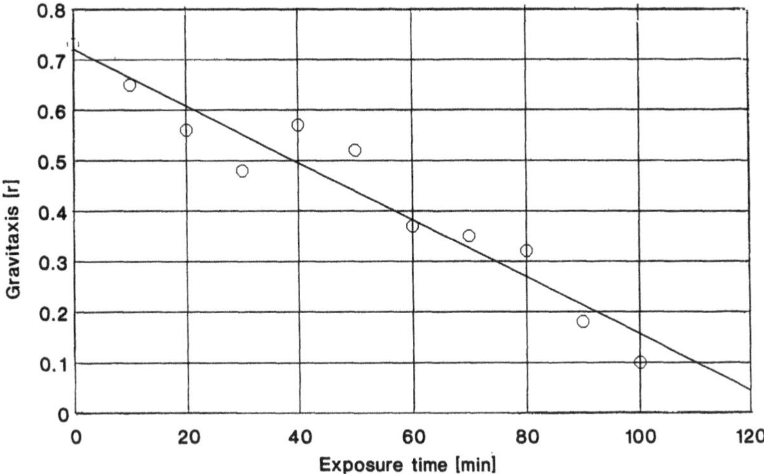

Fig. 4. Inhibition of gravitaxis (ordinate, r-value) in *Euglena gracilis* by exposure to ultraviolet radiation (abscissa, exposure time, min).

In a natural habitat the result of antagonistic oriention mechanisms is the accumulation of the population in a suitable band within the water column. These bands fluctuate with the changing environmental conditions, and many planktonic organisms are known to undergo daily vertical movements of up to 12 m (Burns and Rosa, 1980). These movement patterns have been followed in vertical transparent Plexiglas columns (Häder and Griebenow, 1988) immersed in a pond or lake (Fig. 5). Samples were taken at regular time intervals from the column using equidistantly spaced outlets along the length of the column and the cell densities were determined using an automatic image analysis system capable of handling large numbers of data (Häder and Griebenow, 1987). Currently a 3 m column is used to follow the movement patterns of natural marine phytoplankton organisms in their environment.

Development and Physiology

Some microorganisms undergo a more or less complicated developmental cycle. During some stages of the cycle the organisms are more sensitive to solar UV-B radiation than during others. This can be due to a reduced tolerance or due to a more exposed position in the water column. Any inhibition of one of the developmental stages has long lasting effects on the productivity and fecundity of the whole population. *Dictyostelium* amoebae have been found to be strongly impaired by ultraviolet radiation (Ohnishi and Nozu, 1979; Häder, 1983b); as a consequence the developmental cycle is not continued and pseudoplasmodia are not formed or only at a greatly retarded pace (Häder and Häder, 1989c). Also UV-treated spores were not viable (Ohnishi et al., 1982; Ford and Deering, 1979).

General Metabolism

Below lethal doses UV-B has been found to affect growth and the endogenous rhythm exhibited by many microorganisms (Worrest, 1982; Mitchell, 1990). Higher doses affect the ion permeability of cellular membranes and induce irreversible damage and eventually cause death. Often the direct effects of UV-B are aggravated by additional stress factors such as salinity and temperature (Döhler, 1984).

UV-B Effects on Photosynthesis

Enhanced UV-B radiation has been found to bleach the photosynthetic pigments and thus to impair photosynthesis; this in turn decreases the production of ATP and NADPH which limits the CO_2 incorporation. Furthermore, UV-B radiation damages the photosystem II reaction center accompanied by a structural changes in the membranes and a decrease in the lipid content. A UV-B induced decline in nucleic acids and protein content affects enzyme activity and production. Specific enzyme inhibition was found in some marine diatoms and cyanobacteria (Döhler et al. 1986). Photosynthetic oxygen production is affected by unfiltered solar radiation within a few minutes of exposure in a number of phytoplankton organisms (Fig. 6). After an initial enhancement there is a drastic decline in the O_2 emission. Of cause, in addition to UV-B also other bands in the solar spectrum contribute to this effect.

Nitrogen Assimilation

Prokaryotes are the only organisms capable of utilizing atmospheric nitrogen and convert it into a form which can be utilized by higher plants. Thus, e.g., cyanobacteria play an important role in providing nitrogen for instance for tropical rice paddies. The annual nitrogen fixation by only this group of prokaryotes has been assumed to amount

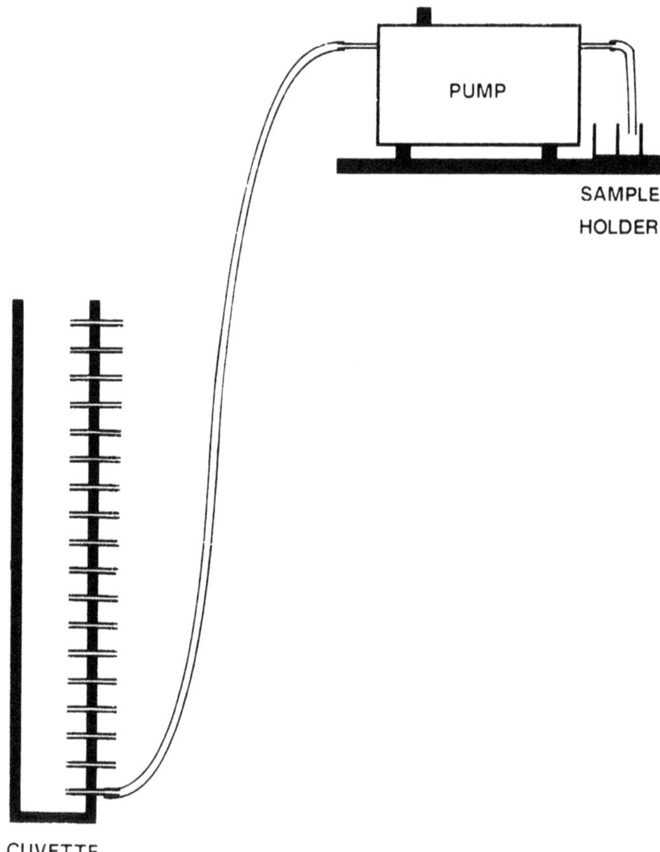

Fig. 5. Schematic diagram of a Plexiglas column to determine the vertical movements of phytoplankton organisms by taking samples at regular time intervals from the 18 outlets along the length of the column.

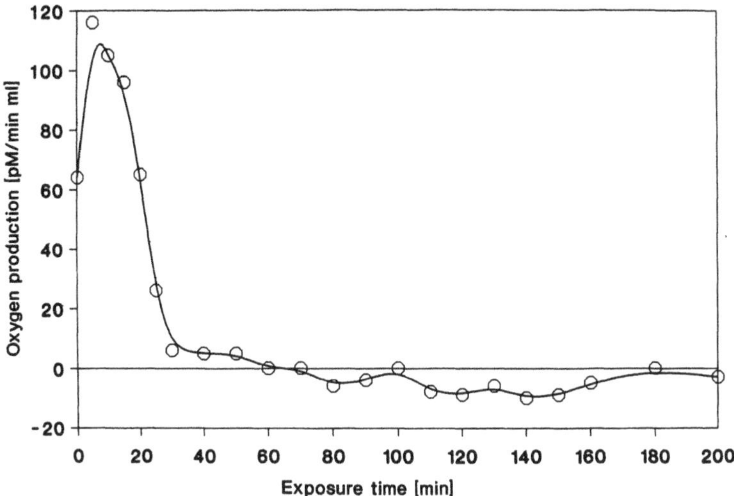

Fig. 6. Inhibition of photosynthetic O_2 production by unfiltered solar radiation in the cyanobacterium *Phormidium uncinatum* (isolated from Lake Baikal).

to 35 million tons anually which exceeds the 30 million tons of artificial nitrogen fertilizer produced each year with the Haber-Bosch process.

One of the main processes for growth is nitrogen fixation since it determines the rate of protein synthesis. The key enzyme for nitrogen assimilation is the nitrogenase, which is activated by light and inhibited by UV-B radiation (Döhler, 1985). This and other enzymes incorporate nitrogen and produce amino acids and proteins. This process has been found to be drastically affected by UV-B radiation in a number of ecologically important phytoplankton organisms (Döhler et al. 1987).

Pigmentation and Fluorescence

In contrast to land plants adapted to full sunlight, most phytoplankton organisms are not capable of tolerating excessive solar radiation. When exposed to unfiltered sunlight the photosynthetic pigments are bleached (Nultsch and Agel, 1986; Häder et al., 1988) and the organisms die within a few days. When exposed to radiation in a UV-B transmitting quartz cuvette, the marine *Cryptomonas maculata* showed a visible color change within a few minutes. This bleaching can be quantified by measuring absorption spectra of the cells at regular time intervals during exposure to artificial ultraviolet radiation (Häder and Häder, 1990a). When the absorption spectrum of the unexposed control is substracted from all the subsequently measured spectra, the resulting difference spectra indicate where the absorption decreases and thus which pigments are destroyed first (Fig. 7).

From these difference spectra bleaching kinetics can be determined for the individual pigments. The accessory pigments in *Cryptomonas*, which operate in light harvesting for photosynthesis, are bleached first upon exposure to full sunlight. The carotenoids, which are thought to have a protective role against excessive radiation, are damaged later and the chlorophylls usually have a lifetime of several hours at constant exposure (Fig. 8). Using artificial UV-B radiation has shown that the photosynthetic pigments are also bleached by the UV-B component alone, indicating the detrimental effect of this radiation on photosynthesis.

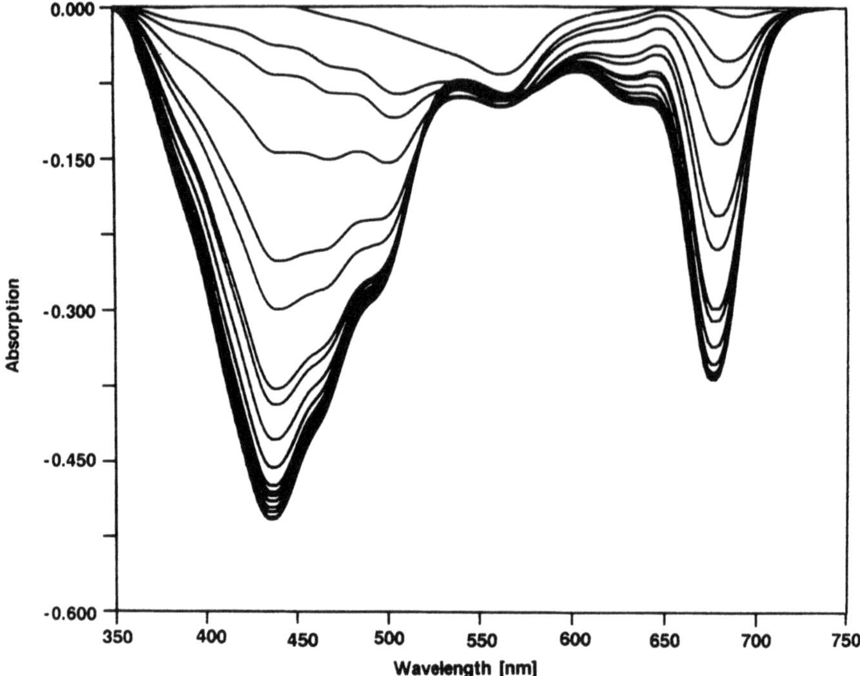

Fig. 7. Absorption spectra of a freshwater *Cryptomonas* species measured after increasing times of exposure to artificial ultraviolet radiation (from top to bottom: 0, 145, 350, 755, 1560, 2915, 3840, 4380, 5615, 6355, 7760, 9205, 10605, 11685, 12775, 13960, 14630, 16320, 17175, and 18990 min) measured as difference spectra between the irradiated sample minus the untreated control.

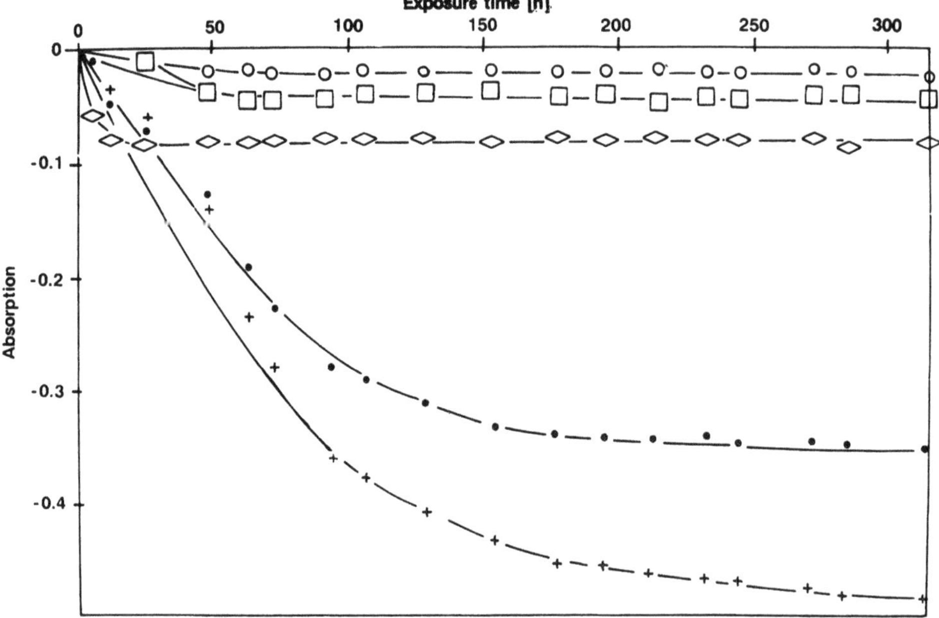

Fig. 8. Decrease in absorption at key wavelengths representing the major photosynthetic pigments of a freshwater *Cryptomonas* species due to UV bleaching after increasing times of exposure: solid circles 680 nm, squares 480 nm, rhomboids 560 nm, open circles 460 nm, crosses 440 nm.

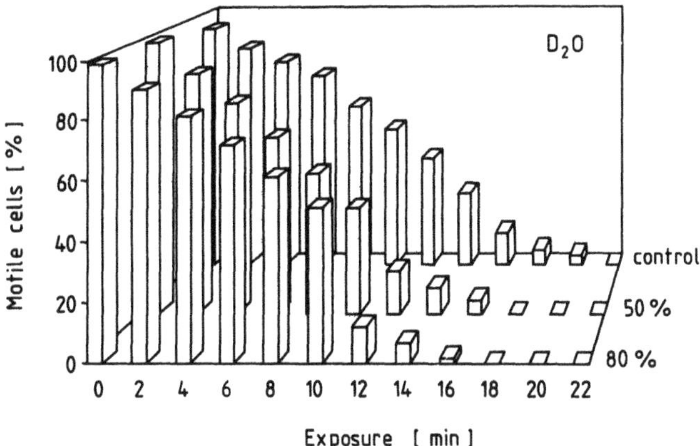

Fig. 9. Effect of short-term exposure to broad-band artificial UV-B radiation on the percentage of motile *Euglena gracilis* in dependence of the exposure time (rear columns, control) and in the presence of 50% (middle columns) and 80% (front columns) D_2O.

After a short exposure to UV-B radiation an increase in fluorescence can be detected in many photosynthetic pigments, which indicates that photosynthetic light is not used effectively by the photosynthetic apparatus but rather wasted in form of fluorescence reimission (Schreiber et al., 1989). After longer exposure times the fluorescence decreases again indicating that the absorbing pigments are increasingly destroyed by the ultraviolet radiation. Pigment concentration can be determined in laboratory samples by extraction. This is also done in natural plankton communities, but on a larger scale the pigment concentration can only be assayed using spectroscopic data from satellite measurements (Jeffrey and Humphrey, 1975).

UV-B Targets

Short wavelength UV-B radiation below 300 nm is most effective in most systems and responses as indicated by the action spectra measured so far. These results demonstrate that any long-term losses in the ozone layer, which results in an increase in UV-B radiation, are bound to have adverse effects. DNA has been found to be a major target in UV-B mediated damages in animals and microorganisms (Yammamoto et al., 1983). This idea was supported by the close resemblance between the DNA and the measured action spectra also in phytoplankton organisms. However, recent research has revealed a host of vital components within the cells with similar absorption characteristics.

In the study of a green flagellate, *Euglena gracilis*, serving as a model organism, DNA seems not to be the major target for UV-B radiation (Häder and Häder, 1988a). This can be concluded from the very fast effect of the radiation on motility and from the fact that no photorepair has been observed in light which is supposed to activate the cellular photolyase system, which removes the lesions produced by high energetic UV radiation and is activated by long wavelength UV-A or visible radiation (Hirosawa and Miyachi, 1983). After inducing UV-B damages no recovery could be found in dim white light. These results argue against an involvement of DNA in these organisms. The same result was found in gliding cyanobacteria (Häder et al., 1986).

Another mechanism by which high energy radiation can damage cells and tissues is by so-called photodynamic responses (Ito, 1983): UV-B radiation is absorbed by a suitable photoreceptor molecule; if it cannot pass the energy on to a photochemical reaction or loose it in the form of another relaxation mechanism, the energy can be accepted by triplet oxygen in the ground state which leads to the production of singlet oxygen (Maurette et al., 1983). Alternatively free radicals are being formed (Spikes, 1977). Both these substances are very aggressive and destroy membranes and other cellular components.

In some studies, however, photodynamic reactions could be excluded as the main mechanism by which UV-B radiation damages plankton organisms. Using diagnostic reagents and quenchers for singlet oxygen and free radicals did not increase the survival time under solar or artificial radiation (Häder et al., 1986; Häder and Häder, 1988b). Likewise, addition of D_2O, which increases the lifetime of singlet oxygen by a factor of 10 did not aggravate the inhibitory effect of the radiation as expected if a type II mechanism were involved (Fig. 9).

In those organisms and their responses, in which neither DNA is the primary target nor photodynamic responses are involved, it can be speculated that specific proteins or other components in the motor apparatus or the photoperception organelle or its sensory transduction are damaged by UV-B radiation. Recently, it could be shown that the proteins which comprise the paraflagellar body, the presumed photoreceptor in *Euglena gracilis*, are damaged by solar or artificial UV radiation. Likewise, similar damages by solar radiation have been found in preliminary studies in other flagellated organisms.

Consequences of Increased Solar UV-B Radiation

Due to the enormous size of the phytoplankton ecosystems in the oceans even small UV-B radiation induced decreases will amount to large losses in biomass production. These losses will be relayed through the whole intricate biological food web. It has been estimated that about 30% of the world's animal protein for human consumption comes from the sea, thus a noticeable reduction in the phytoplankton productivity will certainly affect the global food supply (U.S. Environmental Protection Agency, 1987). Evidence indicates that increased UV-B irradiance could also result in fishery losses through indirect effects on the planktonic food web. Several authors have suggested that fishery yield decreases in a power law fashion with decreases in primary production (Nixon, 1988). Thus, using the equation that fisheries yield increases as productivity raised to the 1.55 power, a 5% decrease in primary production (estimated for a 16% ozone depletion) will cause reductions in fish yield of approximately 6 to 9%. A 7% reduction in fish yield, if it occurred on a global basis, would then represent a loss of about 6 million tons of fish per year.

Furthermore, though all phytoplankton microorganisms studied so far are sensitive to UV-B radiation, there may be considerable differences between species (Kelly, 1986). The consequence will be a pronounced shift in the species composition in the primary producers as well as in the consumers. The generation time of marine phytoplankton is in the range of hours to days; whereas the potential increase to ambient levels of solar UV-B irradiance will occur in the range of decades. The question remains as to whether the gene pool within species is variable enough to adapt during this relatively gradual change in exposure to UV-B radiation. Indirect effects may also occur in the form of altered patterns of predation, competition, diversity, and trophic dynamics if species resistant to UV-B radiation were to replace sensitive species.

The global carbon dioxide cycle has been estimated to involve about 200 gigatons. In addition to the natural CO_2 production and emission, anthropogenic sources, such as fossil fuel consumption and burning of tropical rain forests add another 20 and 8 gigatons, respectively. The phytoplankton in the oceans is a major sink for atmospheric carbon dioxide. Any decrease in the phytoplankton populations will decrease the sink capacity for atmospheric CO_2; an estimated 10% decrease in the populations would equal the net CO_2 increase due to fossil fuel burning, which is not yet accounted for by the current climate change (greenhouse effect) models. Currently, no quantitative assessment of the loss in biomass production in aquatic systems is available. However, even a small loss is bound to have a noticeable adverse effect.

The role of cyanobacteria in nitrogen fixation especially in tropical rice paddies has been mentioned above. A significant decrease in the nitrogen fixation by prokaryotic microorganisms is bound to affect growth and productivity of higher plants. The amounts of artificial nitrogen fertilizer necessary to compensate for a substantial loss will certainly stress the capabilities of third world countries.

References

Batschelet, E., 1965, Statistical methods for the analysis of problems in animal orientation and certain biological rhytms, *in*: "Animal Orientation and Navigation," Galles, S. R., Schmidt-Koenig, K., Jacobs, G. J., and Belleville, R. F., eds., Washington, NASA, pp. 61.

Batschelet, E., 1981, "Circular Statistics in Biology," Academic Press, London.

Bean, B., 1984, Microbial geotaxis, *in*: "Membranes and Sensory Transduction," Colombetti, G., and Lenci, F., eds., Plenum Publishing Corporation, pp. 163.

Björn, L. O., and Murphy, T. M., 1985, Computer calculation of solar ultraviolet radiation at ground level, *Physiol. Vèg.*, 25:555.

Block, J., Briegleb, W., Sobick, V., and Wohlfarth-Bottermann, K. E., 1986, Confirmation of gravisensitivity in the slime mold *Physarum polycephalum* under near weightlessness, *Adv. Space Res.*, 6:143.

Briegleb, W., and Block, J., 1986, Classification of gravity effects on "free" cells, *Adv. Space Res.*, 6:15.

Brinkmann, K., 1968, Keine Geotaxis bei *Euglena*, *Z. Pflanzenphysiol.*, 59:12.

Burns, N. M., and Rosa, F., 1980, *In situ* measurements of the settling velocity of organic carbon particles and ten species of phytoplankton, *Limnol. Oceanogr.*, 25:855.

Caldwell, M. M., Madronich, S., Björn, L. O., and Ilyas, M., 1989, Ozone reduction and increased solar ultraviolet radiation, *in*: "UNEP Environmental Effects Panel Report," pp. 1.

Diehn, B., 1973, Phototaxis and sensory transduction in *Euglena*, *Science*, 181:1009.

Döhler, G., 1984, Effect of UV-B radiation on the marine diatoms *Lauderia annulata* and *Thalassiosira rotula* grown in different salinities, *Marine Biology*, 83:247.

Döhler, G., 1985, Effect of UV-B radiation (290-320 nm) on the nitrogen metabolism of several marine diatoms, *J. Plant Physiol.*, 118:391.

Döhler, G., Biermann, I., and Zink, J., 1986, Impact of UV-B radiation on photosynthetic assimilation of ^{14}C-bicarbonate and inorganic ^{15}N-compounds by cyanobacteria, *Z. Naturforsch.*, 41c:426.

Döhler, G., Worrest, R. C., Biermann, I., and Zink, J., 1987, Photosynthetic $^{14}CO_2$ fixation and ^{15}N-ammonia assimilation during UV-B radiation of *Lithodesmium variabile*, *Physiol. Plantarum*, 70:511.

Ekelund, N. G. A., and Björn, L. O., 1990, Ultraviolet radiation stress in dinoflagellates in relation to targets, sensitivity and radiation climate, *in*: "Proceedings of Workshop," Scripps Institution of Oceanography, University of California, San Diego La Jolla.

Ford, W. T. Jr., and Deering, R. A., 1979, Survival, spore formation and excision repair of UV-irradiated developing cells of *Dictyostelium discoideum* NC-4, *Photochem. Photobiol.*, 30:653.

Foster, K. W., and Smyth, R. D., 1980, Light antennas in phototactic algae, *Microbiol. Rev.*, 44:572.

Häder, D.-P., 1983a, Inhibition of phototaxis and motility by UV-B irradiation in *Dictyostelium discoideum* slugs, *Plant Cell Physiol.*, 24:1545.

Häder, D.-P., 1983b, Effects of UV-B irradiation on sorocarp development of *Dictyostelium discoideum*, *Photochem. Photobiol.*, 38:551.

Häder, D.-P., 1984, Effects of UV-B on motility and photoorientation in the cyanobacterium, *Phormidium uncinatum*, *Arch. Microbiol.*, 140:34.

Häder, D.-P., 1985, Effects of UV-B on motility and photobehavior in the green flagellate, *Euglena gracilis*, *Arch. Microbiol.*, 141:159.

Häder, D.-P., 1986, Effects of solar and artificial UV irradiation on motility and phototaxis in the flagellate, *Euglena gracilis*, *Photochem. Photobiol.*, 44:651.

Häder, D.-P., 1987a, Polarotaxis, gravitaxis and vertical phototaxis in the green flagellate, *Euglena gracilis*, *Arch. Microbiol.*, 147:179.

Häder, D.-P., 1987b, Effects of UV-B irradiation on photomovement in the desmid, *Cosmarium cucumis*, *Photochem. Photobiol.*, 46:121.

Häder, D.-P., 1988, Ecological consequences of photomovement in microorganisms, *J. Photochem. Photobiol. B: Biol.*, 1:385.

Häder, D.-P., and Griebenow, K., 1987, Versatile digital image analysis by microcomputer to count microorganisms, *EDV Med. Biol.*, 18:37.

Häder, D.-P., and Griebenow, K., 1988, Orientation of the green flagellate, *Euglena gracilis*, in a vertical column of water, *FEMS Microbiol. Ecol.*, 53:159.

Häder, D.-P., and Häder, M. A., 1988a, Inhibition of motility and phototaxis in the green flagellate, *Euglena gracilis*, by UV-B radiation, *Arch. Microbiol.*, 150:20.

Häder, D.-P., and Häder, M., 1988b, Ultraviolet-B inhibition of motility in green and dark bleached *Euglena gracilis*, *Current Microbiol.*, 17:215.

Häder, D.-P., and Häder, M., 1989a, Effects of solar UV-B irradiation on photomovement and motility in photosynthetic and colorless flagellates, *Env. Exp. Bot.*, 29:273.

Häder, D.-P., and Häder, M., 1989b, Effects of solar radiation on photoorientation, motility and pigmentation in a freshwater *Cryptomonas*, *Botanica Acta*, 102:236.

Häder, D.-P., and Häder, M., 1989c, Effects of solar radiation on development in the cellular slime mold, *Dictyostelium discoideum*, *Photochem. Photobiol.*, 50:557.

Häder, D.-P., and Häder, M., 1990a, Effects of solar and artificial UV radiation on motility and pigmentation in the marine *Cryptomonas maculata*, *J. Photochem. Photobiol.*, 5:105.

Häder, D.-P., and Häder, M., 1990b, Effects of solar radiation on motility, photomovement and pigmentation in two strains of the cyanobacterium, *Phormidium uncinatum*, *Acta Protozool.*, in press.

Häder, D.-P., Häder, M., Liu, S.-M., and Ullrich, W., 1990, Effects of solar radiation on photoorientation, motility and pigmentation in a freshwater *Peridinium*, *BioSystems*, 23:335.

Häder, D.-P., and Liu, S.-M., 1990a, Effects of artificial and solar UV-B radiation on the gravitactic orientation of the dinoflagellate, *Peridinium gatunense*, *FEMS Microbiol. Ecol.*, 73:331.

Häder, D.-P., and Liu, S.-M., 1990b, Motility and gravitatic orientation of the flagellate, *Euglena gracilis* impaired by artifical and solar UV-B radiation, *Curr. Microbiol.*, 21:161.

Häder, D.-P., Watanabe, M., and Furuya, M., 1986, Inhibition of motility in the cyanobacterium, *Phormidium uncinatum*, by solar and monochromatic UV irradiation, *Plant Cell Physiol.*, 27:887.

Häder, D.-P., Worrest, R. C., and Kumar, H. D., 1989, Aquatic ecosystems, *in*: "UNEP Environmental Effects Panel Report," 39.

Halldal, P., 1963, Zur Frage des Photoreceptors bei der Topophototaxis der Flagellaten, *Ber. Dtsch. Bot. Ges.*, 76:323.

Halldal, P., 1964, Ultraviolet action spectrum of photosynthesis and photosynthetic inhibition in a green and a red alga, *Physiol. Plant.*, 17:414.

Haupt, W., 1959, Die Phototaxis der Algen, *in*: "Handbuch der Pflanzenphysiologie," vol. XVII, 1, Ruhland, W., ed., Springer-Verlag Berlin, Göttingen, Heidelberg, pp. 318.

Haupt, W., 1965, Perception of environmental stimuli orienting growth and movement in lower plants, *Ann. Rev. Plant Physiol.*, 16:267.

Hirosawa, T., and Miyachi, S., 1983, Inactivation of Hill reaction by long-wavelength ultraviolet radiation (UV-A) and its photoreactivation by visible light in the cyanobacterium, *Anacystis nidulans*, *Arch. Microbiol.*, 135:98.

Ito, T., 1983, Photodynamic agents as tools for cell biology, *in*: "Photochemical and Photobiological Reviews," Vol. 7, Smith, K. C., ed., Plenum, New York, pp. 141.

Jagger, J., 1983, Effects of near-UV radiation on bacteria, *in*: "Photochemical and Photobiological Reviews," Smith, K. C., ed., Plenum, New York, pp. 1.

Jeffrey, S. W., and Humphrey, G. F., 1975, New spectrophotometric equations for determining chlorophylls *a, b, c1* and *c2* in higher plants, algae and natural phytoplankton, *Biochem. Physiol. Pflanzen*, 167:191.

Jensen, P., 1893, Über den Geotropismus niederer Organismen, *Pflüger's Arch. ges. Phys.*, 53:428.

Jerlov, N. G., 1970, Light - general introduction, *in*: "Marine Ecology," Vol. 1, Kinne, O., ed., pp. 95.

Kelly, J. R., 1986, How might enhanced levels of solar UV-B radiation affect marine ecosystems? *in*: "Proceedings of EPA/UNEP International Conference on Health and Environmental Effects of Ozone Modification and Climate Change," July, 1986.

Kessler, J. O., 1986, The external dynamics of swimming microorganisms, *in*: "Progress in Phycological Research," vol. 4, Round, F. E., and Chapman, D. J., eds., Biopress Ltd. Bristol, pp. 258.

Kuroda, K., Kamiya, N. M. J. A., Yoshimoto, Y., and Hiramoto, Y., 1986, *Paramecium* behavior during video centrifuge-micrscopy, *Proc. Japan. Cad.* Ser. B, 62:117.

Madronich, S. Frederick, J. Brasseur, G., and Caldwell, M. M., 1989, Predicted changes in surface UV radiation, *in*: "UNEP Atmospheric Sciences Panel Report."

Mardia, K. V., 1972, "Statistics of Directional Data," Acad. Press, London.

Maurette, M.-T., Oliveros, E., Infelta, P. P., Ramsteiner, K., and Braun, A. M., 1983, Singlet oxgen and superoxide: experimental differentiation and analysis, *Helv. Chim. Acta*, 66:722.

Mitchell, B. G., 1990, Action spectra of ultraviolet photoinhibition of Antarctic phytoplankton and a model of spectral diffuse attenuation coefficients, *in*: "Proceedings of Workshop," Scripps Institution of Oceanography, University of California, San Diego La Jolla.

Nixon, S. W., 1988, Physical energy inputs and the comparative ecology of lake and marine ecosystems, *Limnol. Oceanogr.*, 33:1005.

Nultsch, W., 1974, Movements, *in*: "Algal Physiology and Biochemistry," Stewart, W. D. P., ed., Blackwell Scientific Publications, Oxford, London, Edinburgh, Melbourne, pp. 864.

Nultsch, W., and Agel, G., 1986, Fluence rate and wavelength dependence of photobleaching in the cyanobacterium *Anabaena variabilis*, *Arch. Microbiol.*, 144:268.

Nultsch, W., Häder, D.-P., 1979, Photomovement of motile microorganisms. Photochem. Photobiol. 29:423.

Nultsch, W., and Häder, D.-P., 1988, Photomovement in motile microorganisms II, *Photochem. Photobiol.*, 47:837.

Ohnishi, T., and Nozu, K., 1979, Ultraviolet effects on killing, fruiting body formation and the spores of *Dictyostelium discoideum*, *Photochem. Photobiol.*, 29:615.

Ohnishi, T., Hazama, M., Okaichi, K., and Nozu, K., 1982, Formation of non-viable spores of *Dictyostelium discoideum* by UV-irradiation and caffeine, *Photochem. Photobiol.*, 36:355.

Schneider, S. H., 1989, The changing climate, *Sci. Am.*, 261:38.

Schreiber, U., Neubauer, C., and Klughammer, C., 1989, Devices and methods for room-temperature fluorescence analysis, *Phil. Trans. R. Soc. Lond. B*, 323:241.

Smith, R. C., 1989, Ozone, middle ultraviolet radiation and the aquatic environment, *Photochem. Photobiol.*, 50:459.

Spikes, J. D., 1977, Photosensitization, *in*: "The Science of Photobiology," Smith, K. C., ed., Plenum, New York, pp. 87.

Spikes, J. D., and Straight, R., 1981, The sensitized photooxidation of bimolecules, an overview., *in*: "Oxygen and Oxyradicals in Chemistry and Biology," Rodgers, M. A. J., and Powers, E. L., eds., Academic Press, New York, pp. 421.

Stolarski, R. S., 1988, The antarctic ozone hole, *Sci. Am.*, 258:20.

Taneda, K., Miyata, S., and Shiota, A., 1987, Geotactic behavior in *Paramecium caudatum*. II. Geotaxis assay in a population of the specimens, *Zool. Sci.*, 4:789.

Tevini, M., Braun, J., Grusemann, P., and Ros, J., 1989a, UV-Wirkungen auf Nutzpflanzen, *Laufener Sem. Beitr.*, 3:38.

Tevini, M., Teramura, A. H., Kulandaivelu, G., Caldwell, M. M., and Björn,L. O., 1989b, Terrestrial plants, *in*: "UNEP Environmental Effects Panel Report".

USEPA (U. S. Environmental Protection Agency), 1987, An assessment of the effects of ultraviolet-B radiation on aquatic organisms, *in*: "Assessing the Risks of Trace Gases That Can Modify the Stratosphere," EPA 400/1-87/001C, pp. (12)1.

van der Leun, J. C., 1989, Human health, *in*: "UNEP Environmenal Effects Panel Report".

Walsby, A. E., 1968, Mucilage secretion and the movements of blue-green algae, *Protoplasma*, 65:223.

Wolke, A., Niemeyer, F., and Achenbach, F., 1987, Geotactic behavior of the acellular myxomycete *Physarum polycephalum*, *Cell. Biol. Inter. Rep.*, 11:525.

Worrest, R. C., 1982, Review of literature concerning the impact of UV-B radiation upon marine organisms, *in*: "The Role of Solar Ultraviolet Radiation in Marine Ecosystems," Calkins, J., ed., Plenum Publishing Corp., pp. 429.

Yammamoto, K. M., Satake, M., Shinagawa, H., and Fujiwara, Y., 1983, Amelioration of the ultraviolet sensitivity of an *Escherichia coli* recA mutant in the dark by photoreactivating enzyme, *Mol. Gen. Genet.*, 190:511.

On the Trail of the Photoreceptor for Phototropism in Higher Plants

Timothy W. Short, Markus Porst,
and Winslow R. Briggs

Department of Plant Biology
Carnegie Institution of Washington
Stanford, CA. 94305
USA

Introduction

Much of the photomorphogenesis of multicellular green plants is mediated by the red, far red-reversible pigment phytochrome (Hendricks and Van der Woude, 1983). By contrast, only a few selected microorganisms respond to red light signals. However, both plants and microorganisms exhibit a wide range of responses to blue and ultraviolet light (Gressel and Rau, 1983; Senger, 1987), almost certainly mediated by several different blue light photoreceptors (Briggs and Iino, 1983; Gressel and Rau, 1983; Iino, 1988; Palit et al., 1989). A class of photoreceptors showing action spectra that one would expect for flavoproteins are found both in higher plants and fungi (Briggs and Iino, 1983). This class is frequently given the general name "cryptochrome". We are presently working with a higher plant photoreceptor that we suspect to be in this class. We will describe our current studies on this pigment system here in the hopes that at least some of what we have found may be helpful in elucidating responses to blue light in microorganisms.

The effect we are investigating involves the phosphorylation state of a plasma membrane protein, so a brief introduction to some of the known effects of protein phosphorylation is in order. It is well known that protein phosphorylation can play an important role in regulating protein function (Boyer and Krebs, 1986) and there are now accumulating a number of examples from higher plants (Budde and Chollet, 1988; Ranjeva and Boudet, 1987). Several plant enzymes have been subjects of intensive investigations, and it is clear that in some cases (e. g., the pyruvate dehydrogenase complex; Randall et al., 1990) phosphorylation inhibits enzymatic activity. In other cases (e. g., phosphoenol pyruvate carboxylase; Chollet, 1990; Nimmo, 1990) phosphorylation activates the enzyme. In still other cases (sucrose phosphate synthase; Huber and Huber, 1990) phosphorylation may either activate or inactivate the enzyme depending upon which sites on the enzyme are phosphorylated.

We have recently described a dramatic light-induced change in the phosphorylation of a plasma membrane protein obtained from rapidly elongating regions of the epicotyls of etiolated pea (*Pisum sativum*) L.) seedlings (Gallagher et al., 1988; Short and Briggs, 1990), and have hypothesized that this phosphorylation change represents an early step in the transduction chain for phototropism. The aim of our present re-

Biophysics of Photoreceptors and Photomovements in Microorganisms
Edited by F. Lenci *et al.*, Plenum Press, New York, 1991

search is to characterize this light-sensitive system in as much detail as possible and to determine its precise function in the cell. While we can still only speculate as to what its role might be, we have made considerable progress in its characterization. Since the material we have published has already been reviewed elsewhere (Briggs and Short, 1991; Short et al., 1990), we shall only briefly summarize it here before discussing more recent unpublished results related to the *in vitro* characterization of the system.

In vivo Studies

In an investigation of the capacity of various proteins in crude membrane fractions from etiolated pea stems to be phosphorylated upon addition of γ-labeled ^{32}P-ATP, Gallagher noticed a dramatic effect of light on a 120 kDa polypeptide. When the membranes were prepared from etiolated seedlings, this protein became heavily phosphorylated *in vitro*. On the other hand, when the membranes were obtained from etiolated seedlings treated to a few hours of white light, phosphorylation of a protein at 120 kDa was virtually undetectable.

Following this initial observation, Gallagher et al. (1988) began a preliminary characterization of the system. They found, first, that the change occurred in response to blue rather than to red light, eliminating the possible direct involvement of the plant photoreceptor, phytochrome. Second, by using sucrose gradient centrifugation and appropriate enzyme markers, they showed that the 120 kDa protein was localized to the plasma membrane. Finally, they showed that the association of the protein with the membrane was not ionic - as it could not be removed by high salt. It was also not the consequence of the entrapment of a soluble protein within membrane vesicles formed during membrane isolation, as it was released neither by sonication nor by hypo-osmotic shock. Only by treatment with detergents at concentrations that began to solubilize the membranes was it released to solution. Hence the association is strongly lipophilic.

Two physiological responses of etiolated pea seedlings have recently been described in some detail: a rapid inhibition of growth (Laskowski and Briggs, 1988) and phototropism in response to unilateral illumination (Baskin, 1986a, b). These two responses to blue light have somewhat different photobiological properties. In particular, they have fluence-response curves covering very different fluence ranges from threshold to saturation, with phototropism being over an order of magnitude more sensitive. Short and Briggs (1990) therefore examined the photobiological properties of the light-inducible phosphorylation change to determine whether this change could lie along one or the other of the transduction chains leading to the physiological responses, or to neither.

Short and Briggs (1990) first found that the reaction was extremely rapid. It goes to completion within a fraction of a second, even if the tissues are irradiated after being placed in a chilled mortar on ice and ground as rapidly as possible following a brief (0.3 s) saturating blue light pulse. They next found that if the tissue sections were incubated in darkness at 25° C for various times following irradiation and prior to membrane extraction, the capacity of the protein for *in vitro* phosphorylation gradually returned between 10 and 60 minutes. The amount of detectable phosphorylation at 120 kDa was strongest in the most rapidly elongating tissue of the epicotyl, declining basipetally. It was also completely absent from membranes prepared from bud tissue. Wherever it could be found in the epicotyl, however, it was lost if the tissue sections were treated with saturating blue light prior to membrane isolation. Finally, the threshold fluence for reducing phosphorylation was close to 10^{-1} μmol m^{-2} with saturation at about $10^{2.5}$ μmol m^{-2}, and within the limits tested, the photoreaction obeyed the reciprocity law. In summary, the light-altered protein is in the photosensitive tissue for both of the physiological responses (phototropism and rapid suppression of growth) mentioned above; the photoreaction is complete long before the end of the lag period for either of them; and it obeys the reciprocity law as do each of the physiological responses. However, The fluence requirements for the phosphorylation change match those for phototropism, and not those for the rapid inhibition of growth. Hence, Short and Briggs (1990) conclude that the observed reaction occurs early in the transduction chain for phototropism.

When SDS (sodium dodecyl sulfate) polyacrylamide gel electrophoresis of crude membranes from dark control sections is followed by Coomassie staining, a protein band is clearly visible at 120 kDa. However, if the sections have been given saturating blue light treatment prior to membrane extraction, this band can no longer be seen. A detailed study of this protein shows that several of its properties closely parallel those of the phosphorylation change. Specifically, the tissue distribution of this protein, the fluence-response relationships for its disappearance, and the kinetics for its reappearance on dark incubation prior to homogenization are all identical with those for the phosphorylation change. On the basis of these strong correlations, Short and Briggs (1990) therefore conclude that this protein is the substrate for the phosphorylation reaction.

In vitro Studies

The Fate of the 120 kDa Protein Following Irradiation

The disappearance of a Coomassie-stained band at 120 kDa in membranes from irradiated sections has several potential explanations. The putative phosphorylated protein could be lost from the membrane, it could remain with the membrane but show different mobility, its affinity for the Coomassie stain could be reduced, or it could show some combination of these possible changes. Recent unpublished work with polyclonal antibodies has allowed us to assess these possibilities. Plasma membranes were purified from crude membrane fractions by the polymer phase separation method described by Widdell and Larsson (1987), and the membrane proteins were separated by SDS polyacrylamide gel electrophoresis. The protein at 120 kDa was excised, run again on SDS gels, and the band electroeluted and injected into a rabbit using poly(A)-poly(U) as adjuvant (Hovanessian et al., 1988).

With this antibody we did Western blots of plasma membrane proteins from dark controls and from light-treated sections. The results clearly indicated that a) the protein was still with the plasma membrane, b) its mobility was slightly lowered, c) it was spread out over a broader molecular weight range, and, d) its affinity for the Coomassie stain was significantly lower in comparison with the protein from the dark control sections. *In vivo* irradiation caused no immunochemically detectable change in the small amount normally found remaining with the endomembranes after most of the plasma membrane has been separated away, nor was there an increase over the small amount normally detected in the soluble fraction. The changes seen here on phosphorylation were not unlike those reported by Guilfoyle et al. (1990) for the large subunit of RNA polymerase II. Phosphorylation of the large subunit also lowered its mobility on SDS gels and greatly reduced its affinity for Coomassie stain.

The Light Reaction

When Short and Briggs (1990) irradiated isolated membranes instead of tissue sections prior to the *in vitro* phosphorylation, they obtained a strong enhancement of phosphorylation. Short et al. (1990) have advanced a preliminary hypothesis to account for the opposite effects of *in vivo* and *in vitro* irradiation. They suggest that irradiation somehow alters the conformation of the protein to allow phosphorylation to occur. When the irradiation is *in vivo*, endogenous ATP promptly phosphorylates the exposed sites. Hence there are none available for subsequent *in vitro* phosphorylation. When the irradiation is *in vitro*, however, endogenous ATP is no longer present, and the sites are now available for phosphorylation by the radiolabeled exogenous ATP. When the irradiation treatment is *in vitro*, however, we did not always observe the mobility shift on phosphorylation seen with *in vivo* irradiation. Hence, the actual reaction is probably more complex than this model indicates. However, the model serves as a useful starting concept.

As was the case with crude membrane preparations, irradiation of purified plasma membrane led to a significant increase in the capacity of the 120 kDa protein for *in vitro* phosphorylation, indicating that the plasma membrane contained all three elements needed for phosphorylation: the 120 kDa substrate, the kinase activity, and the photoreceptor moiety. At present, we do not known whether one, two, or three or more polypeptides are necessary. However, plasma membranes can be dissolved in 1% Triton X-100 and retain their photoactivity with undiminished quantum efficiency, indicating that if several polypeptides are required for photoactivity, they must be tightly associated and retain complete functionality in Triton.

Since this membrane-associated system remains functional following detergent solubilization, it is accessible to soluble reagents, and one can address the long-standing question regarding the chemical nature of the photoreceptor - the principle candidates being flavins and carotenoids (Galston, 1959). Two kinds of reagents can affect photoexcited flavins. The first includes ions such as iodide and azide which react with photoexcited flavins to depopulate their excited states (Heelis et al., 1978). The second, including molecules such as phenylacetic acid, reacts with photoexcited flavins by undergoing an oxidative decarboxylation, both reducing the flavin and forming a covalent bond with it (Hemmerich et al., 1967). Both of the ions and phenylacetic acid specifically inhibit the light-inducible increase in phosphorylation of the 120 kDa protein at concentrations that do not affect phosphorylation in the dark control samples. Hence it seems likely that the photoreceptor in this case is a flavin moiety.

With an *in vitro* assay available, we could address another question: How long following irradiation can the consequences of photo-excitation be stored prior to phosphorylation? Put differently, how long does the light-altered state of the protein persist without phosphorylation? Preliminary experiments with purified plasma membranes indicate that the protein retains the capacity for enhanced phosphorylation for at least 2-3 minutes *in vitro* at 30° C (the normal temperature at which the phosphorylation reaction is carried out), and then gradually returns to something approaching the dark state over the next twenty to thirty minutes. Since the phosphorylation capacity of the dark controls also declines somewhat over the same time period, there is almost certainly some degradation occurring during the more lengthy incubations.

Preliminary Characterization of the Kinase Activity

With the demonstration that both the phosphorylation itself and the light-inducible increase in phosphorylation of the 120 kDa protein could occur after plasma membranes were solubilized in 1% Triton X-100, we undertook a series of studies to characterize the reaction. First, we examined the detailed time course for phosphorylation both in irradiated and control preparations from solubilized membranes. In all of our previous work (Gallagher et al., 1988; Short and Briggs, 1990), we permitted the phosphorylation reaction to go on for 30 seconds at 30° C before stopping the reaction either by adding SDS and heating to 100° C or by adding EDTA and EGTA to remove the calcium it requires. In the present experiments, phosphorylation of the protein increases rapidly to a maximum at about two minutes, in both irradiated and dark control samples. There is clearly some turnover, as detectable phosphorylation drops gradually thereafter, although significant radioactivity persists even after 20 minutes. We found no difference either in rate of onset or turnover between light and dark samples, and at each time point, the irradiated sample showed significantly more phosphorylation than the dark control. For all subsequent experiments, we carried out the phosphorylation reaction for two minutes. All phosphorylation experiments we have published to date utilized a standard reaction mixture at pH 7.0 (Gallagher et al., 1988). We recently investigated the pH dependency for the reaction in Triton X-100-solubilized plasma membrane preparations. The kinase activity clearly shows a broad optimum near pH 7.5. The optimum is the same whether or not the sample was irradiated prior to phosphorylation. There is still considerable activity at pH 9.0, but it drops sharply below about pH 6.0 so the pH dependency curve is quite asymmetric. As all protein phosphorylation in the plasma membrane, not just that of the protein at 120 kDa shows a steep decline between pH 6.0 and 5.5, the sharp drop is probably unrelated to the kinase activity, and may reflect the buffers used.

The normal reaction mixture includes 5 mM Mg^{++}, and omission of the magnesium results in greatly reduced phosphorylation whether or not the sample has been irradiated (although the light-dark difference persists). However, high magnesium concentrations (either 0.5 M $MgCl_2$ or 0.5 M $MgSO_4$ are also strongly inhibitory. NaCl is also inhibitory, with a concentration of 250 mM dramatically reducing the extent of phosphorylation. This reaction is not entirely irreversible, however, as some activity can be recovered upon removal of the high salt concentration. Throughout all of this as yet incomplete series of experiments with different salt concentrations, whenever any phosphorylation of the 120 kDa protein is detectable, a light-dark difference is also detectable. Hence the salt effects are on the kinase activity itself, and not on the light reaction and subsequent steps leading to the change in capacity for phosphorylation. We have also investigated the nucleotide specificity for the kinase reaction. In preliminary competition experiments, addition of cold ATP to the reaction mixture greatly reduced phosphorylation of the 120 kDa protein. At the concentrations tested, addition of cold GTP, UTP, and CTP had little or no effect. These experiments are currently being extended over a wider range of NTP concentrations to obtain better quantitative information, but there appears to be reasonable specificity for ATP.

We next addressed the question as to the minimum possible number of phosphorylation sites. The phosphorylation reaction was carried out with labeled ATP on purified plasma membranes, and the 120 kDa protein purified by SDS gel electrophoresis as before. The band was cut out and treated with sufficient SV-8 protease to complete proteolytic cleavage. Subsequent gel electrophoresis showed three distinct phosphorylated bands. Hence the minimum number of sites available for phosphorylation is three. Finally, we asked which amino acid (or amino acids) were phosphorylated. Gel-purified protein was acid-hydrolysed and the hydrolysate run on thin layer chromatography following the procedure of Cooper et al. (1983). Comparison of the mobility of the radioactivity with that of authentic phosphotyrosine, phosphothreonine, and phosphoserine, indicated that the 120 kDa protein was exclusively phosphorylated on serine residues.

Tissue and Species Distribution

As mentioned above, the highest level of the 120 kDa protein was found in the most rapidly elongating region of the epicotyl, declining down into the more mature regions of the stem. However, we were able to detect neither a protein band at 120 kDa nor phosphorylation at that position on the gel in membrane preparations from bud tissues. As noted above, the protein isolated from irradiated stem sections was virtually undetectable by Coomassie staining, presumably as a consequence of its phosphorylation. This observation, then, raises an obvious question concerning the buds: Is the protein really absent from plasma membranes isolated from these tissues or is it present but already fully phosphorylated? If the latter were the case, we could fail to detect it by Coomassie staining. Western blots from SDS gels of pea bud membranes indicate that if the protein is present at all, it is present in very low concentrations.

Western blots of gels prepared from crude membrane fractions from roots indicate the presence of a protein at 120 kDa that is recognized by the antibody. However, no *in vitro* phosphorylation could be obtained either from membranes from dark controls or irradiated tissues, or from membranes from dark controls that were subsequently irradiated *in vitro*. It is at present not known whether roots of the age tested show phototropic curvature in response to unilateral blue light. Clearly the root system requires further investigation.

If the protein we have been studying is indeed involved in the transduction of light signals in phototropism, it should be detectable over a wide range of species. We used two different approaches to address this matter. First, we probed Western blots of crude membrane preparations from carrot cell suspension cultures, from rapidly elongating stem tissues of etiolated zucchini, *Arabidopsis*, sunflower, tomato, and fava bean seedlings, and from apical regions of maize coleoptiles with antibody against the pea protein. Although weak cross-reactivity could be detected in some cases, the results were not dramatic. We had also attempted a limited proteolysis of the phosphorylated pea protein to obtain fragments for microsequencing, and found that the antibody did not recognize any proteolytically produced fragments, although autoradiography and Coomassie staining indicated a very clean pattern of proteolysis. Evidently the antibody recognizes only a small number of epitopes, and these are lost on proteolytic cleavage. Hence it is perhaps not surprising that the antibody does not cross-react well with proteins from plants that are not fairly close relatives of pea.

Another test for the possible ubiquity of the light-mediated phosphorylation change is to investigate whether irradiation of membranes isolated from a number of different plant species affects the subsequent capacity of any particular protein for phosphorylation. We investigated the same membranes we had utilized for the Western blotting experiments, and detected significant light enhancement of phosphorylation of one protein in every case. The molecular weights ranged from 114 kDa (maize) to about 130 kDa (sunflower), again suggesting some heterogeneity among species. Especially dramatic is the response in maize where phosphorylation is almost undetectable in membranes from dark control seedlings, and is intense if the seedlings have been irradiated with blue light. It should be cautioned that the phosphorylation change found did not always correspond with a protein band detectable by Western blotting.

Conclusions

From the results of the work described above, we believe that we have identified an early step in the signal transduction chain for phototropism. An increase in the capacity of a membrane protein for phosphorylation on irradiation of isolated membranes has been detected now in membrane preparations from elongating maize coleoptile tissue and from elongating stem tissues of all etiolated dicot seedlings thus far examined. The proteins from different species are somewhat divergent, at least based on immunological criteria, and vary by more than 10% in their molecular weights. Since a detailed study of the properties of the phosphorylation reaction has only been carried out in pea to date, it is premature to conclude that it is involved in phototropism in other species, though this seems a reasonable possibility.

There remain a number of problems to be addressed. First, we have not yet determined whether a single polypeptide or several polypeptides are required for the light-driven reaction to go to completion. Are we studying a photoreceptor protein that can undergo autophosphorylation under the right conditions or are there separate kinase, substrate, and photoreceptor proteins? We expect purification efforts currently under way to resolve this problem. Second, we have as yet only indirect evidence as to the nature of the photoreceptor pigment. Before we can state with certainty that the photoreceptor is a flavin, we need direct spectral evidence - either from absorption or fluorescence spectroscopy. Third, we have no information as yet bearing on the possible function of the 120 kDa protein. We are in the process of isolating and sequencing the gene and the corresponding protein, and it is possible that sequence homology with other genes of known function may provide some leads. Finally, we have strong correlative evidence in pea that we are looking at an element in the phototropism signal transduction chain. However, it will require the tools of genetics and molecular biology to allow us to establish or refute a definitive association of this reaction with phototropism.

Acknowledgements

The research from the authors' laboratory was supported by National Science Foundation Grant DCB-88 19137 to W.R.B. T.W.S. was supported by a National Science Foundation Predoctoral Fellowship. The authors gratefully acknoledge the able technical assistence of Ann McKillop.

References

Baskin, T. I., 1986a, "Phototropism: Light and Growth," Ph. D. Thesis, Stanford University.

Baskin, T. I., 1986b, Redistribution of growth during phototropism and nutation in the pea epicotyl, *Planta*, 169:406.

Boyer, P. D., and Krebs, E. G., eds., 1986, "The Enzymes, Ed. 3, XVII, Control by Phosphorylation," Academic Press, Orlando, Florida.

Briggs, W. R., and Iino, M., 1983, Blue light-absorbing photoreceptors in plants, *Phil. Trans. Roy. Soc. B Biol. Sci.*, 303:347.

Briggs, W. R., and Short, T. W., 1991, The transduction of light signals in plants, in: "Phytochrome Properties and Biological Actions," Thomas, B., and Johnson, C., eds., Springer-Verlag, Berlin, in press.

Budde, R. J. A., and Chollet, R., 1988, Regulation of enzyme activity in plants by reversible phosphorylation, *Physiol. Plantarum*, 72:435.

Chollet, R., 1990, Light/dark modulation of C_4-photosynthesis enzymes by regulatory phosphorylation, *in*: "Current Topics in Plant Biochemistry and Physiology 9," Randall, D. D., and Blevins, D. G., eds., University of Missouri, Columbia, Missouri, pp. 232.

Cooper, J. A., Sefton, B. M., and Hunter, T., 1983, Detection and quantification of phosphotyrosine in proteins, *Methods in Enzymology*, 99:387.

Gallagher, S., Short, T. W., Ray, P. M., Pratt, L. H., and Briggs, W. R., 1988, Light-mediated changes in two proteins found associated with plasma membrane fractions from pea stem sections, *Proc. Natl. Acad. Sci. U.S.A.*, 85:8003.

Galston, A. W., 1959, Phototropism of stems, roots, & coleoptiles, *in:* "Handbuch der Pflanzenphysiologie," Ruhland, W., ed., Vol. XVII 1, Springer-Verlag, Berlin, p. 492.

Gressel, J., and Rau, W., 1983, Photocontrol of fungal development, *in:* "Encyclopedia of Plant Physiology, New Series, 16B," Shropshire, W. Jr., and Mohr, H., eds., Springer-Verlag, Berlin, p. 603.

Guilfoyle, T. J., Dietrich, M. A., Prenger, J. P., and Hagen, G., 1990, Phosphorylation/dephosphorylation of the carboxyl terminal domain of the largest subunit of RNA polymerase II, *in:* "Current Topics in Plant Biochemistry and Physiology 9," Randall, D. D., and Blevins, D. G., eds., University of Missouri, Columbia, Missouri, p. 299.

Heelis, P. F., Parsons, B. J., Phillips, G. O., and McKellar, J. F., 1978, A laser flash photolysis study of the nature of flavin mononucleotide triplet states and the reactions of the neutral forms with amino acids, *Photochem. Photobiol.*, 28:169.

Hemmerich, P., Massey, V., and Weber, G., 1967, Photoinduced benzyl substitution of flavins by phenylacetate: a possible model for flavoprotein catalysis, *Nature*, 213:728.

Hendricks, S. B., and Van der Woude, W. J., 1983, How phytochrome acts - perspectives on the continuing quest, *in:* "Encyclopedia of Plant Physiology, New Series, 16A," Shropshire, W. Jr., and Mohr, H., eds., Springer-Verlag, Berlin, p. 3.

Hovanessian, A. G., Galabru, J., Riviere, Y., and Montagnier, L., 1988, Efficiency of poly(A)poly(U) as an adjuvant, *Immunol. Today*, 9:161.

Huber, S. C., and Huber, J. L. A., 1990, Regulation of spinach leaf sucrose-phosphate synthase by multisite phosphorylation, *in:* "Current Topics in Plant Biochemistry and Physiology 9,", Randall, D. D., and Blevins, D. G., eds., University of Missouri, Columbia, Missouri, p. 329.

Iino, M., 1988, Pulse-induced phototropisms in oat and maize coleoptiles, *Plant Physiol.*, 88:823.

Laskowski, M., and Briggs, W. R., 1989, Regulation of pea epicotyl elongation by blue light, *Plant Physiol.*, 89:293.

Nimmo, H. G., 1990, Regulation of phosphoenolpyruvate carboxylase by reversible phosphorylation in C_4 and Crassulacean Acid Metabolism Plants, *in:* "Current Topics in Plant Biochemistry and Physiology 9," Randall, D. D., and Blevins, D. G., eds., University of Missouri, Columbia, Missouri, p. 357.

Palit, A., Galland, P., and Lipson, E. D., 1989, High- and low-intensity photosystems in *Phycomyces* phototropism: Effects of mutations in genes *mad*A, *mad*B, and *mad*C, *Planta*, 177:547.

Randall, D. D., Miernyk, J. A., David, N. R., Budde, R. J. A., Schuller, K. A., Fang, T. K., and Gemel, J., 1990, Phosphorylation of the leaf mitochondrial pyruvate dehydrogenase complex and inactivation of the complex in the light, *in:* "Current Topics in Plant Biochemistry and Physiology 9," Randall, D. D., and Blevins, D. G., eds., University of Missouri, Columbia, Missouri, p. 313.

Ranjeva, R., and Boudet, A., 1987, Phosphorylation of proteins in plants: regulatory effects and potential involvement in stimulus/response coupling, *Annu. Rev. Plant Physiol.*, 38:73.

Senger, H., ed., 1987, "Blue Light Responses: Phenomena and Occurrence in Plants and Microorganisms," Vol. I and II, CRC Press, Boca Raton, Florida.

Short, T. W., and Briggs, W. R., 1990, Characterization of a rapid, blue light-mediated change in detectable phosphorylation of a plasma membrane protein from etiolated pea (*Pisum sativum* L.) seedlings, *Plant Physiol.*, 92:179.

Short, T. W., Gallagher, S., and Briggs, W. R., 1990, Protein phosphorylation as a possible signal transduction step for blue light-mediated phototropism in pea (*Pisum sativum* L.) epicotyls, *in:* "Current Topics in Plant Biochemistry and Physiology 9,", Randall, D. D., and Blevins, D. G., eds., University of Missouri, Columbia, Missouri, p. 232.

Widdell, S., and Larsson, C., 1987, Plasma membrane purification, *in:* "Blue Light Responses: Phenomena and Occurrence in Plants and Microorganisms," Vol. II, CRC Press, Boca Raton, Florida, p. 99.

LIAC Activity in Higher Plants

Han Asard and Roland Caubergs

Department of Botany
University of Antwerp (RUCA)
Groenenborgerlaan 171
B-2020 Antwerp
Belgium

Introduction

Numerous examples of blue light responses are known for plants, fungi and microorganisms. In the context of photomovements, the main topic of this Advanced Study Institute, we can mention phototropism, growth responses, phototaxis and chloroplast movement as well known phenomena. From several typical blue light responses action spectra are available. However, due to the intrinsic limitations, this technique has not been successful in unambiguously identifying the nature of the photoreceptor involved (see for example Briggs and Iino, 1983). As suggested in this paper, flavins are favored as most likely photoreceptors involved in typical UVA/blue light responses. As became clear in the course of this workshop, at least two other chromophores, pterins and rhodopsins, could possibly be involved in certain blue light phenomena, eventually in concert with a flavin.

Progress in blue light perception is seriously hampered by the lack of an appropriate assay to test the involvement of the blue light photoreceptor in a light-induced physiological response. In this respect phytochromists are 'spoiled' by the easy red-far red assay indicative for the participation of phytochrome. To distinguish the flavoprotein that acts as a blue light photoreceptor from the other flavins, the so-called blue light inducible absorbance changes or LIAC's have been proposed as a promising test.

An Introduction to LIAC Activity

Light-inducible absorbance changes (LIAC), induced by blue light (maximum around 460 to 470 nm) or UVA radiation (around 360 nm) were first noticed by Berns and Vaughn (1970). Mycelium and sporangiophores of different strains of *Phycomyces* were used. All of these showed an absorbance change caused by irradiation with 354 nm radiation, with a maximum at 460 nm. Except for one particular mutant (mad) this reaction reversed within a few minutes.

A link between the phenomenon of LIAC activity and blue light photoperception was suggested by Butler, Poff and Munoz (Poff et al., 1973; Poff and Butler, 1974; Munoz and Butler, 1975). They recorded blue light-mediated absorbance changes in

Biophysics of Photoreceptors and Photomovements in Microorganisms
Edited by F. Lenci *et al.*, Plenum Press, New York, 1991

cell suspensions and mycelia of *Dictyostelium*, *Phycomyces*, and *Neurospora*. These LIAC's, obtained after oxygen starvation of the tissue, indicated the concomitant photoreduction of a flavoprotein and a *b*-type cytochrome. According to the authors, these results represented the reflection of the primary action of a universal blue light photoreceptor. This view was supported by the fact that the action spectra for the LIAC in fungi and higher plants closely resembled the action spectrum of many 'typical' blue light phenomena (Munoz and Butler, 1975; Widell et al., 1983).

The hypothesis that redox activity upon blue light perception induces the typical so-called cryptochrome responses, contributed in the long standing discussion about the nature of the blue light photoreceptor. Indeed, action spectroscopy does not discriminate between a flavin or a carotenoid as possible candidates. However, carotenoids are never involved in redox phenomena and therefore, if LIAC activity is linked to blue light perception, preference should be given to flavins as responsible photoreceptor. Comparison of spectral and photochemical properties of flavins and carotenoids support this view. As explained by Song (1980) carotenoids show such a short lifetime for the excited singlet states that virtually no biological activity can be induced. Flavins on the other hand are strongly fluorescent, indicating a potency to induce photodynamic action. Other indirect evidence that favors flavins is available. A detailed description, however, seems exhaustive and the reader is advised to consult appropriate literature (Senger 1980, 1984; Galland and Senger, 1988 and references therein).

Although at first glance LIAC activity seems to be a good specific assay for the action of the blue light photoreceptor, it was rightfully stressed that a similarity in action spectra guaranties no causal relationship between LIAC and cryptochrome responses. Since flavins are ubiquitous in the cell and in all organisms it is prudent to give up the original concept of one common blue light photoreceptor whose primary action is reflected by LIAC activity.

Starting at its discovery in the early 1970s, LIAC activity is recorded in several organisms by different groups of researchers. Studies on higher plants revealed essentially the same blue light-induced absorbance change. For example in a microsomal fraction of corn coleoptiles Brain et al. (1977) showed a LIAC that was comparable to that of *Neurospora*. However, it was often unclear from the difference spectra whether aside from reduction of a *b*-type cytochrome also a flavin was involved. Obtaining stable and reproducible light- mediated cytochrome reductions has been proven to be more difficult with many samples.

To end this introductory overview, it has to be mentioned that the physiological relevance of the LIAC measurements is still a matter of debate (Widell, 1987; Schmidt, 1987). One of the most relevant criticisms concerns the specificity of the LIAC. The reader is reminded that free flavins can act as potent sensitizers. Particularly in those experiments, where cell disruption takes place, an artefactual adherence of released flavins and even cytochromes to membraneous structures cannot be excluded.

It is the aim of the work described below to characterize both the LIAC and the *b*-type cytochrome in more detail. Evidence will also be discussed indicating that the participating cytochrome and flavoprotein are native constituents of a particular membrane type, the plasma membrane.

Detecting LIAC Activities

Current methods to detect LIAC activities either involve a single-beam or dual-beam (dual-wavelength) spectrophotometer setup. With the single-beam configuration the sample absorbance spectrum is recorded before (baseline) and after an actinic irradiation with blue light. However, due to the very large absorbance differences of most samples in the blue and red wavelength regions, sensitivity and reproducibility of these measurements are often limited. Much better resolutions are obtained operating

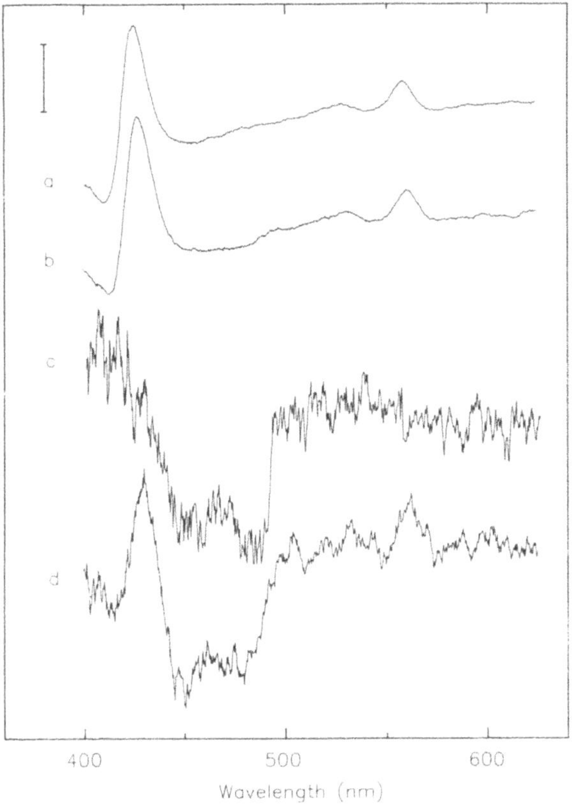

Fig. 1: Reduction of *b*-type cytochromes in purified plasma membrane fractions from bean hypocotyls. Curve a: dithionite-reduced minus oxidized difference spectrum. Curve b: light-minus-dark difference spectrum (LIAC) obtained in the presence of GOD/glucose and EDTA. Curve c: after the first irradiation in the absence of GOD and EDTA a flavin photoreduction is apparent. Curve d: upon slow oxygen starvation the reduction of the *b*-type cytochrome becomes detectable ("new"-LIAC). The scale bar represents: 0.02 OD (a), 0.005 OD (b), 0.0005 OD (c, d). Protein concentration: 1.14 mg ml^{-1}.

the single-beam spectrophotometer in a 'split-beam' (double beam) mode. Mechanically splitting the measuring beam with a high speed light-chopper results in the recording of the (small) optical difference between the sample and the reference cuvette as a baseline. After irradiation of the sample (while shielding the reference), a second spectrum is recorded. Subtraction of these measurements by computer results in the typical light-minus-dark difference spectra as shown in Figure 1 (curve b). This type of measurement allows the identification of the light reduction of a specific *b*-type cytochrome.

In dual-wavelength mode the time dependent change of the absorbance difference between two wavelengths is directly recorded. From the single-beam measurements it follows that the kinetics of the blue LIAC are most easily detected at the cytochrome Soret-band (maximum around 430 nm) or the α-band (maximum around 560 nm) with appropriate reference wavelengths set at 410 nm or 580 nm, respectively. Although the α-band absorbance maximum is about six times smaller, the interference of the actinic blue light is much smaller at this wavelength. Therefore, choosing these wavelengths and using specific cut-off filters allows the recording of the light-on kinetics (during illumination) of the LIAC activity.

In most samples tested so far the actual absorbance changes induced by blue light are rather small and range between 0.001 and 0.05 absorbance units if measured at about 430 nm. As a result high concentrations of tissue (mycelium) or membrane suspension are often used. However, as pointed out by Butler (1962) special care should be taken to perform measurements in turbid samples. In order to efficiently collect a high portion of the intensely scattered light, a large diameter photomultiplier tube is mounted as close to the sample as possible in combination with the 'opal glass method'. Higher resolutions can sometimes also be obtained through mathematically averaging a series of recordings, provided that a complete dark reversion is obtained and that sequential illumination of the sample does not result in altered light efficiencies. LIACs obtained in membrane fractions from higher plants are usually stable and reproducible after two illuminations.

Reproducible and large LIACs were only detected if the intact tissue (mycelium) was kept in the cuvette for some time apparently to develop partial anaerobiosis. Similarly, Goldsmith et al. (1980) found that stable LIACs in corn membrane fractions were only obtained in the presence of an oxygen scavenging system such as glucose oxidase or galactose oxidase with their respective sugar substrates. In addition to lowering the oxygen tension, also the addition of an external electron donor such as EDTA was necessary. LIAC activity in cauliflower membrane fractions was strongly stimulated by the addition of catalase. Apparently catalase prevented the dark reoxidation which could be restored by hydrogen peroxide. Clearly these data suggest the involvement of oxygen radicals in the LIAC reaction.

Characteristics of LIAC Activity

The light-mediated reduction of the *b*-type cytochrome by blue light is particularly sensitive to KI and phenylacetic acid (PAA). Half maximal inhibitions in corn microsomal fractions were obtained with 50 mM and 40 mM, respectively (Caubergs et al., 1979). Both agents are known to interfere with flavin- mediated reactions either by quenching the excited triplet state or by formation of a stable photoadduct (Hemmerich et al., 1967) preventing electron transfer. Although the involvement of a flavin was not clear from the difference spectra, these effects strongly support the flavin role in the LIAC reaction. Other highly effective inhibitors are the phenols salicyl hydroxamic acid (SHAM) (Caubergs et al., 1979; Borgeson and Bowman, 1985) and ferulic acid (Askerlund et al., 1987). Micromolar concentrations generally result in more than 50% inhibition. It has been suggested by Rich et al. (1978) that SHAM specifically binds close to the heme group of peroxidases. A similar mechanism could possibly operate at the cytochrome. On the other hand a possible relation between LIAC-inhibition and a phenol-stimulated NADH-dehydrogenase activity at the plant plasma membrane has been put forward (Askerlund et al., 1987).

Addition of external flavins has a remarkably strong stimulatory effect on LIACs. Blue light-mediated cytochrome reduction in cauliflower plasma membrane fractions was increased three-fold by 1-3 μM riboflavin. Similar effects were also demonstrated by Goldsmith et al. (1980) and Borgeson and Bowman (1985) in corn coleoptile and *Neurospora* membrane fractions, respectively. On the other hand the addition of NADH or NADPH (mM concentrations) has no detectable effect on the light signal (Asard et al., 1987). Stimulation of LIACs by flavins is observed both in the presence and absence of the oxygen scavenging system (see below). The possible significance of these observations will be discussed later.

A particularly interesting aspect of LIAC activity concerns its subcellular localization. Cell fractionation studies were carried out using differential centrifugation

(Brain et al., 1977) and density gradient centrifugation on sucrose and Renografin (Leong and Briggs, 1981; Caubergs et al., 1983). The significant correlation between LIAC activity and the marker enzyme glucan synthetase II indicated a localization at the plasma membrane. The LIAC in cauliflower also co-sedimented with a specific vanadate-sensitive plasma membrane ATPase (Caubergs et al., 1986) on sucrose gradients. Further evidence was obtained using an aqueous polymer phase-partitioning technique to obtain a highly purified plasma membrane fraction. A blue-LIAC was demonstrated and characterized in these fractions isolated from several higher plant tissues (Widell et al., 1983; Caubergs et al., 1983; Asard et al., 1987, 1989). The localization of the LIAC at the plasma membrane level is consequent with predictions based on physiological evidence. Blue light responses often involve directional changes (movement or growth) that suggest a photoreceptor localization at a stable subcellular structure. Also studies with polarized light (Jesaitis, 1974) and bio- physical arguments related to the optical properties of the *Phycomyces* sporangiophore (Steinhardt et al., 1989) suggest a dichroic oriented photoreceptor localized at the plasma membrane. The possible presence of LIAC activity in other plant membranes remains somewhat controversial. Using a combined centrifugation and phase-partitioning technique Widell (1987) suggested the presence of a blue light-mediated cytochrome *b* reduction in cauliflower endoplasmic reticulum. LIAC activity in *Neurospora* endoplasmic reticulum fractions was also much more sensitive to the addition of exogenous flavins as compared to the plasma membrane fraction (Borgeson and Bowman, 1985). In our hands, however, no significant LIAC activities were found in the endoplasmic reticulum from cauliflower or mung beans. LIAC was also absent in Golgi and tonoplast membranes (Caubergs et al., 1986).

Identification and Partial Purification of the Cytochrome

The establishment of the idea that a specific *b*-type cytochrome localized in the plant plasma membrane was involved (Jesaitis et al., 1977; Widell et al., 1983), initiated further research to characterize this heme-protein. From spectral data it was clear that a very similar cytochrome is involved in many systems. α-Band maxima are always detected between 558 and 561 nm. In a few cases LIAC measurements were also performed at higher resolution by freezing the sample in liquid nitrogen (Widell et al., 1983). These spectra generally show one single α-band at 556 nm.

In order to more accurately describe the *b*-type cytochrome investigations were started to determine the redox potential of the protein in highly purified plasma membrane fractions. Using a redox titration technique as described by Van Wielink et al. (1982) a major cytochrome component with a redox potential between +120 to +160 mV was detected in cauliflower, bean and zucchini (Caubergs et al., 1986; Asard et al., 1989). The α-band wavelength maximum was identified to be at 560.7 nm. The concentration of this cytochrome varied between the tested species and also low concentrations of additional cytochrome components were detected (Asard et al., 1989; Askerlund et al., 1989). Comparisons with literature data revealed that none of the *b*-type cytochromes found in chloroplasts and mitochondria was identical in both wavelength and midpoint potential to the major plasma membrane component.

Due to its electrochemical properties the high potential cytochrome in the plant plasma membranes is irreversibly reduced by Na-ascorbate. This reduction completely inhibited LIAC activity, indicating the involvement of the same cytochrome. Other arguments supporting this hypothesis are presented elsewhere (Caubergs et al., 1988).

Following the characterization of the cytochrome the next step was to solubilize and purify the protein (Lambrechts, in preparation). Several detergents were tested for their ability to disrupt the membrane structure and release the cytochrome. One major problem encountered in this was the non-covalent association between the protein moiety and the heme-group. Since detection of the cytochrome was based on the spectrophotometric measurement of the chemically reduced heme-protein (with Na-dithionite or Na-ascorbate); apparently very low yields were obtained. A partial purification of the solubilized fraction was obtained through affinity chromatography on a Con A sepharose (Pharmacia) system. Approximately 65-75% of the total recovered cytochrome was retained on the column due to binding with the immobilized lectins. As a consequence the cytochrome is preliminarily identified as a glycoprotein. Further attempts to purify the cytochrome through FPLC are in progress.

Detecting a "New" LIAC?

In recent experiments highly purified plasma membrane fractions from 5 days old etiolated bean hypocotyls were used to investigate the LIAC activity. In the presence of glucose oxidase (GOD), glucose and EDTA about 25% of the total amount of cytochromes present in the membrane were reduced by blue light (Fig. 1, curve b). Irradiation of the sample in the absence of EDTA, GOD and glucose resulted in a distinct reversible absorbance change with minima around 450 nm and 470-475 nm (Fig. 1, curve c). The profile of this curve suggests the photoreduction of a flavin-chromophore. With the repeated irradiation (1 min pulses with about 4 min dark intervals) of the same sample, a second absorbance change became apparent (Fig. 1, curve d). This LIAC is again identified as the reduction of a *b*-type cytochrome superimposed on the absorption decrease of curve c. For comparison the chemical reduction of all cytochromes in the sample by Na-dithionite is shown in curve a. Two further observations illustrate that this "new" LIAC is obtained after a slow oxygen depletion in the cuvette during measurements. First, when leaving the sample undisturbed for about 5-7 min, the cytochrome reduction is obtained immediately, together with the flavin-reduction. And second, gently shaking the sample after LIAC measurements reverts the signal to the initial observation of the flavin reduction without cytochrome reduction.

The LIAC reactions in the absence of the GOD/glucose system (= "new" LIAC) are again highly sensitive to the addition of exogenous flavins. Concentrations below 1 μM (Fig. 2) result in about 5-fold stimulations. The inhibitors described in a previous section are at least equally effective in this "new" LIAC or even much more effective in the case of PAA. Finally, the irreversible chemical reduction of *b*-type cytochromes by the addition of Na-ascorbate apparently reduces the level of cytochrome reduction without significantly affecting the initial flavin-bleach.

Discussion

The question how LIAC could fit into a reaction chain leading to a still unknown physiological response, remains unresolved. Several different hypotheses have been formulated. Illustrative for one of them is the model by Rau and co- workers (Rau, 1980). In this scheme it is assumed that photoreduction of a flavin photoreceptor causes a concomitant oxidation of a yet hypothetical compound "X_{red}". The resulting "X_{ox}" is stabilized rapidly by subsequent reactions yielding a photooxidation product. The reduced flavin photoreceptor may be reoxidized by transferring electrons either to a cytochrome (LIAC) and finally to oxygen. Similar models have been presented in relation to other blue light phenomena (Senger, 1980). This scheme is not fundamentally changed if one of the steps in the redox chain is of an enzymatic nature. A particular interesting case is the one of nitrate reductase (Ninnemann, 1987 for review). Its activity was found to be correlated under certain conditions with the photostimulation of conidiophore formation. This complex enzyme is composed of a large and a small subunit. The former contains FAD and cytochrome *b*. The latter is a molybdopterin complex. These different chromophoric groups render the enzyme suitable for photoregulation.

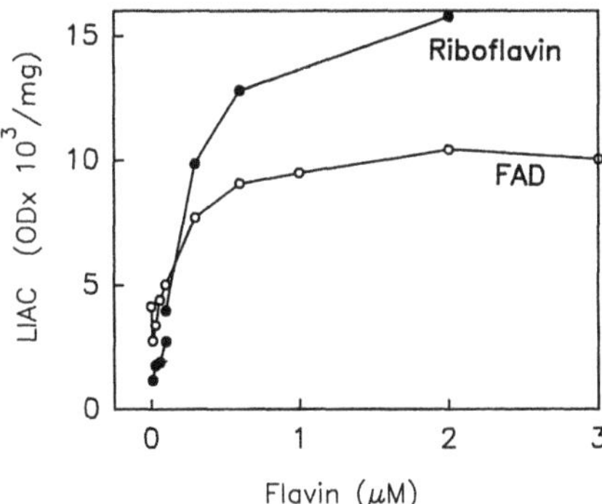

Fig. 2. The effect of varying concentrations of flavins on the cytochrome reduction induced by blue light ("new"-LIAC) in bean hypocotyl plasma membranes.

Induction of large ion fluxes with some blue light responses strongly suggests the involvement of membrane transport systems in the primary action of the photoreceptor. Using a patch-clamp technique to directly detect K^+-fluxes in guard cell plasma membranes, Assman et al. (1985) were able to demonstrate blue light-enhanced ion transport. Three models were designed by Zeiger (1984) that account for a direct or indirect activation of vectorial transport properties. A direct activation of the plasma membrane H^+-transport ATPase has been proposed (Shimazaki et al., 1986). However, no blue light effect on the *in vitro* enzyme activity could be demonstrated (Caubergs, unpublished). Indirect activation of membrane transport (and/or ATPase) could possibly be mediated by a plasma membrane electron transport chain capable of creating an electrochemical potential gradient. LIAC activity most likely reflects the operation of a (short) electron transport reaction. Flavin mediated cytochrome reductions in artificial membrane systems have been used to demonstrate blue light-mediated vectorial transport of redox equivalents (Schmidt, 1984). Recent work from several laboratories has also established the existence of plasma membrane oxido-reductase enzymes using NAD(P)H as a substrate (reviews in Crane et al., 1988). Whether this system is also involved in transmembrane proton or electron translocation is still a matter of debate. Although several models have been hypothesized combining the redox enzyme and the *b*-type cytochrome, we were unable to proof a direct relation between these components (Asard et al., 1987).

Finally, we would like to briefly draw the attention to an interesting hypothesis formulated by different authors (e.g., Hertel, 1980; Ninnemann, 1983). Free (soluble) flavins are present at micromolar concentrations in many plant and fungal cells. Upon binding with a specific receptor protein, these molecules are possible candidates for a physiologically active blue light photoreceptor. A saturable and reversible flavin binding site has been demonstrated in membrane vesicles from corn, zucchini (Hertel et al., 1980), *Phycomyces* (Dohrmann, 1983), cauliflower and bean (Asard, unpublished). Gradient centrifugation studies of this activity suggest an association with at least two different membrane types, one of which is positively identified as the plasma membrane. Work to further characterize this protein is in progress.

Sufficient arguments are presented to safely state that the plasma membrane contains a flavin-cytochrome protein complex that specifically reacts upon blue light irradiation. Consequently, the hypothesis that LIAC activity is an indication of flavin activity due to blue light perception still holds. However, the concept that LIAC activity is a general assay for all cryptochrome phenomena is difficult to support. As already mentioned, it is more likely that flavins, possibly with the additional action of other pigments, control different blue light inducible phenomena.

Acknowledgements

This work was supported by a grant from the National Science Department (87/92-119).

References

Asard, H., Venken, M., Caubergs, R., Reijnders, W., Oltmann, F. L., and De Greef, J. A., 1989, *b*-Type cytochromes in higher plant plasma membranes, *Plant Physiol.*, 90:1077.

Asard, H., Caubergs, R., Renders, R., and De Greef, J. A., 1987, Duroquinone-stimulated NADH oxidase activity and *b*-type cytochromes in the plasma membrane of cauliflower and mung beans, *Plant Science*, 53:109.

Askerlund, P., Larsson C., Widell, S., and Moller I. M., 1987, NAD(P)H oxidase and peroxidase activities in purified plasma membranes from cauliflower inflorescences, *Physiol. Plant.*, 71:9.

Askerlund, P., Larsson, C., and Widell, S., 1989, Cytochromes of plant plasma membranes. Characterization by absorbance difference spectrophotometry and redox titration, *Physiol. Plant.*, 76:123.

Assmann, S. M., Simoncini, L., and Schroeder, J. I., 1985, Blue light activates electrogenic ion pumping in guard cell protoplasts of *Vicia faba*, *Nature*, 318:285.

Berns, D. S., and Vaughn, J. R., 1970, Studies on the photopigment system in *Phycomyces*, *Biochem. Biophys. Res. Comm.*, 39:1094.

Borgeson, C. E., and Bowman B. J., 1985, Blue light-reducible cytochromes in membrane fractions from *Neurospora crassa*, *Plant Physiol.*, 78:433.

Brain, R. D., Freeberg, J. A., Weiss, C. V., and Briggs, W. R., 1977, Blue light-induced absorbance changes in membrane fractions from corn and *Neurospora*, *Plant Physiol.*, 59:948.

Briggs, W. R., and Iino, M., 1983, Blue-light-absorbing photoreceptors in plants, *Phil. Trans. R. Soc. Lond. B.*, 303:347.

Butler, W. L., 1962, Absorption of light in turbid materials, *J. Opt. Soc. Am.*, 52:292.

Caubergs, R. J., Goldsmith, M. H. M., and, Briggs W. R., 1979, Light inducible cytochrome reduction in membranes from corn coleoptiles: fractionation and inhibitor studies, *Carnegie Inst. Wash. Year Book*, 78:121.

Caubergs, R., Widell, S., Larsson, C., and De Greef, J. A., 1983, Comparison of two methods for the preparation of a membrane fraction of cauliflower inflorescences containing a blue light reducible *b*-type cytochrome, *Physiol. Plant.*, 57:291.

Caubergs, R. J., Asard, H. H., De Greef, J. A., Leeuwerik, F. J., and Oltmann, F. L., 1986, Light-inducible absorbance changes and vanadate-sensitive ATPase activity associated with the presumptive plasma membrane fraction from cauliflower inflorescences, *Photochem. Photobiol.*, 44:641.

Caubergs, R. J., Asard, H., De Greef, J. A., 1988, *b*-Type cytochromes, light- and NADH-dependent oxidoreductase activities in plant plasma membranes, *in*: "Plasma Membrane Oxidoreductases in Control of Animal and Plant Growth," Crane, F. L., Morre, D. J., and Low, H., eds., NATO ASI series, vol. 157, Plenum Press, New York, London, p. 273.

Crane, F. L., Morre, D. J., and Low, H., 1988, "Plasma Membrane Oxidoreductases in Control of Animal and Plant Growth," NATO ASI series, vol. 157, Plenum Press, New York, London.

Dohrmann, U., 1983, *In-vitro* riboflavin binding and endogenous flavins in *Phycomyces blakesleanus*, *Planta*, 159:357.

Galland P., and Senger H., 1988, The role of flavins as photoreceptors, *J. Photochem. Photobiol.*, 1:277.

Goldsmith, M. H. M., Caubergs, R. J., and Briggs, W. R., 1980, Light-inducible cytochrome reduction in membrane preparations from corn coleoptiles, *Plant Physiol.*, 66:1067.

Hertel, R., 1980, Phototropism of lower plants, *in*: "Photoperception and Sensory Transduction in Aneural Organisms," Lenci, F., Colombetti, G., and Song, P. S., eds., NATO ASI series, vol. 89, Plenum Press, New York, London, p. 89.

Hertel, R., Jesaitis, A. J., Dohrmann, U., Briggs, W. R., 1980, *In vitro* binding of riboflavin to subcellular particles from maize coleoptiles and *Cucurbita* hypocotyls, *Planta*, 147:312.

Hemmerich, P., Massey, V., and Weber, G., 1967, Photo-induced benzyl substitution of flavins by phenylacetate: a possible model for flavoprotein catalysis, *Nature*, 213:728.

Jesaitis, A. J., 1974, Linear dichroism and orientation of the *Phycomyces* photopigment, *J. Gen. Physiol.*, 63:1.

Jesaitis, A. J., Heners, P. R., Hertel, R., and Briggs, W. R., 1977, Characterization of a membrane fraction containing a *b*-type cytochrome, *Plant Physiol.*, 59:941.

Leong, T. Y., and Briggs, W. R., 1981, Partial purification and characterization of a blue light sensitive cytochrome flavin complex from corn membranes, *Plant Physiol.*, 67:1042.

Munoz, V., and Butler, W. L., 1975, Photoreceptor pigment for blue light in *Neurospora crassa, Plant Physiol.*, 55:421.

Ninnemann, H., 1983, Reversible absorbance changes and modulation of biological activities by blue light, *in*: "Molecular models of photoresponsiveness," Montagnoli, G., and Erlanger, B. F., eds., NATO ASI series, vol. 68, Plenum Press, New York, London, pp. 133.

Ninnemann, H., 1987, Photoregulation of eukaryotic nitrate reductase, *in*: "Blue Light Responses: Phenomena and Occurrence in Plants and Microorganisms," Senger, H., ed., CRC Press, Boca Raton, pp. 17.

Poff, K. L., Butler, W. L., and Loomis, W. F., 1973, Light induced absorbance changes associated with phototaxis in *Dictyostelium, Proc. Natl. Acad. Sci. USA*, 57:813.

Poff, K. L., and Butler, W. L., 1974, Absorbance changes induced by blue light in *Phycomyces blakesleanus* and *Dictyostelium discoideum, Nature*, 248:799.

Rau, W., 1980, Blue light-induced carotenoid biosynthesis in microorganisms, *in*: "The Blue Light Syndrome," Senger, H., ed., Springer-Verlag, Berlin, pp. 283.

Rich, P. R., Wiegand, N. K., Blum, H., Moore, A. L., and Bonner, Jr, W. D., 1978, Studies in the mechanism of inhibition of redox enzymes by substituted hydroxamic acids, *Biochim. Biophys Acta.*, 525:325.

Schmidt, W., 1984, Blue light-induced, flavin-mediated transport of redox equivalents across artificial bilayer membranes, *J. Membr. Biol.*, 82:113.

Schmidt, W., 1987, Primary reactions and optical spectroscopy of blue light photoreceptors, *in*: "Blue Light Responses: Phenomena and Occurrence in Plants and Microorganisms," Senger, H., ed., CRC Press, Boca Raton, pp. 19.

Senger, H., 1984, "Blue Light Effects in Biological Systems," Springer-Verlag, Berlin.

Senger, H., 1980, "The Blue Light Syndrome," Springer-Verlag, Berlin.

Shimazaki, K., Iino, M., and Zeiger, E., 1986, Blue light- dependent proton extrusion by guard-cell protoplasts of *Vicia faba, Nature*, 319:324.

Song, P. S., 1980, Spectroscopic and photochemical characterization of flavoproteins and carotenoproteins as blue light photoreceptors, *in*: "The Blue Light Syndrome," Senger, H., ed., Springer Verlag, Berlin, Heidelberg, New York, pp. 157.

Steinhardt, A. R., Popescu, T., and Fukshansky, L., 1989, Is the dichroic photoreceptor for *Phycomyces* phototropism located at the plasma membrane or at the tonoplast? *Photochem. Photobiol.*, 49:79.

Van Wielink, J. E., Oltmann, L. F., Leeuwerik, F. J., De Hollander, J. A., and Stouthamer, A. H., 1982, A method for *in situ* characterization of *b*- and *c*-type cytochromes in *E. coli* and in complex III from beef heart mitochondria by combined spectrum deconvolution and potentiometric analysis, *Biochim. Biophys Acta.*, 681:177.

Widell, S., Caubergs, R. J., and Larsson, C., 1983, Spectral characterization of light-reducible cytochrome in a plasma membrane-enriched fraction and in other membranes from cauliflower inflorescences, *Photochem. Photobiol.*, 38:95.

Widell, S., 1987, Membrane-bound blue light receptors - Possible connection to blue light photomorphogenesis, *in*: "Blue Light Responses: Phenomena and Occurrence in Plants and Microorganisms," Senger, H., ed., CRC Press, Boca Raton, pp. 89.

Zeiger, E., 1984, Blue light and stomatal function, *in*: "Blue Light Effects in Biological Systems," Senger, H., ed., Springer-Verlag, Heidelberg, New York, Tokyo, pp. 485.

Electrophysiology of Photomovements in Flagellated Algae

Oleg A. Sineshchekov

Biology Dept.
Moscow State University
Moscow
U.S.S.R.

Present address
Biology Dept.
Philipps-Univ.
Marburg
F.R.G.

Introduction

Photo-induced motile responses in microorganisms differ greatly by their appearance and mechanisms (see Nultsch and Häder, 1988 for comprehensive review). The traditional classification of photomovements has long been based on the behavioral principles (Diehn et al., 1977). Accordingly, the definitions used (*photokinesis, photophobic responses* and *phototaxis*) described the final results of light stimulation of a cell, reflecting mostly the strategy of light-induced behavior. However, for the analysis of a sensory transduction mechanism of the photo-induced responses it seems more useful to differentiate them on the basis of the probable nature of their photoreceptors and primary photobiological events in a cell. Three main photobiological types of photoreception could be suggested from this point of view (Sineshchekov and Litvin, 1974; 1982).

In the first group the motile responses are directly or indirectly related to the *photodynamic action* of the light stimulus. The photoreceptor pigments in these cases are not the elements of the evolutionarily developed photobiological apparatus, whose primary function is light energy conversion. Even an artificial photodynamic pigment can serve as a receptor for these responses. Such light-induced motile behavior can be considered as a response of a cell to "underthreshold" photodamage.

The photomotile responses of the second group are obviously linked to photoenergetics of the organism. The same pigment systems are used for both light energy conversion and photoreception. The cell only senses the changes in the level of photosynthesis. These *energy-dependent* photomovements are mainly observed in prokaryotic cells, in which the photosynthetic apparatus is not isolated from other cell structures by a chloroplast envelope. But they were also found in eukaryotic organisms and play an important role in the strategy of photomotile behavior.

Many microorganisms possess the third *specialized* type of photosensory mechanisms. These responses show very high light sensitivity, despite the extremely low concentration of the photoreceptor molecules. This suggests the existence of efficient amplification steps in the signal transduction. Additionally, in the photosynthetic flagellates specialized photoreception provides the cell with the most effective sensing of the direction to the light source, because of the complex structural organization of the photoreceptor. An increasing amount of evidence strongly suggests that specialized photoreception in Chlamydomonaceae (which will be the subject of this paper) could be considered as a primitive analog of vision in higher organisms (Foster and Smyth, 1980; Foster et al., 1984; Hegemann et al., 1988, and this volume; Sineshchekov, 1991; Sineshchekov and Litvin, 1982; Sineshchekov et al., 1990; Uhl and Hegemann, 1990).

Methods

The key role of bioelectric phenomena in photomovements in microorganisms has been proposed a number of times since the 1950's (Sineshchekov and Litvin, 1974, and refs. therein) and remained the subject of intensive discussion later (Nultsch, 1983; Lenci et al., 1984). However, traditional *intracellular microelectrode* measurements gave no clear indication of the existence of specific photoelectric responses involved in photoreception in unicellular algae. As it had become clear later (Sineshchekov, 1978; Sineshchekov et al., 1976, 1978), this was due to the high sensitivity of the sensory transduction processes of phototaxis to the mechanical damage of the cell by microelectrode insertion. Nevertheless, this method can be used for investigation of more stable photosynthetically driven electrical responses at both plasma and chloroplast membranes, which are involved in the interrelation between photoenergetic and photosensory systems of a cell.

The first indication of the existence of a blue light-induced electric response in flagellates was obtained by using *extracellular microelectrodes* not penetrating the cell but placed closely adjacent to the cell surface (Sineshchekov et al., 1976). This "patch" technique (although without gigaseal and voltage clamping, and thus difficult to interprete) allowed us also to find strictly periodic local changes in the state of the cell membrane, involved in klinokinesis and sign control of photoorientation (Sineshchekov et al., 1976, 1984; Sineshchekov and Litvin, 1988).

Investigation of photoelectric phenomena in phototaxis was significantly advanced with the development of the non-damaging *measuring suction micropipette* technique (Sineshchekov, 1978; Sineshchekov et al., 1978; Litvin et al., 1978). In this method the cell is sucked into the tip of the micropipette, causing two parts of the cell surface (those inside and outside the pipette) to become electrically isolated by the glass. If an electrical response is local and takes place only in one of these two parts of the membrane, a potential difference appears between the inside and the outside of the pipette (in the "current clamp" version of measurements). When a voltage is clamped between the inside of a pipette and a chamber, part of the asymmetric photo-induced current can be measured directly. Both voltage and current signals have exactly the same origin and kinetics. This method allows photoelectric responses to be measured for several hours without damaging the cell as well as a number of other advantages highlighted below. The most amenable for exploring this technique were relatively big objects with elastic cell walls such as *Haematococcus pluvialis* and *H. lacustris* and a wall-less mutant of *Chlamydomonas reinhardtii* all showing similar photoelectric responses (Sineshchekov, 1988).

Recently, we have developed a technique for the measurement of photoelectric responses in phototaxis from a population of cells in suspension (Sineshchekov et al., in preparation). The idea of this measurement was based upon the previously obtained data that phototactic electrical responses are asymmetric and the photoreceptor for

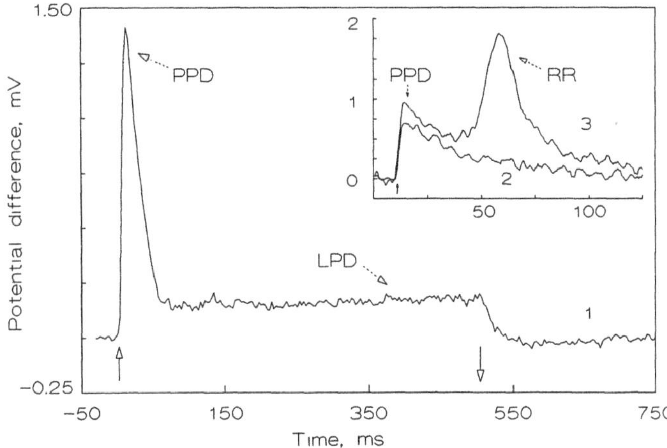

Fig. 1. Blue light-induced electric responses in *Haematococcus pluvialis* registered by measuring suction pipette technique: 1, continuous light; 2, flash of under-threshold (for eliciting RR) intensity; 3, flash of high intensity. Light on, off and the moment of the flash are indicated by arrows.

these signals has a directional sensitivity. Due to this fact, upon the unilateral illumination of the suspension of randomly oriented cells, extracellular potential differences of cells oriented at that moment by their photoreceptors toward the light source are not fully compensated by the smaller potential differences of cells with the opposite orientation. The difference between these signals can thus be registered by *macroelectrodes*. The application of this method allowed us to extend the electrical measurements to small objects with non-elastic cell walls like wild type *Chlamydomonas* and its mutants.

The Cascade of Bioelectric Processes Involved in Photophobic Responses and Phototaxis

In cells of both *Haematococcus* and *Chlamydomonas* light induces a complex of asymmetric bioelectric processes (Litvin et al., 1978; Sineshchekov, 1978, 1988; Sineshchekov et al., 1978, 1990) which can be measured by either micropipettes or macroelectrodes in a cell suspension. The electrical signal can be divided into three main parts which differ by their kinetics and features (Fig. 1). A transient *primary potential difference* (PPD) appears in the millisecond time range upon the intense flash or continuous light stimulus. It is followed by the permanent *late potential difference* (LPD), which has a lower amplitude and dissipates with a time constant of 15 to 40 ms upon switching off the light. Both responses depend on light intensity in a graded fashion and above the threshold level become superimposed by a nearly all-or-none *regenerative response* (RR) (Fig. 1, insert).

Only short-wavelengths light is effective in the generation of these responses (Fig. 2). The absence of any electrical signal upon the red light flash (at least in this time range) indicates that photosynthetically driven photopotentials do not participate in the formation of the described asymmetric electrical responses.

The spectral sensitivity undoubtedly relates the observed electrical responses to phototaxis (Litvin et al., 1978; Sineshchekov, 1978, 1991; Sineshchekov and Litvin, 1988; Sineshchekov et al., 1978). Action spectra of potential generation and photomotile behavior coincide with each other even in their complex fine structure (Fig. 3).

Photo-induced electric responses are generated only when the stigma region of the cell is illuminated with a microbeam (Sineshchekov, 1978; Ristori et al., 1981). This proves the widely accepted view that the photoreceptor is located in the part of the

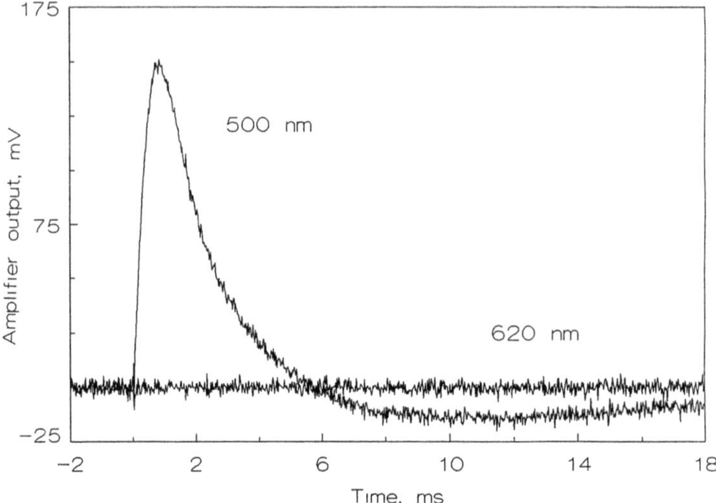

Fig. 2. Photoelectric signals in response to laser flashes of equal energies (about 1 mJ) of 500 nm and 620 nm measured in suspension of *Chlamydomonas reinhardtii* cells.

membrane covering the stigma (Foster and Smyth, 1980; Nultsch, 1983). By changing the position of the cell in the tip of the micropipette it was possible to determine the localization of the electrical generators responsible for the electrical responses. The sign of PPD is changed from positive to negative only when the stigma region of the cell is transferred during suction from outside to the inside of the pipette. It means that the photoreceptor potential is generated in the same place where light is absorbed and it has a depolarizing sign (movement of positive charges into the cell) (Sineshchekov, 1988, 1991; Sineshchekov et al., 1990).

In experiments with nanosecond laser excitation and improved time resolution it was found that PPD has no time delay (at least in the microsecond time range) and even when the amplitude of the signal is greatly suppressed by cooling the cell

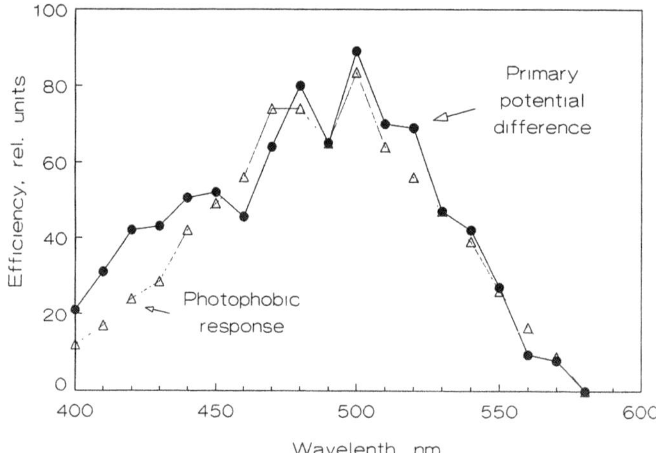

Fig. 3. Action spectra for the rate of photoreceptor potential generation (single cell measurements) and photophobic response of *Haematococcus pluvialis* freely motile cell. The reciprocal of the quantum requirement for equal (33 %) probability of the stop response was chosen as a criterium of light efficiency.

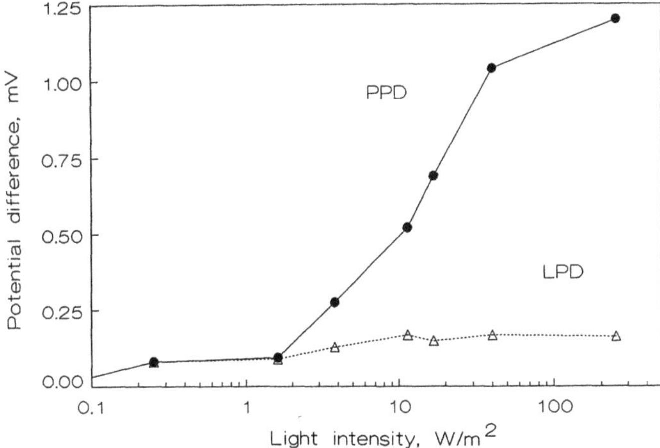

Fig. 4. Light intensity dependence of the amplitudes of PPD and LPD under continuous actinic light (500 nm) in *Haematococcus pluvialis* (single cell measurements). The LPD amplitude was measured after 1 s of illumination.

(Sineshchekov et al., 1990). It has a very high level of light saturation, compared to the later electrical events and is not saturated up to 500 W/m^2 (Fig. 4).

The primary potential difference is very stable. Transient responses, although progressively lower in amplitude, can be measured in a single cell for more than 12 h, a long time after the slower electrical signals have disappeared. The fast electrical signal is insensitive to a great extent to the variation of the ion composition of the medium and could not be completely inhibited by different ion channel blockers.

All these data indicate that at least a part of the fast electrical signal precedes the biochemical steps of the light signal transduction. It most probably reflects the charge movement in pigment molecules and can be thus considered as the *early receptor potential of phototaxis* (ERP) (Sineshchekov, 1991; Sineshchekov et al., 1990).

Light activation of photoreceptor molecules leads to the generation of the second more physiological electric response. By analogy with the processes in vision it can be named the *late receptor potential of phototaxis* (LRP). It is registered as the permanent asymmetric potential under continuous illumination. But the existence of the second component can be observed already in both rising and decaying parts of the transient electrical response to a laser flash (Sineshchekov et al., 1990). LRP has the same localization as ERP (in the stigma region) and has also a depolarizing sign. But it differs from ERP in the following features. It has a lag-period which varies from 150 to 400 ms, depending on the light stimulus intensity, and the saturation of LRP under the continuous illumination begins at a very low light intensity (Fig. 4). This suggests the existence of limiting biochemical steps, preceding the generation of LRP.

At intensities of 0.1 W/m^2 the calculated current through the photoreceptor which is responsible for the LRP can reach 10^7 unit charges/s and the amount of quanta absorbed at this intensity by the photoreceptor (assuming approxumately 30000 molecules per cell according to Smyth et al., 1989) should be not more then 10^3 per second. Thus, an amplification of at least 4 orders of magnitude is involved in the photosensory chain of phototaxis (Sineshchekov, 1991).

The ionic nature of the late photoreceptor current remains unknown. Although the photoreceptor potential in *Haematococcus* is to some extent calcium-dependent (Litvin et al., 1978), it is clear that not only calcium ions are involved. Both PPD and the permanent photoreceptor current demonstrate very weak sensitivity to the removal of traces of Ca^{2+} from the medium and are not markedly inhibited by such Ca^{2+} channel blockers as ruthenium red or verapamyl, although the effectiveness of these treatments on Ca^{2+} transport is obvious from the disappearance of the electrical RR and the motile responses. In gametes of the high light-sensitive strain of *Chlamydomonas reinhardtii* the dependence of PPD on the Ca^{2+} concentration is more obvious, but still even 3 times washing of the suspension in 2 mM EGTA does not lead to a complete inhibition of the photoreceptor potential (Fig. 5). Thus, light-sensitive ion channels in the region of the photoreceptor are not strictly calcium selective, or channels selective for other ions are involved in LRP generation.

The late receptor potential of phototaxis as well as the photomotile responses are inhibited by l-*cis*-diltiazem - the blocker of c-GMP-dependent ion channels at the usually effective concentrations. This fact and the highly efficient signal amplification mentioned above allow the suggestion that the light signal transduction chain in phototaxis involves a biochemical cascade probably similar to that in vision (Kaupp and Koch, 1986).

The following steps in the light signal transduction chain are most obviously represented by the threshold, an almost all-or-none, *regenerative electric response* (Fig. 1, insert). Independently of the location of the stigma in the pipette, the sign of this signal depends upon whether the flagella are located outside the pipette (positive) or inside the pipette (negative) (Sineshchekov, 1978, 1988, 1991). Thus, the RR takes place in the region of the flagella and has a depolarizing direction.

Opposed to the preceding electrical events in the stigma region, the RR is strictly calcium dependent and disappears upon its removal by EGTA (Fig. 5) or by addition of low concentrations of calcium channel blockers. It has a light intensity-dependent lag-period in the range of 5 to 40 ms. There is a linear dependence between the rate of photoreceptor potential generation and the reciprocal of RR lag-period, which suggests that the signal transduction from the photoreceptor to the motor apparatus is based most probably on the electrotonic spread of a potential difference change along the cell membrane (Sineshchekov, 1988, 1991; Sineshchekov et al., 1990). Indeed, it was found that the area under the curve of the light-induced potential difference before the beginning of RR is constant for different light intensities, even those close to saturation. This shows that an equal amounts of charge should enter the cell to initiate the regenerative response. According to our calculation the membrane must be discharged by several tens of millivolts to open the potential-dependent Ca^{2+} channels and initiate the threshold electrical response.

Direct microscopic observations revealed that flagella reorientation takes place only when the RR is registered. This led us to conclusion (Sineshchekov, 1978;

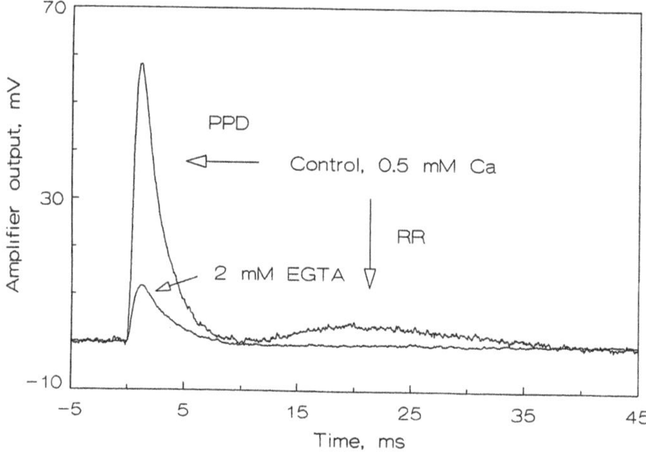

Fig. 5. The effect of Ca^{2+} ion removal on PPD and RR in *Chlamydomonas reinhardtii* (suspension measurements).

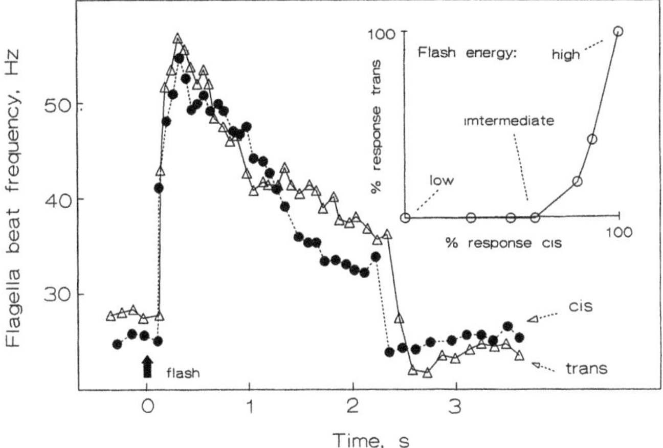

Fig. 6. The kinetics of photophobic responses of the *cis*- and *trans*-stigma flagella of a *Haematococcus pluvialis* cell fixed on the tip of the micropipette to intense blue flash stimulation. Insert: different probability of phobic responses of the *cis*- and *trans*-flagella for the rising intensities of the flashes near the threshold.

Sineshchekov et al., 1978) that the RR can be considered as the direct driving force of the photophobic behavior of the cell - the motile response which appears upon the abrupt change in light intensity and prevents the organism from crossing the light/dark boundary. This was confirmed by the findings of close similarity between the features of RR and photophobic response of freely motile organisms or flagella, as the cell was fixed on the tip of micropipette. Indeed, both electric and motile responses are strictly Ca^{2+}-dependent, nearly all-or-none threshold phenomena with similar time delay. They have identical spectral sensitivity and depend in the same way on red background illumination and other experimental conditions. Thus, the massive Ca^{2+} influx during the described RR response leads to the photophobic response of flagella.

The quantitative analysis of the *motile responses* in individual flagella based upon the red microbeam probe monitoring of their beat frequencies shows that the behavior of the two flagella are not identical (Sineshchekov, 1983, 1991; Sineshchekov and Litvin, 1988). It differs in the kinetics of photophobic responses upon a high intensity flash stimulus (Fig. 6). Moreover, as the stimulus rises in the narrow light intensity range near the threshold the response of only one flagellum could be observed. Independent of the position of the cell in the pipette, the flagellum on the stigma bearing side (*cis*-stigma flagellum) is more sensitive to illumination (Fig. 6, insert). This observation indicates that each flagellum behaves like an independent excitable organelle. Thus, the calcium channels responsible for RR (mediating the photophobic behavior) are most probably located in the membrane covering each flagellum. This is confirmed by the observation that a pair of electric RR responses could be registered under intermediate stimulus intensities, each of them probably reflecting the excitation in individual flagella.

The functional *in vivo* difference between the two flagella found in these experiments provides the basis for the second main type of photomotile behavior - photo-orientation, or true phototaxis. According to a widely accepted hypothesis a cell uses the spatially sensitive antenna for its phototaxis (Foster and Smyth, 1980). We have checked this assumption directly by the rotation of the measuring pipette with the cell so that the different parts of the cell surface were exposed to light. The maximum photo-induced potential was observed when the cell was oriented toward the light source with its stigma bearing side. Maximum difference between the electrical signals of the two opposite orientations of the cell was found at the wavelength actively absorbed by stigma and chloroplast (Sineshchekov, 1988, 1991).

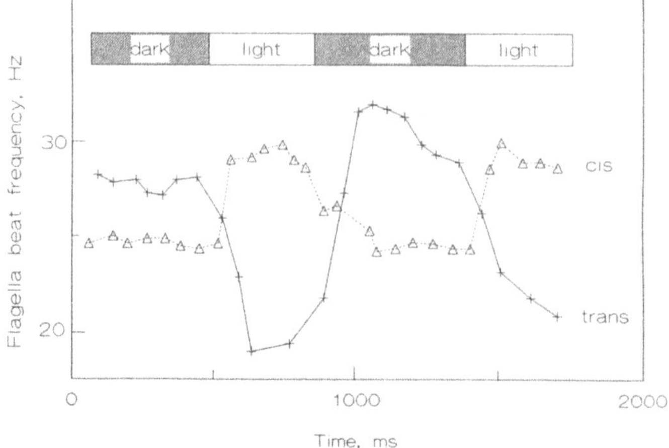

Fig. 7. Opposed responses of the *cis*- and *trans*-flagella of *Haematococcus pluvialis* cell fixed on the tip of micropipette to periodic changes in light intensities.

According to these data in freely motile organism the light incident on the photoreceptor is modulated (due to the cell rotation) with about one second period and three times in intensity. These conditions were simulated for the fixed cell (Sineshchekov and Litvin, 1988). Due to the existence of the refractive period of the RR and the relatively low amplitude of light intensity variation only a gradual electrical response was observed under these conditions. The motile responses of individual flagella were found to be also smooth, gradual and without changing the undulation from the cilia- to the flagella-type. The two flagella show not only different, but in most cases opposite motile behavior (Fig. 7). The *cis*-stigma flagellum reversibly increases its beat frequency during the illuminated period and the *trans*-stigma flagellum decreases it. Parallel with the changes in beat frequencies, smooth reorientation of both flagella were observed: more out of the cell body with an increased frequency and closer to it,

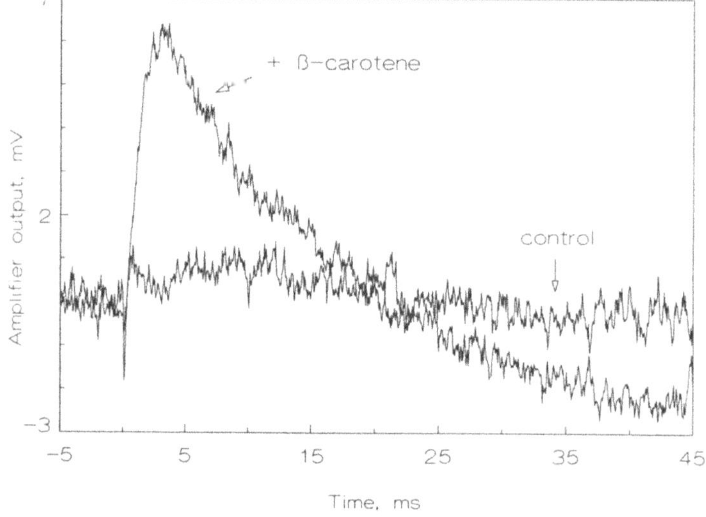

Fig. 8. The induction of photoreceptor potential generation in a carotenoidless mutant of *Chlamydomonas reinhardtii* by addition of ß-carotene (suspension measurements).

when the beat frequency is decreased. This reorientation adds to the unbalance in the driving force of two flagella, which causes the turn of the cell.

In *Chlamydomonas*, which differs from *Haematococcus* by a mostly synchronous beat of the two flagella (Rüffer and Nultsch, 1990), the unbalance of the motive forces of the two flagella is achieved by the difference in the amplitude and orientation of the two flagella strokes (Rüffer and Nultsch, in press). But in both bi-flagellated algae the correction of cell orientation during the phototaxis is based on the opposite effect of light on the activity of two morphologically identical flagella and takes place twice during one rotation cycle.

The periodic modulation in the intensity of flagella beating is most probably due to the small under-threshold variations of divalent cation concentration in each flagellum, since the true phototaxis like the photophobic response also needs calcium ions (Nultsch, 1979). Gradual late receptor potential spread on the flagella membrane, could fulfill this even without active modulation of the ion conductivity of the flagella membranes. The ion fluxes between axonema and outer medium should be modulated already as a direct consequence of the changes in the electrical component of the electrochemical motive force across the flagellar membrane.

Photoreceptor Features and Characteristics Revealed by Photoelectric Response Measurements

Since the photoreceptor potential represents the most early detectable effect of the light action on the cell, its measurements provides a powerful tool for the investigation of photoreceptor functioning *in vivo*. Indeed, most of the conclusions about the primary steps of light signal transduction obtained earlier were based on the observation of the final motile responses of the cell or even cell population. Obviously, these criteria could, and in same cases even should, reflect the features of not only the photoreceptor itself, but also of the later transduction chain elements or other processes involved in the complex motile behavior of the cell. Photoreceptor potential registration, particularly in a single cell, excludes these uncertainties and allows us to get new information about the photoreceptor and to confirm more correctly previously made conclusions.

The basic idea of the retinal nature of the photoreceptor molecule in phototaxis of Chlamydomonaceae (Foster et al., 1984) was approved by the direct observation of primary electric events in the photoreceptor (Sineshchekov et al., in preparation). The early receptor potential is absent in a carotenoid-less mutant of *Chlamydomonas reinhardtii*, but appears after addition of exogenous ß-carotene (Fig. 8). In agreement with earlier experiments on *Chlamydomonas* motile behavior (Hegemann et al., 1988), hydroxylamine inhibits phototaxis also in *Haematococcus*. But light-dependent inhibition of the receptor potential, which we found recently in both organisms, allows us to exclude the possible effect of this agent on the sensory transduction chain or the motor apparatus and to attribute its action directly to the photoreceptor pigment (Govorunova et al., in press).

The action spectrum for the photoelectric response gives the spectral characteristics of the receptor molecules, which are not disturbed by shading devices or secondary effects due to the control of phototaxis by photosynthesis (Sineshchekov et al., 1989). A complex fine structure of this spectrum was found (Litvin et al., 1978; Sineshchekov et al., 1978), which points to the high rigidity of the chromophore in the protein moiety (Fig. 3).

Recently, a retinal-protein was isolated from *Chlamydomonas reinhardtii* which has a very similar but less obvious fine structure of its absorption spectrum (Hegemann, this volume). As one of the possible explanations of the pronounced fine structure in the action spectra (compared to the absorption spectrum) we can assume that only the

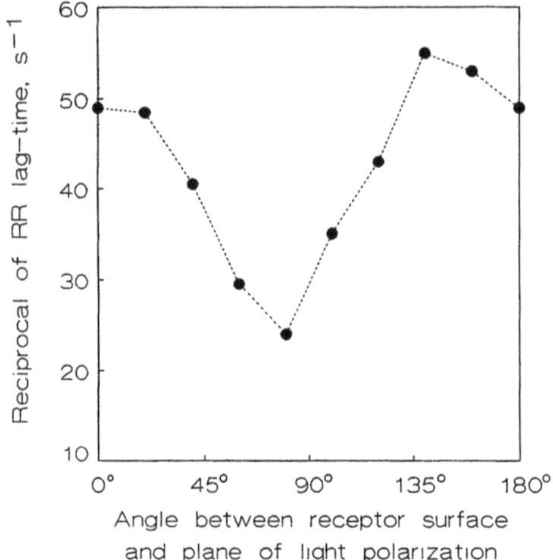

Fig. 9. Action dichroism of actinic light (500 nm) for eliciting electric response in *Haematococcus pluvialis* (single cell measurements).

molecules which absorb a light quantum being at that moment in a photoactive substate with stronger interaction between the chromophore and the protein moiety (and hence more rigid fixation of the former) can efficiently undergo physiologically significant conformational changes in the protein part of the molecule.

Even assuming this model, the spectral characteristics of one pigment could hardly account for six bands in the action spectra. This led us to the conclusion that probably more than one pigment participates in the absorption of phototactically active light in flagellates (Sineshchekov, 1988; Sineshchekov and Litvin, 1982, 1988) as it was firmly established for photoreception in *Halobacterium halobium* (Spudich and Bogomolni, 1988).

The quantitative analysis of light intensity dependence of the photoreceptor potential shows that the data fit with good accuracy to the exponential function. If we assume that the high light intensity saturated component of PPD is limited only by the product of quantum yield and optical cross-section of the photoreceptor molecule, the experimental data give the value of approximately 0.8 $Å^2$ for this product, which is reasonable for a single retinal-protein molecule (Sineshchekov, 1988, 1991).

The maximum efficiency of the polarized actinic light in eliciting the electrical responses in a single cell was found to be when the plane of polarization was parallel to the stigma surface (Fig. 9). This indicates that the receptor molecules are oriented so that their transition dipole moments lie in the plane of the cell membrane (Sineshchekov, 1988, 1991). This orientation is typical for retinal-proteins and provides the additional basis for the directional sensitivity of light antenna of phototaxis (Sineshchekov, Litvin, 1982).

The early receptor potential recovers after the saturation flash within several hundred milliseconds (Sineshchekov, 1988, 1991). This "refractive period" most probably reflects the turn-over time of the light-induced cycle in the photoreceptor molecule (at least gives the short limit for this time) and it also is in the order of magnitude typical for rhodopsins.

Conclusion

The data presented show that the electric phenomena are deeply involved in the sensory transduction chain of specialized photoreception in unicellular green algae from its very beginning up to the final motile responses of the flagella, and provide the basis for many steps in this chain. The great similarities in photoelectric responses in *Haematococcus* and *Chlamydomonas* cells allow us to suggest that the cascade of electrical processes and the scheme of their involvement in photomotions described above could be universal to some extent and one could expect them to exist also in other related phototactic algae.

The results obtained by the direct electrical measurements provide further evidence for a close similarity between the photoreception in flagellates and vision in multicellular organisms. The proposed techniques may serve as a promising experimental approach to the evaluation of the mechanisms of sensory reception in microorganisms.

It should be emphasized that the strategy and mechanisms of the final optimization of the light conditions in the photosynthetic organisms can not be understood solely on the basis of the data on "blue light" transduction chain. The feed-back control of photoreception from the photosynthetic apparatus and the energetic state of the cell should be taken into consideration. This control, which also involves electrical processes (photosynthetically driven and rhythmically spontaneous) will be discussed elsewhere.

Acknowledgements

This work was partly supported during the preparation by the Alexander von Humboldt Foundation, F.R.G. I thank Prof. W. Nultsch and Prof. F. Lenci for their help in my participation in NATO ASI and work on the manuscript.

References

Diehn, B., Feinleib, M., Haupt, H., Hildebrand, E., and Nultsch, W., 1977, Terminology of behavioral responses of motile microorganisms, *Photochem. Photobiol.*, 26:559.

Foster, K. W., and Smyth, R. D., 1980, Light antennas in phototactic algae, *Microbiol. Rev.*, 44:572.

Foster, K. W., Saranak, L., Patel, N., Zarilli, G., Okabe, M., Kline, T., and Nakanishi, K., 1984, A rhodopsin in the functional photoreceptor for phototaxis in the unicellular eukaryote *Chlamydomonas, Nature*, 311:756.

Hegemann, P., Hegemann, U., and Foster, K. W., 1988. Reversible bleaching of *Chlamydomonas reinhardtii* rhodopsin *in vivo, Photochem. Photobiol.*, 48:123.

Kaupp, U. B., and Koch, K.-W., 1986, Mechanism of photoreception in vertebrate vision, *Trends Biochem. Sci.*, 11:43.

Lenci, F., Häder, D.-P., and Colombetti, G., 1984, Photosensory responses in freely motile microorganisms, *in*: "Membranes and Sensory Transduction," Colombetti, G., and Lenci, F., eds., Plenum Press, New York, p. 199.

Litvin, F. F., Sineshchekov, O. A., and Sineshchekov, V. A., 1978, Photoreceptor electric potential in the phototaxis of the alga *Haematococcus pluvialis, Nature*, 271:476.

Nultsch, W., 1979, Effect of external factors on phototaxis of *Chlamydomonas reinhardtii*. III: Cations, *Arch. Microbiol.*, 123:93.

Nultsch, W., 1983, The photocontrol of movement of *Chlamydomonas*, *in*: "The Biology of Photoreception," Cosens, D. J., and Vince-Prue, D., eds., Soc. Exp. Biol. Symp., vol. 36, pp. 521.

Nultsch, W., and Häder, D.-P., 1988, Photomovement in motile microorganisms - II, *Photochem. Photobiol.*, 47:837.

Ristori, T., Ascoli, C., Banchetti, R., Parrini, P., and Petracci, D., 1981, Localization of photoreceptor and active membrane in the green alga *Haematococcus pluvialis*, Proceedings of the Sixth International Congress on Protozoology, Warsaw, p. 314.

Rüffer, U., and Nultsch, W., 1990, Flagellar responses of *Chlamydomonas* cells held on micropipettes: change in flagellar beat frequency, *Cell Mot. Cytoskel.*, 15:162.

Sineshchekov, O. A., 1978, The investigation of the photoelectrical processes in phototaxis in green alga (in Russian), Ph. D. Thesis. Moscow State University.

Sineshchekov, O. A., 1983, Phototaxis in microorganisms, in Russian, Proceeding of the All-Union Conference "Application of lasers in biology," Moscow Univ. Press, Moscow, pp. 91.

Sineshchekov, O. A., 1988, Phototaxis in microorganisms and its role in photosynthesis regulation (in Russian), *in*: "Phototrophic microorganisms," Gogotov, I. N., ed., Acad. Sci. USSR, Puschino, p. 11.

Sineshchekov, O. A., 1991, Photoreception in unicellular flagellates: bioelectric phenomena in phototaxis, *in*: "Light in Biology and Medicine," vol. II, Douglas, R. H., ed., Proceedings of the III Congress of the European Society for Photobiology, Budapest, 1989, Plenum Press, in press.

Sineshchekov, O. A., and Litvin, F. F., 1974, Phototaxis in microorganisms, its mechanism and relation to photosynthesis (in Russian), *Usp. Sovr. Biol.*, 78:57.

Sineshchekov, O. A., and Litvin, F. F., 1982, Photoregulation of movement in microorganisms (in Russian), *Usp. Sovr. Mikrobiol.*, 17:62.

Sineshchekov. O. A., and Litvin, F. F., 1988, The mechanisms of phototaxis in microorganisms (in Russian), *in*: "Molecular mechanisms of Biological Action of Optic Radiation," Rubin, A. B., ed., *Nauka, Moscow*, p. 412.

Sineshchekov, O. A., Kurella, G. A., Andrianov, V. A., and Litvin, F. F., 1976, Bioelectric phenomena in unicellular flagellated alga, their relation to phototaxis and photosynthesis (in Russian), *Fiziol. Rastenij*, 23:229.

Sineshchekov, O. A., Sineshchekov, V. A., and Litvin, F. F., 1978, Photo-induced bioelectric responses in phototaxis of unicellular flagellated alga (in Russion), *Dokl. Akad. Nauk S. S. S. R.*, 239:471.

Sineshchekov, O. A., Sudnitzin, V. V., and Litvin, F. F., 1984, Periodic electrical activity in unicellular flagellated alga and its possible relation to kliniphotokinesis (in Russian), *Biofizika*, 29:643.

Sineshchekov, O. A., Litvin, F. F., and Keszthelyi, L., 1990, Two components of photoreceptor potential in phototaxis of the flagellated green alga *Haematococcus pluvialis*, *Biophys. J.*, 57:33.

Smyth, R. D., Saranak, J., and Foster, K. W., 1989, Algal visual systems and their photoreceptor pigments, *Prog. Phycol. Res.*, 6:255.

Spudich, J. L., and Bogomolni, R. A., 1988, Sensory rhodopsins of Halobacteria, *Ann. Rev. Biophys. Biophys. Chem.*, 17:193.

Uhl, R., and Hegemann, P., 1990, Probing visual transduction in plant cell, *Biophys. J.*, in press.

Phototaxis and Gravitaxis in Euglena gracilis

Donat-P. Häder

Institut für Botanik und
Pharmazeutische Biologie
Friedrich-Alexander-Universität
Staudtstr. 5
D-8520 Erlangen
Germany

Introduction

Like many other motile microorganisms, the photosynthetic unicellular flagellate, *Euglena gracilis*, orients in its habitat using a number of external chemical and physical parameters (Häder, 1988; Nultsch and Häder, 1988). While some microorganisms have been found to move and orient in the water column with the aid of chemical (Berg, 1985; Macnab, 1985) and thermal (Mizuno et al., 1984; Poff, 1985) gradients, the magnetic field of the earth (Ofer et al., 1984; Frankel, 1984; Esquivel and de Barros, 1986) and even electrical currents (Mast, 1911), *Euglena* mainly orients with respect to light (Diehn et al., 1977; Häder et al., 1981, 1986b; Colombetti et al., 1982; Lenci et al., 1983) and gravity (Brinkmann, 1968; Häder, 1987a). Recent unpublished experiments failed to demonstrate responses to thermal gradients and the magnetic field of the earth.

In addition to step-up and step-down photophobic responses (Diehn, 1969; Doughty and Diehn, 1980, 1983, 1984; Shimmen, 1980) and a weak photokinetic effect (Wolken and Shin, 1958), the green flagellate, *Euglena gracilis*, controls its position in its habitat by phototactic (Bancroft, 1913; Jennings, 1904, 1906; Häder et al., 1986) and gravitactic orientation (Häder, 1987). Movement toward the light source (positive phototaxis) has been found at low white light fluence rates up to 150 - 400 lx, but the degree of orientation is rather low (Colombetti et al., 1982; Häder et al., 1981; Häder 1987). At higher fluence rates the cells orient very precisely away from the light source (negative phototactically). In the absence of a light stimulus the cells move upward (negative gravitaxis) and the antagonism between negative gravitaxis (supported by positive phototaxis under weak light conditions) and negative phototaxis causes the cells to accumulate in a band of suitable light conditions (Häder and Griebenow, 1988). This behavior has important ecological consequences, since cellular pigments of phytoplankton organisms have been found to be easily bleached by the bright light intensity at the surface of the water column (Nultsch and Agel, 1986; Häder et al., 1987, 1988; Ekelund and Häder, 1988; Rhiel et al., 1988a, b). Furthermore, the UV-B component of solar radiation has been found to affect both motility and photoorientation in *Euglena* (Häder, 1985, 1986; Häder and Häder, 1988a, b) and other photosynthetic and

Biophysics of Photoreceptors and Photomovements in Microorganisms
Edited by F. Lenci *et al.*, Plenum Press, New York, 1991

non-photosynthetic microorganisms (Häder, 1984, 1987b; Häder et al., 1986a; Häder and Häder, 1989).

Tracking Many Microorganisms in Real Time Simultaneously

While many earlier experiments on the orientation of microorganisms have been performed using population techniques, where the displacement of a whole population is monitored, real time image analysis allows to track individual organisms and to determine their deviation from the stimulus direction as well as their individual velocity.

Recent attempts allowed tracking one organism at a time over a short track segment during a predefined period of time (Häder and Lebert, 1985). Recently we have developed a new technique by which up to several hundred organisms can be tracked simultaneously (Häder and Vogel, in press).

The image of the moving organisms is recorded by a b/w CCD camera mounted on top of a dark field microscope operating with an infrared monitoring beam. Each image is digitized in real time (40 ms per frame) using 256 possible gray levels ranging from 0 (dark) to 255 (white) (Preston, 1983). The digitizer card is accomodated in an IBM AT compatible microcomputer (Tatung CS 8000, Taipei, Taiwan) and can store up to four images of 512 x 512 pixels each which can be recorded and accessed individually. A look-up table (LUT) is used to manipulate the image before storage in memory, which allows to e.g. enhance the contrast or produce a binary image. Before digital/analog (D/A) conversion the stored gray levels can again be altered using sets of LUTs which allow pseudocolor presentation of the image by mapping certain gray levels to either of the three basic color channels displayed on a suitable analog monitor.

The program, written in the computer language C, handles the input and output routines as well as the mathematical analysis of the data. All time consuming procedures such as screening, manipulation of image pixels, outline detection and calculation of centroid and area are written in ASSEMBLY Language (80286). Since for the purpose of tracking no cellular details are of interest the image is regarded as binary: all pixels above (in dark field microscope images) a predefined threshold are organism pixels, all others are regarded as background.

The algorithm starts by taking four snapshots at 80 ms intervals. Then the first of the four digitized and stored images is scanned starting from the top left corner. The outline routine (Fig. 1) is based on the chain code scheme (Freeman, 1961). The direction vectors of the eight neighbors of a pixel P under consideration are assigned numbers from 0 to 7 in a counterclockwise fashion starting with 0 to the right of the pixel. In order to find the next edge position on the circumference the neighboring pixels are tested for being edge pixels. The vector leading to the next edge pixel is stored as an element in the growing chain code. With this routine performed iteratively the whole outline is scanned until the first edge point is found again (Fig. 2). The advantage of this robust technique is that each area can be described uniquely by its edge vectors. In addition, there are several algorithms to calculate the circumference, area and centroid (Freeman, 1974, 1980). During the chain code determination all edge pixels are arbitrarily set to 255 (bright white) and serve as markers for already found organisms.

Starting from the first found edge pixel of the previously calculated organism the search is continued until the whole frame has been analyzed and all organisms have been found. The centroids and areas of all organisms are stored in an array. Upper and lower limits for the area allow to distinguish cells from debris or noise in the image.

The next frame is digitized and stored in memory about 80 ms later and the positions of all organisms are determined in a similar fashion as described above; however the image is not scanned sequentially but the positions of the organisms are searched for starting from the previous centroids. If no organism is found in the position

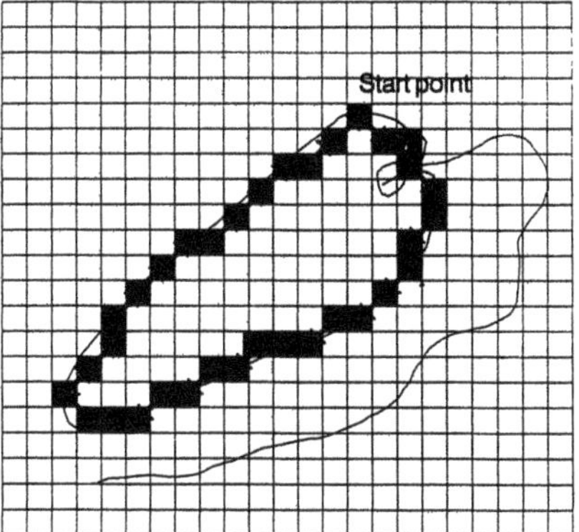

Traced contour of a cell

Chain code:

$A = a_1, a_2, \ldots a_n$

A = 7, 0, 6, 7, 6, 5, 6, 5, 5, 4, 5, 4, 4, 5, 4, 5
5, 4, 4, 1, 1, 2, 1, 1, 1, 0, 1, 1, 1, 0, 1, 1

Fig. 1. Sketch to exemplify the chain code technique to analyze the outline of an organism.

of the previous centroid the close surrounding is searched. If this is also without success the cell is regarded as lost and equally all organism colliding with the image boundaries are discarded. The positions and areas of all found organisms are stored in a second array and the process is repeated for the third and fourth digitized image, overwriting the second array. When two organisms meet, the direction of movement is no longer defined; therefore a sudden increase in the area of an organism from one image to the next is taken as a signal to discard this organism.

From the data stored in the two arrays, the movement vector can be determined for each individual organism and the value can be stored in terms of the deviation angle Á from a predetermined stimulation direction (defined as 0°). The individual speed of movement v of all cells can be calculated from the time elapsed between the first and final frame, using the built-in hardware clock of the computer. Under optimized conditions the system analyzes about 500 organisms per minute.

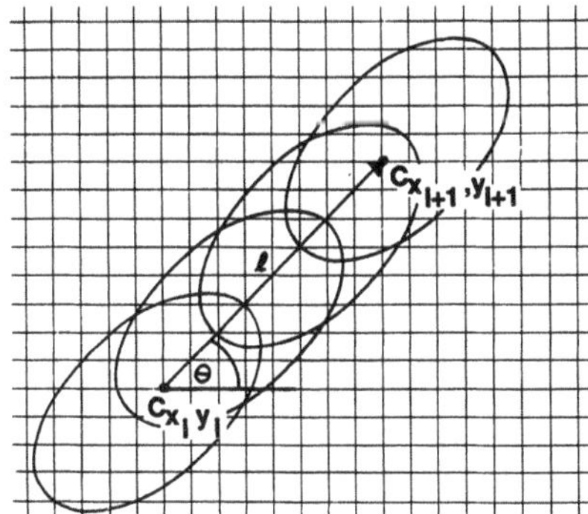

Fig. 2. Subsequent analysis of an organism outline starting from the previous centroid. The movement vector is determined from the first and final .i.centroids;.

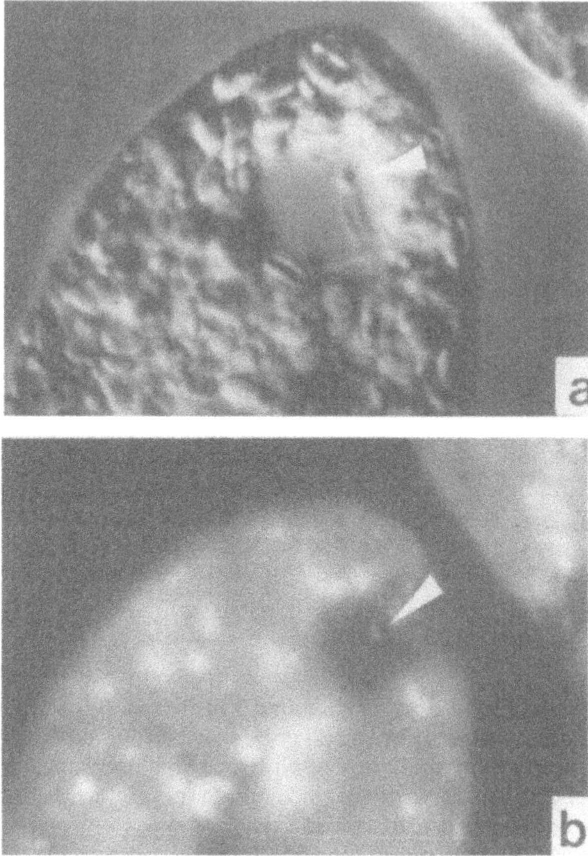

Fig. 3. Light microscopic photographs (a) of *Euglena gracilis* cells showing the PFB and the two flagellar bases. At the same place a bluish fluorescent spot is seen in fluorescence microscopy (b).

Subsequent programs are used to calculate circular histograms for the direction distribution as well as the velocity distribution using a resolution of 64 sectors of 5.6° each (Häder, 1985). In addition, a Rayleigh test is performed to determine the directedness of the moving organisms (Batschelet, 1965, 1981; Mardia, 1972). The degree of orientation for a unimodular distribution is quantified by an r-value in the range between 0 (random distribution) and 1 (precise orientation of all organisms in a single direction). The direction of movement can also be determined using a Fast Fourier Analysis which also allows smoothing of the raw histograms and Inverse Fourier Analysis reconstructs the smoothed histograms (Häder and Lipson, 1986).

Phototaxis

Mechanism of Orientation

The photoreceptor pigments for phototactic orientation in *Euglena gracilis* are thought to be localized in a swelling at the basis of the emerging flagellum, called the paraflagellar body (PFB) (Benedetti and Checcucci, 1975; Ghetti et al., 1985). The second flagellum is too short to leave the reservoir, and its tip is attached to the longer flagellum at the position of the PFB. High magnification microscopy is capable of demonstrating the paraflagellar body (PFB) even in living cells which have been immobilized by a short UV radiation (Fig. 3a). When studied by fluorescence microscopy a bluish fluorescence was found at the location of the PFB distinct from other fluorescing particles within the cell (Fig. 3b). After PFB preparation (see below), however, the fluorescence was much weaker, though still visible. When excited with a blue light beam a green flavin fluorescence can be observed.

Previous hypotheses assumed that the mechanism of orientation is based on a periodic shading (Buder, 1917; Jennings, 1906; Mast, 1911, 1941; Mast and Johnson, 1932) in which the stigma (a collection of lipid droplets stained with carotenoids, located in the cytoplasm adjacent to the reservoir) casts a shadow on the PFB. During forward locomotion the cell rotates with a frequency of about 1 or 2 Hz around its long axis (Diehn, 1969). Therefore the stigma periodically intercepts a lateral light beam before it falls on the PFB. This sudden decrase in the fluence rate experienced by the photoreceptor causes the flagellum to swing out triggered by a series of sensory transduction steps which causes the front end of the cell to turn by a certain angle toward the light source. Following this hypothesis, phototaxis is the result of repetitive photophobic reactions which after a number of iterations lead to the reorientation of the cell with respect to the light source until the front end faces the light source and the stigma no longer casts a shadow on the PFB. During negative phototaxis the reasoning is similar only that the rear end faces the light source and the intracellular organelles (plastids, mitochondria and nucleus) provide an additional shade.

However, this mechanism was found not to be valid in *Euglena*, since agents affecting the sensory transduction chain of the phobic response (Doughty and Diehn, 1982) had no effect on phototactic orientation (Häder et al., 1987). In addition, if the periodic shading mechanism were responsible for phototactic orientation, the cell population should be expected to move on the resultant between two perpendicularly oriented light beams. The experimental evidence, however, is that the population splits in two subpopulations and the cells move towards either light source (Häder et al., 1986). Furthermore, stigmaless mutants of *Euglena* were found to be capable of phototactic orientation in the light field, which strongly contradicts the shading hypothesis.

The alternative hypothesis for the orientation mechanism is based on a dichroic orientation of the photoreceptor pigments in the paraflagellar body. Electron micrographs of thin sections show a paracrystalline array in the PFB (Kivic and Vesk, 1972). At low fluence rates (2 mW m^{-2}) at 472 nm) Creutz and Diehn (1976) observed the cells to swim perpendicular to the plane of polarization.

In polarized light from above the cells move at an angle of 30° clockwise to the plane of polarization (Fig. 4) in a flat horizontal cuvette which restricts the swimming to a horizontal plane and prevents the cells from moving phototactically (Häder, 1987). When the cells were tracked in a vertical cuvette under a polarized actinic beam from above unexpected results were obtained: At some polarization planes with respect to the axis of the cuvette the cells moved downward, guided by the negative phototaxis to the strong incident light beam. However, at other polarization planes the cells moved upwards (Fig. 5). The explanation of this mysterious result is that the flat cuvette did not allow a reorientation of the cells when the flagellar plane was perpendicular to the glass planes and thus the cells were prevented from reorienting away from the light beam. At other orientation angles reorientation was mechanically not prevented. With this trick it was possible to determine the angular deviation of the main absorbing vector within the paraflagellar body from the flagellar plane to be 60° counterclockwise. Combining all results with polarized light reveals the three dimensional orientation of the absorbing vector of the photoreceptor pigments with respect to the cells' axes. Currently we develop a computer simulation of a swimming Euglena cell which involves the receptor pigment orientation and which is intended to respond to external light stimuli.

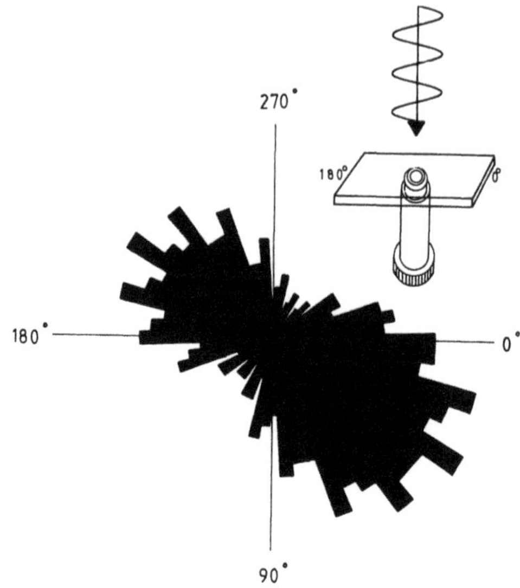

Fig. 4. Histogram of *Euglena gracilis* moving in a horizontal cuvette in linearly polarized light (0° - 180°) from above. The cells orient 30° clockwise to the polarization plane.

The photoreceptor is thought to be a flavoprotein arranged in a paracrystalline array in the paraflagellar body (PFB), a swelling of the emerging flagellum inside the reservoir (Ghetti et al. 1985; Doughty and Diehn 1980; Benedetti and Checcucci 1975). A number of action spectra have been published for the phototactic orientation of *Euglena*; however, some of them have been measured using a so-called phototaxigraph

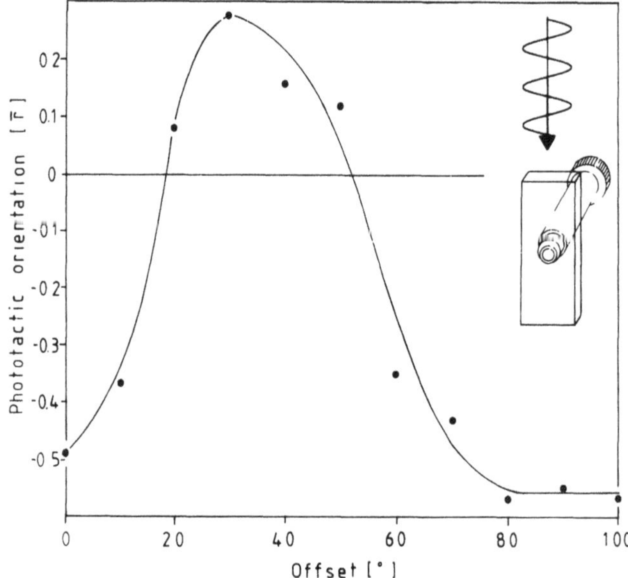

Fig. 5. Dependence of the direction of movement and degree of orientation (ordinate, r-value) of *Euglena gracilis* in a vertical cuvette on the polarization angle of the actinic light (150 W m^{-2}) impinging from above. Abscissa: clockwise offset (seen from above) of the polarization plane from the cuvette plane.

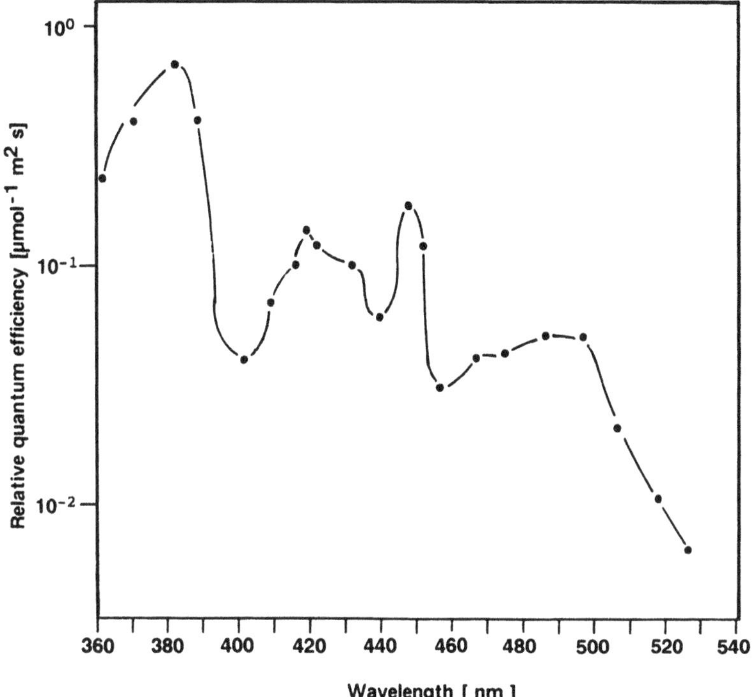

Fig. 6. Action spectrum for positive phototaxis in green *Euglena gracilis* based on fluence rate response curves determined for individual cell tracks.

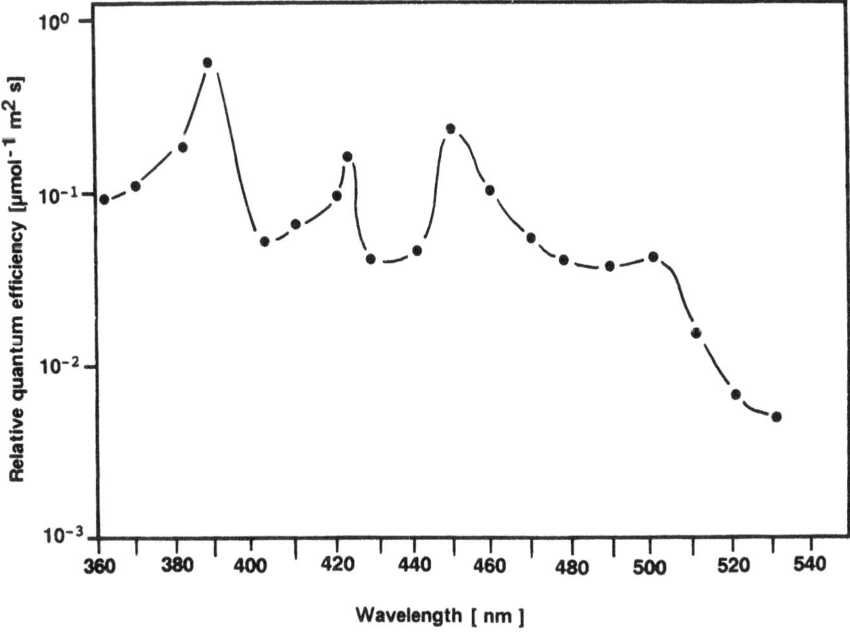

Fig. 7. Action spectrum for positive phototaxis in dark bleached *Euglena gracilis* based on fluence rate response curves determined for individual cell tracks.

(Bound and Tollin 1967; Diehn 1969) which - despite of its name - measures photoaccumulations of cells in a light beam which can be brought about by photokinesis, phobic responses, preferential settling on the glass wall in light in addition to phototaxis. All action spectra determined so far are based on mass movement techniques, assaying the behavior of a population rather than individual cells (Gössel 1957; Checcucci et al. 1975, 1976).

In order to circumvent these problems we measured an action spectrum for the positive phototactic orientation by determining the tracks of individual cells. It is based on fluence rate response curves at individual wavelengths. In a random orientation 50 % of the cells move in the two quadrants towards the light source while in perfect orientation 100 % of all cells move within these two quadrants; therefore a mean value of 75 % was chosen to calculate the action spectrum (Fig. 6). It has a major maximum at 380 nm and a second at 450 nm. Two additional maxima are found at 420 and 495 nm. In order to determine the role of the chloroplast pigments in photoperception or shading, the action spectrum was also measured for dark bleached cells (Fig. 7). It is similar to the one for green *Euglena* cells, but there is a striking difference: dark grown cells need lower fluence rates for the same response in the visible part of the spectrum. This does not hold for the ultraviolet region, where the cells need the same fluence rates as green *Euglena*.

Biochemical and Spectroscopic Analysis of the Paraflagellar Body

The chromophoric group of the photoreceptor pigment has been supposed to be a flavin based on the action spectra for photophobic responses (Doughty and Diehn, 1980), phototaxis and photoaccumulations (Diehn, 1969). In addition, microspectroscopic studies and fluorometric measurements indicated the presence of flavins (Benedetti and Lenci, 1977). Recently, Gualtieri et al. (1986) refined a method originally described by Rosenbaum and Child (1967) to isolate the complete flagella with the PFBs still attached (Gualtieri et al., 1989) from the cells by swelling the cells osmotically before the flagella are removed by a cold and calcium shock. Based on this procedure, PFBs can be isolated in large quantities to study the protein content and the chromophoric groups.

An excitation of the PFB preparation at a wavelength of 450 nm resulted in a fluorescence emission an about 520 nm indicative for the presence of flavins, which support the hypothesis described above. However, when excited at 360 nm, we observed both an emission at 520 nm plus a second peak at 450 nm indicative for pterins in their oxidized state (Galland et al., 1990). Judging from these results there are two photoreceptor chromophores in the PFB: flavins and pterins. These pigments could be arranged in sequence so that the shorter wavelength absorbing pigment operates as an antenna pigment and transfers its absorbed energy to the flavins, which in turn are the photochemically active chromophore.

Using the same isolation technique for PFBs allowed to prepare enough material for a detailed biochemical analysis: The proteins contained in the PFBs were separated by FPLC (Fast protein liquid chromatography): 500 μl samples were solubilized by 9 M urea, 2 % emulfogen, 8 mM PMSF, 0.8 % pharmalyte, 2 % mercaptoethanol at pH 8.7 and additionally sonicated (Häder and Brodhun, 1990). Isoelectric titration curves had shown previously that the isoelectric points of most flagellar proteins are above a pH of 8; therefore the sample was separated on an anion exchange column. After binding, the proteins were eluated from the column by an NaCl gradient in the range from 0 to 0.5 M and collected in 70 samples in a fraction collector.

The FPLC separation of the solubilized flagella yielded four major and a number of minor components (Fig. 8). Fluorescence emission spectra were run for all these fractions separately at the key wavelengths indicative for flavin and pterin excitation and emission, respectively (Häder and Brodhun, 1990). Three of the fractions showed

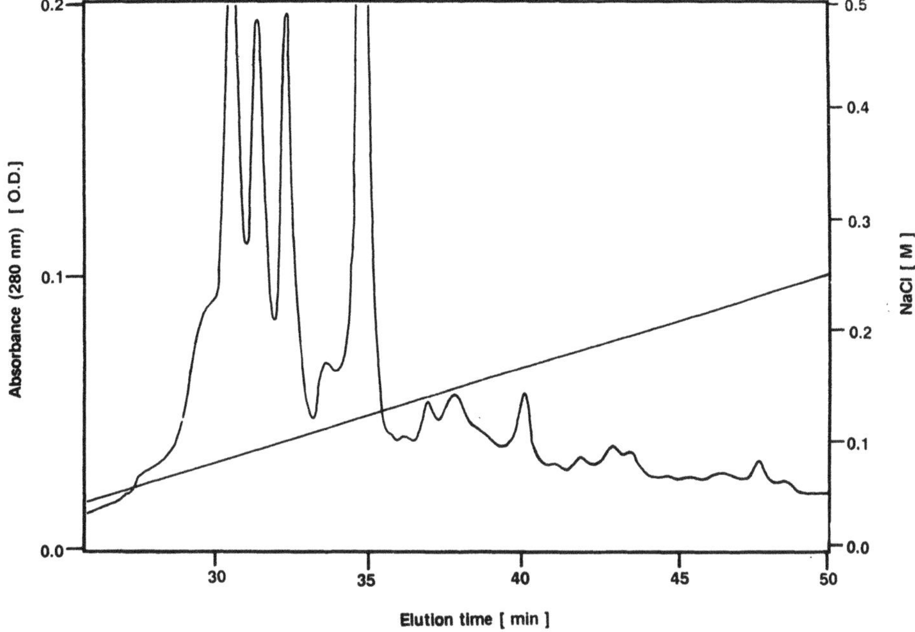

Fig. 8. Elution diagram of the FPLC separation of the solubilized PFB proteins in an anionic exchange column (MonoQ). The rising line indicates the increasing NaCl concentration to eluate the proteins from the column. Abscissa elution time [min], left ordinate relative absorbance (280 nm), right ordinate molar concentration of NaCl.

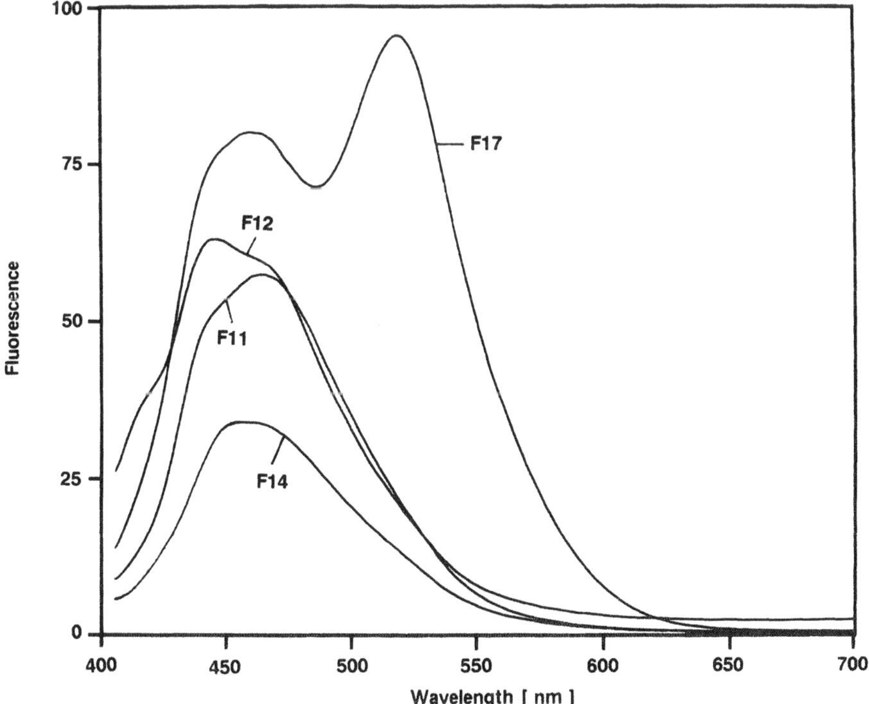

Fig. 9. Fluorescence emission spectra of the four fractions separated on the FPLC anionic exchange column excited at a wavelength of 380 nm.

maximal emission peaks at around 450 nm when excited at 380 nm; only on fraction had an additional, even higher peak at 520 nm (Fig. 9). The flavin nature of the last fraction was more obvious when excited at 450 nm: only the latter fraction showed a noticeable fluorescence at 520 nm while the other fractions did not (Fig. 10). The same results were obtained when excited at 470 nm (data not shown). These findings were supported by the excitation spectra with an emission wavelength at 520 nm where only the latter fraction showed a major absorption in the range between 400 and 500 nm as well as in the UV-A range indicative of flavins (Fig. 11).

The separated protein samples from the fractions described above were separated by gel electrophoresis (gradient 10 to 22.5% polyacrylamide) and stained by an enhanced silver staining. Two of the fractions showed major bands at apparent molecular weights of 27 and 31.6 kDa, respectively (Fig. 12). One fraction could not be analyzed by gel electrophoresis for lack of sufficient material and the remaining fraction showed two bands at 27.5 kDa and 33.5 kDa. The band with an apparent molecular weight of 27.5 kDa is identical with those of the other two fractions. The second band with a molecular weight of 33.5 kDa is not present in the two other ones and represents the protein which carries the flavin chromophore.

In order to demonstrate that the pterin fluorescence in the PFB is not mimicked by flavin degradation products which show a similar fluorescence, the extracted chromophores were compared with lumiflavin, lumichrome and riboflavin in thin layer chromatography (Fig. 13). The PFB sample consisted of two components, the R_f values of which significantly differed from those of the flavin derivatives. When compared with commercially available pterins, xanthopterin and biopterin, one component had a similar but not identical R_f value as biopterin.

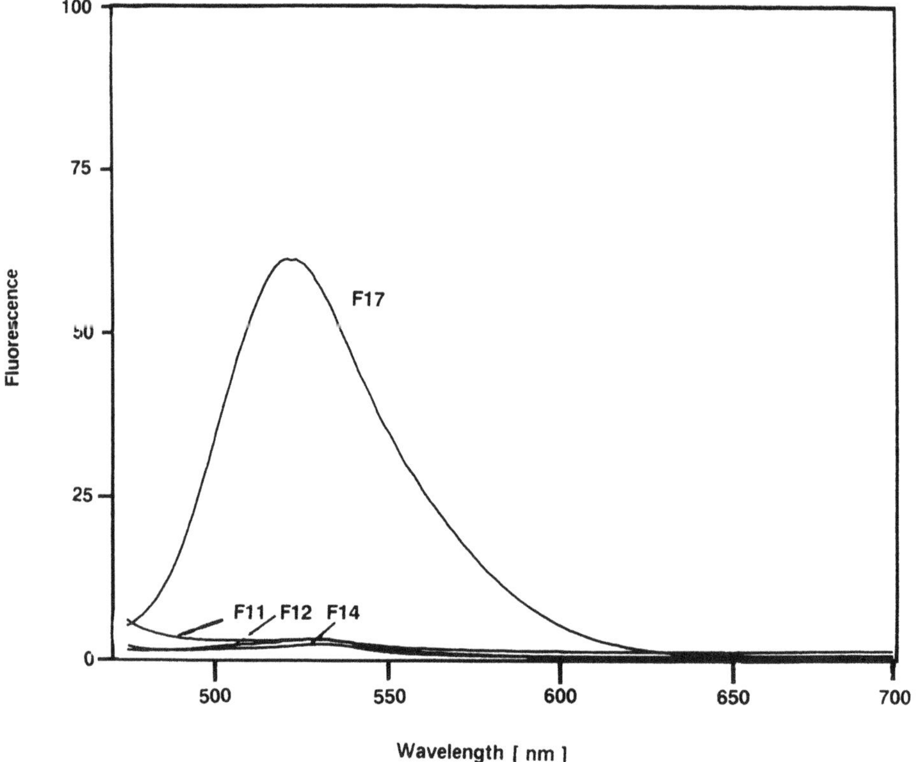

Fig. 10. Fluorescence emission spectra of the four fractions separated on the FPLC anionic exchange column excited at a wavelength of 450 nm.

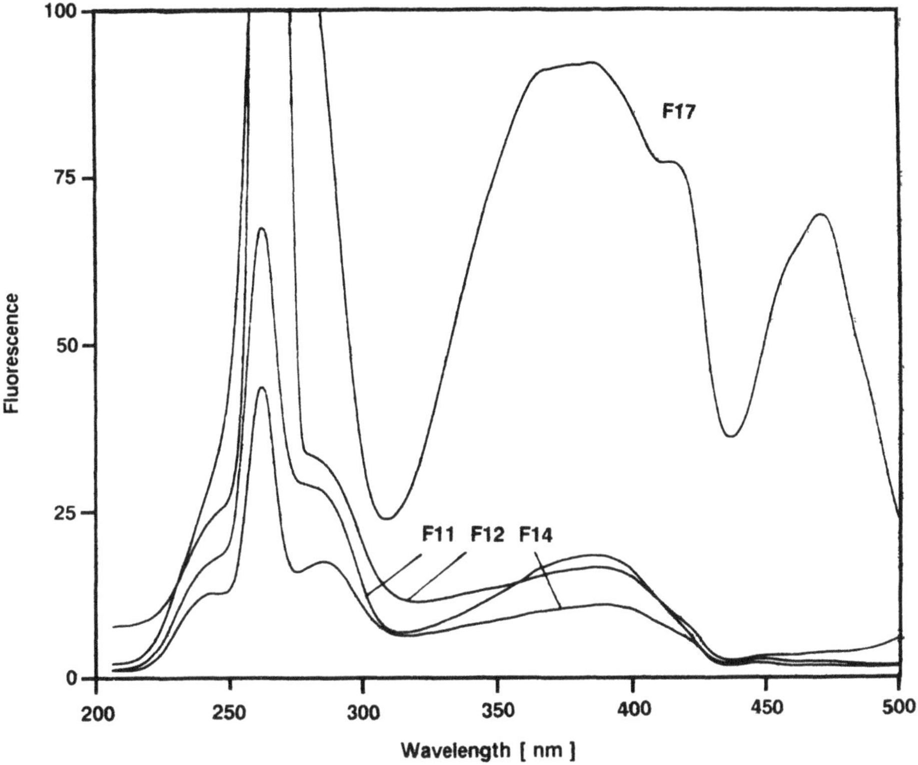

Fig. 11. Fluorescence excitation spectra of the four fractions separated on the FPLC anionic exchange column at an emission wavelength of 520 nm.

Gravitaxis

As indicated above, *Euglena* cells show a pronounced negative gravitaxis in a vertical cuvette in the absense of light. The upward component can be compensated by a light stimulus from above, to which the cells orient negative phototactically. By varying the fluence rate, the two responses can be titrated against each other (Fig. 14). The compensation point is at about 30 W m^{-2}, which is the fluence rate at the band where the cells accumulate in the water column in their natural habitat.

In contrast to the photoreceptor which has been analyzed biochemically, the gravireceptor has not been found in these cells. It has even been speculated that the graviorientation is brought about by a passive physical mechanism: when the center of gravity is in the rear, the cells will orient with their flagella upwards so that they swim to the water surface unless reoriented by other stimuli (Brinkmann, 1968). However, gravitactic orientation can be impaired by ultraviolet radiation: When exposed to a strong ultraviolet radiation the degree of orientation decreased significantly even after short exposure times (10 min) and the cells moved almost randomly after an exposure of about 90 min. This result argues in favor of an active physiological gravireceptor which is impaired by UV radiation rather than a passive physical orientation mechanism.

In order to prove that the upward movement in the water column in the absence of light is due to an orientation in the gravity vector of the earth - and not to, e.g., a temperature or gaseous gradient in the water - a sample of the cells was flown in a sounding rocket experiment (TEXUS) under microgravity conditions. During the flight the sample was under constant observation by a CCD camera and the image was

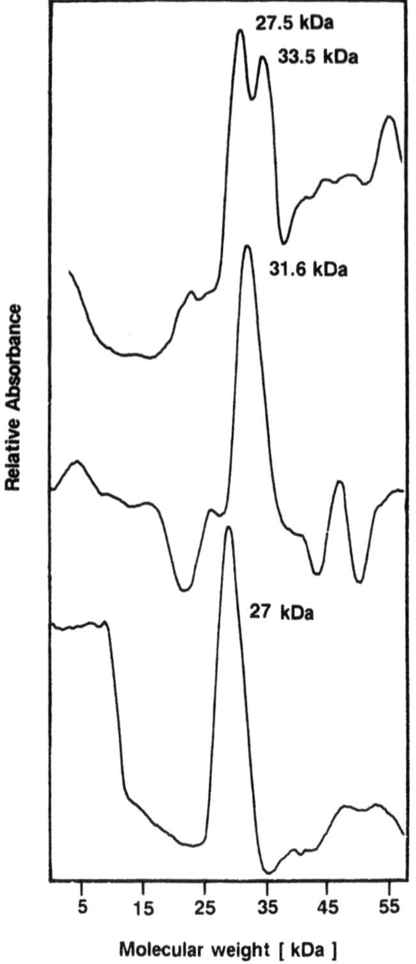

Fig. 12. Gelscans of three FPLC fractions run on SDS polyacrylamid gradient gel electrophoresis.

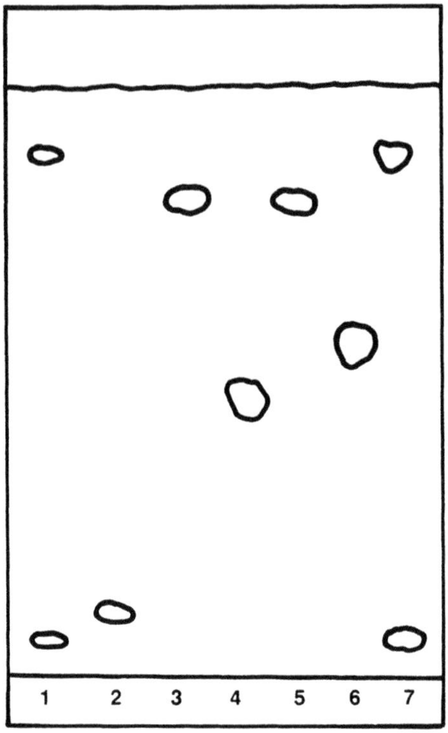

1 2 3 4 5 6 7

Fig. 13. Thin layer chromatography of PFBs extracted in methanol (1,7), biopterin (2), xanthopterin (3), riboflavin (4), lumiflavin (5) and lumichrome (6).

transmitted by a video link to a ground station where it was analyzed using the image analysis system described above. Immediately after the despin of the rocket, which stabilized the flight during its ascent, the cells were found to move randomly in the cuvette (Fig. 15). This behavior could be observed even during the first minute, indicating that there is no adapatation period. After the flight the rocket was returned to the lab and the sample module retrieved. The cells showed normal gravitaxis as before the flight.

In order to analyze the gravitactic behavior of the cells in hypergravity we used a slow rotating centrifuge microscope (NIZEMI), developed by Dornier on behalf of the BMFT (German minister for research and technology) and DLR (German aerospace organisation). This instrument is based an a Zeiss Axioplan oriented horizontally on a rotation platform. It is tested for centrifugal accelerations of up to 5 g and allows to observe the image of the moving cells by a CCD camera which transmits its image to an external screen, where it can also analyzed by real time image analysis.

Under the acceleration force in the NIZEMI the cells were capable of orienting even at higher g values. At 4.5 g most cells were still capable of moving against the ac-

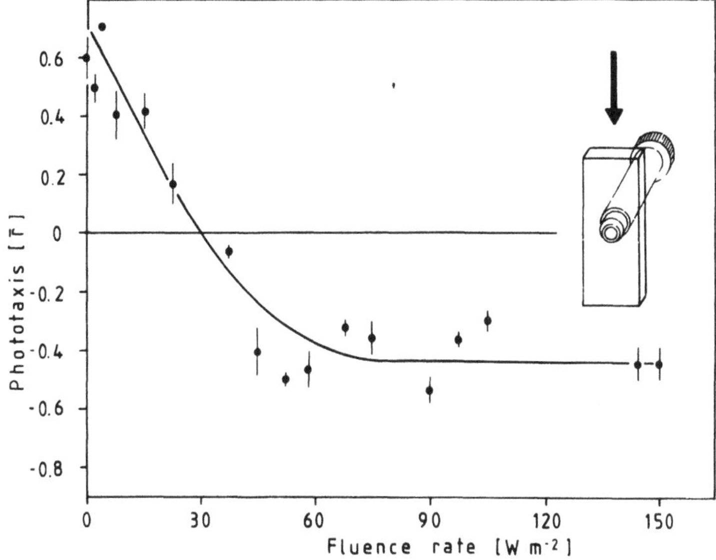

Fig. 14. Direction and degree of orientation of *Euglena gracilis* moving in a vertical cuvette under the antagonistic stimulation of the gravitational force and the light beam from above in dependence of the fluence rate.

celeration vector though the degree of orientation was rather low (Fig. 16), while at 5 g most cells moved in the direction of the centrifugal force (Fig. 17).

Even under 1 g conditions, cells swimming against the gravity vector move slower than those moving in any other direction. This effect is even more pronounced at higher acceleration forces up to 4.5 g (Fig. 18). Cells which still swim against the acceleration vector at 5 g are even slower. In contrast, under microgravity conditions in the

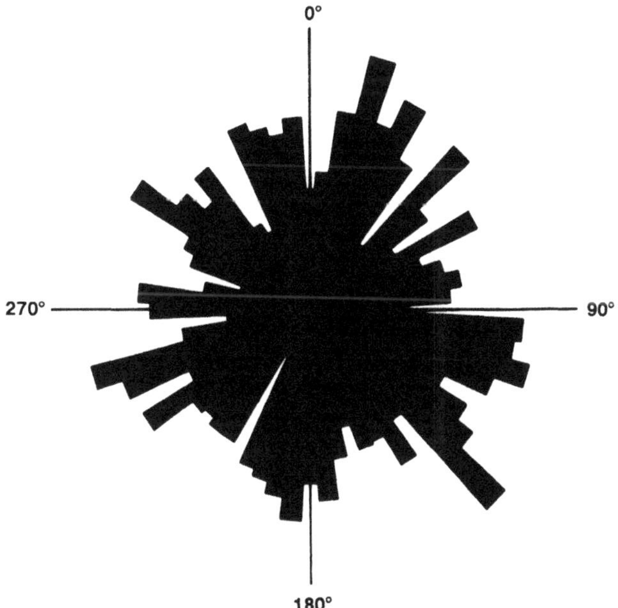

Fig. 15. Circular histogram of the movement vectors of *Euglena gracilis* during microgravity conditions.

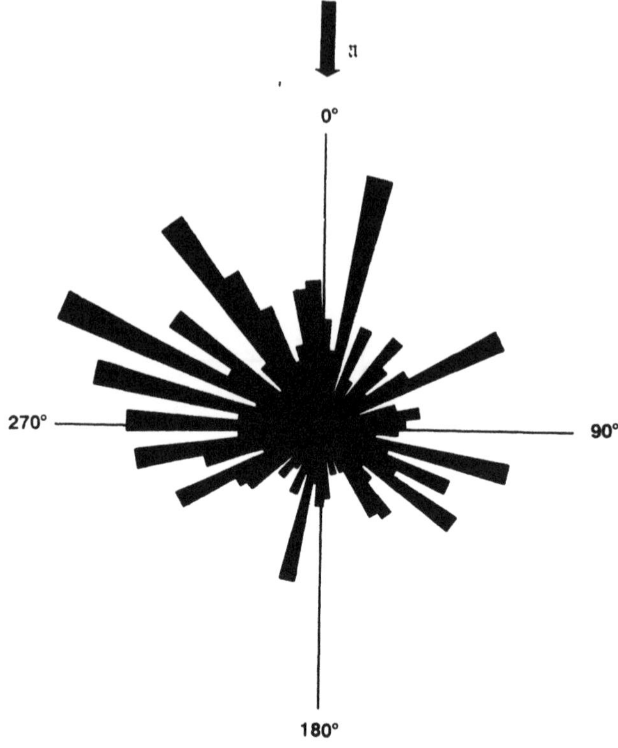

Fig. 16. Circular histogram of the movement vectors of *Euglena gracilis* of a population at an acceleration force of 4.5 g in the NIZEMI.

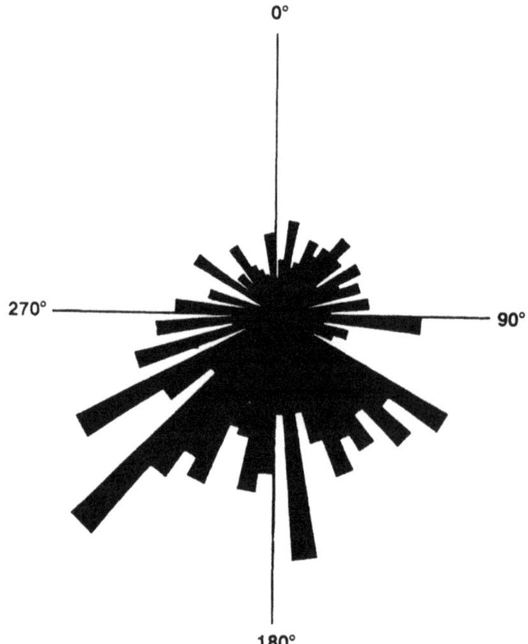

Fig. 17. Circular histogram of the movement vectors of *Euglena gracilis* of a population at an acceleration force of 5 g in the NIZEMI.

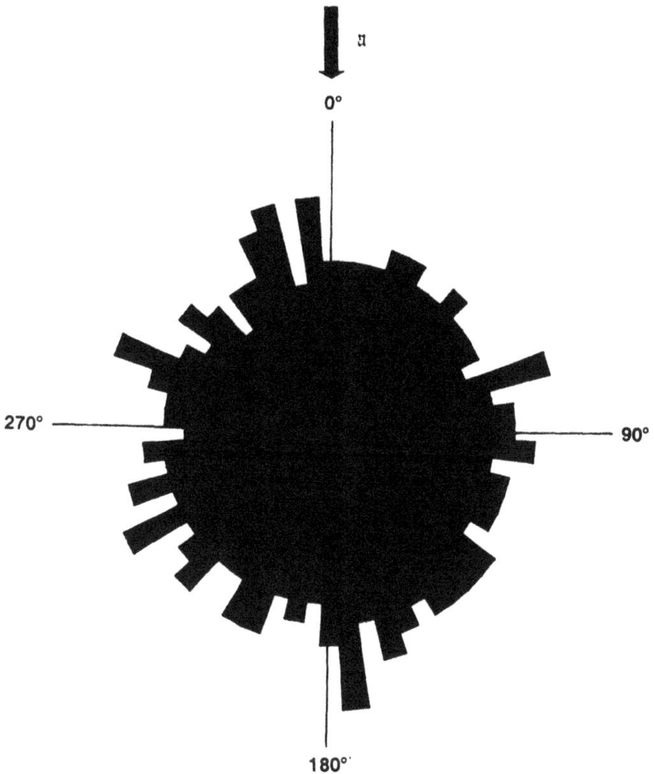

Fig. 18. Circular histogram of the velocity distribution in a population of *Euglena gracilis* at an acceleration force of 5 g in the NIZEMI.

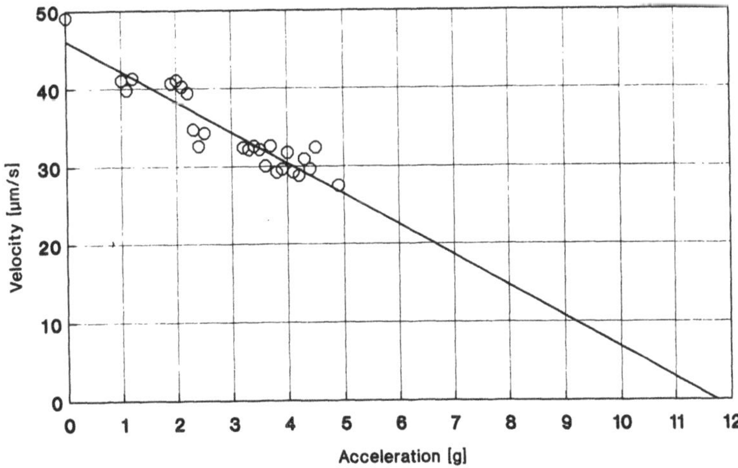

Fig. 19. Linear velocity of *Euglena gracilis* moving against the acceleration force deviating no more than ± 30° from the acceleration vector in dependence of the acceleration force.

TEXUS experiment the cells moved about 20 % faster than in the control. Figure 19 plots the linear velocity of the cells moving in a sector deviating no more than ± 30° from the direction against the gravity vector in dependence of the acceleration force.

The observed dependence of the linear velocity of the cells can be employed for an estimation of the cell's energy requirements for locomotion. The frictional component the cells encounter during forward propulsion is equal at all acceleration forces. In contrast, the cells have to supply additional energy to move against the acceleration vector which depends on the acceleration force in each experiment. The specific weight of the cells has been determined to be 1.04 g/ml and the power necessary to overcome the acceleration can be calculated from the access weight above buoyancy (0.04 g/ml) multiplied by the volume of the cell, which has been calculated to be about 2450 μm^3 multiplied by the applied acceleration force. At 1 g the energy required to swim against the gravity vector amounts to about 9.7×10^{-13} N and at 5 g to 4.9×10^{-12} N. An extrapolation of the velocity curve to zero suggests that the total energy requirement for movement at 1 g amounts to about 3.0×10^{-10} N including both the frictional component and the energy necessary to overcome the gravitational force.

Acknowledgements

This work was in part supported by financial support from the Bundesminister für Forschung und Technologie and by the Deutsche Forschungsgemeinschaft. The authors gratefully acknowledge the skilfull technical assistance of B. Brodhun, D. Grothe, O. Joop, H. Koch, P. Kyr, W. Lork, R. Treichl, I. Meyer, J. Schäfer, U. Schleier, S. Seeler and H. Vieten.

References

Bancroft, F. W., 1913, Heliotropism, differential sensibility and galvanotropism in *Euglena*, *J. Exp. Zool.*, 15:383.

Batschelet, E., 1965, Statistical methods for the analysis of problems in animal orientation and certain biological rhytms, *in*: "Animal Orientation and Navigation," Galles, S. R., Schmidt-Koenig, K., Jacobs, G. J., and Belleville, R. F., eds., Washington, NASA, pp. 61.

Batschelet, E., 1981, "Circular Statistics in Biology," Academic Press, London.

Benedetti, P. A., and Checcucci, A., 1975 Paraflagellar body (PFB) pigments studied by fluorescence microscopy in *Euglena gracilis, Plant. Sci. Lett.,* 4:47.

Benedetti, P. A., and Lenci, F., 1977, *In vivo* microspectro-fluorometry of photoreceptor pigments in *Euglena gracilis, Photochem. Photobiol.,* 26:315.

Berg, H. C., 1985, Physics of bacterial chemotaxis. *In*: "Sensory Perception and Transduction in Aneural Organisms," Colombetti, G., Lenci, F., and Song, P.-S., eds., Plenum Press, New York, London, pp. 19.

Bound, K. E., and Tollin, G., 1967, Phototactic response of *Euglena gracilis* to polarizied light, *Nature*, 216:1042.

Brinkmann, K., 1968, Keine Geotaxis bei *Euglena*, *Z. Pflanzenphysiol.,* 59:12.

Buder, J., 1917, Zur Kenntnis der phototaktischen Richtungsbewegungen, *Jahrb. Wiss. Bot.,* 58:105.

Checcucci, A., Favati, L., Grassi, S., and Piaggesi, T., 1975, The measurement of phototactic activity in *Euglena gracilis, Klebs. Monit. Zool. Ital.,* 9:83.

Checcucci, A., Colombetti, G., Ferrara, R., and Lenci, F., 1976, Action spectra for photoaccumulation of green and colorless *Euglena*: evidence for identification of receptor pigments, *Photochem. Photobiol.,* 23:51.

Colombetti, G., Häder, D.-P., Lenci, F., and Quaglia, M., 1982, Phototaxis in *Euglena gracilis*: effect of sodium azide and triphenylmethyl phosphonium ion on the photosensory transduction chain, *Curr. Microbiol.,* 7:281.

Creutz, C., and Diehn, B., 1976, Motor responses to polarized light and gravity sensing in *Euglena gracilis*, *J. Protozool.*, 23:552.

Diehn, B., 1969, Action spectra of the phototactic responses in *Euglena*, *Biochim. Biophys. Acta*, 177:136.

Diehn, B., Feinleib, M., Haupt, W., Hildebrand, E., Lenci, F., and Nultsch, W., 1977, Terminology of behavioral responses of motile microorganisms, *Photochem. Photobiol.*, 26:559.

Doughty, M. J., and Diehn, B., 1980, Flavins as photoreceptor pigments for behavioral responses, *Structure and Bonding*, 41:45.

Doughty, M. J., and Diehn, B., 1982, Photosensory transduction in the flagellated alga, *Euglena gracilis*. III. Induction of Ca^{2+}-dependent responses by monovalent cation ionophores, *Biochim. Biophys. Acta*, 682:32.

Doughty, M. J., and Diehn, B., 1983, Photosensory transduction in the flagellated alga, *Euglena gracilis*. IV. Long term effects of ions and pH on the expression of step-down photobehavior, *Arch. Microbiol.*, 134:204.

Doughty, M. J., and Diehn, B., 1984, Anion sensitivity of motility and step-down photophobic responses of *Euglena gracilis*, *Arch. Microbiol.*, 138:329.

Ekelund, N., and Häder, D.-P., 1988, Photomovement and photobleaching in two *Gyrodinium* species, *Plant Cell Physiol.*, 29:1109.

Esquivel, D. M. S., and de Barros, H. G. P. L., 1986, Motion of magnetotactic microorganisms, *J. Exp. Biol.*, 121:153.

Frankel, R. B., 1984, Magnetic guidance of organisms, *Annu. Rev. Biophys. Bioeng.*, 13:85.

Freeman, H., 1961, On the Encoding of Arbitrary Geometric Configurations, *IRE Trans EC*-10:260.

Freeman, H., 1974, Computer Processing of Line-Drawing Images, *Computing Surveys*, 6:57.

Freeman, H., 1980, "Analysis and Manipulation of Lineal Map Data. Map Data Processing," Academic Press, pp. 151.

Galland, P., Keiner, P., Dörnemann, D., Senger, H., Brodhun, B., and Häder, D.-P., 1990, Pterin- and flavin-like fluorescence associated with isolated flagella of *Euglena gracilis*, *Photochem. Photobiol.*, 51:675.

Ghetti, F., Colombetti, G., Lenci, F., Campani, E., Polacco, E., and Quaglia, M., 1985, Fluorescence of *Euglena gracilis* photoreceptor pigment: an *in vivo* microspectrofluorometric study, *Photochem. Photobiol.*, 42:29.

Gössel, I., 1957, Über das Aktionsspektrum der Phototaxis chlorophyllfreier Euglenen und über die Absorption des Augenflecks, *Arch. Mikrobiol.*, 27:288.

Gualtieri, P., Barsanti, L., and Rosati, G., 1986, Isolation of the photoreeptor (paraflagellar body) of the phototactic flagellate *Euglena gracilis*, *Arch. Microbiol.*, 145:303.

Gualtieri, P., Passarelli, V., and Barsanti, L., 1989, *In vivo* microspectrophotometric investigation of *Blepharisma japonicum*, *J. Photochem. Photobiol. B: Biol.*, 3:379.

Häder, D.-P., 1984, Effects of UV-B on motility and photoorientation in the cyanobacterium, *Phormidium uncinatum*, *Arch. Microbiol.*, 140:34.

Häder, D.-P., 1985, Effects of UV-B on motility and photobehavior in the green flagellate, *Euglena gracilis*, *Arch. Microbiol.*, 141:159.

Häder, D.-P., 1986, Effects of solar and artificial UV irradiation on motility and phototaxis in the flagellate, *Euglena gracilis*, *Photoch. Photobiol.*, 4:651.

Häder, D.-P., 1987a, Polarotaxis, gravitaxis and vertical phototaxis in the green flagellate, *Euglena gracilis*, *Arch. Microbiol.*, 147:179.

Häder, D.-P., 1987b, Effects of UV-B irradiation on photomovement in the desmid, *Cosmarium cucumis*, *Photochem. Photobiol.*, 46:121.

Häder, D.-P., 1988, Ecological consequences of photomovement in microorganisms, *J. Photochem. Photobiol. B: Biol.*, 1:385.

Häder, D.-P., and Brodhun, B. 1990, Photoreceptor proteins and pigments in the paraflagellar body of the flagellate, *Euglena gracilis*, *Photochem. Photobiol.*, 52:865.

Häder, D.-P., and Griebenow, K., 1988, Orientation of the green flagellate, *Euglena gracilis*, in a vertical column of water, *FEMS Microbiol. Ecol.*, 53:159.

Häder, D.-P., and Häder, M. A., 1988a, Inhibition of motility and phototaxis in the green flagellate, *Euglena gracilis*, by UV-B radiation., *Arch. Microbiol.*, 150:20.

Häder, D.-P., and Häder, M., 1988b, Ultraviolet-B inhibition of motility in green and dark bleached *Euglena gracilis*, *Curr. Microbiol.*, 17:215.

Häder, D.-P., and Häder, M. A., 1989, Effects of solar UV-B irradiation on photomovement and motility in photosynthetic and colorless flagellates, *Env. Exp. Bot.*, 29:273.

Häder, D.-P., and Lebert, M., 1985, Real time computer-controlled tracking of motile microorganisms, *Photochem. Photobiol.*, 42:509.

Häder, D.-P., and Lipson, E., 1986, Fourier analysis of angular distributions for motile microorganisms, *Photochem. Photobiol.*, 44:657.

Häder, D.-P., and Vogel, K., 1990, Simultaneous tracking of flagellates in real time by image analysis, *J. Math. Biol.*, in press.

Häder, D.-P., Colombetti, G., Lenci, F., and Quaglia, M., 1981, Phototaxis in the flagellates, *Euglena gracilis* and *Ochromonas danica*. *Arch. Microbiol.*, 130:78.

Häder, D.-P., Lebert, M., and DiLena, M. R., 1986a, New Evidence for the mechanism of phototactic orientation of *Euglena gracilis*, *Current Microbiol.*, 14:157.

Häder, D.-P., Watanabe, M., and Furuya, M., 1986b, Inhibition of motility in the cyanobacterium, *Phormidium uncinatum*, by solar and monochromatic UV irradiation, *Plant Cell Physiol.*, 27:887.

Häder, D.-P., Rhiel, E., and Wehrmeyer, W., 1987, Phototaxis in the marine flagellate *Cryptomonas maculata*, *J. Photochem. Photobiol.*, 1:115.

Häder, D.-P., Rhiel, E., and Wehrmeyer, W., 1988, Ecological consequences of photomovement and photobleaching in the marine flagellate *Cryptomonas maculata*, *FEMS Microbiol. Ecol.*, 53:9.

Jennings, H. S., 1904, Reactions to light in ciliates and flagellates, *in*: "Contributions to the study of the behavior of microorganisms," Carnegie Inst. Washington, Washington, pp. 29.

Jennings, H. S., 1906, "The behavior of lower organisms," 1962 edn., Indiana Univ. Press, pp. 134.

Kivic, P. A., and M. Vesk, 1972, Structure and function in the euglenoid eyespot apparatus: the fine structure, and response to environmental changes, *Planta*, 105:1.

Lenci, F., Colombetti, G., and Häder, D.-P., 1983, Role of flavin quenchers and inhibitors in the sensory transduction of the negative phototaxis in the flagellate, *Euglena gracilis*, *Current Microbiol.*, 9:285.

MacNab, R. M., 1985, Biochemistry of sensory transduction in bacteria, *in*: "Sensory Perception and Transduction in Aneural Organisms," Colombetti, G., Lenci, F., and Song, P.-S. eds., Plenum Press, New York, London, pp. 31.

Mardia, K. V., 1972, "Statistics of Directional Data," Academic Press, London.

Mast, S. O., 1911, "Light and Behavior of Organisms," John Wiley & Sons, New York; Chapman & Hall, London.

Mast, S. O., 1941, Motor response in unicellular organisms, *in*: "Protozoa in Biological Research," Calkins, G. N., and Summer, F. M., eds., Columbia University Press, New York, pp. 271.

Mast, S. O., and Johnson, P. L., 1932, Orientation in light from two sources and its bearing on the function of the eyespot, *Z. Vergl. Physiol.*, 16:252.

Mizuno, T., Maeda, K., and Imae, Y., 1984, Thermosensory transduction in *Escherichia coli*, *in*: "Transmembrane Signaling and Sensation," Oosawa, F., Yoshioka, T., and Hayashi, H., eds., Japan Scientific Society Press, Tokyo and VNU Sci. Press BV, Netherlands, pp. 147.

Nultsch, W., and Agel, G., 1986, Fluence rate and wavelength dependence of photobleaching in the cyanobacterium *Anabaena variabilis*, *Arch. Microbiol.*, 144:268.

Nultsch, W., and Häder, D.-P., 1988, Photomovement in motile microorganisms II, *Photochem. Photobiol.*, 47:837.

Ofer, S., Nowik, I., Bauminger, E. R., Papaefthymiou, G. C., Frankel, R. B., and Blakemore, R. P., 1984, Magnetosome dynamics in magnetotactic bacteria, *J. Biophys.*, 46:57.

Poff, K. L., 1985, Temperature sensing in microorganisms, *in*: "Sensory Perception and Transduction in Aneural Organisms," Colombetti, G., Lenci, F., and Song, P.-S., eds., Plenum Press, New York, London, pp. 299.

Preston, K. Jr., 1983, Gray level image processing by cellular logic transforms. IEEE transactions on pattern analysis and machine intelligence, Vol. Pami-5:55.

Rhiel, E., Häder, D.-P., and Wehrmeyer, W., 1988a, Photo-orientation in a freshwater *Cryptomonas* species, *J. Photochem. Photobiol. B: Biol.*, 2:123.

Rhiel, E., Häder, D.-P., and Wehrmeyer, W., 1988b, Diaphototaxis and gravitaxis in a freshwater *Cryptomonas*, *Plant Cell Physiol.*, 29:755.

Rosenbaum, J. L., and Child, F. M., 1967, Flagellar regeneration in protozoan flagellates, *J. Cell. Biol.*, 34:345.

Shimmen, T., 1980, Quantitative studies on step-down photophobic responses of *Euglena* in an individual cell, *Protoplasma*, 106:37.

Wolken, J. J., and Shin, E., 1958, Photomotion in *Euglena gracilis*. I. Photokinesis, II. Phototaxis, *J. Protozool.*, 5:39.

Photoreception in Chlamydomonas

Peter Hegemann

Max-Planck-Institut für Biochemie
8033 Martinsried
Germany

Introduction

When the unicellular photosynthetic alga *Chlamydomonas* is faced with changes in direction, wavelength or intensity of the ambient light, it exhibits distinct behavioral responses, which are direction changes if the changes are small or stop responses upon larger changes. At continuous irradiation *Chlamydomonas* orients to or away from the light source (positive or negative phototaxis). These movement responses have been under investigation for more than 100 years (Pfeffer, 1904; Nultsch and Häder, 1988) but neither the photoreceptor nor the signal transduction process have been characterized in detail.

Rhodopsin Serves as the Photoreceptor for Photomovement Responses

Until very recently the nature of the photoreceptor controlling photomovement responses in *Chlamydomonas* has been discussed exclusively on the basis of action spectra. When constructed from phototaxis measurements at high photon irradiance (equal response spectra) they exhibit two maxima at 503 and 440 nm (Nultsch et al., 1971) leading to the suggestion that flavins or carotenoids may be the functional photoreceptor.

Foster and Smyth (1980) argued on the basis of physical principles that low intensity action spectra much better reveal the absorption of the functional photoreceptor as do high intensity spectra. Threshold action spectra for phototaxis are suggestive of rhodopsin and peak at 503 nm, very similar to the absorption spectra of rhodopsin from higher animals. The fact that the maximum at 440 nm is missing may be caused either by a second photoreceptor present in much smaller amounts or by a photochromism of only one photoreceptor. Further evidence for the rhodopsin has been provided from experiments with FN68 mutant cells, which are almost blind if grown in darkness and phototactic only at very high photon irradiance. Full phototactic sensitivity of these blind cells can be reconstituted, however, by supplementation of the cells with retinal or analog compounds (Foster et al., 1984).

Biophysics of Photoreceptors and Photomovements in Microorganisms
Edited by F. Lenci *et al.*, Plenum Press, New York, 1991

More than 30 analogs have been tested for their ability to restore phototaxis in this mutant. In the first series of experiments the authors showed that 11-*cis* and all-*trans* retinal result in action spectra maxima very similar to that of wild type cells (505 and 501 nm) whereas the action spectrum in the presence of 9-*cis* retinal is slightly shifted to the blue (488 nm). Retinal analogs with shorter conjugated polyene chain (dihydro compounds) shifted the action spectrum more drastically to shorter wavelength. The retinal dependence of the phototactic sensitivity together with the shift of the action spectrum maximum by retinal analogs led to the conclusion that rhodopsin is the functional photoreceptor for phototaxis (Foster et al., 1984).

Later phototaxis was tested by using retinal analogs that can not isomerise around certain double bonds or have very short polyene chains (Foster et al., 1988, 1989). Surprisingly most of the compounds readily restored phototaxis. From these reconstitution experiments several conclusions were derived about the properties of *Chlamydomonas* rhodopsin and its light activation: A) The aldehyde function is essential for the formation of an active rhodopsin. B) If isomerization is required for rhodopsin activation, it must be non-regiospecific. C) Neither detachment of the chromophore from the binding site (bleaching) nor protonation of a C=N bond is required for restoration of phototaxis. D) Phototaxis is restored by aldehydes, which have only one or no C=C bond as they are hexenal or hexanal. Since it is well known that specific double bound isomerizations accompany ion pumping in bacteriorhodopsin and halorhodopsin (Oesterhelt et al., 1985) and signal transduction in the visual process, the possibility that mechanisms other than isomerization may activate a rhodopsin make the *Chlamydomonas* rhodopsin especially interesting. An attractive hypothesis was that *Chlamydomonas* rhodopsin activation might be triggered by charge redistribution in the excited state of the chromophore and not by isomerization (Nakanishi et al., 1989).

Originally 11-*cis* retinal was favored as the natural retinal isomer, because the final phototactic sensitivity with 11-*cis* was about three times higher than that with all-*trans* retinal (Foster et al., 1984). However, by using retinal analogs the all-*trans* isomers showed the same and sometimes even higher sensitivity as the 11-*cis* isomers (Nakanishi et al., 1989), so that the question about the natural isomer cannot be answered from these reconstitution experiments alone.

The rhodopsin is easily accessible from the extracellular medium by hydroxylamine, which suppresses phototaxis in wild type cells in a dose-dependent manner. Phototaxis, however, can be restored upon washing the cells and addition of retinal (Hegemann et al., 1988)

Later flash-induced movement responses as they are direction changes or stop responses have also been analyzed in detail (Hegemann and Marwan, 1988; Uhl and Hegemann, 1990). They are retinal dependent, they show rhodopsin action spectra with a maximum at 495 nm and are sensitive to hydroxylamine, suggesting that flash-induced responses and phototaxis are controlled by the same or very similar rhodopsin type photoreceptors.

The physiological experiments described established rhodopsin as the functional photoreceptor for photomovement responses in *Chlamydomonas* but several questions about the mechanism of rhodopsin activation arose and have to be answered in the future: 1) Why is the phototactic sensitivity in cells that have been reconstituted with naphthylretinals higher than in those reconstituted with normal retinals? Why is the sensitivity with the short chained analogs as high as with retinal itself although a chromophore with the former should have a much lower extinction coefficient? 2) Does retinal isomerize in *Chlamydomonas* rhodopsin under physiological conditions although this isomerization seems not to be essential for its function?. 3) Are stop response and phototaxis controlled by the same rhodopsin or by two different rhodopsin species?

Retinal Analysis

Since wild type cells contain large amounts of carotenoids in the eyespot and in the photosynthetic membranes (Krinsky and Levine, 1964) the identification of retinoids in wild type cells has been always difficult. The blind mutant FN68, however, is virtually free of carotenoids, the retinoid deficiency being caused by a defect in the carotenoid pathway at an early state. Since the phototactic sensitivity can be recovered by irradiation with green light (Foster et al., 1988a), FN68 cells were ideal for retinal extraction. All-*trans* retinal but no 11-*cis* isomer has been identified in extracts of these resensitized cells (Beckmann and Hegemann, 1991). This finding strongly favors all-*trans* as the functional isomer in the rhodopsin-binding site and not 11-*cis* as in animal rhodopsin. The amount of retinal increased up to 30000 molecules per cell before the sensitivity had been fully restored. Later all-*trans* retinal has also been identified in wild type strain 806 cells that have been grown at low light in order to reach the maximal phototactic sensitivity (Hegemann et al., 1988). The HPLC diagram of wild type cells is distorted by large amounts of carotenoids, but retinal is detectable by online absorption spectroscopy. The final concentration again corresponded roughly to 30000 molecules per cell (Hegemann et al., in preparation).

In Vitro Identification of Rhodopsin

For the spectroscopic identification of rhodopsin, cells from a 20 l culture were collected and subsequently lysed under nitrogen pressure. The membranes were separated on a succrose step gradient and three yellow and one dominant green fraction were collected for further purification. Only one yellow fraction (S2) contained significant amounts of retinal. This S2 fraction was washed under mild conditions with ß-octylglycoside. The resulting membranes were free of carotenoids and photosynthetic material, the absorption resembled that of a rhodopsin and showed a maximum at 495 nm. The absorption in the visible range, however, had a fine structure which is extremely unusual for a rhodopsin. It has been observed only very recently in phoborhodopsin, a photoreceptor for the repellent photomovement responses in Halobacteria (Takahashi et al., 1990). The rhodopsin in *Chlamydomonas* membranes can be bleached by continuous green light (500 nm) leading to the appearence of free retinal. The corresponding bleaching difference spectra still exhibit fine structure. The clear isosbestic point suggested that the absorption difference results from a single rhodopsin species. The rhodopsin is not stable in these detergent-washed membranes and decays during a few days even at 4° C. However, it can be reconstituted by addition of all-*trans* retinal. From these reconstitution experiments the maximal extinction coefficient was determined to be near 50000 M^{-1} cm^{-1}. Some reconstitution was also found with 11-*cis* retinal, whereas 13-*cis* and 9-*cis* were almost uneffective (Beckmann and Hegemann, unpublished observation). These experiments again favor all-*trans* as the natural isomer.

Reconstitution with ^{3}H-retinal followed by reduction with cyanoborohydride identified only one labelled retinal protein with a molecular weight of 32 kDa in the final membrane praparation (Beckmann and Hegemann, 1991). We concluded that this is the photoreceptor that functions as the photoreceptor for photomovement responses in *Chlamydomonas.*

Properties of the Rhodopsin Chromophore

In contrast to smooth bell-shaped absorption spectra, fine structured spectra allow certain conclusions about the retinylidene conformation. A detailed analysis has been presented for phoborhodopsin by Takahashi et al. (1990). The major general arguments should hold true also for *Chlamydomonas* rhodopsin: It is generally accepted that the major causes of spectroscopic broadening in retinal and related compounds in solution are a distribution of vibronic activity among a large number of excited state vibrational modes and/or a conformational heterogeniety of the ground state. Both arise primarily from a distorted *cis* conformation around the C6-C7 single bond (Warshel and Karplus, 1974). Therefore the development of fine structure may serve as an indication for a forced coplanar conformation around the C6-C7 bond. A planar conformation is expected to be 6-s-*trans* because a steric hindrance argues against a planar 6-s-*cis* geometry. Assuming that the spectroscopically identified rhodopsin is the photoreceptor for phototaxis, strong support for the ring/chain coplanarity may be derived from phototaxis experiments with "naphthylretinal" (3,7-dimethy-12,4,6-heptatrienal) in which the bond corresponding to the C6-C7 bond in retinal is already fixed to form a planar conformation. As already mentioned, all-*trans* and 11-*cis* naphthylretinal regenerate the phototaxis to a level above that reached with normal retinals (Nakanishi et al., 1989). The planar binding site apparently accelerates naphthylretinal incorporation and explains the high light sensitivity in their presence. In visual pigments further distortion of the chromophore planarity arises from the steric interaction between the C13 methyl group and C10 proton resulting from the C11=C12 *cis* retinal configuration (Warshel and Karplus, 1974; Smith et al., 1987). Therefore fine-structured absorption spectra point to a retinal chromophore with all-*trans* conformation. This argument lends further support to the notion that the natural chromophore of *Chlamydomonas* is all-*trans* as already concluded from retinal analysis and reconstitution experiments.

Although an all-*trans* and 6-s-*trans* conformation is a prerequisite for a rigid chromophore, it does not necessarily imply the existence of a fine structured absorption spectrum. Bacteriorhodopsin, which meets these two requirements (van der Steen et al., 1986), is an example. In bacteriorhodopsin (maximum at 568 nm) as well as in the light-driven chloride pump halorhodopsin (maximum at 578 nm) and in sensory rhodopsin I (maximum at 588 nm), all from Halobacteria, the strong red shift (opsin shift) indicates heavy chromophore-opsin interactions. These are responsible for the inhomogenious broadening of the absorption spectra. Since the opsin shift of *Chlamydomonas* rhodopsin is much lower, the protein-chromophore interaction has to be weak similar to that in phoborhodopsin. The location of the negative charges provided by the protein has been probed by action spectroscopy with a series of dihydroretinals. The sharp change in the opsin shift observed between trienals and dienals (7,8- and 9,10-dihydroretinal) suggests that one negative charge is near the N-terminus of the conjugated system) and no charge is located near the C5 position of the retinal ring (Foster et al., 1988b).

Calcium is Essential for Signal Transduction

Phototaxis measurements and analysis of the stop response (Schmidt and Eckert, 1976; Dolle et al., 1986, Hegemann and Bruck, 1989) together with observations on permeabilized cells (Kamyia and Witman, 1984) at various calcium concentrations gave conclusive evidence that calcium directly controls the beating pattern of the flagella and that calcium plays an important role for the signal transduction from the photoreceptor to the flagella in *Chlamydomonas*. The role of calcium found further support by the suppression of phototaxis and stop responses in the presence of calcium channel inhibitors (Nultsch et al., 1986; Hegemann et al., 1990). Calcium, however, not only plays a role in the primary signal transduction process but also during adaptation of the cells to continuous light or to repetitive flash stimulation (Hegemann and Bruck, 1989). Direct measurements of the ^{45}Ca uptake have been precluded in wild type cells by the cell wall but have been possible in cell wall defective 806 strain cells (Hegemann et al., 1990). Verapamil, a selective inhibitor for L-type calcium channels completely suppres-

Tab. 1 Properties of *Chlamydomonas* rhodopsin

maximal absorbance	495 nm
extinction coefficient	50000 M^{-1} cm^{-1}
functional isomer	all-*trans*
C6-C7 bond	planar and *trans*
negative charge	close to the C=N-bond
molecular mass	32 KDa

sed the calcium uptake in light even below the dark level. ^{45}Ca measurements, however, suffer from low time resolution, and the calcium uptake during a stop response is too small to be detected.

Electrical Recording of Ion Channel Activation

The disadvantages of ^{45}Ca have been overcome by application of the cell attached patch clamp methodology, first to *Haematococcus* (Litvin et al., 1978), and recently to *Chlamydomonas* (Harz and Hegemann, 1991). Upon stimulation with green flashes two clearly separated calcium currents are detectable in two small distinct areas of the cell. The first, appearing between 0.1 and 5 ms after a flash, is localized in the eyespot region and is called "photoreceptor current". Its amplitude can be titrated with light and shows a rhodopsin action spectrum centered at 494 nm, which is very close to the absorption spectrum of the rhodopsin in membrane preparations. The second photocurrent originates in the flagellar region and follows the photoreceptor current with a delay of several milliseconds. This delay decreases at high flash energies whereas the amplitude stays unchanged. Below a certain threshold the flagellar current disappears abruptly similar to an action potential. How the two currents are functionally related is completely unclear at present.

The Ionic Composition of the Medium Controls the Cellular Sensitivity

Besides these direct current measurements, more indirect conclusions about the ionic requirements for movement responses can be drawn from light scattering measurements that detect with micro-second time resolution structural changes in the cell and alterations of the flagellar beating pattern (Uhl and Hegemann, 1990): Using this assay it can be demonstrated that potassium also plays an important role in transmembrane signalling. Potassium is the major ion carrying the membrane potential. At least 10 μM potassium have to be present during cell growth to maintain the intracellular calcium level and a stable membrane potential, but during the light response po-

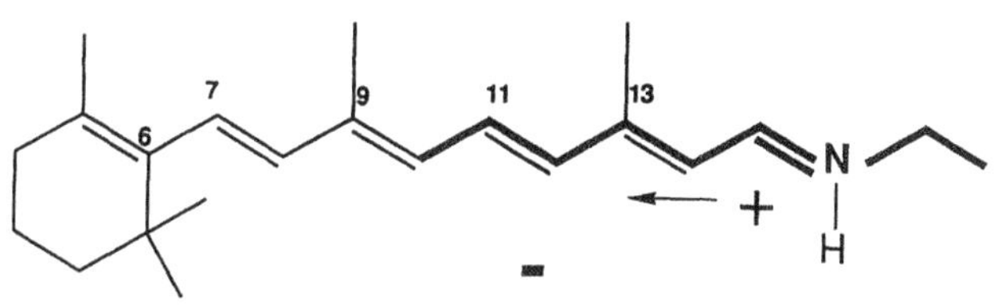

Fig. 1. Structure of the *Chlamydomonas* rhodopsin chromophore as derived from action and absorption spectroscopy. The bold part indicates structural features that have been present in all retinal compounds that form a functional chromophore.

tassium is not required. Monovalent, divalent and especially tri- or tetravalent cations have a strong influence on the cellular surface potential thus drastically reducing the light sensitivity at nM to mM concentrations. Calcium channel inhibitors also diminish excitability, whereas potassium channel inhibitors lower the amplitude of the light response but leave the sensitivity unchanged (Hegemann et al., in preparation). A 50 KDa protein that binds dihydropydridine type calcium channel inhibitors was identified by photolabelling with the dihydropydridine derivative azidopine (Hegemann et al., 1990). This protein might function as a rhodopsin regulated calcium channel in *Chlamydomonas.*

The mechanism of calcium channel activation, the relation between both currents and the regulation of flagellar beating is not understood at the moment. The participation of a G-proteins like in the visual system of higher animals is presently under study (Korolkov et al., 1990).

References

Beckmann, M., and Hegemann, P., 1991, In vitro identification of rhodopsin in the green alga *Chlamydomonas*, *Biochemistry*, submitted.

Dolle, R., Pfau, J., and Nultsch, W., 1986, Role of calcium ions on motility and phototaxis on *Chlamydomonas reinhardtii*, *J. Plant Physiol.*, 126:467.

Foster, K. W., Saranak, J., Derguini, F., Rao, J., Zarrilli, G. R., Okabe, M., Fang, J.-M., Shimizu, N., and Nakanishi, K., 1988b, Rhodopsin activation: A novel view suggested by *in vivo Chlamydomonas* experiments, *J. Am. Soc.*, 110:6588.

Foster, K. W., Saranak, J., Derguini, F., Zarrilli, G., Johnson, R., Okabe, M., and Nakanishi, K., 1989, Activation of *Chlamydomonas* rhodopsin *in vivo* does not require isomerization of retinal., *Biochemistry*, 28:819.

Foster, K. W., Saranak, J., Patel, N., Zarrilli, G., Okabe, M., Kline, T., and Nakanishi, K., 1984, A rhodopsin is the functional chromophore for phototaxis in the unicellular alga *Chlamydomonas*, *Nature*, 311:756.

Foster, K. W., Saranak, J., and Zarrilli, G. R., 1988a, Autoregulation of rhodopsin synthesis in *Chlamydomonas reinhardtii*, *Proc. Natl. Acad. Sci. USA*, 85:6379.

Foster, K. W., and Smyth, R. D., 1980, Light antennas in phototactic algae, *Microbiol. Rev.*, 44:572.

Harz, H., and Hegemann, P., 1991, Rhodopsin-regulated ion channel activation in *Chlamydomonas*, *Biophys. J.*, submitted.

Hegemann, P., and Bruck B., 1989, The light-induced stop response in *Chlamydomonas*. Occurence and adaptation phenomena, *Cell Motil.*, 14:501.

Hegemann, P., Hegemann, U., Foster, K. W., 1988, Reversible bleaching of *Chlamydomonas reinhardtii* rhodopsin *in vivo*, *Photochem. Photobiol.*, 48:123.

Hegemann, P., Neumeier, K., Hegemann, U., and Kuehnle, E., 1990, The role of calcium in *Chlamydomonas* movement responses as analysed by calcium channel inhibitors, *Photochem. Photobiol.*, 52:575.

Kamyia. R., and Witman, G., 1984, Submicromolar levels of calcium control the balance of beating between the two flagella in demembrenated models of *Chlamydomonas.*, *J. Cell Biol.*, 98:97.

Korolkov, S. N., Garnovskaya, M. N., Basov, A. S., Chunaev, A. S., and Dumler, I. L., 1990, The detection and characterisation of G-proteins in the eyespot of *Chlamydomonas reinhardtii*, *FEBS Lett.*, 270:132.

Krinski, N. I., and Litvin, R. P., 1964, Carotenoids of wild type and mutant strains of the green alga *Chlamydomonas reinhardi.*, *Plant Physiol*,. 48:680.

Litvin, F. F., Seneshchekov, O. A., and Seneshchekov, V. A., 1978, Photoreceptor electric potential in the phototaxis in the alga *Haematococcus pluvialis*, *Nature*, 271:476.

Nakanishi, K., Derguini, F., Rao, J., Zarrilli, G., Okabe, M., Lien, T., Johnson, R., Foster, K. W., and Saranak, J., 1989, Theory of rhodopsin activation: probable charge redistribution of excited state chromophore, *Pure Appl. Chem.* 61:361.

Nultsch, W., and Häder, D.-P., 1988, Photomovement in motile microorganisms II, *Photochem. Photobiol.*, 47:837.

Nultsch, W., Pfau, J., and Dolle, R., 1986, Effect of calcium channel blockers on phototaxis and motility of *Chlamydomonas reinhartii*, *Arch. Microbiol.*, 144:393.

Nultsch, W., Throm, G., v. Rimscha, I., 1971, Phototaktische Untersuchungen an *Chlamydomonas reinhardtii* Dangeard in homo-kontinuierlicher Kultur, *Arch. Microbiol.* 80:351.

Oesterhelt, D., Hegemann, P., Tavan, P., and Schulten, K., 1986, *Trans-cis* isomerization of retinal and a mechanism for ion translocation in halorhodopsin, *Eur. J. Biophys.*, 14:123.

Pfeffer, W., 1904, "Pflanzenphysiologie," Leipzig.

Schmidt, J. A., and Eckert, R., 1976, Calcium couples flagellar reversal to photostimulation in *Chlamydomonas reinhardtii.*, *Nature*, 262:713.

Smith, S. O., Paling, I., Copie, V., Raleigh, D. P., Courtin, J., Pardoen, J. A., Lugtenburg, J., Mathies, R. A., and Griffin, R. G., 1987, Low temperature solid state ^{13}C NMR studies of retinal chromophore in rhodopsin, *Biochemistry*, 26:1606.

Stavis, R. L., and Hirschberg, R., 1973, Phototaxis in *Chlamydomonas reinhardtii*, *J. Cell Biol.*, 59:367.

Takahashi, T., Yan, B., Mazur, P., Derguini, F., Nakanishi, K., and Spudich, J., 1990, Color regulation in the archibacterial phototaxis receptor phoborhodopsin (sensory rhodopsin II), *Biochemistry*, in press.

Uhl, R., and Hegemann, P., 1990a, Adaptation of *Chlamydomonas* phototaxis: I. A light scattering apparatus for measuring the phototactic rate of microorganisms with high time resolution, *Cell Motil.*, 15:230.

Uhl, R., and Hegemann, P., 1990b, Probing visual transduction in a plant cell. Optical recording of rhodopsin-induced structural changes from *Chlamydomonas*, *Biophys. J.*, in press.

van der Steen, R., Biesheuvel, P. L., Mathies, R. A., and Lugtenburg, J., 1986, Retinal analogs with locked 6-7 conformation show that bacteriorhodopsin requires the 6-s-*trans* conformation, *J. Am. Chem. Soc.*, 108:6410.

Warshel, A., and Karplus, M., 1974, Calculation of $\pi\pi^*$ excited state conformations and vibronic structure of retinal and related molecules, *J. Am. Chem. Soc.*, 96:5677.

Processing of Photosensory Signals in Halobacterium halobium. Common Features of the Bacterial Signalling Chain and of Information Processing in Higher Developed Organisms

Angelika Schimz
and Eilo Hildebrand

Institut für Biologische
Informationsverarbeitung
Forschungszentrum Jülich
D-5170 Jülich
Germany

Introduction

Halobacteria recognize changes of light intensity, process the information within the cell, and respond by a particular locomotor behavior. The cell can integrate signals from different photoreceptors and chemoreceptors, and it adapts to constant environmental conditions. The bacteria are already equipped with highly developed devices for acquisition and processing of information, and the question arises, whether the mechanisms which they use can be found as basic components of complex signalling processes in multicellular organisms.

Swimming Behavior and Responses to Light Stimuli

Halobacteria swim by means of polarly inserted flagella. According to a change of the sense of flagellar rotation, the cells spontaneously reverse their swimming direction about every 10 s on the average (Alam and Oesterhelt, 1984). They can detect changes of light intensity in different spectral regions by means of membrane bound receptor proteins, which carry retinal as the chromophore (Dencher and Hildebrand, 1979; Sperling and Schimz, 1980; Bogomolni and Spudich, 1982; Schimz et al., 1983; Takahashi et al., 1985).

Light stimuli transiently disturb this pattern of spontaneous behavior. Depending on the wavelength and on the sign of the intensity change, the response is either a prolonged swimming interval, the attractant response (Fig. 1), or a shortened interval, the repellent response (Hildebrand and Dencher, 1975; Spudich and Stoeckenius, 1979; Hildebrand and Schimz, 1983). A step-like change of light intensity alters one swimming interval only; thereafter the cell resumes its autonomous behavior (Schimz and Hildebrand, 1985).

In a similar way halobacteria respond to gradients of light intensity (Schimz and Hildebrand, 1989), and their light-dependent behavior allows them to accumulate in

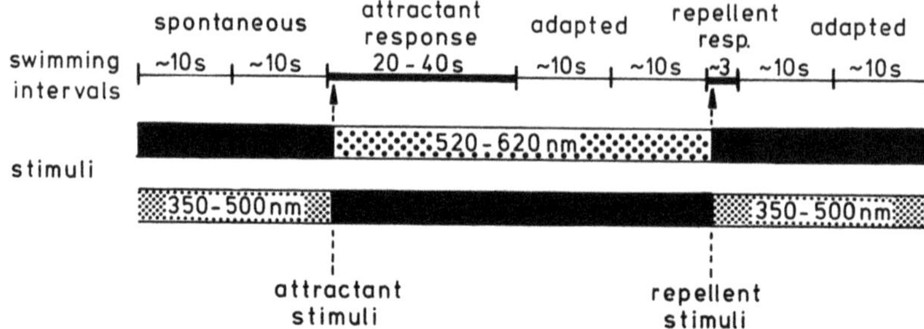

Fig. 1. Spontaneous and stimulus-controlled behavior of *Halobacterium halobium*. Each vertical bar indicates a reversal of the swimming direction. After one stimulus-determined swimming interval (attractant or repellent response) the behavior adapts, and the next intervals are not significantly different from spontaneous ones.

regions, which are optimal for light-energy conversion, and to avoid damaging UV radiation.

Control of Reversal Events by an Endogeneous Oscillator

An important question was, whether reversal events occur stochastically or in a regulated manner, and how their occurrence is influenced by stimuli.

Phase-response Curves

A first indication for a regulated process was the observation of a phase-relationship between the stimulus and the response (Fig. 2). When an attractant stimulus was given at different moments during a swimming interval, it turned out that the magnitude of the response, i.e., the resulting interval length, was not constant but increased and decreased in a reproducible manner (Schimz and Hildebrand, 1985). Such a phase-response curve is usually interpreted as an indication, that a particular state of the cell, which is necessary to trigger an event, rises and falls periodically. One possibi-

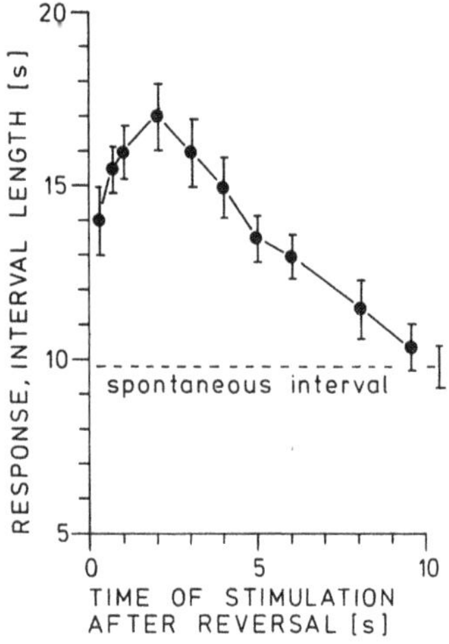

Fig. 2. Phase-response curve for the attractant response. Change of the magnitude of the response to a constant attractant stimulus in dependence on the moment of stimulation during a swimming interval. Each data point is the mean ± SEM of 30 measurements.

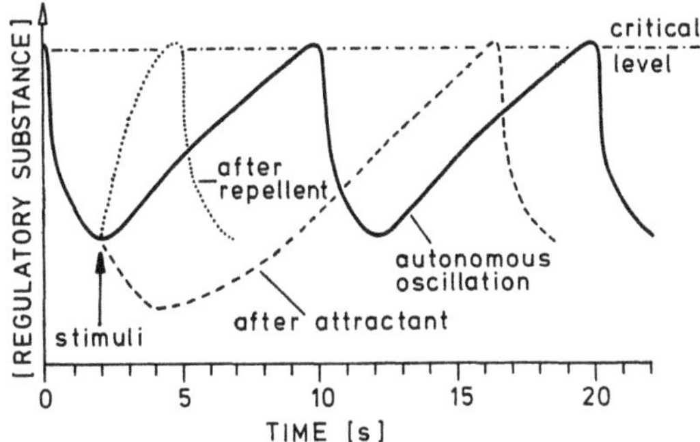

Fig. 3. Oscillator model. It is assumed that a still unknown regulatory substance changes rhythmically in its concentration, and, at a critical level, triggers a reversal of the flagellar motor. The shape of free-running oscillation has been deduced from the phase-response curve in Fig. 2. Light stimuli are thought to either enhance (repellent) or slow down (attractant) the rise of the regulatory substance.

lity is that the concentration of a regulatory substance oscillates and thereby controls periodic behavior. This is outlined in our working hypothesis for *Halobacterium* (Fig. 3).

Periodic Stimulation Induces Phase-locking, Period Doubling, and Deterministic Chaos

If the assumption of an endogeneous clock, i.e., a cellular oscillator, holds true, the system is expected to fulfil certain criteria. The behavior of a non-linear oscillator upon exposure to periodic perturbation is predictable. Within a certain frequency range, the oscillation becomes entrained at the frequency of perturbation; this is called "phase-locking". Above this frequency range, complex phenomena begin to develop. They are initiated by period doubling, i.e., bifurcation into a long and a short period occurring alternatingly. Further increase of the frequency results in the occurrence of strongly varying period lengths without any obvious regularity. Such a "period doubling route" is the result of a change from a simple limit cycle of periodic oscillation into a "strange attractor" of a complex geometrical structure (Fig. 4), which gives rise to deterministic chaos (for review see Olsen and Degn, 1985).

We subjected halobacteria to periodic repellent stimuli (pulses of UV light) and measured successive swimming intervals of a single bacterium. At a certain frequency of stimulation, we observed phase-locking, i.e., the reversal frequency became identical with the stimulation frequency (Fig. 5b). With increasing stimulation frequency, we induced period doubling (Fig. 5c), and finally aperiodic behavior (Fig. 5d).

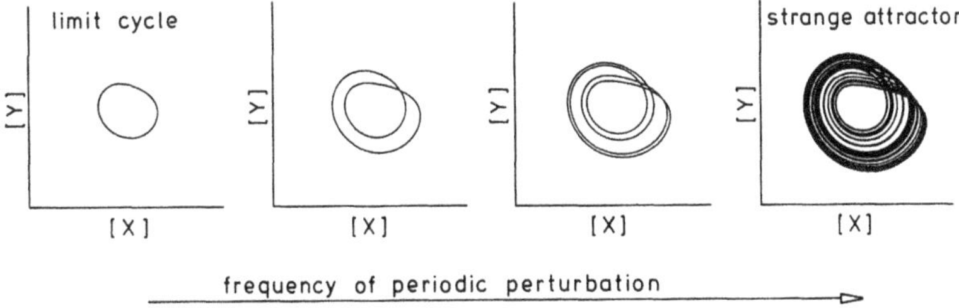

Fig. 4. Simulation of the behavior of an oscillator under periodic perturbation by non-linear differential equations. X and Y are the concentrations of the two substances involved. The frequency of periodic perturbation is increased from left to right. (After Rössler, 1976, from Olsen and Degn, 1985).

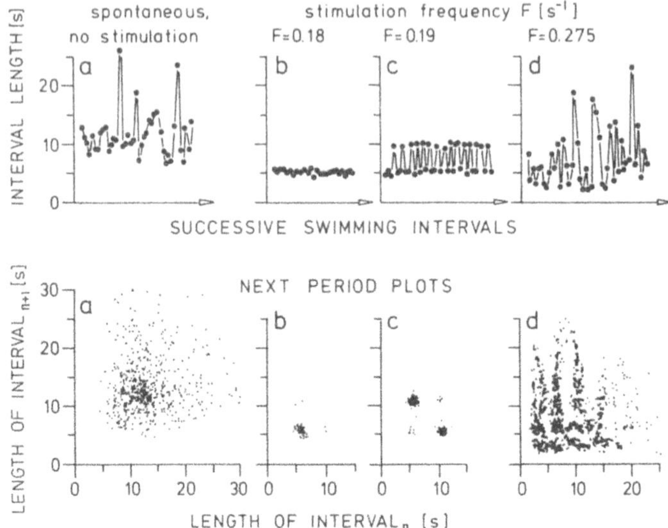

Fig. 5. Behavior of *Halobacterium* during periodic stimulation by repellent stimuli. Upper panels: successive swimming intervals of single bacteria at different stimulation frequencies. Lower panels: next period plots (interval n + 1 vs interval n); number of intervals were: a) 500, b) 110, c) 510, d) 730. For further explanation see text.

In order to diagnose whether the aperiodic events represent chaotic motion, one can see if there is a functional relationship between successive intervals by plotting each interval length as a function of the preceding one. If this "next period plot" shows a non-random structure, the system is most probably governed by a deterministic dynamical law (Olsen and Degn, 1985).

The narrow spots in the next period plots (Fig. 5b, c) indicate predictable periodic oscillation, the clearly discernible structure in Fig. 5d suggests deterministic chaos. A model calculation for chaos as a result of periodic perturbation of a limit cycle oscillation yields a similar structure, when a small amount of noise is added (see Schimz and Hildebrand, 1989). These results confirm our hypothesis, that an endogeneous oscillator controls the occurrence of reversals in *Halobacterium*.

Biochemistry of the Oscillator

The biochemical nature of the oscillator is still unknown. Any oscillator requires at least two components, which are coupled to each other through feedback mechanisms. Input signals would alter the level of a tuning parameter, which in turn varies the frequency of the oscillator over a certain range. In *Halobacterium* the length of spontaneous intervals can be altered in opposite directions by varying the concentration of cGMP or calcium, respectively. Also the shape of the phase-response curve is changed in a characteristic manner (Fig. 6; Schimz and Hildebrand, 1987). By uncovering the nature of these cGMP- and calcium-sensitive processes it may be possible to identify key components of the oscillator.

Integration of Sensory Signals

Signals generated by the different photoreceptors are integrated in a step prior to the oscillator. A particular property of the system helped us to draw this conclusion: The cell does not respond to repellent stimuli, when they are given within 0.5 s after a

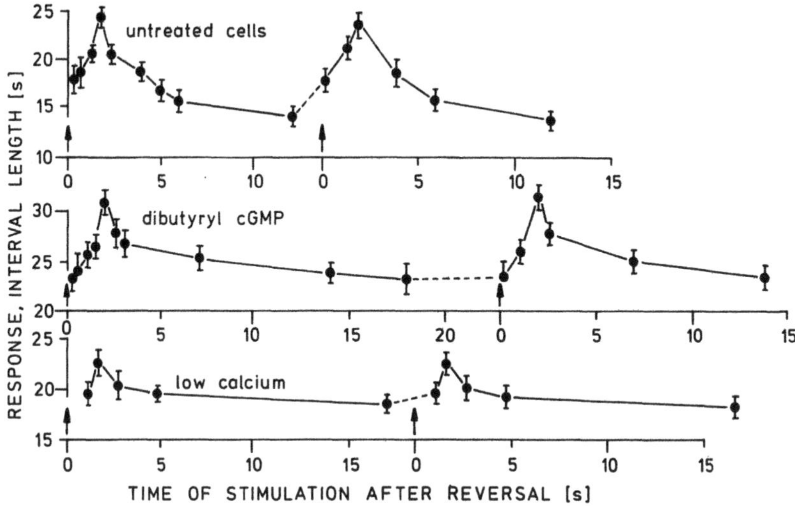

Fig. 6. Phase response curves at increased cGMP and at lowered Ca^{2+} concentration. The experiments were done as in Fig. 2. Arrows indicate spontaneous reversals of the swimming direction.

reversal (Fig. 7), whereas at the same time it responds to attractant stimuli. But even during the period of absolute refractoriness, a repellent signal is generated and can cancel the effect of a simultaneously evoked attractant signal (Schimz and Hildebrand, 1985; Hildebrand and Schimz, 1986). Several observations led us to conclude that refractoriness is a property related to the oscillator, and hence signal integration must occur in an earlier step which lies between the receptors and the oscillator (Schimz and Hildebrand, 1985; 1987; Hildebrand and Schimz, 1986; 1987).

G Protein

A protein with homology to the nucleotide binding site of most known G proteins could be detected immunologically. Measurements of the hydrolysis of cGMP in membrane vesicles in the presence of G protein activators indicate that this G protein activates a cGMP-phosphodiesterase (Schimz et al., 1989). Our hypothesis is that the G protein interacts with the excited photoreceptors or with an additional transducer protein to transmit the excitatory signal, and that the G protein or the cGMP-pool is the site of signal integration. The integrated signal is assumed to act on the oscillator.

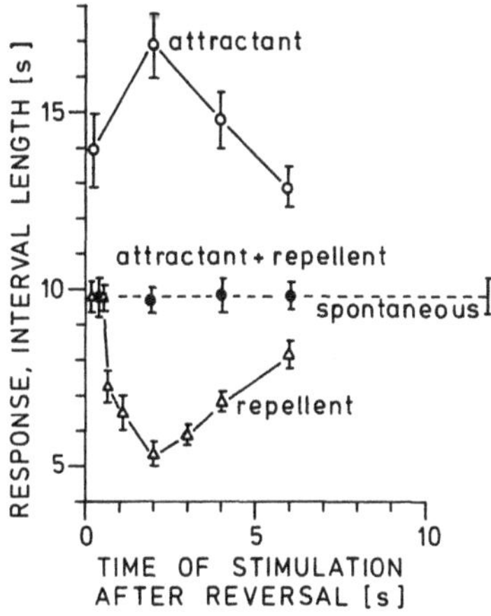

Fig. 7. Integration of an attractant and a repellent signal evoked at different times after a spontaneous reversal. Response to the attractant stimulus alone (open circles); response to the repellent stimulus alone (open triangles); attractant and repellent stimulus simultaneously applied (closed circles).

Lifetime of Excitatory Signals and Protein Methylation as a Device for Adaptation

Signal Lifetimes

To determine the lifetime of a repellent signal, although not knowing its nature, we used the following approach: We gave a repellent stimulus during the refractory period, when the signal is transmitted to the site of integration but does not lead to a motor response. An attractant stimulus was applied with increasing delay thereafter (Fig. 8). By comparing these responses with the responses obtained upon the attractant stimuli alone, we found that the repellent signal weakens the response to the following attractant only as long as the delay between the two stimuli does not exceed 1.2 s. This shows that the repellent signal must be extinguished 1.2 s after its generation, and we regard this time as the total lifetime of a repellent signal at the site of integration (Hildebrand and Schimz, 1990). With a particular stimulation program the decay of the signal could be monitored. The kinetics cannot be described by a single exponential equation. We rather have to postulate at least two rate constants (Hildebrand and Schimz, 1990).

By use of a modified stimulation program, we could also measure the lifetime of an attractant signal, which is longer than that of the repellent signal, namely about 4 s (Hildebrand and Schimz, 1990).

Protein Methylation and Decay of Excitatory Signals

As a consequence of attractant or repellent stimuli, certain membrane proteins in the range of 67-100 kDa become transiently methylated or demethylated, respectively (Schimz, 1981; 1982; Alam et al., 1989; Spudich et al., 1989; Hildebrand and Schimz, 1990). The methylation reaction can be inhibited in living cells by addition of homocysteine to the medium. Under these conditions the decay of sensory signals is slowed down. The lifetime of a repellent signal could be extended up to 3.5 s, and the lifetime of an attractant signal increased up to about 7 s (Schimz and Hildebrand, 1987; Hildebrand and Schimz, 1990).

Protein methylation is not required for excitatory signalling (Schimz and Hildebrand, 1987). Our results suggests that it is the device to extinguish excitatory signals and thus allows the cell to adapt to constant light conditions. Results obtained by simultaneous stimulation of the blue-light and the long-wavelength light photoreceptor indicate that the methylation system is regulated by an integrated signal (Spudich et al., 1989). We assume that the feedback signal to control methylation initiates at the site of

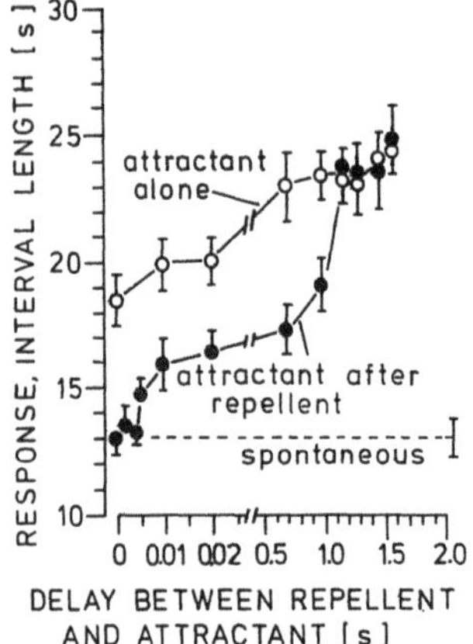

Fig. 8. Determination of the lifetime of the repellent signal. A repellent stimulus was applied during the time of absolute refractoriness, and an attractant with various delays thereafter. A comparison of the obtained curve with the responses to the attractant stimulus alone shows that the repellent signal does not influence the attractant response any longer, when the delay between the two stimuli exceeds 1.2 s. (The increase of the curves results from the phase-response relationship shown in Fig. 2).

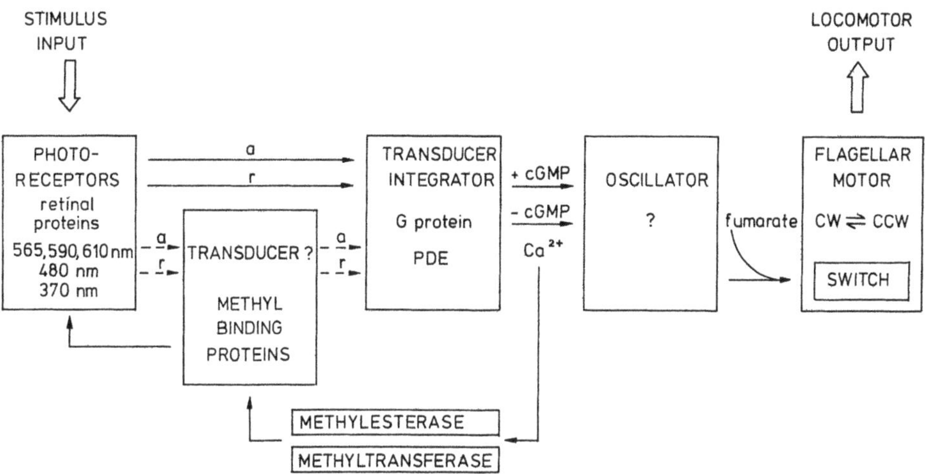

Fig. 9. Hypothetical sensory signalling chain of *Halobacterium*. For details see text.

signal integration. Calcium may be involved in this process, because we found that the methyltransferase is inhibited by calcium, whereas the methylesterase is activated (Schimz, 1982; Hildebrand and Schimz, 1990).

Working Hypothesis on the Photosensory Signalling Chain of Halobacterium

Our results are summarized in the following model. Reversals of the direction of flagellar rotation are controlled by a cellular oscillator (Fig. 9). When a critical level of the concentration of a regulatory substance is reached, the flagellar switch is triggered to induce a reversal of the motor. Marwan et al. (1990) showed that fumarate is involved in the switching process. The photoreceptors generate repellent (r) or attractant (a) signals. We assume that the excited receptors interact with a G protein, which transduces the signal by activating a cGMP-phosphodiesterase. The G protein can be the site of signal integration. The integrated excitatory signal shifts the phase of the oscillator and thereby the moment, at which a flagellar reversal occurs. A feedback signal, which controls protein methylation, most probably initiates at the site of signal integration. We propose that the methyl binding proteins interact with the excited receptors and thereby stop signalling. When the excitatory signals have decayed, the oscillator resumes its autonomous rhythm, and the behavior of the cell adapts. There is evidence that the methylation reaction is not required for excitatory signalling. The methyl binding proteins, however, may have signalling domains, through which they act as transducers between the receptors and the G protein.

Ubiquitous Phenomena in Biological Signalling

Biological systems must respond to internal and external signals such as the concentration of nutrients, the levels of hormones, and changes of physical parameters as light, temperature, or pressure. For the purpose of information processing within the cell, some basic mechanisms seem to be realized in all types of organisms from the simplest to the most complex.

Biological Oscillators

Autonomous periodicities occur in a large variety of biological systems. The periods can range from 30 ms, e.g., in the firing of thalamic sensory neurons (Poggio and Viernstein, 1964) to 24 h circadian rhythms or even monthly or annual rhythms (Bünning, 1973). Oscillations occur in all major biological processes: movement, secretion, reproduction, growth and development, and in information processing in sensory and nervous systems (for reviews see Berridge and Rapp, 1979; Rapp, 1987). There is no common molecular mechanism that generates all biological oscillations. Oscillations in the concentrations of metabolites, or of cyclic nucleotides and calcium, which can interact with each other through feedback loops, have been implicated in many different systems; a large number of processes are characterized by oscillations in the membrane potential (for review see Berridge and Rapp, 1979). What are the functional advantages of an oscillating system? Several aspects have been discussed by Rapp (1987).

Efficiency

Periodic signalling can increase the efficiency of intercellular communication. The signal for aggregation of *Dictyostelium* cells is the periodic release of cAMP from the cells forming the center. Although periodic signalling is not essential for aggregation, an analysis of the parameters involved led to the conclusion, that periodic signalling is more efficient than signalling at a constant rate in that sense that less molecules are required to give the same effective signalling radius (Nanjundiah, 1973).

Precision of Control

Frequency-encoded information is generally more resistent to distortion by background noise than amplitude-encoded information. The best known examples of frequency encoding in biology occur in neural systems. A frequency modulation by external signals in two directions, which we have shown for halobacteria, is realized, e.g., in the vestibular receptors of the inner ear, which are responsible for the sense of equilibrium: The nerve fires spontaneously 20 times per second; a torsion of the body in one direction increases the rate, a torsion in the opposite direction decreases the rate.

Temporal and Spacial Organization

The capacity of biological oscillators for self-synchronization and entrainment provides a mechanism for coordinating biological processes in time. An important example is the synchronization among embryonic heart cells (DeHaan, 1967). Synchronization of spontaneous membrane potential oscillations in nerve cells may be essential for the coordination of central nervous system activity (Pinsker and Willis, 1980). Entrainment by a periodic external force, which we have shown for *Halobacterium*, is a dynamical process in which a recipient oscillator is driven to oscillate at the frequency of the input signal. A well known example for entrainment in higher organisms is the periodic signal from the sino-atrial node in the heart, which acts as a pacemaker and entrains the entire heart to oscillate at the same frequency (Noble, 1975).

Gradients of oscillator frequencies can serve to organize events in space. In the human small intestine, the frequency of muscle contractions decreases from the duodenum towards the ileum and thus unidirectional contraction waves are generated (Linkens, 1979).

Reversible Covalent Modification of Proteins

We have shown that in *Halobacterium* reversible methylation of membrane proteins is involved in the turnoff of excitatory signals. The same mechanism operates in eubacteria (for review see Koshland et al., 1988). The strategy of signal turnoff by covalent modification is also realized in the vertebrate photoreceptor (Wilden et al., 1986) or in the ß-adrenergic receptor (Benovic et al., 1988); in both cases phosphorylation of the receptor is involved in adaptation. In eubacteria, successive phosphorylation of signalling proteins is required to process the excitatory signal (for review see Parkinson, 1988). Covalent modification of enzymes has also been identified with control in carbohydrate and fat metabolism or protein synthesis. In sensing, it is important that the turnon and turnoff of signals be sensitive to small changes in effector concentration. It has been shown by Goldbeter and Koshland (1981) that there is an enhanced sensitivity inherent in covalent modification mechanisms, when at least one enzyme component operates outside the region of first-order kinetics, i.e., in the region of saturation with respect to a substrate. Covalent systems can generate a sensitivity equivalent to allosteric enzymes with a high Hill coefficient, i.e., an output response that is much more sensitive to a change in input signal than the hyperbolic Michaelis-Menten system.

Signal Coupling by G Protein

GDP-/GTP-binding proteins are involved in a large variety of signal-transducing systems, and the number of newly detected G proteins is still incrasing (Gilman, 1987; Casey et al., 1988). The G protein, which we found in *Halobacterium*, is the first example of a G protein in bacteria. It seems to be a widespread mechanism, therefore, that G proteins, which carry highly conserved nucleotide binding sites, transduce an excitatory signal by interacting with an excited receptor and by transmitting the signal to a subsequent enzyme in the signalling chain. In many cases, G proteins represent the site of integration of signals from different receptors.

Conclusions

As we have seen, *Halobacterium* shares many molecular mechanisms with signal processing systems of higher complexity. The use of cellular oscillators seems to be a general tool to archieve high sensitivity and precision of fine tuning. G proteins are common transducer molecules in sensory and other receptive cells, and reversible covalent modification of proteins is frequently used for the turnon and turnoff of signalling events. Therefore, it seems justified to regard *Halobacterium* a most suitable model system to investigate basic molecular mechanism of signal transduction and further processing with respect also to the understanding of complex systems.

References

Alam, M., Lebert, M., Oesterhelt, D., and Hazelbauer, G. L., 1989, Methyl-accepting taxis proteins in *Halobacterium halobium*, *EMBO J.*, 8:631.

Alam, M., and Oesterhelt, D., 1984, Morphology, function and isolation of halobacterial flagella, *J. Mol. Biol.*, 176:459.

Benovic, J. L., Bouvier, M., Caron, M. G., and Lefkowitz, R. J., 1988, Regulation of adenylyl cyclase-coupled ß-adrenergic receptor, *Ann. Rev. Cell Biol.*, 4:405.

Berridge, M. J., and Rapp, P. E., 1979, A comparative survey of the function, mechanism and control of cellular oscillators, *J. Exp. Biol.*, 81:217.

Bogomolni, R. A., and Spudich, J. L., 1982, Identification of a third rhodopsin-like pigment in phototactic *Halobacterium halobium*, *Proc. Natl. Acad. Sci. USA*, 79:6250.
Bünning, E. , 1973, "The physiological clock. Circadian Rhythms and Biological Chronometry," Rev. 3rd ed., The English Universities Press, London.
Casey, P. J., Graziano, M. P., Freissmuth, M., and Gilman, A. G., 1988, Role of G proteins in transmembrane signalling, *Cold Spr. Harbor Symp. Quant. Biol.*, 53:203.
DeHaan, R. L., 1967, Regulation of spontaneous activity and growth of embryonic chick heart cells in tissue culture, *Devel. Biol.*, 16:216.
Dencher, N. A., and Hildebrand, E., 1979, Sensory transduction in *Halobacterium halobium*: Retinal protein pigment controls UV-induced behavioral response, *Z. Naturforsch.*, 34c:841.
Gilman, A. G. , 1987, G proteins: transducers of receptor-generated signals, *Ann. Rev. Biochem.*, 56:615.
Goldbeter, A., and Koshland, D. E., Jr., 1981, An amplified sensitivity arising from covalent modification in biological systems, *Proc. Natl. Acad. Sci. USA*, 78:6840
Hildebrand, E., and Dencher, N., 1975, Two photosystems controlling behavioural responses of *Halobacterium halobium*, *Nature*, 257:46.
Hildebrand, E., and Schimz, A., 1983, Photosensory behavior of a bacteriorhodopsin-deficient mutant, ET-15, of *Halobacterium halobium*, *Photochem. Photobiol.*, 37:581.
Hildebrand, E., and Schimz, A., 1986, Integration of photosensory signals in *Halobacterium halobium*, *J. Bacteriol.*, 167:305.
Hildebrand, E., and Schimz, A., 1987, Role of the response oscillator in inverse responses of *Halobacterium halobium* to weak light stimuli, *J. Bacteriol.*, 169:254.
Hildebrand, E., and Schimz, A., 1990, The lifetime of photosensory signals in *Halobacterium halobium* and its dependence on protein methylation, *Biochim. Biophys. Acta*, 1052:96.
Koshland, D. E., Jr., Sanders, D. A., and Weis, R. M., 1988, Roles of methylation and phosphorylation in the bacterial sensing system, *Cold Spr. Harbor Symp. Quant. Biol.*, 53:11.
Linkens, D. A., 1979, Modelling of gastro-intestinal electric rhythms. *In*: "Biological Systems Modelling and Control," Linkens, D. A., ed., pp. 202, Peregrinus, Stevenage.
Marwan, W., Schäfer, W., and Oesterhelt, D., 1990, Signal transduction in *Halobacterium* depends on fumarate, *EMBO J.*, 9:355.
Nanjundiah, V., 1973, Chemotaxis, signal relaying and aggregation morphology, *J. Theor. Biol.*, 42:63.
Noble, D., 1975, "The Initiation of the Heartbeat," Clarendon Press, Oxford.
Olsen, L. F., and Degn, H., 1985, Chaos in biological systems, *Q. Rev. Biophys.*, 18:165.
Parkinson, J. S., 1988, Protein phosphorylation in bacterial chemotaxis, *Cell*, 53:1
Pinsker, H. M., and Willis, W. D., eds., 1980, "Information Processing in the Nervous System," Raven Press, New York.
Poggio, G. F., and Viernstein, L. J., 1964, Time series analysis of impulse sequences of thalamic somatic sensory neurons, *J. Neurophysiol.*, 27:517.
Rapp, P. E., 1987, Why are so many biological systems periodic? *Progress Neurobiol.*, 29:261.
Rössler, O. E., 1976, An equation for continuous chaos, *Phys. Lett.*, 57A:397.
Schimz, A., 1981, Methylation of membrane proteins is involved in sensory behavior of *Halobacterium halobium*, *FEBS Lett.*, 125:205.
Schimz, A., 1982, Localization of the methylation system involved in sensory behavior of *Halobacterium halobium* and its dependence on calcium, *FEBS Lett.*, 139:283.
Schimz, A., and Hildebrand, E., 1985, Response regulation and sensory control in *Halobacterium halobium* based on an oscillator, *Nature*, 317:641.
Schimz, A., and Hildebrand, E., 1987, Effects of cGMP, calcium and reversible methylation on sensory signal processing in halobacteria, *Biochim. Biophys. Acta*, 923:222.
Schimz, A., and Hildebrand, E., 1989, Periodicity and chaos in the response of *Halobacterium* to temporal light gradients, *Eur. Biophys. J.*, 17:237.
Schimz, A., Hinsch, K.-D., and Hildebrand, E., 1989, Enzymatic and immunological detection of a G-protein in *Halobacterium*, *FEBS Lett.*, 249:59.
Schimz, A., Sperling, W., Ermann, P., Bestmann, H. J., and Hildebrand, E., 1983, Substitution of retinal by analogues in retinal pigments of *Halobacterium halobium*. Contribution of bacteriorhodopsin and halorhodopsin to photosensory activity, *Photochem. Photobiol.*, 38:417.

Sperling, W., and Schimz, A., 1980, Photosensory retinal pigments in *Halobacterium halobium*, *Biophys. Struct. Mech.*, 6:165.

Spudich, J. L., and Stoeckenius, W., 1979, Photosensory and chemosensory behavior of *Halobacterium halobium*, *Photobiochem. Photobiophys.*, 1:43.

Spudich, E. N., Takahashi, T., and Spudich, J. L., 1989, Sensory rhodopsins I and II modulate a methylation/demethylation system in *Halobacterium halobium* phototaxis, *Proc. Natl. Acad. Sci. USA*, 86:7746.

Takahashi, T., Tomioka, H., Kamo, N., and Kobatake, Y., 1985, A photosystem other than PS 370 also mediates the negative phototaxis of *Halobacterium halobium*, *FEMS Microbiol. Lett.*, 28:161.

Wilden, U., Hall, S. W., and Kühn, H., 1986, Phosphodiesterase activation by photoexcited rhodopsin is quenched when rhodopsin is phosphorylated and binds the intrinsic 48-kDa protein of rod outer segments, *Proc. Natl. Acad. Sci. USA,* 83:1174.

Color Discriminating Pigments in Halobacterium halobium

John L. Spudich

Albert Einstein College of Medicine
Bronx, NY 10461
U.S.A.

Introduction

As in the world of higher organisms, light is a key carrier of information in the microbial world, guiding the motion of single-celled prokaryotic and eukaryotic organisms. The archaebacterial species *Halobacterium halobium* uses chromoproteins chemically similar to our visual pigments (rhodopsins) to sense light and to distinguish its color (Spudich and Bogomolni, 1988). The purpose of this chapter is to review briefly progress on the molecular properties of the bacterial sensory rhodopsins since the first of these pigments (sensory rhodopsin I (SR-I)) came to light in 1982, and to present some recent work from our laboratory.

H. halobium cells are flagellated and their motility is modulated by light or chemicals. The motion of halobacteria in isotropic conditions consists of linear paths interrupted by a roughly 180° reversal of swimming direction every 5-30 seconds. This zigzag motility pattern is biased by light intensity gradients in that reversals are suppressed or enhanced when the cell experiences a favorable or unfavorable change, respectively, in light intensity. Increases (decreases) in orange light suppress (enhance) reversals, and thereby orange light is an attractant. Increases (decreases) in blue light, on the other hand, enhance (suppress) reversal probability, thereby repelling the cells from high intensity regions of blue light. We quantitate these motility responses by computerized cell tracking and motion analysis, capable of processing on the order of a thousand cells per stimulus at 67 msec resolution. Advances in computer hardware have improved on the rate of processing of data, so that with RISC technology essentially real-time assessment of stimulus-induced reversal frequency responses is now available.

The migration of cells in light and chemical gradients is generally referred to as phototaxis and chemotaxis, respectively, by investigators of bacterial behavior, although phototaxis in a stricter sense implies sensing of the direction of light, which has not been observed in halobacteria.

Sensory Rhodopsins I & II

The color-discriminating behavior of *H. halobium* was shown to require retinal, the vitamin A-derived prosthetic group of visual pigments, by Hildebrand and Dencher

Biophysics of Photoreceptors and Photomovements in Microorganisms
Edited by F. Lenci *et al.*, Plenum Press, New York, 1991

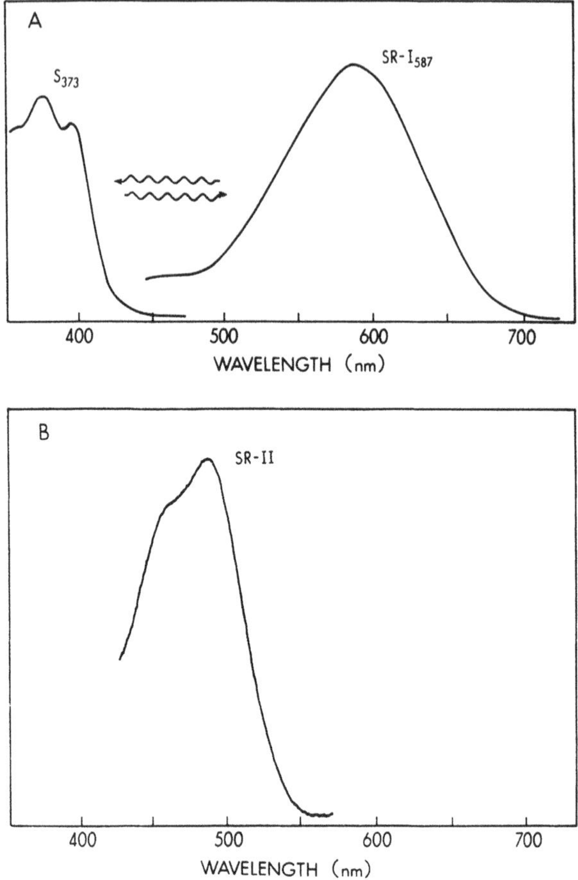

Figure 1. Upper panel: The photointerconvertible attractant (SR-I_{587}) and repellent (S_{373}) receptor forms of SR-I. Lower panel: SR-II absorption spectrum with absorption maximum at 487 nm. The structured shape of SR-II is due to vibrational fine structure (Takahashi et al., 1990).

(1975). At that time a membrane protein in *H. halobium* with a retinal chromophore was known to function in photoenergy conversion: bacteriorhodopsin (BR, absorption maximum 568 nm), an electrogenic light-driven proton pump (Oesterhelt and Stoeckenius, 1973). Later a second retinal-containing light-driven ion pump catalyzing electrogenic chloride transport was found, halorhodopsin (HR, absorption maximum 578 nm) (as reviewed in Lanyi, 1986). Each of the pumps consists of a single polypeptide (26 kDa for BR and 27 kDa for HR) which folds in the membrane, forming an internal pocket where the chromophore all-*trans* retinal is bound (Khorana, 1988; Henderson et al., 1990).

H. halobium mutants in light-activated ion transport (Flx mutants), which are deficient in BR and HR, are still fully phototactic (Spudich and Spudich, 1982). Studies of Flx mutants revealed two retinal-containing phototaxis receptors in the membranes: sensory rhodopsin I (SR-I, absorption maximum 587 nm (Bogomolni and Spudich, 1982) and sensory rhodopsin II (SR-II, also called phoborhodopsin and P480, absorption maximum 487 nm (Takahashi et al., 1985)) (Fig. 1). Each of the sensory rhodopsins undergoes a cyclic series of photochemical reactions, which can be monitored by flash spectroscopy. Their photocycles are nonelectrogenic; i.e., they do not catalyze active ion transport, but rather their light-induced conformational changes generate signals to the flagellar motor switch. Like transport by BR and HR, photocycling and phototaxis signaling by both sensory receptors require all-*trans*/13-*cis* isomerization of the retinal chromophore (Yan et al., 1990a).

SR-I is unusual among known photoreceptors in that it exists in two spectrally distinct forms, each of which can be photoactivated to generate sensory signals (Spudich and Bogomolni, 1984). The thermally stable species $SR\text{-}I_{587}$ (Fig. 1) is an attractant receptor; its photoexcitation suppresses directional reorientation (swimming reversals) thereby favoring cell migration into higher intensity regions of orange-red light. A long-lived species (S_{373}) in the SR-I photocycle accumulates in a photostationary state maintained by photoexcitation of $SR\text{-}I_{587}$ (Fig. 1). S_{373} is photochemically reactive and functions as a repellent receptor; its photoexcitation induces swimming reversals, repelling cells from regions of high UV-blue light intensity. The opposing signals from the attractant and repellent receptor forms of SR-I add algebraically (i.e., mutually cancel). Therefore, the photochromic mixture consisting of these two spectrally distinct forms is sensitive to the wavelength of light. The activation of as few as 1-2 S_{373} molecules is sufficient to elicit a response (Spudich and Bogomolni, 1984). Sensory rhodopsin II is a distinct chromoprotein which mediates repellent responses in the blue-green spectral region (Fig. 1) (Takahashi et al., 1985, 1990).

Receptor Proteins

Incorporation of radiolabeled retinal into the chromophoric pockets of SR-I and SR-II allowed identification of their apoproteins in denaturing gels as distinct polypeptides at 25 and 23 kDa M_r, respectively (Spudich et al., 1986a), similar in size to BR. The retinal binding pockets of BR and SR-I have been shown to be closely similar in their electrostatic and hydrophobic interactions with the chromophore by studies of retinal analog effects on absorption (Spudich et al., 1986b; Baselt et al., 1989; Fodor et al., 1989; Yan et al., 1990a; 1990b). The SR-I apoprotein gene (*sopI*) has been cloned and the deduced amino acid sequence bears substantial homology to BR (Blanck et al., 1989), with 29% identity in the BR transmembrane regions. From the above considerations, a structural model for SR-I can be developed based on the more established models of BR, and the resulting folded structure fits well the hydropathy predictions of transmembrane helical regions in SR-I (Blanck et al., 1989; Henderson et al., 1990).

SR-I is not an ion pump and membrane polarization changes are not involved in sensory signaling. We have identified another protein in the membrane associated with SR-I signaling (Spudich et al., 1988, 1989). This is a methylated protein of 97 kDa (methyl-accepting phototaxis protein I, MPP-I) which is likely to be the receiver for the signal transmitted by the SR-I retinal-binding protein.

MPP-I: A Methyl Accepting Protein Associated with SR-I

Some discussion of the mechanism of phototaxis adaptation is necessary to present this finding. Photoexcitation of SR-I and SR-II modulates swimming behavior by controlling the frequency of directional changes by swimming cells. The reversal frequency is altered within 200-800 ms after photoactivation and in seconds returns to the prestimulus value. In other words, in the presence of continued sensory rhodopsin photocycling the signal modulating the motor switch is attenuated and in time cancelled. In the parlance of investigators of bacterial sensory behavior, the cells "adapt" to continuous illumination (and similarly respond and adapt to chemical stimuli). In chemotactic eubacteria (e.g., *Escherichia coli*) one mechanism of adaptation is by carboxylmethylation of transmembrane chemoreceptors.

The effects of methionine starvation and methylation inhibitors implicated a methylation system in *H. halobium* taxis as well, and protein methylation has been detected in *H. halobium* membranes (Schimz, 1981). To sort out the involvement of this process in *H. halobium* taxis, *in vivo* methyl-radiolabeling methods developed to visualize the *E. coli* chemotaxis methyl-accepting proteins have been applied (Spudich et al., 1988; Alam et al., 1989). These procedures reveal altered patterns of methylation among *H. halobium* taxis mutants (Sundberg et al., 1990), and a methyl-labeled protein of M_r 97 kDa required for SR-I function (Spudich et al., 1988, 1989). SR-II also activates a methylation system in the cells and evidently uses a methyl-accepting component distinct from that of SR-I. Our working model is that photoactivated SR-I and SR-II relay the signal to methyl-accepting proteins in the membrane, probably by protein/protein interaction.

SR-I Phototransduction

A crucial step in photosensory transduction by SR-I is the coupling of retinal isomerization to the generation of a conformational change in the protein. An important clue to the mechanism of this step has recently been found by Bing Yan in his doctoral thesis work (in a collaborative effort between my laboratory and that of Koji Nakanishi, Columbia University, New York). Bing Yan compared the retinal binding sites of SR-I apoprotein (SOP-I) and bacterioopsin (BOP) by incorporating retinal isomers and retinal analogs. Unlike BOP, SOP-I does not form a retinylidene pigment with 13-*cis* retinal, as assessed by absorption measurements and SDS-polyacrylamide gel electrophoresis, isolation of SOP-I after [^{3}H]13-*cis* retinal addition and $NaCNBH_3$ reduction. Also unlike BOP (Gartner et al., 1983), SOP-I does not thermally isomerize all-*trans* 13-desmethyl retinal to 13-*cis*. Most importantly, unlike BR the corresponding all-*trans* analog SR-I pigment does not exhibit photochemical reactivity (1 ms resolution). This indicates that photoactivation of SR-I requires a specific steric interaction between the protein and the 13-methyl group of the retinal (Yan et al., in preparation). This requirement differs from that of BR but resembles that of rhodopsin, in which steric interaction between the protein and the retinal 9-methyl group is crucial for photoactivation (Ganter et al., 1989). As noted below ("signaling states"), there is an additional interesting analogy of SR-I to rhodopsin. Meta-II_{380} is the rhodopsin photoproduct which is active in signal transduction (Bennett et al., 1982), and the meta-II_{380}-like intermediate S_{373} appears to be the physiologically active conformation in the SR-I single photon cycle (see next section).

Signaling States of SR-I and SR-II

A general approach to localize the signaling process to specific conformations of a receptor is to (i) define stimulus-induced conformations by monitoring structural alterations with spectroscopic reporter groups, (ii) modulate the lifetime of these conformational states by modification of the receptor or its effector, and (iii) assess receptor signaling efficiency by physiological measurements. Correlation of signaling efficiency with the time-averaged concentration of a particular stimulus product (determined over the interval of time over which the physiological response is assessed) can be taken as evidence this product is a signaling conformation of the receptor. We have applied this procedure to SR-I and SR-II using retinal analogs to modulate photocycle intermediate lifetimes.

Many retinal analogs which retard the photocycles of SR-I and SR-II enhance the sensitivity of the cells to sR-I_{587}-mediated attractant and SR-II-mediated repellent responses, respectively. Our interpretation of this effect is that slowing steps of the photocycle with retinal analogs increases the lifetime of conformations of the receptors which modulate the flagellar motor switch, thereby increasing the effectiveness of receptor photoactivation. The analogs therefore provide a way to identify these active conformations ("signaling states") by comparing behavioral sensitivity with lifetimes of photocycle intermediates in the analog photoreceptors as outlined in the previous paragraph.

Photoexcitation of SR-I or SR-II is followed by a sequence of thermal steps which return the molecules to their original state (photocycling). In the SR-I photocycle only one long-lived intermediate, S_{373}, has been detected. Analogs which increase the lifetime of S_{373} increase the effectiveness of the attractant receptor form of the molecule (SR-I_{587}). This result indicates S_{373} (the repellent receptor form of SR-I) is the attractant signaling state.

SR-II exhibits two long-lived intermediates (Tomioka et al., 1986): S-II_{350} forms in < 1 ms after photon absorption. The decay of S-II_{350} ($t_{1/2}$ 140 ms, 23° C) is paralleled by the rise of a red shifted species S-II_{530} which thermally returns to SR-II_{487} ($t_{1/2}$ 300 ms, 23° C) completing the cycle. Analogs which extend the lifetime of the later intermediate S-II_{530} increase receptor signaling efficiency (Yan et al., in preparation). Using 7 retinal-related chromophores, a clear relationship is found: receptor efficiency exhibits an exponential dependence on the average lifetime of the S-II_{530} intermediate. Our results show S-II_{530} is a signaling conformation of SR-II. We cannot exclude that S-II_{350} also may be effective in signaling.

References

Baselt, D. R., Fodor, S. P. A., van der Steen, R., Lugtenburg, J., Bogomolni, R. A., and Mathies, R. A., 1989, Halorhodopsin and sensory rhodopsin contain a C_6C_7-s-*trans* retinal chromophore, *Biophys. J.*, 55:193.

Bennett, N., Michel-Villaz, M. and Kuhn, H., 1982, Light-induced interaction between rhodopsin and the GTP-binding protein. Metarhodopsin II is the major photoproduct involved, *Eur. J. Biochem.*, 127:97.

Blanck, A., Oesterhelt, D., Ferrando, E., Schegk, E.S. and Lottspeich, F., 1989, Primary structure of sensory rhodopsin I, a prokaryotic photoreceptor, *EMBO J.*, 8:3963.

Bogomolni, R. A. and Spudich, J. L., 1982, Identification of a third rhodopsin-like pigment in phototactic *Halobacterium halobium*, *Proc. Natl. Acad. Sci. USA*, 79:6250.

Fodor, S. P. A., Gebhard, R., Lugtenburg, J., Bogomolni, R. A. and Mathies, R. A., 1989, Structure of the retinal chromophore in sensory rhodopsin I from resonance Raman spectroscopy, *J. Biol. Chem.*, 264:18280-18283.

Ganter, U. M., Schmid, E. D., Perez-Sala, D., Rando, R. R., and Siebert, F., 1989, Removal of the 9-methyl group of retinal inhibits signal transduction in the visual process. A Fourier transform infrared and biochemical investigation, *Biochemistry*, 28:5954.

Gartner, W., Towner, P., Hopf, H., and Oesterhelt, D., 1983, Removal of methyl groups from retinal controls the activity of bacteriorhodopsin, *Biochemistry*, 22:2637.

Henderson, R., Baldwin, J. M., Ceska, T. A., Zemlin, F., Beckmann, E., and Downing, K. H., 1990, Model for the structure of bacteriorhodopsin based on high-resolution electron cryo-microscopy, *J. Mol. Biol.*, 213:899.

Hildebrand, E. and Dencher, N., 1975, Two photosystems controlling behavioral responses of *Halobacterium halobium*, *Nature*, 257:46.

Khorana, H. G., 1988, Bacteriorhodopsin, a membrane protein that uses light to translocate protons, *J. Biol. Chem.*, 263:7439.

Lanyi, J. K., 1986, Halorhodopsin: a light-driven chloride ion pump, *Annu. Rev. Biophys. Biophys. Chem.*, 70:319.

Oesterhelt, D. and Stoeckenius, W., 1973, Functions of a new photoreceptor membrane, *Proc. Natl. Acad. Sci. USA*, 70:2853.

Schimz, A., 1981, Methylation of membrane proteins is involved in chemosensory and photosensory behavior of *Halobacterium halobium*, *FEBS Lett.*, 125:205.

Spudich, E. N. and Spudich, J. L., 1982, Control of transmembrane ion fluxes to select halorhodopsin-deficient and other energy-transduction mutants of *Halobacterium halobium*, *Proc. Natl. Acad. Sci. USA*, 79:4308.

Spudich, E. N., Hasselbacher, C. A. and Spudich, J. L., 1988, A methyl-accepting protein associated with bacterial sensory rhodopsin I, *J. Bacteriol.*, 170:4280.

Spudich, E. N., Sundberg, S. A., Manor, D. and Spudich, J. L., 1986a, Properties of a second sensory rhodopsin in *Halobacterium halobium*, *Proteins* 1:239.

Spudich, E. N., Takahashi, T., and Spudich, J.L., 1989, Sensory rhodopsins I and II modulate a methylation/demethylation system in *Halobacterium halobium* phototaxis, *Proc. Natl. Acad. Sci. USA*, 20:7746.

Spudich, J. L. and Bogomolni, R. A., 1988, Sensory rhodopsins of halobacteria, *Annu. Rev. Biophys. Biophys. Chem.*, 17:193.

Spudich, J. L. and Bogomolni, R. A., 1984, The mechanism of colour discrimination by a bacterial sensory rhodopsin, *Nature*, 312:509.

Spudich, J. L., McCain, D. A., Nakanishi, K., Okabe, M., Shimizu, N., Rodman, H., Honig, B. and Bogomolni, R. A., 1986b, Chromophore/protein interaction in bacterial sensory rhodopsin and bacteriorhodopsin, *Biophys. J.*, 49:479.

Stoeckenius, W. and Bogomolni, R. A., 1982, Bacteriorhodopsin and related pigments of halobacteria, *Annu. Rev. Biochem.*, 52:587.

Sundberg, S. A., Alam, M., Lebert, M., Spudich, J. L., Oesterhelt, D. and Hazelbauer, G. L., 1990, Characterization of mutants of *Halobacterium halobium* defective in taxis, *J. Bacteriol.*, 172:2328.

Takahashi T., Tomioka, H., Kamo, N. and Kobatake, Y., 1985, A photosystem other than PS370 also mediates the negative phototaxis of *Halobacterium halobium*, *FEMS Microbiol. Lett.*, 28:161.

Takahashi, T., Yan, B., Mazur, P., Derguini, F., Nakanishi, K., and Spudich, J. L., 1990, Color regulation in the archaebacterial phototaxis receptor phoborhodopsin (sensory rhodopsin II), *Biochemistry*, 29:8567.

Tomioka, H., Takahashi, T., Kamo, N., and Kobatake, Y., 1986, Flash spectrophotometric identification of a fourth rhodopsin-like pigment in *Halobacterium halobium*, *Biochem. Biophys. Res. Commun.*, 139:389.

Yan, B., Takahashi, T., Johnson, R., Derguini, F., Nakanishi, K. and Spudich, J. L., 1990a, All-*trans*/13-*cis* isomerization of retinal is required for phototaxis signaling by sensory rhodopsins in *Halobacterium halobium*, *Biophys. J.*, 57:807.

Yan, B., Takahashi, T., McCain, D. A., Rao, V. J., Nakanishi, K., and Spudich, J. L., 1990b, Effects of modifications of the retinal beta-ionone ring on archaebacterial sensory rhodopsin I, *Biophys. J.*, 57:477.

Absorption and Action Spectroscopy of Phoborhodopsin (Sensory Rhodopsin II)

Tetsuo Takahashi

Faculty of Pharmaceutical Sciences
Hokkaido University
Sapporo 060 Japan

Introduction

The presence of the second photosensory pigment (Takahashi et al., 1985a) called phoborhodopsin (pR, Tomioka et al., 1986b) or sensory rhodopsin II (sR-II, Spudich et al., 1986a) was first suggested during the experiment on action spectroscopy of a *Halobacterium halobium* (*H. halobium*) mutant (Tomioka et al., 1986a). *H. halobium* is an extremely halophilic bacterium in which four retinal-containing membrane proteins (often referred to as bacterial rhodopsins) have been discovered so far. Among them, bacteriorhodopsin (bR) and halorhodopsin (hR) utilize light energy to pump out protons from the interior of the cell, or to transport chloride ions inwardly across the cell membrane, respectively (for review, see Henderson et al., 1990; Khorana, 1988; Lanyi, 1990). The first photosensory pigment was found in a both bR and hR-defective strain, which is deficient in light-driven transmembrane ion movement but retains the photobehavioral response (Bogomolni and Spudich, 1982; Spudich and Spudich, 1982). The pigment was called sensory rhodopsin or sensory rhodopsin I (sR or sR-I; Spudich and Bogomolni, 1984). This was the best candidate for receptors of both photoattractant and photorepellent responses which have been shown in halobacteria (Hildebrand and Dencher, 1975).

It was of course important to attribute the photobehavioral responses to the function of the newly discovered pigment on the basis of behavioral studies. The repellent response of *H. halobium* to near UV light apparently corresponded to the absorption of an intermediate (λ_{max}, 373 nm) in the photoreaction cycle of sR-I, since it was observable only under background light which accumulates the long-lived intermediate (Spudich and Bogomolni, 1984, Takahashi et al., 1985c). On the other hand, the attractant response often showed an action maximum at 565 nm (Hildebrand and Dencher, 1975; Schimz et al., 1982; Hildebrand and Schimz, 1983; Dencher, 1983), which is substantially different from the absorption maximum of sR-I (587 nm; Spudich and Bogomolni, 1983). It is possible that bR and/or hR contribute to the photoattractant response in a manner mediated by light-induced changes in the proton motive force of the cell (Bibikov and Skulachev, 1989). We therefore were interested in the action spectrum of the photoattractant response of the bR$^-$, hR$^-$ strain. In a sense,

Biophysics of Photoreceptors and Photomovements in Microorganisms
Edited by F. Lenci *et al.*, Plenum Press, New York, 1991

we intended to demonstrate the pure action spectrum of sR-I. The observed spectrum, however, was distorted at wavelength shorter than 580 nm, suggesting the presence of a second sensory pigment which mediates additional repellent response (Tomioka et al., 1986a). Furthermore, a mutant was obtained which did not show the photoattractant response but show a repellent response to blue-green light. The mutant also lacked the background-depended near UV response.

The action spectrum of the repellent response of the mutant cells to blue-green light was measured. The shape of the spectrum is unique and the spectral bandwidth is not so broad as those expected for a retinal-containing pigment, whereas the behavioral response apparently requires retinal (Takahashi et al., 1987). Recently, by means of retinal analog incorporation into apo-membranes of the mutant strain, the absorption property of pR was elucidated (Takahashi et al., 1990). This paper provides a short review of halobacterial action spectroscopy, the action spectrum of sensory rhodopsins, and the spectral property of pR.

Action Spectroscopy of Halobacteria

H. halobium is a rod shaped, polarly flagellated bacterium of 0.5-1.0 μm wide by 3-4 μm in length. The cell swims in almost saturated brine with a velocity of 2-6 μm/s. A period of smooth swimming of a cell lasts 3-15 s, interrupted by an abrupt stop which is usually followed by a reversal of the swimming direction. A light stimulus modulates the frequency of the interruption of the smooth swimming. The response to a repellent light can be classified as a photophobic response since the cells show only a stop or a reversal of the swimming direction. (Hildebrand and Dencher, 1975; Hildebrand and Schimz, 1990). A step-down stimulus of repellent light or step-up of attractant light (λ > 530 nm) causes a kind of klinokinesis (Diehn et al., 1977), i.e., the cell prolongs the period of smooth swimming (Spudich and Stoeckenius, 1979).

For measurement of the photobehavioral responses, the cells in complex medium are put between a slide and a cover glass set under a microscope objective. Either dark field illumination or phase-contrast optics can be used with a 32-40 x objective. Computer-linked video image analysis greatly reduced the time and the labor in tracking individual cells (Takahashi and Kobatake, 1982; Sundberg et al., 1985). The actinic light illuminates the cells vertically through the microscope optics either via the objective or via the condenser lens. Because the thickness of the cell suspension is less than 0.2 mm, and normally the cell density is as low as 10^6/ml, the light absorbed or scattered by the cell suspension is negligible. Therefore, no correction for the observed action spectrum is necessary in our experimental system. Either the time-lag between a light stimulus and the response or the fraction of the reversed cells within a given period after a light stimulus can be used as a measure of the magnitude of a response.

The principle of action spectroscopy is based on the Grotthus-Draper's law. The response is a function of the number of photons absorbed:

$$R = F(N_p) \tag{1}$$

where R is the magnitude of the behavioral response of the cell, and N_p the number of photons absorbed by the cell.

From the Lambert-Beer's law, we can calculate the photon absorption by a material having a very short path length.

$$-\Delta I = \log_e 10 \, I \, \epsilon \, c \Delta l \tag{2}$$

where I is the intensity of the incident light, ΔI the change in the intensity of the light due to absorption by the material, c the concentration of the pigment in the material, Δl the path length of the material, and ϵ the molar extinction coefficient of the pigment. Note that $c\,\Delta l$ indicates a density of the pigment molecule in a unit area of a plane perpendicular to the beam axis.

If we imagine that the light beam is so narrow that it illuminates only the entire body of a single bacterial cell, and the thickness of the cell can be assumed to be Δl, then $c\,\Delta l$ becomes the number of the pigments in the cell divided by both the cross-sectional area of the cell and Avogadro's number, i.e.,

$$c\,\Delta l = p/N_A\,s \tag{3}$$

where p is the number of the pigments in the cell, N_A the Avogadro's number, and s the cross-sectional area of the cell projected to a plane perpendicular to the light beam.

On the other hand, the number of photons absorbed by the cell in a unit time is represented as:

$$N_p = -\Delta I\,s \tag{4}$$

Using equation (2) and (3), we obtain

$$N_p = \log_e 10\,I\,p\,\epsilon/N_A \tag{5}$$

The term $\log_e 10\,\epsilon/N_A$ is called capture cross section of the pigment, denoted by σ. And,

$$\sigma = 3.81 \times 10^{-21}\,\epsilon\ [\mathrm{cm}^2] \tag{6}$$

Equation (5) is the basis for action spectroscopy of *H. halobium*. This indicates that if the product of I and ϵ is kept constant at various wavelengths, then $R = F(N_p)$ becomes constant at any of these wavelengths. All we have to do is to plot the reciprocal of the light intensity at which the cell shows constant response probability. The plot against wavelengths then becomes the spectrum of ϵ, i.e., the absorption spectrum.

Fluence Rate-Response Curves

Practically, to find the light intensity that causes the defined response, we first measure the fluence rate-response curves in which the magnitude of the response is plotted against the intensity of actinic light. The shape or the steepness of the fluence rate-response curves is a good measure for the validity of the action spectroscopy. If a single photoreceptor species is concerned, and no other term is dependent on the wavelength than the molar extinction coefficient, equations (1) and (5) indicate that these curves become parallel with each other when the magnitude of the response is plotted against the logarithm of the light intensity.

A typical example is our measurement of the photoattractant response of bR^-, hR^- mutant strain, where the fluence rate-response curves run in parallel to each other only at wavelengths longer than 580 nm (Tomioka et al., 1986a). We measured the step-down response and obtained the light intensity at which 50 % of the cells showed the reversal response. However, a step-down of the actinic light at wavelengths shorter than 560 nm never resulted in more than 50 % reversal of the cells. At 565 nm, the fraction of reversing cells increases monotonously with increasing intensity of the actinic light to less than 10^{14} photons/mm^2 s. At intensities higher than 2×10^{14} photons/mm^2 s, however, the fraction steeply decreased with increasing the intensity of the actinic light. Simultaneous measurement of step-down and step-up responses indicated that high intensity actinic light caused a step-up photophobic response at wavelengths shorter than 580 nm. This not only indicates the existence of two, step-down and step-up, photophobic systems, but means that the fluence rate-response relationship is different between the two photosystems.

A detailed discussion of the shape of fluence rate-response curves of halobacterial photobehavior is given by Marwan et al. (1988).

Action Spectrum of Halobacterial Sensory Rhodopsins

A genuine action spectrum of sensory rhodopsin (sR-I) was not yet obtained. As described above, *H. halobium* strain Flx3 (bR^-, hR^-, sR^+) contained the second sensory pigment pR (Takahashi et al., 1985a; Tomioka et al., 1986a). In contrast to sR-I, pR mediates only a step-up photophobic response and a step-down photoattractant response (klinokinesis) of the cells to blue-green light. When the magnitude of step-up and step-down responses of Flx_3 cells were plotted against the wavelength - one in the positive direction of the longitudinal axis and the other in the negative direction - the curve intersected the abscissa at 550 nm (Takahashi, unpublished result). Using a population method, Wolff et al. (1986) also observed a similar result with their strain Flx_3 km1. They also reported that their Flx_3 strain contained only a small amount of pR, as determined by photobehavioral measurements. Even in the same strain, the pigment content varies depending on the growth condition (Takahashi et al., 1985a). Furthermore, it may vary between different laboratories since the halobacterial genome has a remarkably high mutation rate (Pfeifer, 1986), and the pigment content may change after repetitive inoculation and growth. Spudich et al. (1986a) reported that strain Flx_3R, a retinal deficient derivative of Flx_3, showed no detectable amount of pR (sR-II) as judged by $[^3H]$-retinal labeling. Pure action spectrum of sR-I may be obtained with this strain.

On the other hand, a pure action spectrum of pR has been determined using an sR-I deficient derivative of Flx_3 (Nakamori et al., in preparation). Fluence rate-response curves are parallel to each other from 350 nm to 550 nm. Light intensity that causes 50 % reversal ranges from 2×10^{13} photons/mm^2 s at 485 nm to 10^{15} photons/mm^2 s at 550 nm. A prominent feature of the observed action spectrum is that it shows a shoulder at 460 nm in addition to the main peak. No absorption spectrum of a retinal-containing protein has been reported exhibiting such a structure at ambient temperature.

Absorption Spectroscopy of pR

Flash Spectroscopy

A cyclic photoreaction occurs in pR after bleaching by a flash (Tomioka et al., 1986b; Scherrer et al., 1987). An intermediate absorbing at $\approx$360 nm appears and decays with a half time of 150-200 ms. At natural pH, the decay of the UV intermediate (P_{350}) is accompanied by formation of another intermediate with a bathochromic shift (P_{530}). The life-time of P_{530} is about 300 ms. Apparently, these intermediates correspond to the M and O intermediates of bR (Stoeckenius et al., 1982), though the detection of intermediates having a fast decay time ($t_{1/2} < 1$ ms) is difficult due to the low concentration of the pigment in native membranes. Shichida et al., (1988) observed a bathochromic

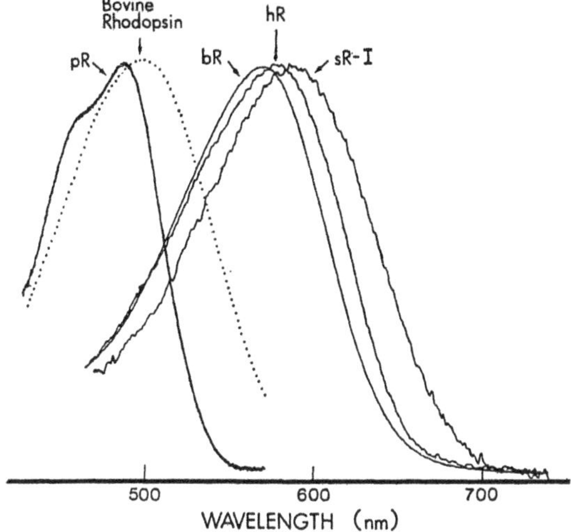

Fig. 1. Absorption spectrum of pR (from Takahashi et al., 1990) compared to those of bovine rhodopsin (from Koutalos et al., 1989) and halobacterial rhodopsins (from Spudich and Bogomolni, 1983).

intermediate (P_{520}) before P_{350} by low temperature spectroscopy. This presumably corresponds to the K intermediate of bR.

The qualitative similarity of pR photocycle to that of bR may suggest a similar geometry around the active site. However, even when pR pumps protons, the detection would be difficult because of both the slow cycling rate of pR and the low pigment content in the vesicle preparations.

A flash-induced difference spectrum between pR and P_{350} also showed a shoulder at 460 nm (Tomioka et al., 1986b).

Absorption Spectrum

Even when the concentration of a retinal-protein chromophore in the native membrane is low, the absorption spectrum of the chromophore can be measured as a difference between the absorption in the apo-membrane suspension and that in the membrane reconstituted with exogenous chromophores (Spudich and Bogomolni, 1983). Care has to be taken when this method is applied to pR, because the absorbance of pR in these natural membrane suspension yields only 10-30 mOD (Takahashi et al., 1990). The low pigment content indicates the relatively high quantum efficiency in the function of pR (Marwan et al., 1988).

The observed absorption spectrum of pR is compared to those of other retinal-containing pigments in Figure 1 (the maximum height of absorbance in each pigment is normalized). The absorption maximum of pR is isolated from those of other bacterial rhodopsins, and a shoulder at 460 nm in the pR spectrum is obvious. A number of data suggest that the shoulder is due to a vibrational structure. Note that neither retinal itself in solution nor so far known retinal-containing proteins exhibit such a structured absorption (Christensen and Köhler, 1973; Birge et al., 1982).

First, the bandwidth of the pR spectrum is narrower than those expected for retinal-containing proteins (Dartnall, 1953; Ebrey and Honig, 1977). If the pR spectrum is composed of the spectra of two retinal-protein chromophores, it should be wider than a normal spectrum of retinal protein. Secondly, the energy difference between the main peak and the shoulder matches the vibronic spacing of the chromophore, i.e., the 1550

cm^{-1} difference in the pR spectrum is close to those C-C stretching frequencies observed in bacterial and visual rhodopsins by resonance Raman method (Fodor et al., 1989). We know that any UV-visible absorption band of a compound consists of a number of vibronic transitions. Sometimes we see a typical three peaked vibrational structure in the absorption spectrum of a compound. Whether it exhibits some structure or not is dependent on the relation between the vibronic spacing and the intrinsic width of each vibronic band. A simulation of the PR spectrum assuming an intrinsic width of 760 cm^{-1} reproduced the observed spectrum well (Takahashi et al., 1990).

A diffuse spectrum of retinal in solution is observed even at liquid nitrogen temperature (Christensen and Köhler, 1973). This is explained by the distorted *cis* conformation of the C_6-C_7 single bond (Warshel and Karplus, 1974; Birge et al., 1982; Meyers et al., 1983). Therefore, a major cause of the structure in pR can be assumed to be a planar C_6-C_7 conformation as first suggested by Schreckenbach et al. (1977) for the chromophore of bR. It is further supported by the spectral change of a compound called C_{18}-keton that the chromophore is forced to have a planar conformation at the binding pocket in pR, along with the fact that naphthyl retinal binds to the apoprotein (Takahashi et al., 1990).

Mechanism of Opsin Shift and the Fine Structure

The absorption properties of retinal-containing pigments reconstituted with various analog chromophores provide information about the retinal binding site (Derguini and Nakanishi, 1986). A most intriguing, presumably the most important one is about the mechanism of color regulation (Honig et al., 1979; Nakanishi et al., 1980; Spudich et al., 1986b). This approach has been applied to pR, using a series of dihydro analogues of retinal (Takahashi et al., 1990). The external point charge (or a dipole-dipole interaction; Kakitani et al., 1985), a dominant factor regulating the colors of visual and bacterial rhodopsins, seemed not important in pR. Rather, a planar conformation of the chromophore contributes about one third of the spectral shift from a model retinal-Schiff's base chromophore to the pigment. Also, a tight interaction between the Schiff base and its counter ion explains why λ_{max} of pR is much shorter than those of bacterial rhodopsins, in addition to the lack of an external point charge. Note that both the lack of an external point charge and the tight interaction between the Schiff's base and the counter ion potentially make the intrinsic width of vibronic bands narrow, thereby causing the fine structure in pR.

Acknowledgements

Part of the work described was performed in Y. Kobatake's laboratory in Sapporo, and in J. L. Spudich's laboratory at the Albert Einstein College of Medicine, New York, in collaboration with K. Nakanishi's group at Columbia University, New York. I wish to thank B. Yan, H. Tomioka, and Y. Nakamori as well as Y.K., J.L.S. and K.N.

References

Bibikov, S. I., and Skulachev, V. P., 1989, Mechanism of phototaxis and aerotaxis in *Halobacterium halobium, FEBS Letters*, 243:303.

Birge, R. A., Bocian, D. F., and Hubbard, L. M., 1982, Origins of inhomogeneous broadening in the vibronic spectra of visual chromophores and visual pigments, *J. Am. Chem. Soc.*, 104:1196.

Bogomolni, R. A., and Spudich J. L., 1982, Identification of a third rhodopsin-like pigment in phototactic *Halobacterium halobium, Proc. Natl. Acad. Sci. USA*, 79:6250.

Christensen, R. L., and Kohler, B. E., 1973, Low resolution optical spectroscopy of retinyl polyenes: low lying electronic levels and spectral broadness, *Photochem. Photobiol.*, 18:293.

Dartnall, H. J. A., 1953, The interpretation of spectral sensitivity curves, *Br. Med. Bull.*, 9:24.

Dencher, N., 1983, The five retinal-protein pigments of halobacteria: bacteriorhodopsin, halorhodopsin, P 565, P 370, and slow-cycling rhodopsin, *Photochem. Photobiol.*, 38:753.

Derguini, F., and Nakanishi, K., 1986, Synthetic rhodopsin analogs, *Photobiochem. Photobiophys.*, 13:259.

Diehn, B., Feinleib, M., Haupt, W., Hildebrand, E., Lenci, F., and Nultsch, W., 1977, Terminology of behavioral responses of motile microorganisms, *Photochem. Photobiol.*, 26:559.

Ebrey, T. G., and Honig, B., 1977, New wavelength dependent visual pigment nomograms, *Vision Res.*, 17:147.

Fodor, S. P. A., Gebhard, R., Lugtenburg, J., Bogomolni, R. A., and Mathies, R. A., 1989, Structure of the retinal chromophore in sensory rhodopsin I from resonance Raman spectroscopy, *J. Biol. Chem.*, 264:18280.

Henderson, R., Baldwin, J. M., Ceska, T. A., Zemlin, F., Beckmann, E., and Downing, K. H., 1990, Model for the structure of bacteriorhodopsin based on high-resolution electron cryo-microscopy, *J. Mol. Biol.*, 213:899.

Hildebrand E., and Dencher, N., 1975, Two photosystems controlling behavioral responses of *Halobacterium halobium, Nature*, 257:46-48.

Hildebrand, E., and Schimz, A., 1983, Photosensory behavior of a bacteriorhodopsin-deficient mutant, ET-15, of *Halobacterium halobium, Photochem. Photobiol.*, 37:581.

Hildebrand, E., and Schimz, A., 1990, The lifetime of photosensory signals in *Halobacterium halobium, Biochim. Biophys. Acta*, 1052:96.

Honig, B., Dinur, U., Nakanishi, K., Balogh-Nair, V., Gawinowicz, M. A., Arnaboldi, M., and Motto, M. G., 1979, An external point-charge model for wavelength regulation in visual pigments, *J. Am. Chem. Phys.*, 101:7084.

Kakitani, H., Kakitani, T, Rodman, H, and Honig, B., 1985, On the mechanism of wavelength regulation in visual pigments, *Photochem. Photobiol.*, 41:471.

Khorana, H. G., 1988, Bacteriorhodopsin, a membrane protein that uses light to translate proteins, *J. Biol. Chem.*, 263:7439.

Koutalos, Y., Ebrey, T. G., Tsuda, M., Odashima, K., Lien, T., Park, M. H., Shimizu, N., Derguini, F., Nakanishi, K., Gilson, H. R., and Honig, B., 1989, Regulation of bovine and octopus opsins *in situ* with natural and artificial retinals, *Biochemistry*, 28:2732.

Lanyi, J. K., 1990, Halorhodopsin, a light-driven electrogenic chloride-transport system, *Physiol. Rev.*, 70:319.

Marwan, W., Hegemann, P., and Oesterhelt, D., 1988, Single photon detection by an archaebacterium, *J. Mol. Biol.*, 199:663.

Meyers, A. B., Harris, R. A., and Mathies, R. A., 1983, Resonance Raman excitation profiles of bacteriorhodopsin, *J. Chem. Phys.*, 79:603.

Nakanishi, K., Balogh-Nair, V., Arnaboldi, M., Tsujimoto, K., and Honig, B., 1980, An external point-charge model for bacteriorhodopsin to account for its purple color, *J. Am. Chem. Soc.*, 102:7945.

Pfeifer, F., 1986, Genetics of Halobacteria, *in*: "Halophilic Bacteria, Volume II," F. Rodriguez-Valera, ed., CRC Press, Boca Raton, pp. 105.

Schreckenbach, T., Walckoff, B., and Oesterhelt, D., 1977, Studies on the retinal-protein interaction in bacteriorhodopsin, *Eur, J. Biochem.*, 76:499.

Scherrer, P., McGinnis, K., and Bogomolni, R. A., 1987, Biochemical and spectroscopic characterization of the blue-green photoreceptor in *Halobacterium halobium, Proc. Natl. Acad. Sci. USA*, 84:402.

Schimz, A., Sperling, E., Hildebrand, E., and Köler-Hahn, D., 1982, Bacteriorhodopsin and the sensory pigment of the photosystem 565 in *Halobacterium halobium, Photochem. Photobiol.*, 36:193.

Shichida, Y., Imamoto, Y, Yoshizawa, T., Takahashi, T., Tomioka, H., Kamo, N., and Kobatake, Y., 1988, Low-temperature spectrophotometry of phoborhodopsin, *FEBS Letters*, 236:333.

Spudich, J. L., and Stoeckenius, W., 1979, Photosensory and chemosensory behavior of *Halobacterium halobium, Photobiochem. Photobiophys.*, 1:45.

Spudich, E. N., and Spudich, J. L., 1982, Control of transmembrane ion fluxes to select halorhodopsin-deficient and other energy-transduction mutants of *Halobacterium halobium*, *Proc. Natl. Acad. Sci. USA.*, 79:4308.

Spudich, J. L., and Bogomolni, R. A., 1983, Spectroscopic discrimination of the three rhodopsinlike pigments in *Halobacterium halobium*, *Biophys. J.*, 43:243.

Spudich, J. L., and Bogomolni, R. A., 1984, Mechanism of color discrimination by a bacterial sensory rhodopsin, *Nature*, 312:509.

Spudich, E. N., Sundberg, S. A., Manor, D., and Spudich, J. L., 1986a, Properties of a second sensory receptor protein in *Halobacterium halobium*, *PROTEINS: Structure Function and Genetics*, 1:239.

Spudich, J. L., McCain, D. A., Nakanishi, K., Okabe, M., Shimizu, N., Rodman, H., Honig, B., and Bogomolni, R. A., 1986b, Chromophore/protein interaction in bacterial sensory rhodopsin and bacteriorhodopsin, *Biophys. J.*, 49:479.

Stoeckenius, W., and Bogomolni, R. A., 1982, Bacteriorhodopsin and related pigments of halobacteria, *Annu. Rev. Biochem.*, 52:587.

Sundberg, S. A., Bogomolni, R. A., and Spudich, J. L., 1985, Selection and properties of phototaxis-deficient mutants of *Halobacterium halobium*, *J. Bacteriol.*, 164:282.

Takahashi, T., and Kobatake, Y., 1982, Computer-linked automated method for measurement of the reversal frequency in phototaxis of *Halobacterium halobium*, *Cell Struct. Funct.*, 7:183.

Takahashi, T., Mochizuki, Y., Kamo, N., and Kobatake, Y., 1985c, Evidence that the long-lifetime photointermediate of s-rhodopsin is a receptor for negative phototaxis in *Halobacterium halobium*, Biochem. Biophys. Res. Commun., 127:99.

Takahashi, T., Tomioka, H., Kamo, N., and Kobatake, Y., 1985a, A photosystem other than PS370 also mediates the negative phototaxis of *Halobacterium halobium*, *FEMS Microbiol. Lett.*, 28:161.

Takahashi, T., Watanabe, M., Kamo, N., and Kobatake, Y., 1985b, Negative phototaxis from blue light and the role of third rhodopsinlike pigment in *Halobacterium cutirubrum*, *Biophys. J.*, 48:235.

Takahashi, T., Tomioka, H., Nakamori, Y., Kamo, N., and Kobatake, Y., 1987, Phototaxis and the second sensory pigment in *Halobacterium halobium*, *in*: "Primary Processes in Photobiology," T. Kobayashi, ed., Springer-Verlag, Berlin, Hiderberg, pp. 101.

Takahashi, T., Yan, B., Mazur, P., Delguini, F., Nakanishi, K., and Spudich, J. L., 1990, Color regulation in the archaebacterial phototaxis receptor phoborhodopsin (sensory rhodopsin II), *Biochemistry*, 29:8467.

Tomioka, H., Takahashi, T., Kamo, N., and Kobatake, Y., 1986a, Action spectrum of the photoattractant response of *Halobacterium halobium*, *Biochim. Biophys. Acta* 884:578.

Tomioka, H., Takahashi, T., Kamo, N., and Kobatake, Y., 1986b, Flash spectrophotometric identification of a fourth rhodopsin-like pigment in *Halobacterium halobium*, *Biochem. Biophys. Res. Commun.*, 139:389.

Warshel, A., and Karplus, M., 1974, Calculation of $\pi\pi^*$ excited state conformations and vibronic structure of retinal and related molecules, *J. Am. Chem. Soc.*, 96:5677.

Wolff, E. K., Bogomolni, R. A., Scherrer, P., Hess, B., and Stoeckenius, W., 1986, Color discrimination in halobacteria: Spectroscopic characterization of a second sensory receptor covering the blue-green region of the spectrum, *Proc. Natl. Acad. Sci. USA*, 83:7272.

Photoreception and Photomovements in Blepharisma japonicum

Francesco Ghetti

C.N.R. Istituto di Biofisica
via San Lorenzo 26
56127 Pisa
Italy

Death and Light are everywhere, always and they begin, end, strive, attend into and upon the Dream of the Nameless, that is the World...
Roger Zelazny "Lord of Light"

Introduction

From the beginning of the century, the red heterotrichous ciliate *Blepharisma* has been an interesting subject of studies in the field of photobiology because of its endogenous pigment blepharismin. In fact, if *Blepharisma* is exposed to relatively strong light, in the presence of oxygen, blepharismin acts as a strong photosensitizer and readily causes its death. This almost unique behavior of an organism producing a pigment that photosensitizes not only its predators but the organism itself, is a puzzle for protozoologists studying evolutionary processes.

In spite of this clear effect of light on the very survival of *Blepharisma*, its motile reactions to light stimuli have been studied in detail only since a few years.

Blepharismin and Blepharisma Photobiology

Blepharisma lives at the bottom of ponds and ditches, usually under plant debris, where it can find food (bacteria, but also small ciliates and even other individuals of its own species) and can shield itself from the lethal action of light.

The cell is generally pear-shaped and its body length ranges from 150 to 500 μm, depending on the variety and the nutrition. If cultivated in the dark or in dim light (the natural condition), it appears red or pink, but in non-controlled conditions its color markedly depends on the illumination.

Concerning the physiology of *Blepharisma* and the basic processes of its photodynamic sensitization, the reader is referrd to Giese (1973, 1981). Some basic points are, however, worthwhile to be reminded:

a) the action spectrum for photodynamic immobilization of *Blepharisma* spans the visible spectrum up to 600 nm and it suggests blepharismin as the photosensitizing pig-

Biophysics of Photoreceptors and Photomovements in Microorganisms
Edited by F. Lenci *et al.*, Plenum Press, New York, 1991

ment; the photokilling action only occurs in the presence of oxygen for light intensities above about 30 W/m^2.
b) Under relatively low intensity irradiation (about 3.5 W/m^2), the red form of blepharismin progressively converts into a gray-blue form, apparently neither toxic nor phototoxic for the cell; this transformation too requires the presence of oxygen. If the cells are exposed to slightly higher light intensities for several days, they expel most of the pigment. Keeping the cells in the dark or under dim light in a nutrient medium, red blepharismin is re-synthetized within two days.
c) Blepharismin is mainly localized in subpellicular membrane-limited granules of about 500 nm in diameter; they are arranged in strings parallel to the ciliary rows and spread all over the cellular body; electron microscopy shows an internal organization in these organella.

Sevenants (1965) developed a procedure to obtain acetone crude extracted pigment from *Blepharisma* and he demonstrated, by means of optical and IR spectroscopy, the analogy between blepharismin and hypericin (see Fig. 1), a *meso*-naphtodianthrone pigment synthetized in species of the genus *Hypericum* and quite similar to the chromophore of stentorins, the photoreceptor proteins of *Stentor coeruleus* (Song, 1983; Kim et al., 1990; Song et al., 1990). Actually a wide variety of molecules based either on the anthraquinone or the *meso*-naphtodianthrone skeleton are present in plants and even in some species of insects (Banks et al., 1976).

Sevenants (1965) proposed a structural formula for blepharismin in which the two methyls in the 2 and 2' positions (see Fig. 1) had been substituted by two unspecified residues. Moreover, he suggested that blepharismin might form aggregates due to a hydrogen bonding network.

Interestingly, the optical absorption spectrum of crude extracts of blue blepharismin definitely resembles that of hypericin, suggesting that blepharismin might be a reduced form of hypericin (see Fig. 2).

Even though a precise determination of the intracellular blepharismin concentration is not possible, it seems quite high. In fact the *in vivo* optical density, measured by means of a computer-controlled micro-spectrophotometer, reaches the maximum value of about 0.75 at 580 nm (Gualtieri et al., 1989). The molar extinction coefficient of blepharismin can be assumed about $3x10^4$ at this wavelength, that of hypericin being

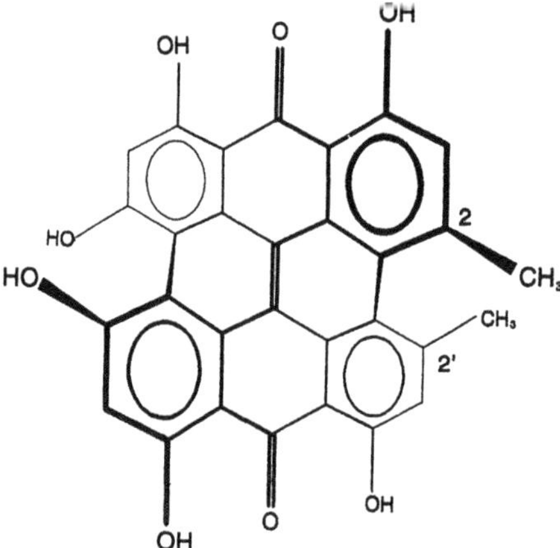

Fig. 1. Molecular structure of hypericin. The two anthraquinonic parts of the molecule are connected by a central double bond and two single bonds. As the double bond is shorter, the molecule is not planar but twisted

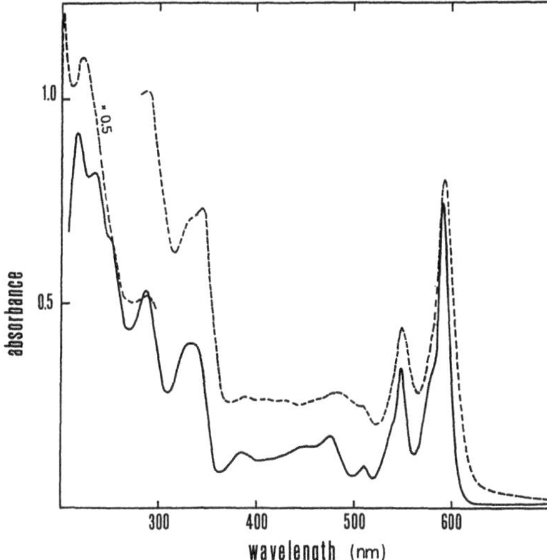

Fig. 2. Optical absorption spectra of hypericin (solid line) and of the blue form of blepharismin (broken line)

$2.7x10^4$. With this figures and an estimated optical path of the order of 1 μm, a value for the blepharismin concentration of about 0.25 M can be calculated.

The general photophysical and photochemical properties of the anthraquinone derivatives are described in a recent review by (Navas Diaz, 1990). Besides being efficient photosensitizers through the triplet state, yielding, for example, singlet molecular oxygen (Spikes, 1989), these molecules are able to participate in many photochemical reactions, like keto-enol tautomerism, photoreduction, photooxidation and other processes requiring oxygen. Moreover, the acidity of the hydroxy-anthraquinones is greater in the singlet excited state, so that photoexcitation makes them more efficient proton donors.

As a final observation on one of the possible functions of blepharismin, it is worthwhile to mention a recent work of Miyake et al. (1990) that describes the offense-defense interaction between *Blepharisma* and the predator ciliate *Dileptus margaritifer*. Using normal red cells, bleached cells and albino mutant cells it was possible to demonstrate the toxic effect of blepharismin on predator cells and hence its defensive function.

Phenomenology of Blepharisma Photomovements

Blepharisma is known to accumulate in shaded regions (Giese, 1981), and this behavior was described as negative phototaxis, resulting from two different photomotile responses: step-up photophobic response and positive photokinesis (Matsuoka, 1983b).

Actually it has been reported that *Blepharisma* is not able to perceive light direction (Kraml and Marwan, 1983; Passarelli et al., 1984; Scevoli et al., 1987). So the accumulation of cells in non-illuminated areas is the result of a trial-and-error process, assisted by the change in velocity: dark and light may be seen as the analogues of the attractant and the repellent signals controlling frequency of reversal movement in *Halobacterium*.

Blepharisma's velocity increases when the cells are irradiated with continuous light, the rise time and the the saturation level of velocity depending on light intensity

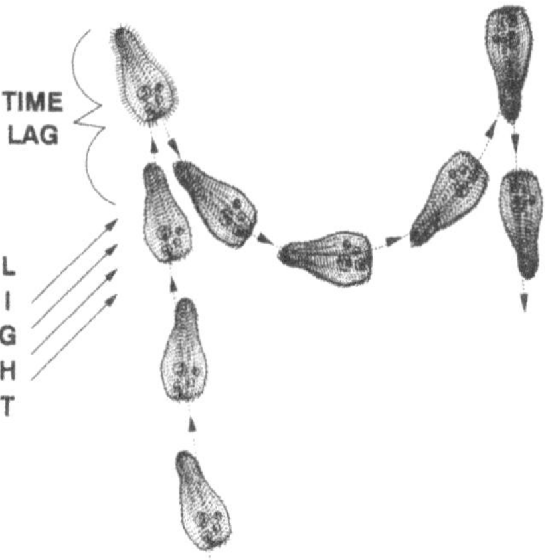

Fig. 3. Step-up photophobic reaction of *Blepharisma*. The time-lag represents the delay between the onset of the light stimulus and the occurrence of the stop response

(Matsuoka, 1983b; Kraml and Marwan, 1983). The maximum velocity value is about 0.4 mm/s with white light intensities up to 16 W/m^2 according to Matsuoka (1983b), whereas it gets to about 1.5 mm/s at a fluence rate of 2.5 W/m^2 (537 nm) according to Kraml and Marwan (1983). Thus, the change in speed is an efficient means for *Blepharisma* to implement its strategy to stay in the dark and to escape from illuminated areas.

The step-up photophobic response of *Blepharisma*, elicited by sudden increases in light intensity, consists of a stop of forward movement, followed by backward swimming along a bent trajectory, so that the cell's longitudinal axis changes its direction (see Fig. 3). If the rise in light intensity results from entering into an illuminated area (a light trap) and the cell, moving backward, attains the dark, it simply resumes forward swimming in a new direction (Kraml and Marwan, 1983). Under continuous illumination the light avoiding maneuver is followed by a tumbling and it is repeated until the cell adapts to the new illumination conditions (Matsuoka, 1983b; Checcucci et al., in preparation).

In dark-adapted cells, the response is elicited by light intensities above about 10-20 mW/m^2 at the most effective wavelength (Kraml and Marwan, 1983; Ghetti et al., unpublished results). The time-lag between the application of the stimulus and the appearance of the reaction, depends on the light intensity, and its minimum value is reported to be not less than 1 s for a gap in white light intensity of the order of 4 W/m^2 (Matsuoka, 1983b) or 1.4 s for a light intensity of the order of 2.2 W/m^2 at 400 nm (Kraml and Marwan, 1983). Also the adaptation time seems to increase with increasing light intensity (Checcucci et al., in preparation).

It is interesting to note that in a population of *Blepharisma*, even in the absence of any light stimulation, a percentage of cells show motile maneuvers similar to photophobic reactions. This "dark response", probably due to mechanical or chemical stimuli, is exhibited by 10-15% of the cells (Matsuoka, 1983a; Matsuoka, 1983b; Ghetti et al., unpublished results).

In addition to the above described photomotile reactions, an elongation of the cell body caused by light intensities higher than 1.5 W/m^2 was observed. This phenomenon had a time-lag of about two minutes and attained its maximum within six minutes. The elongation was light-intensity dependent, cells attaining 1.5 times their normal length at about 15 W/m^2. Ultrastructural studies of colchicine treated cells suggest the involvement of vacuole-associated microtubules in the change of shape (Matsuoka, 1983b; Matsuoka and Shigenaka, 1985).

Positive photokinesis might actually be partly due to a hydrodynamic effect dependent on the elongation of the cell body, but a simple calculation on the possible reduction of water resistance indicates that this contribution is not sufficient to produce the observed swimming acceleration. The change in shape seems rather aimed to achieve the expulsion of pigment granules to avoid the light sensitized killing.

Identification of Receptors for Photomovement

The identification of the pigment(s) triggering the photomotile responses of *Blepharisma*, mainly based on the determination of action spectra, is a debated issue. As a general consideration, blepharismin, that pervades the entire cell body and that is the prime reason for the cell to avoid even moderate light intensities, is a good candidate for the role, also taking into account that the related pigment stentorin is the photoreceptor for photomovements of the related ciliate *Stentor*.

Kraml and Marwan (1983) presented an action spectrum of the photophobic reaction using the light dependent time-lag to quantify the response. It showed a large band peaked at 400 nm, with a shoulder around 500 nm; wavelengths higher than 550 nm were quite ineffective to elicit the response. These authors suggested a blue light photoreceptor, identified with a yellow pigment, with absorption maximum at 420 nm, extracted from the cell by means of thin layer chromatography (TLC). However, on the basis of a comparison of the IR spectrum of this pigment with that of hypericin, they did not exclude this chromophore as a derivative of hypericin. In that same set of experiments, Kraml and Marwan, to prompt a very pronounced positive photokinesis, used light of 537 nm, a wavelength at which blepharismin absorbs, but not the "yellow pigment" responsible for photophobic reactions. Whereas these authors mentioned the existence of different sensory areas for the two photoresponses, they did not put forward any hypothesis on the possible role of blepharismin in regulating *Blepharisma* photobehavior.

According to our results, *Blepharisma* does exhibit photophobic reactions in the red region of the visible spectrum up to 600 nm, at fluence rates of the same order of those effective around 400 nm (Scevoli et al., 1987; Checcucci et al., in preparation).

Scevoli et al. (1987) determined two action spectra for the photophobic response, both ranging from 390 nm to 615 nm. An action spectrum, based on the response time-lag, was determined fitting at each wavelength the data points for the time-lag versus the fluence rate by means of a straight line; the absolute value of its slope was chosen as a measure of the effectiveness of the light at that wavelength.

For the second action spectrum, the percentage of responding cells (P) versus the fluence rate (I) was plotted and the best fitting function was the two-parameter exponential curve:

$$P = 1 - b \cdot \exp(-a \cdot I)$$

Analogously with the time-lag spectrum, the slope in the origin ($a \cdot b$) was chosen to quantify the action.

The two spectra are very similar and both show three maxima at about 480 nm, 540 nm and 580; these data are consistent with the identification of blepharismin as the photoreceptor for the photophobic response.

This hypothesis is further supported by the results of independent measurements of fluorescence lifetime and quantum yield performed on cold extruded granules and on crude extracted pigments. Fluorescence lifetimes, measured by means of a phase-shift fluorometer, were 0.62 ns and 1.08 ns, respectively; the fluorescence quantum yields, determined relative to rhodamine B, were 0.018 and 0.036, respectively.

From these data it was possible to estimate the rate constant for the radiative decay to ground level. In both cases the calculated value was about 3×10^7 s^{-1}, and the sum of the rate constants for radiationless deexcitation pathways was $1.6x10^9$ s^{-1} in the granules and $0.9x10^9$ s^{-1} in free blepharismin.

These figures indicate that the probability of processes alternative to fluorescence for releasing light energy is higher in the natural environment than in free molecules, and support the hypothesis that blepharismin is the actual photoreceptor pigment. In any phototransduction chain, in fact, the first step is likely to be not trivial radiative dissipation but non-radiative transformations yielding biochemical/biophysical signals for the cell (Scevoli et al., 1987; Lenci, this volume).

The existence of different sensory areas for photophobic response and for photokinesis was demonstrated, by means of microbeam irradiation of different parts of the cell. The anterior part was found to be mainly responsive to light stimuli eliciting the photophobic reaction, whereas the rear end seems to be the sensory zone for photokinesis (Matsuoka, 1983a; Kraml and Marwan, 1983).

According to microspectrofluorometric measurements, blepharismin photoreceptors are uniformly distributed over the whole cell body, so the hypothesis was formulated that the existence of specific sensory areas depends on an irregular distribution of Ca^{2+} channels in the plasmatic membrane (Colombetti and Lenci, unpublished results).

Hypotheses on Transduction Mechanisms

Whereas the role of blepharismin as photoreceptor pigment triggering the molecular events which end in the photophobic response can be thought as well established, at present a complete and experimentally well supported scheme of signal transduction in *Blepharisma* is unavailable.

It was demonstrated that the Ca^{2+}-blocking agent Ruthenium Red at $5x10^{-7}$ M reduced the percentage of reacting cells and completely inhibited the response at higher concentrations, without significantly affecting cell motility and unstimulated swimming pattern (Passarelli et al., 1984). So, as in other ciliates, the ciliary reversal, that is the final step of the transduction chain, seems to be prompted by an intracellular increase in Ca^{2+} concentration (Song, 1981).

Furthermore, experiments with the protonophore carbonyl-cyanide-chlorophenylhydrazone (CCCP) showed that the phobic response is reduced to 20% for CCCP at 5 μM and completely inhibited by CCCP at 50 μM. Also in this case no effect of CCCP on the motile behavior in the absence of a light stimulus was observed (Passarelli et al., 1984).

On the basis of what has been proposed for other ciliates, and particularly for *Stentor coeruleus* (Song, 1981), the following scheme for the photosensory transduction chain in *Blepharisma* was suggested: light absorbed by the photoreceptor pigment evokes a transient proton gradient that, in turn, generates, across the membrane, a pH gap or a steep potential variation that favors the above-mentioned Ca^{2+} influx responsible for ciliary beating reversal (Passarelli et al., 1984; Lenci and Ghetti, 1989; Lenci et al., 1989) (see Fig. 4).

The fundamental point to investigate is the actual occurrence of a proton release from blepharismin singlet excited state and its role of primary trigger for the transduction process. The proposed chemical structures for blepharismin (Sevenants, 1965) and for similar molecules are compatible with the hypothesis that changes in the electronic structure due to light absorption bring about an increase in the acidity of the hydroxyl groups (Navas Diaz, 1990). Thus, blepharismin may serve as a proton source.

Experiments on blepharismin crude extracts were performed to check this hypothesis. In pure ethanol, regardless of the excitation wavelength, the fluorescence emission spectrum of blepharismin showed a single band centered around 600 nm. Increasing the alkalinity of the medium by addition of NaOH, a second band centered at

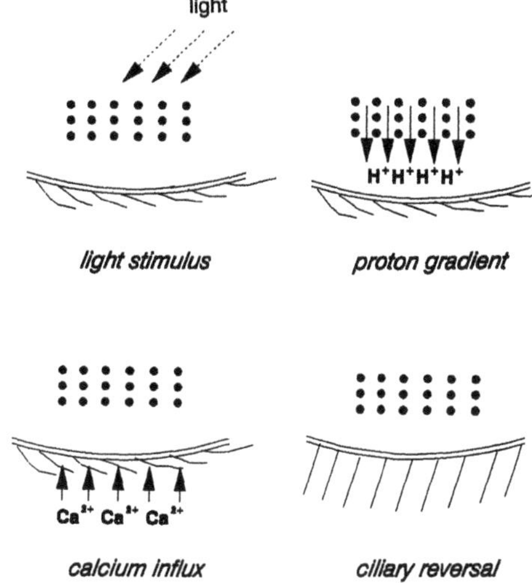

Fig. 4. Schematic representation of the "proton translocation hypothesis" for the photophobic reaction of *Blepharisma*. Light absorbed by the photoreceptor pigment in the granules evokes a transient proton gradient that, in turn, generates, across the membrane, a pH gap or a steep potential variation that favors a Ca^{2+} influx, responsible for ciliary beating reversal.

about 650 nm appeared, while the fluorescence emission at 600 nm diminished; only the 650 nm emission band was detectable at an NaOH concentration around 200 mM. Under the same experimental conditions, only minor shifts in absorption spectral maxima were observed.

These emission bands at 600 nm and 650 nm were attributed to the first excited singlet states of the neutral and of the anionic form of blepharismin, respectively, and the hypothesis was put forward that the charged species resulted from deprotonation of a hydroxyl group of the chromophore in its first excited singlet state (Lenci et al., 1989).

More recently, fluorescence lifetimes and time-gated spectra of blepharismin crude extracts have been measured in ethanol and aqueous solutions, at various OH^- concentrations (Cubeddu et al., 1990; Cubeddu et al., this volume).

In pure ethanol two molecular species, both emitting around 600 nm, with lifetimes 1 ns (relative amplitude ≈ 80%) and 0.5 ns (rel. amp. ≈ 20%), respectively, were observed; in aqueous solution (pH = 7) again two molecular species (lifetimes ≈ 0.9 ns and 0.2 ns; rel. amp. ≈ 9% and 91%) were detected. The different relative amounts of these species have been attributed to a solvent depending equilibrium between a phenolic (short lifetime) and a quinonic (long lifetime) neutral form of the pigment.

Upon increasing OH^- concentration, the anionic species, emitting at 660 nm, with a lifetime of 4-6 ns, appeared, its relative amplitude attaining 50% in ethanol (NaOH 100 mM) and 90% in aqueous solution (pH = 13).

By means of time-gated spectra, discriminating the contributions of species emitting with different decay times (Cubeddu et al., this volume), it was possible to demonstrate that, in aqueous solutions at physiological pH, the formation of the anionic species, if any, takes place with a very low yield (Cubeddu et al., 1990).

This means that the occurrence of deprotonation of blepharismin in its first excited singlet state and its role as the primary molecular event in *Blepharisma* photoreception is still a speculation.

An alternative hypothesis to check is the intervention of blepharismin-sensitized photoreactions. In fact singlet oxygen might be produced via photodynamic action also at low light intensities, and its concentration might be the trigger signal for the photophobic response, similar to the proposed case of *Anabaena variabilis* (Nultsch and Schuchart, 1985).

To test this assumption, photobehavioral experiments were performed on cell treated with crocetin, a partially water soluble carotenoid, that is an efficient quencher of singlet oxygen (Schuchart and Nultsch, 1984). The samples were irradiated at 593 nm (10 nm bandwidth) and at 550 nm (40 nm bandwidth) and crocetin concentration ranged from 0.005 mM to 1.4 mM. No detectable reduction of the response was observed in crocetin treated *Blepharisma* at fluence rates eliciting a phobic reaction in about 50% of the cells in control samples (Checcucci et al., in preparation).

Comparing these results with those of Nultsch and Schuchart (1985), who found that 1 mM crocetin inhibited negative phototaxis in *Anabaena*, it seems possible to exclude any significant contribution of photodynamically produced singlet oxygen in inducing *Blepharisma* photoresponses.

On the other hand the efficiency of crocetin in protecting *Blepharisma* cells from the photokilling action was tested when irradiating with white light, 1200 W/m^2, a control culture and a 0.004 mM crocetin containing culture for one minute. Whereas the control cells were severely damaged, those in the presence of crocetin were still alive and motile.

Characterization of Blepharisma Chromoproteins

As mentioned before, all the above reported experiments on *Blepharisma* chromophore were performed on crude extracts from acetone-treated cells and the photoreceptor pigments were not in their physiological state. Work is in progress to isolate and characterize from a spectroscopic point of view intact photoreceptor proteins.

The chromoprotein isolation method mainly consists of hydroxylapatite (HPT) chromatography followed by elution with different buffers. All our elution profiles show one single band for absorption at 280 nm coinciding with blepharismin band for absorption at 580 nm. These profiles closely parallel the elution profile of total protein content.

SDS PAGE of the peak fractions yields deeply red colored bands at the level of the bromophenol blue band, probably due to a partial SDS-induced release of the chromophore from the apoprotein; these bands showed an intense red fluorescence. A red fluorescent band corresponding to a molecular weight of about 68 kDa was also detected.

Under non-denaturating conditions, no release of the chromophore from the apoprotein is observed. Also under these conditions, the red fluorescence from a protein-bound chromophore is revealed.

Optical absorption spectra from peak-fractions of HPT chromatography shows, superimposed to protein absorption, the typical absorption spectrum of the blepharismin chromophore.

Upon excitation at 280 nm, fluorescence emission spectra show a non-structured band centered at about 330 nm and a band around 600 nm typical of the chromophore moiety.

Acknowledgements

The author is sincerely grateful to Prof. Pill-Soon Song and to Prof. Osvaldo Pieroni for critical reading of the manuscript and for helpful discussions.

References

Banks, H. J., Cameron, D. W., and Raverty, W. D., 1976, Chemistry of the Coccoidea. II Condensed polycyclic pigments from two Australian *Pseudococcids* (Hemiptera), *Aust. J. Chem.*, 29:1509.

Cubeddu, R., Ghetti, F., Lenci, F., Taroni, P., and Ramponi, R., 1990, Time-gated fluorescence of blepharismin, the photoreceptor pigment for photomovement of *Blepharisma*, *Photochem. Photobiol.*, 51:567.

Giese, A. C., 1973, "*Blepharisma*: The Biology of a Light-sensitive Protozoan," Stanford University Press, Stanford.

Giese, A. C., 1981, The photobiology of *Blepharisma*, *in*: Smith, K. C., ed., "Photochemical Photobiological Reviews," Plenum, New York, pp. 139.

Gualtieri, P., Passarelli, V., and Barsanti, L., 1989, *In vivo* micro-spectrophotometric investigation of *Blepharisma japonicum*, *J. Photochem. Photobiol. B: Biol.* 3:379.

Kim, I.-H., Rhee, J. S., Huh, J. W., Florell, S., Faure, B., Lee, K. W., Kahsai, T., Song, P.-S., Tamai, N., Yamazaki T., and Yamazaki, I., 1990, Structure and function of the photoreceptors stentorins in *Stentor coeruleus*. I. Partial characterization of the photoreceptor organelle and stentorins, *Biochim. Biophys. Acta*, 1040:43.

Kraml, M., and Marwan, W., 1983, Photomovements responses of the heterotrichous ciliate *Blepharisma japonicum*, *Photochem. Photobiol.*, 37:313.

Lenci, F., and Ghetti, F., 1989, Photoreceptor pigments for photomovement of microorganisms: some spectroscopic and related studies, *J. Photochem. Photobiol. B.: Biol.*, 3:1.

Lenci, F., Ghetti, F., Gioffré, D., Passarelli, V., Heelis, P. F., Thomas, B., Phillips, G. O., and Song, P.-S., 1989, Effects of the molecular environment on some spectroscopic properties of *Blepharisma* photoreceptor pigment, *J. Photochem. Photobiol. B: Biol.*, 3:449.

Matsuoka, T., 1983a, Distribution of photoreceptors inducing ciliary reversal and swimming acceleration in *Blepharisma japonicum*, *J. Exp. Biol.*, 225:337.

Matsuoka, T., 1983b, Negative phototaxis in *Blepharisma japonicum*, *J. Protozool.*, 30:409.

Matsuoka, T., and Shigenaka, Y., 1985, Mechanism of cell elongation in *Blepharisma japonicum*, with special reference to the role of cytoplasmic microtubules, *Cytobios*, 42:215.

Miyake, A., Harumoto, T., Salvi, B., and Rivola, V., 1990, Defensive function of pigment granules in *Blepharisma japonicum*, *Eur. J. Protist.*, 25:310.

Navas Diaz, A., 1990, Absorption and emission spectroscopy and photochemistry of 1,10-anthraquinone derivatives: a review, *J. Photochem. Photobiol., A: Chem.*, 53:141.

Nultsch, W., and Schuchart, H., 1985, A model for phototactic reaction chain of the cyanobacterium *Anabaena variabilis*, *Arch. Microbiol.*, 142:180.

Passarelli, V., Lenci, F., Colombetti, G., Barone, E., and Nobili, R., 1984, The possible role of H^+ and Ca^{2+} in photobehavior of *Blepharisma japonicum*, *in*: Senger, H., ed., "Blue Light Effects in Biological Systems," Springer, Berlin, pp. 480.

Scevoli, P., Bisi, F., Colombetti, G., Ghetti, F., Lenci, F., and Passarelli, V., 1987, Photomotile responses of *Blepharisma japonicum* I: Action spectra determination and time-resolved fluorescence of photoreceptor pigments, *J. Photochem. Photobiol., B: Biol.*, 1:75.

Schuchart, H., and Nultsch, W., 1984, Possible role of singlet molecular oxygen in the control of the phototactic reaction sign of *Anabaena variabilis*, *J. Photochem.*, 25:317.

Sevenants, M. R., 1965, Pigments of *Blepharisma undulans* compared with hypericin, *J. Protozool.*, 12:240.

Song, P.-S., 1981, Photosensory transduction in *Stentor coeruleus* and related organisms, *Biochim. Biophys. Acta*, 639:1.

Song, P.-S., 1983, Protozoan and related photoreceptors: molecular aspects, *Ann. Rev. Biophys. Bioeng.*, 12:35.

Song, P.-S., Kim, I.-H., Florell, S., Tamai, N., Yamazaki, T., and Yamazaki, I., 1990, Structure and function of the photoreceptor stentorins in *Stentor coeruleus*. II. Primary photoprocess and picosecond time-resolved fluorescence, *Biochim. Biophys. Acta*, 1040:58.

Spikes, J. D., 1989, Photosensitization, *in*: Smith, K. C., ed., "The Science of Photobiology," Plenum, New York, pp. 79.

Photoreception and Photomovements in Stentor coeruleus

Pill-Soon Song, Il-Hyun Kim, Jae Seong Rhee,
Jae Wook Huh, Scott Florell, Brigitte Faure, Kit W. Lee
and Tesfamichael Kahsai

Department of Chemistry and Institute for
Cellular and Molecular Photobiology
University of Nebraska
Lincoln, NE 68588
USA

and

Naoto Tamai, Tomoko Yamazaki
and Iwao Yamazaki

Department of Chemical Process Engineering
Hokkaido University
Sapporo
Japan

Introduction

The step-up photophobic and negative phototactic responses of *Stentor coeruleus* to visible wavelength light represent two clear examples of the photosensitivity of aneural motile organisms. The former is a stop/reversal response exhibited by the organism as it crosses from the dark (lower light intensity) to the light side (higher light intensity). Figure 1 shows a videomicroscopic recording of the step-up photophobic response exhibited by two stentor cells.

The negative phototactic response is a directional response of the *S. coeruleus* cell. Thus, the ciliate cell is able to perceive the direction of light propagation and swims in the direction parallel to light propagation, i.e., swimming away from the light source. Since both the intensity (step-up photophobic) and direction (negative phototactic) responses result in a higher population density of the cells in the shaded/weaker intensity area, it is important to discriminate between these two responses unambiguously. This has been done (Song et al., 1980b), as illustrated in Figure 2.

Thus, *Stentor coeruleus* cells exhibit two types of photoresponses, namely, step-up photophobic response (Fig. 1) and negative phototactic response (Fig. 2). These two photoresponses will be discussed in this chapter.

Biophysics of Photoreceptors and Photomovements in Microorganisms
Edited by F. Lenci *et al.*, Plenum Press, New York, 1991

Stentorin as the Photosensor Molecule

The action spectra for the photophobic and phototactic responses in *Stentor coeruleus* suggest that the cerulean pigment of the cell serves as the photosensor/photoreceptor molecule (Song et al., 1980a,b; Kim et al., 1984). The chromophore of stentorin has been identified as a hypericin derivative (Walker et al., 1979; Kim et al., 1990). The chromophore is apparently covalently linked to its apoprotein, although the nature of the linkage remains to be elucidated. Figure 3 shows the chemical structure of hypericin and its absorbance spectrum is compared with that of stentorin I purified by the reverse-phase HPLC.

The pigment granule apparently serves as the photoreceptor organelle in *Stentor coeruleus*, as it contains the photosensor chromoproteins (stentorins) (Kim et al., 1990). Figure 4 shows an electron micrograph of the pigment granules. Note that the core ma-

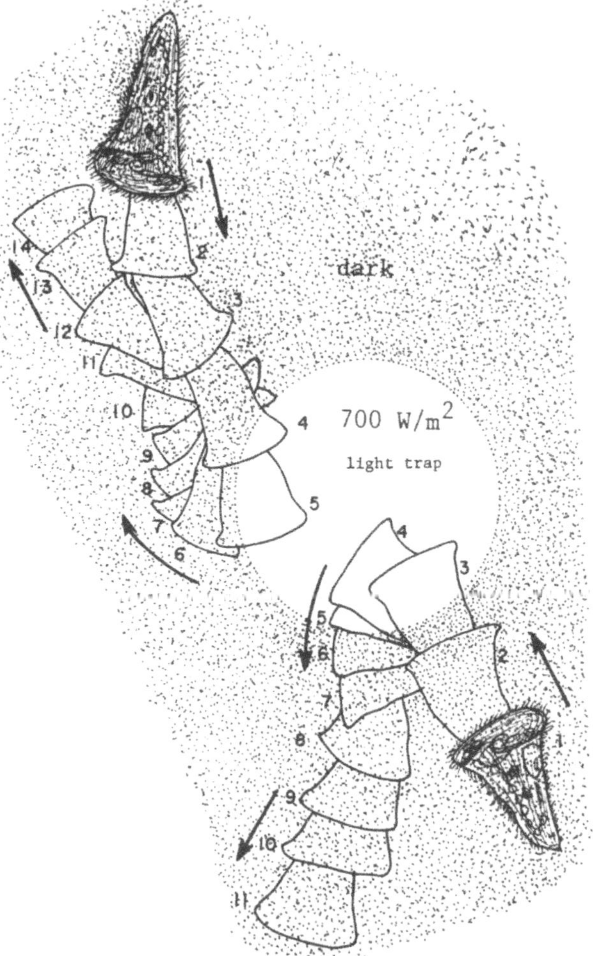

Fig. 1. Step-up photophobic response of *Stentor coeruleus*. The position of each cell shown was videomicroscopically traced every 0.24 s. The stippled area represents darkness, whereas the "light trap" represents an area lighted by white light. (Adapted from Song et al., 1980a; Song and Poff, 1989).

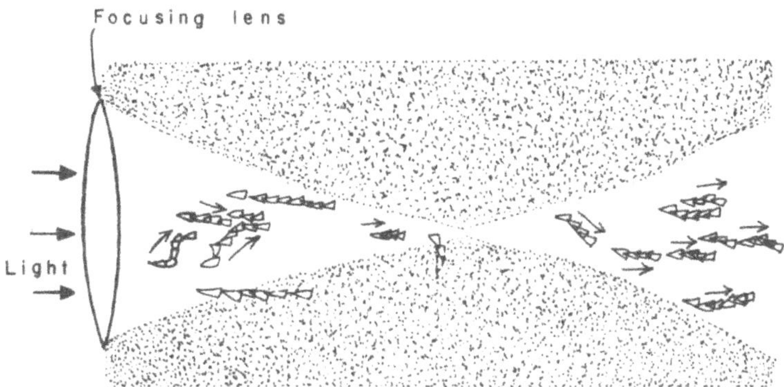

Fig. 2. Traces of the *Stentor coeruleus* cells in a focused light beam, 94 W/m^2. The cells swim away from the light source in both converging (intensity-increasing) and diverging (intensity-decreasing) light beams. The position of the cells is shown at 0.17 s intervals, as traced from a video recording. (After Song et al., 1980b; Song and Poff, 1989).

trix contains a layered structure. The core matrix can be extruded from the pigment granule and the cell upon cold shock treatment of the cell. Whether these ultrastructures represent something similar to the thylakoid membrane stacks in plant chloroplast remains an interesting question. It is possible that stentorins are associated with these structures, although there is no evidence for or against this suggestion at the present time.

The cerulean pigment can be separated into two spectroscopically distinct forms, namely,stentorin I and stentorin II (Kim et al., 1990). Table 1 compares the properties of stentorin I and stentorin II. Stentorin I is strongly fluorescent, whereasstentorin II is only weakly emitting. Both pigment proteins are highly glycosylated, with an indication that the glycan residues may be different between stentorin I and II. The highly glycosylated proteins seem to account for the repeatedly observed anomalies of these proteins on SDS PAGE, as the precise molecular weights of the constituent subunits have not been determined unambiguously on the gel. From the size exclusion HPLC, it appears that stentorin II is a large molecular assembly/complex composed of several subunits.

Which of these stentorin molecules represents the photoactive sensor for the observed action spectra of the step-up photophobic response in *Stentor coeruleus*? First, it is important to establish that both forms exist *in vivo*, rather than as an isolation/purification artifact. The fourth derivative spectrum from the absorbance spectrum (maximum at 618 nm) of the pigment granule did not resolve these two spectrally distinct species (Kim et al., 1990). On the other hand, a recent microspectrofluorometric analysis of the cell indicated that both stentorin I and II fluorescences are present within/in the vicinity of the pigment granule (Deforce, Song and Furuya, unpublished data). Since there is no established assay for the photoreactivity of stentorin I and II as the photosensor molecule for the cellular photosensitivity, we cannot identify the photoactive form of stentorin at the present time. On the basis of available data, i.e., approximate coincidence between the action spectrum and the absorbance spectrum of stentorin II (spectral shape similar to that of stentorin I shown in Figure 3 but red shifted by ca. 10 nm), the low fluorescence quantum yield, long wavelength absorbance maximum in the fourth derivative absorbance spectrum of the pigment granule, and the time-resolved fluorescence spectra (*vide infra*), we tentatively identified stentorin II as the photoactive sensor molecule for the photoresponses of the ciliate cell (Kim et al., 1990; Song et al., 1990).

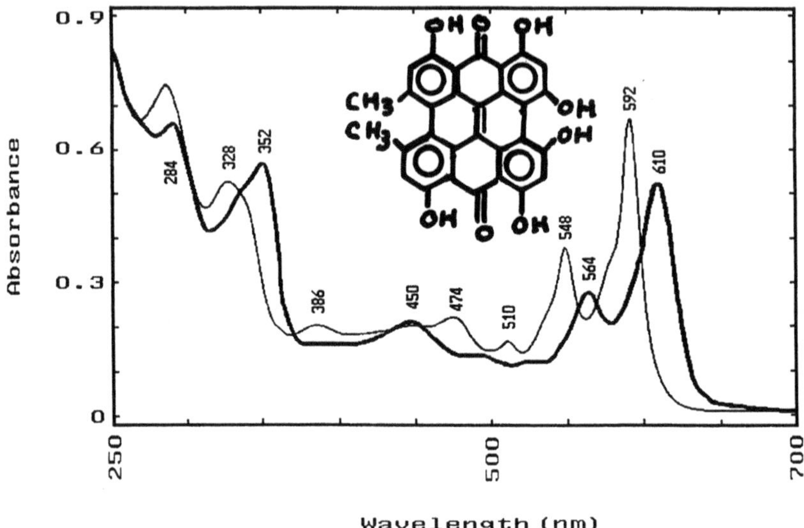

Fig. 3. Absorbance spectra of stentorin I (bold line) purified by the reverse-phase RP-304 HPLC in 10 mM potassium phosphate buffer, pH 7.8, containing 0.02% reduced Triton X-100, and of hypericin (thin line) in ethanol at room temperature. Inset: Chemical structure of hypericin. (Adapted from Kim et al., 1990).

Table 1 Properties of stentorin I and II

Property	Stentorin I	Stentorin II
Solubility	all detergents used	less solubilized by SDS
'Subunits'	2 (?)	(?)
Mol. wt	46 600 (SDS PAGE) to 102 900 (HPLC)	> 805 000-810 000 (HPLC)
Glycan	glycosyl, mannosyl	glycosyl, mannosyl, N-acetylglucosamine
Abs. max.	610 nm	620 nm
Fluorescence max.	618 nm	626 nm
Fluorescence intensity (rel.)	1	0.014
CD	weak	strong
Rotational corr. time	7.6 ns	14.1-15.0 ns
Diameter	3.8 nm	4.8 nm

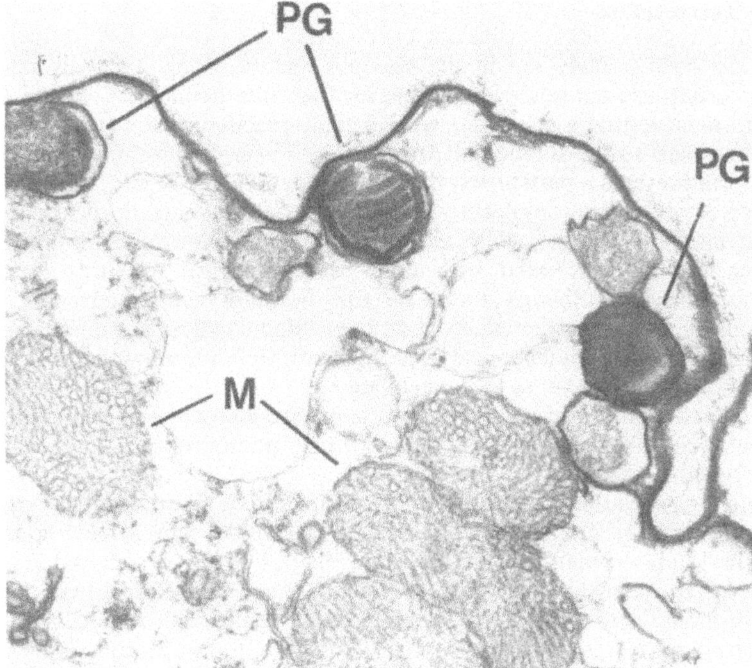

Fig. 4. The electron micrograph of a transverse section of *Stentor coeruleus*. Magnification x 1400; original x 20000. Some layered structure is seen with one of the pigment granules (PG) and mitochondria (M).

Light Signal Transduction

General Scheme

In the previous NATO ASI volume (Song, 1985), a general scheme for the photosensory transduction was briefly discussed. In the signal perception/sensor step of this scheme, the light stimulus/signal is absorbed by a photoreceptor pigment which initiates an ultrafast efficient primary photoprocess. The primary photoreaction leads to the generation of a transducible signal at the cellular level through the functional linkage or coupling between the primary photoreaction and the driving force for signal generation, eg., conformation change. Once the initial signal is generated, it is further amplified to the extent sufficient to elicit the final response such as ciliary beat reversal in the photophobic response of *Stentor coeruleus*. In this lecture, we will examine each of these light signal transduction steps involved in the photoresponse exhibited by the ciliate cell. Figure 5 depicts a general scheme for the light signal transduction in photosensitive cells such as *Stentor coeruleus*.

Light signal → Absorption/Perception by the sensor → Primary photoreaction → Signal generation → Signal amplification → Output (response)

Fig. 5. A general scheme for the light signal transduction in photoresponsive cells.

Primary Photoreaction

As has been pointed out in the previous lecture (Song, 1985), isolated photosensor molecules are not readily assayable for their photobiological activity. It is often difficult to demonstrate a particular type of photoreaction(s) as the signal-generating process that leads to the subsequent transduction chain. Stentorin is no exception. In fact, stentorin presents a particularly difficult problem in this regard, unlike other well-established photoreceptor molecules such as rhodopsin, bacteriorhodopsin, sensory rhodopsin, and phytochrome, which all exhibit readily demonstrable photochemical cycles which can be correlated with their photobiological activities at the cellular/organismic levels. Although studies certainly have not been exhaustive, preliminary experiments, performed under steady state irradiation conditions, indicate no spectroscopically identifiable photochemical cycle in stentorin. Unfortunately, flash photolytic studies on stentorin have yet to be carried out.

If indeed stentorin does not exhibit a spectrophotometrically identifiable photochemical cycle, what then is the possible primary photoprocess? Let us consider the following possible reactions:

Proton dissociation/release. The stentorin chromophore contains several hydroxyl groups (Fig. 3). The average pK_a of these hydroxyl groups is likely to be lowered in the excited state of stentorin (Walker et al., 1979). It is thus expected that proton dissociation and translocation during the excited state can be facilitated via an acid-base network of amino acid residues of the apoprotein, analogous to the charge relay network of serine proteases (Blow and Steitz, 1970). An ultrafast proton dissociation from the excited state of stentorin imbedded in pigment granules generates anionic chromophore species which may undergo local conformational changes. Reassociation of protons therefore can be retarded, resulting in a transient net drop of cytosolic pH.

If the above model is operative, it is necessary but not sufficient that an extremely short fluorescence lifetime is associated with the photoreceptor molecule and that the fluorescence emission from its excited state is time-resolvable due to the emission from the anionic species of the molecule. Stentorin I exhibited nearly exponential fluorescence decay, having two comparable lifetime decay components of 2.53 ns (47%) and 5.95 ns (53%) at pH 7.8, whereas the weakly fluorescent form, stentorin II, showed an ultrafast fluorescence decay component (10 ps, 90%) at pH 7.8 (Song et al., 1990). More suggestively, the former showed no apparent time- resolvable fluorescence emission, whereas the latter exhibited time-resolvable emission centered around 660 nm (Fig. 6). The 660 nm emission has been tentatively assigned to the fluorescence band of anionic species of stentorin II (Song et al., 1990). It is significant that the fluorescence spectra of both stentorin I and II depended on solvent deuterium oxide; thus, the emission band in heavy water is broader than in water. The observed solvent effect may reflect either an effect of hydrogen-deuterium exchange on the chromophore fluorescence or/and a conformational change in the acid-base network of the apoprotein arising from hydrogen-deuterium exchange.

A light-induced transient pH drop proposed for the primary photoprocess of stentorin entails facilitated dissociation and retarded reassociation of protons within the photoreceptor assembly. Proton dissociation from the excited singlet state of an aromatic acid can be accelerated by a specific solvent cluster and can generate a detectable pH drop in the medium. For example, picosecond laser pulses transiently lowered the pH of the 2-naphtholate-6-sulfonate solution from 7 to below 4 (Clark et al., 1979). To produce a significant pH drop under physiologically relevant conditions, a lower light intensity excitation of the acid/base groups of the chromophore is sufficient if the forward proton dissociation is accelerated and the backward proton reassociation is retarded by the matrix or protein in which the chromophore is positioned. For example, in reverse micelles, the proton dissociation rate was accelerated by more than 40 times, whereas proton release from the excited state of naphthylammonium included in the

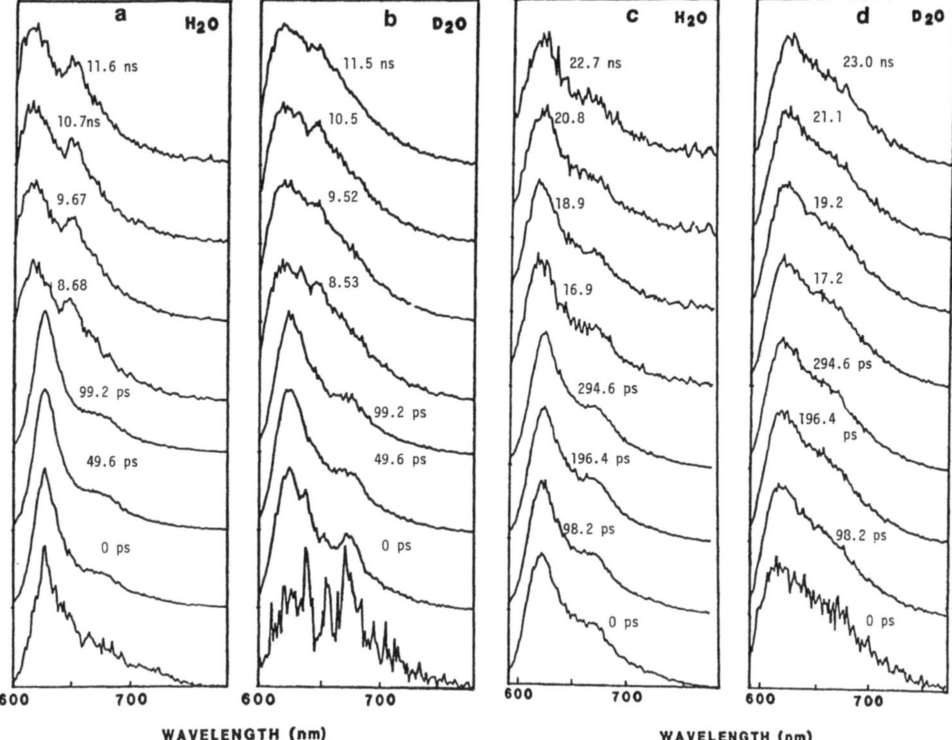

Fig. 6. Time-resolved fluorescence emission spectra of stentorins at 290 K. (a) Stentorin I in Tris buffer, pH 7. 0. (b) Stentorin II in D_2O-Tris buffer, pD 7.0. (c) Stentorin II in Tris buffer, pH 7.0. (d) Stentorin II in D_2O-Tris buffer, pD 7.0.

crown ether was a one-way process, with a very slow rate of proton reassociation (Shizuka et al., 1985, 1986a,b). Similarly, the excited state photoreceptor assembly in the pigment granules of *Stentor coeruleus* may facilitate proton dissociation preferentially over proton reassociation (Song, 1983; Song et al., 1990).

Light-induced electron transfer. Hypericin is an efficient photodynamic sensitizer in nature. It is possible that singlet oxygen is generated in a Type II photosensitization process in solution and *in vivo* (Duran and Song, 1986; Yang et al., 1986). Recent studies have shown that hypericin in its triplet state acts as an electron donor, rather than an acceptor (Lee et al., 1990). For example, hypericin (Hy) photooxidizes ferric ions to ferrous ions in DMSO. Ferric ions can be reduced by hypericin via either Type I or Type II photosensitization route. On the other hand, hypericin and stentorin were found to be unreactive in the presence of an electron donor (dithionite; Kim et al., 1990), suggesting that the hypericin chromophore is a poor electron acceptor.

$$\mathrm{Hy} + h\nu \rightarrow {}^{1}\mathrm{Hy} \rightarrow {}^{3}\mathrm{Hy}$$

$${}^{3}\mathrm{Hy} + \mathrm{Fe}^{3+} \rightarrow \mathrm{Hy}^{+} + \mathrm{Fe}^{2+}$$

The significance of electron transfer as the possible primary photoprocess of stentorin in *Stentor coeruleus* remains to be investigated. The following photocycle can be envisioned as the photoprocess of stentorin (S-OH) that may be coupled to the subsequent transduction chain in the light response of the ciliate:

$$S\text{-}OH + h\nu \rightarrow S\text{-}OH^{*}\ (^{1}S\text{-}OH \text{ and/or } ^{3}S\text{-}OH)$$

$$S\text{-}OH^{*} + A \rightarrow S\text{-}OH^{\cdot +} + AH^{-} \rightarrow S{=}O + AH$$

$$AH + O_2 \rightarrow A + O_2^{-} + H^{+}$$

where A is an electron acceptor. The presumed photocycle can transiently generate either an electrochemical potential across the pigment granule membrane and/or plasma membrane or a pH drop resulting from the dissociation of protons from AH or S-OH upon electron transfer to oxygen or A, respectively. In the above scheme, S=O can be reduced photochemically or enzymatically to regenerate S-OH in an overall cyclic pathway.

Although stentorins in solution showed very little spectrophotometrically detectable photoreactions under aerobic and anaerobic conditions, whole cell suspensions of *Stentor coeruleus* exhibited spectral changes upon prolonged irradiations (R. Dai and P.S. Song, unpublished data, 1990). The photobiological significance of the light-induced spectral changes in whole cells, however, is difficult to ascertain, since these spectral changes were irreversible.

Signal Generation: Ionic Basis

The light signal transduction mechanism for the step-up photophobic response in *Stentor coeruleus* appears to be ionic in nature (Song, 1981, 1983; Kim et al., 1984; Iwatsuki and Song, 1985, 1989). Thus, the following scheme underscores the ionic basis of the light signal transduction in the ciliate cell:

Light signal → Receptor potential (signal generation) → Action & regenerative potentials (signal amplification & transmission) → Mechanoresponse

In this scheme, the receptor potential serves as a trigger signal (Wood, 1976, 1982). Based on circumstantial and indirect evidence from physiological experiments using protonophores and other ionophores and antagonist drugs, it has been proposed that the trigger signal is a transient intracellular pH drop. Whether or not a transient pH change can be identified with the receptor potential remains to be studied. At the present time, such identification is only a conjecture.

A light-induced intracellular pH drop can occur transiently either via proton dissociation from the excited state of stentorin or proton translocation through the electron transfer cycle, *vide supra*. The strong dependencies of the photoresponses of *Stentor coeruleus* on pH and heavy water are necessarily consistent with the proposed role of proton concentration, but such dependencies alone are not sufficient to establish the proposed model as a definitive mechanism (Iwatsuki and Song, 1985).

If the time-resolved emission from stentorin II indeed is attributable to the anionic species of the photosensor molecules (Fig. 6), it is possible to estimate the magnitude of the intracellular pH change under certain assumptions. An approximate cell volume can be calculated to 46 nL. Each Stentor cell contains 1.4×10^{-17} moles of stentorins. Assuming that all stentorin molecules are photoreactive and all the hydroxyl protons of the excited chromophore dissociate due to the marked lowering of their average pK_a in the excited state, a transient intracellular pH drop of about 0.5 can be calculated. Here, we assumed zero buffering capacity and an intracellular pH of 6.8. The assumptions made here are unsubstantiated and difficult to verify. However, regardless of the buffering capacity of the bulk cytosolic medium, heterogeneity of pH (i.e. microscopic pH) in the aqueous cytoplasm or the pH in the proximity of pigment granules located near the cell membrane can lower the local pH within the cell. Therefore, the pH drops estimated are probably sufficient to activate voltage-sensitive calcium channels and trigger an influx of calcium ions in the signal amplification step in *Stentor coeruleus, vide infra.* Work is in progress to electrophysiologically identify the nature of the trigger signal involved in the light signal transduction in the ciliate cell (S. Fabczak and P.S. Song, to be published).

Signal Amplification: Calcium Influx

Assuming that the trigger signal resulting from light excitation of the photosensor apparatus (stentorin plus pigment granule) in *Stentor coeruleus* is a transient proton release from the excited stentorin in the pigment granules to the cytoplasm, a depolarizing effect of the intracellular pH drop (receptor potential) can elicit the opening of calcium channels. This then results in an action potential. A light-induced action potential of the ciliate has been recorded, consistent with the calcium ion influx from the extracellular medium to the cytoplasm when the cell is subjected to the light stimuli. The action potential cannot be induced by light in the presence of protonophores and can be completely quenched by calcium blocking agents, in qualitative agreement with the above scheme.

In addition to the inhibitory effects of specific calcium antagonists on the photophobic response of the ciliate, calimycin and ionomycin stimulated the photophobic sensitivity of the ciliate, apparently due to the enhancement of calcium permeability across the cellular membrane (Song, 1983; Prusti, 1987). Other drugs that facilitate calcium permeability and thus enhance the photosensitivity of the *Stentor coeruleus* cell include caffeine (Prusti et al., 1984) and α-phosphatidic acid (Kim et al., 1984).

Mechanoresponse: Ciliary Beat Reversal

Ciliary beating in the clockwise direction propels the stentor cell to rotate and swim forward. A step-up light intensity gradient abruptly stops the ciliary beating and its direction reverses momentarily. This "gear" shift in the motor apparatus accounts for the stop-light escape behavior of the organism. Figure 7 shows the motor apparatus.

The link between the calcium ion influx and the dynein ATPase activity in the ciliary contractile axoneme has not been established. The contractile mechanism of *km* fibers and myonemes is operative in the cell body contraction of *Stentor coeruleus* upon mechanical and electrical stimulation, and is based on calcium ion influx as the driving force for the contraction phenomenon. However, the role of calcium ions in the contractile axonemes of cilia and ciliary basal bodies remains to be elucidated.

Concluding Remarks: Many Questions Remain Unanswered

The light signal transduction in *Stentor coeruleus* is by no means adequately understood at the mechanistic level, although some aspects of the transduction chain such as the identity of the photosensor molecule and the role of calcium ions in ciliary beat reversal are reasonably well established. As a pointer for further studies in the future, the following unresolved questions remain to be addressed.

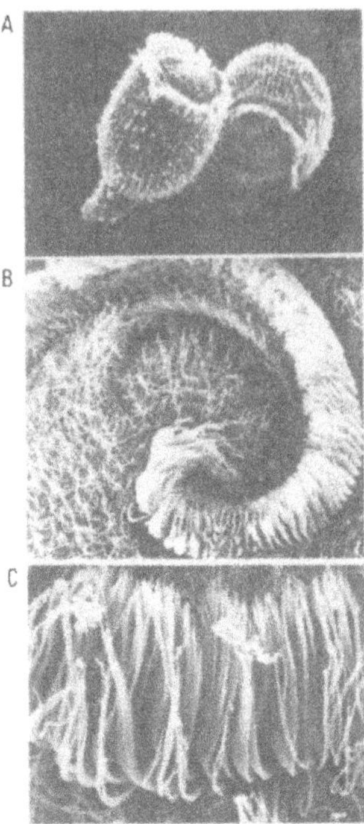

Fig. 7. Scanning electron micrographs of *Stentor coeruleus* and its motor apparatus (cilia). Note that cilia are distributed along the whole cell body, with ciliary bundles preferentially localized around the membranelle band/oral fold. (A) Intact cell. (B) Ciliated membranelle band. (C) Cilia.

Linkage between the chromophore and apoprotein. The linkage appears to be covalent and acid labile (Walker et al., 1979), but its chemical nature is yet to be elucidated. To understand the photochemical basis of the stentorin photoreceptor, it is critical to establish the structure of the linkage.

Apoproteins. Genes for stentorin I and II have not been cloned, and their amino acid sequences are presently unknown. Work is in progress to clone stentorin genes. One major difficulty in characterizing stentorins is the fact that these proteins behave abnormally, most likely due to a high degree of glycosylation (Kim et al., 1990).

Photochemical cycle ? Flash photolytic studies are desirable to reveal whether or not a photobiologically significant photoreaction cycle is operative in *Stentor coeruleus*. Electron transfer as the possible primary photoprocess also awaits a careful investigation.

Pigment granule. The apparent ultrastructure seen in the pigment granule (Fig. 4) is interesting, as it is reminiscent of the thylakoid stacks in chloroplast. It is possible that the photosensor molecules are associated with these structures.

Receptor potential. What is the molecular/ionic nature of the receptor potential preceding the action potential for the light-induced ciliary beat reversal in *Stentor coeruleus*?

Contractile axonemes and mechanism for ciliary beat reversal. The biochemical and biophysical basis of ciliary beat reversal remain largely unexplored.

Mechanisms for photoresponses: Photophobic vs. phototactic response. Is the negative phototaxis in *Stentor coeruleus* a consequence of a series of the course-correcting photophobic response? On the basis of selective inhibition or enhancement of the photophobic response, without a parallel inhibition or enhancement of the negative phototactic response, we postulated that the mechanisms for these two responses are different, i.e. the latter is not a consequence of a trial-error type course-correcting series of photophobic response (Iwatsuki and Song, 1985). Further studies are needed to definitively establish the mechanistic relationship between the two responses.

There are many other questions about the mechanism of the photosensory transduction in *Stentor coeruleus*. This organism offers many challenging opportunities for study by the students of photobiology in the future.

Summary

The unicellular ciliary protozoan, Stentor coeruleus, exhibits photophobic and phototactic responses to visible light stimuli. The pigment granule contains the photoreceptor chromoproteins (stentorins). Stentorins localized in the pigment granules of the cell serve as the primary photoreceptor for the step-up photophobic and negative phototactic responses in this organism. An initial characterization of the pigment granules has been described in terms of size, absorbance spectra and ATPase activity. Two forms of stentorin pigments have been isolated from the pigment granules. Stentorin I has an apparent molecular weight of 68,600 and 52,000 by SDS PAGE (at 10 and 13% gel, respectively) or 102,000 by steric exclusion HPLC, whereas stentorin II is a larger molecular assembly probably composed of several proteins (mol wt greater than 500,000). Stentorin I is composed of at least two heterologous subunits corresponding to apparent mol wts of 46,600 (fluorescent, Coomassie Blue negative) and 52,000 (fluorescent, Coomassie Blue positive) on SDS PAGE (13% gel). However, mobilities of stentorin I and II on polyacrylamide were found to be strongly dependent on the degree of crosslinking in the acrylamide gel, apparently due to the fact that both stentorin I and II are highly glycosylated chromoproteins. Stentorin II appears to be the primary photoreceptor whose absorption and fluorescence properties are consistent with the action spectra for the photoresponses of the ciliate to visible light. How stentorin generates an initial trigger signal that cascades down the sensory transduction chain leading to the ciliary beat reversal required for the step-up photophobic response of the ciliate remains to be established. A photo-induced spectral change or pigment-bleaching cycle for the photoreceptor analogous to the photochemical cycles found in other well-known photoreceptor molecules such as rhodopsin, bacteriorhodopsin, and sensory rhodopsin or slow- cycling rhodopsin is apparently inoperative in stentorin. The hypothesis, the primary photoprocess of stentorin is due to proton dissociation from the excited state of the photoreceptors imbedded in the pigment granules of the ciliate cell, is qualitatively consistent with the transient cytoplasmic pH drop observed upon excitation of the ITStentor cell. In this lecture, we will present results of a comprehensive study on the excited state properties of stentorins I and II in an attempt to elucidate the nature of the primary photoprocess of the photoreceptor molecules. Stentorin II exhibited an ultrafast primary photoprocess, whereas stentorin I did not. It is, therefore, proposed that stentorin II functions as the primary photoreceptor in Stentor coeruleus. The strongly fluorescent form, stentorin I at pH 7. 8, exhibited nearly exponential fluorescence decay monitored at 620 nm, having two comparable lifetime decay components of 2. 53 ns (47%) and 5. 95 ns (53%). Stentorin I showed no significant time-resolved fluorescence emission spectra in the picosecond- nanosecond time scales. The weakly fluorescent form, stentorin II, exhibited an ultrafast fluorescence decay component (10 ps) at emission wavelength of 630 nm and pH 7. 8. The amplitudes of the multi-component fluorescence in stentorin II were found to be slightly emission wavelength-dependent. Furthermore, the fluorescence emission spectrum was time-resolvable in the picosecond time scale. Effects of pH and pD on the fluorescence decay kinetics and time-resolved spectra of stentorins I and II have also been investigated. Results are suggestive of proton dissociation (transient intracellular pH change) as a primary photoprocess from the excited state of stentorin II. Further testing of alternative mechanisms is also suggested. Following the primary step in the signal transduction chain for the photophobic response of Stentor coeruleus, calcium ions play a pivotal role in transducing the stimulus light signal into the mechanical response, i. e., reversal of the ciliary beating direction. The photophobic action spectrum exhibited a maximum peak at ca. 615 nm, corresponding to the absorbance maximum of stentorin. Although the role of calcium ions is reasonably well established, the nature of the primary photochemical event mediated by stentorin awaits further study and direct evidence to substantiate/refute the involvement of transient intracellular pH change as a primary signal is presently lacking.

Acknowledgements

This work was supported in part by grants from NIH (NS15426) and the Center for Biotechnology-UNL.

References

Blow, D. M., and Steitz, T. A., 1970, X-ray diffraction studies of enzymes, *Annu. Rev. Biochem.*, 39:63.

Clark, J. H., Shapiro, S. L., Campillo, A. J., and Winn, K. R., 1979, Picosecond studies of excited-state protonation and deprotontation kinetics. The laser pH jump, *J. Am. Chem. Soc.*, 101:746.

Duran, N., and Song, P. S., 1986, Hypericin and its photodynamic action, *Photochem. Photobiol.*, 43:677.

Iwatsuki, K., and Song, P. S., 1985, Deuterium oxide (D_2O) enhances the photosensitivity of *Stentor coeruleus, Biophys. J.*, 48:1045.

Iwatsuki, K., and Song, P. S., 1989, The ratio of extracellular Ca^{2+} to K^+ ions affects the photoresponses in *Stentor coeruleus, Comp. Biochem. Physiol.*, 92A:101.

Kim, I. H., Rhee, J. S., Huh, J. W., Florell, S., Faure, B., Lee, K. W., Kahsai, M., Song, P. S., Tamai, N., Yamazaki, T., and Yamazaki, I., 1990, Structure and function of the photoreceptor stentorins in *Stentor coeruleus*. I. Partial characterization of the photoreceptor organelle and stentorins, *Biochim. Biophys. Acta*, 1040:43.

Kim, I. H., Prusti, R. K., Song, P. S., Häder, D. P., and Häder, M., 1984, Phototaxis and photophobic responses in *Stentor coeruleus*. Action spectrum and role of Ca^{2+} fluxes, *Biochim, Biophys. Acta*, 798:298.

Lee, T. Y., Yoon, G. J., Ahn, J. S., Chung, J. K., Häder, D. P., Häder, M., and Song, P. S., 1990, Effects of solar radiation on motility in *Stentor coeruleus*. Assessment of photodynamic mechanism by hypericin sensitizers, *Arch. Microbiol.*, submitted.

Prusti, R. K., 1987, *Ph. D. Dissertation*, Texas Tech University, Lubbock, TX.

Prusti, R. K., Song, P. S., Häder, D. P., and Häder, M. 1984, Caffeine-enhanced photomovement in the ciliate, *Stentor coeruleus, Photochem. Photobiol.*, 40:369.

Shizuka, H., Kameta, K., and Shinozaki, T., 1985, Excited-state proton-transfer reactions of naphthylammonium ion-18-crown-6 complexes, *J. Am. Chem. Soc.*, 107:3956.

Shizuka, H., and Serizawa, M., 1986a, Dynamic behavior in the excited state of phenylanthrylammonium ion-18-crown-6 complexes. A one-way proton-transfer reaction, *J. Phys. Chem.*, 90:4573.

Shizuka, H., Serizawa, M., Okazaki, K., and Shioya, S., 1986, Stabilization constants of phenylanthrylammonium ion-18-crown-6 complexes determined by fluorimetry, *J. Phys. Chem.*, 90:4694.

Song, P. S., 1981, Photosensory transduction in *Stentor coeruleus* and related organisms, *Biochim. Biophys. Acta*, 639:1.

Song, P. S., 1983, Protozoan and related photoreceptors. Molecular aspects, *Annu. Rev. Biophys. Bioengin.*, 12:35.

Song, P. S., 1985, Primary molecular events in aneural cell photoreceptors, *in*: "Sensory Perception and Transduction in Aneural Organisms," Colombetti, G., Lenci, F., Song, P.-S., eds., NATO ASI series A. Vol. 89, Plenum, New York. p. 47.

Song, P. S., Häder, D. P., and Poff, K. L., 1980a, Step-up photophobic response in the ciliate, *Stentor coeruleus, Arch. Microbiol.*, 126:181.

Song, P. S., Häder, D. P., and Poff, K. L., 1980b, Phototactic orientation by the ciliate, *Stentor coeruleus, Photochem. Photobiol.*, 32:781.

Song, P. S., Kim, I. H., Florell, S., Tamai, N., Yamazaki, T., and Yamazaki, I., 1990, Structure and function of the photoreceptor stentorins in *Stentor coeruleus*. II. Primary photoprocess and picosecond time-resolved fluorescence, *Biochim. Biophys. Acta*, 1040:58.

Song, P. S., and Poff, P. S., 1989, Photomovement, *in*: "The Science of Photobiology," 2nd Ed., K. C. Smith, ed., Plenum, New York. pp. 305.

Walker, E. B., Lee, T. Y., and Song, P. S., 1979, Spectroscopic characterization of the *Stentor* photoreceptor, *Biochim. Biophys. Acta*, 587:129.

Wood, D. C., 1976, Action spectrum and electrophysiological responses correlated with the photophobic response of *Stentor coeruleus*, *Photochem. Photobiol.*, 24:261.

Wood, D. C., 1982, Membrane permeabilities determining resting, action and mechanoreceptor potentials in *Stentor coeruleus*, *J. Comp. Physiol.*, 146:537.

Yang, K. C., Prusti, R. K., Song, P. S., Watanabe, M., and Furuya, M., 1986, Photodynamic action in *Stentor coeruleus* sensitized by endogenous pigment stentorin, *Photochem. Photobiol.*, 43:305.

Electrophysiology and Photomovement of Stentor

David C. Wood

Department of Behavioral Neuroscience
University of Pittsburgh
Pittsburgh, PA 15260
USA

Introduction

As first reported by Jennings (1906) and Mast (1906) *Stentor coeruleus* swims away from light sources (photodispersal) and collects in the more shaded areas of an unevenly illuminated container. *Stentor* are sensitive to light because they contain a pigment called stentorin (Møller, 1962; Walker et al., 1979), which releases H^+ ions as a result of light absorption (Walker et al., 1981). This rapid proton release is thought to be an initial step in a series of steps which eventuate in photodispersal (Song et al., 1983). This paper summarizes the available evidence describing some of these intervening processes and the physiological mechanisms producing them.

Much of the analysis of these mechanisms has been conducted using the tools of electrophysiology, therefore it is appropriate at the outset to question whether the behaviors underlying photodispersal are completely or even partially dependent on voltage changes produced across the cell's plasma membrane, since this is the variable which electrophysiologists measure. The involvement of transmembrane potentials in the production of photodispersal is supported by electrophysiological recordings from *Stentor* which reveal both receptor potentials and action potentials. Since the receptor potentials are highly correlated with light presentation and the action potentials trigger changes in ciliary beating, they appear to be coupled to sensory input on the one hand and motor output on the other in precisely the fashion expected of intermediates in the production of photodispersal. But are they necessary and sufficient? In a series of studies Song and his colleagues have shown that agents such as ruthenium red (Chang et al., 1981), the Ca^{++} ionophores A23187 and ionomycin (Chang et al., 1981; Prusti and Song, 1986), the protonophores FCCP and TPMP (Song et al., 1980a,b), phospatidic acid (Kim et al., 1984), and La^{+++} block photodispersal up to 90%. All of these agents are thought to act more or less specifically on the surface membrane of cells and thereby disrupt the generation of transmembrane potentials. Alteration of the extracellular concentration of K^+ and Ca^{++} also strongly influences the production of photophobic responses and phototactic behavior in *Stentor* (Colombetti et al., 1982; Iwatsuki and Song, 1989) at concentrations which influence resting and action potentials in the organism (Wood, 1982). However, none of these manipulations completely eliminates photodispersal, hence it cannot be unequivocally stated that transmembrane potential changes control all of the behaviors producing it. Indeed,

Biophysics of Photoreceptors and Photomovements in Microorganisms
Edited by F. Lenci *et al.*, Plenum Press, New York, 1991

such a definitive statement could be made only if a quantitative descriptive model providing a close simulation of the behavior of individual cells undergoing photodispersal was developed, and this model assigned a necessary role to transmembrane potential changes. In the meantime, it is clear that transmembrane potentials are important determinants of at least some of these behaviors.

The Electrical Properties of Resting Cells

Intracellular microelectrodes record a -48 to -56 mV resting potential in *Stentor* when the cells are maintained at 8.5 to 10° C and are in the extended trumpet-shaped form (Wood, 1982). This resting potential is maintained by the penetrated cell throughout several hours of recording and has the same value everywhere inside the cellular compartment. Cells in the pear-shaped form characteristic of swimming cells have more variable resting potentials being frequently depolarized to as little as -30 mV.

In both extended and pear-shaped *Stentor* two separate microelectrodes with their tips positioned 50 to 80 μm apart and penetrating the same cell record essentially identical intracellular potentials. When there are transient changes in intracellular potential both microelectrodes record these changes simultaneously and to the same amplitude (Wood, 1970). These data agree with the data of Eckert and Naitoh (1970) obtained from *Paramecium*. Apparently, the cytoplasmic compartment of both organisms contains sufficient electrolyte to be a relatively good conductor and thereby permit the uniform distribution of charge, and hence voltage, along the cell's plasma membrane.

If a rectangular pulse of current is passed through one of these two intracellular electrodes, the potential recorded by the other microelectrode follows a negatively accelerated exponental time course until an asymptotic voltage deflection is reached. This is the pattern of voltage change that would be expected if the cell membrane contained a resistance and capacitance in parallel. From the rate of rise of this voltage deflection and its magnitude we can determine the magnitude of this membrane resistance and capacitance; the specific membrane resistance of *Stentor* is between 100 and 200 kohm cm^2 and its specific membrane capacitance is between 0.3 and 0.9 $\mu f/cm^2$. Most of the variability in these measures is attributable to uncertainties in the measurement of the cell's surface area. However, the value calculated for the specific membrane capacitance is characteristic of membrane capacitance in other cell types confirming the intracellular location of the recording microelectrodes.

The experimentally determined value for membrane resistance together with an estimate of cytoplasmic resistivity can be used in conjunction with cable theory to calculate the distance along the cell membrane in which a receptor or other potential will decay. These calculations reveal that the distance over which transient potentials should decay in both *Paramecium* (Eckert and Naitoh, 1970) and *Stentor* (Wood, 1982) is 10 to 20 times longer than the length of the cell. Hence, little or no decay in transient potential amplitude is expected from one end of a ciliate to the other in agreement with the previously noted experimental observations. There are two important consequences of these initial observations - one procedural and one theoretical. The procedural consequence derives from the observation that when recording at 20-25° C the recorded intracellular potential invariably decreases to near 0 mV or may even reach values which are positive relative to the extracellular medium within a minute or two after microelectrode penetration. When this happens, it is observed simultaneously that the passage of current through one microelectrode no longer produces a voltage deflection which can be recorded from the other microelectrode. This failure can be explained by assuming that one or both of the microelectrodes has been "encapsulated" by the cell membrane and is no longer intracellular even though they may appear to be intracellular by visual inspection (Wood, 1970). With microelectrodes in this encapsulated configuration, recordings of transient potentials can still be obtained but they are

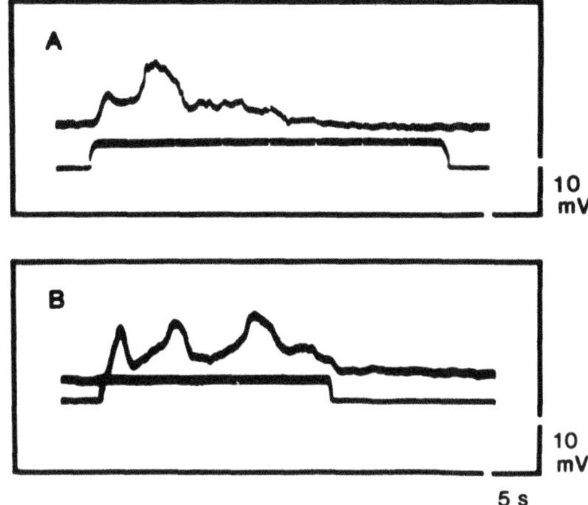

Fig. 1. Two examples of photic receptor potentials recorded from intravacuolar microelectrodes with the cells at 22-24° C. The bottom trace is the output of a monitoring photovoltaic cell. Note that the delay between light onset and receptor potential onset can be seen even at the slow sweep speed used to generate this picture.

much smaller than when recorded from intracellular electrodes and more variable. Both the amplitude and stability of such recordings can be improved by placing both microelectrodes inside large intracellular food vacuoles (Wood, 1970; 1973; 1976). In this configuration the recording locus is sufficiently stable to allow meaningful comparison of potential amplitudes recorded from a given cell at different times. More commonly we circumvent this encapsulation problem entirely by recording from cells chilled to 8.5-10° C, temperatures which retard or prevent encapsulation.

The uniformity of potentials recorded throughout the ciliate cytoplasm also has a theoretical consequence. If electrical signals are necessary intermediates in the production of photic responses in *Stentor*, then cilia in one portion of the cell surface will receive the same photically generated electrical signals and thus behave similarly to cilia in other portions of the cell surface. "One instant" mechanisms of photic orientation require different responses from cilia in different portions of the cell surface at a given instant in time (Feinleib, 1985). Therefore, if the photically generated signals controlling ciliary beating are electrical, photic orientation by ciliates could not occur via a "one instant" mechanism. Of course, chemically mediated photic signals might be localized to small patches of the cell's cilia.

It is convenient and informative to summarize these electrophysiological data in the form of a simple electrical model (Fig. 3). The observations cited above are summarized by drawing a wire (a good conductor) in the cytoplasm and connecting it to the extracellular medium by both a capacitative current pathway and a resistive current pathway with properties characteristic of the resting cell membrane.

Receptor Potentials

When *Stentor* are illuminated, a depolarizing receptor potential is produced (Wood, 1973; 1976) (Fig. 1). This receptor potential is unusual in that it occurs with a discernible delay of 100 to 200 ms after the light onset. The potential then rises in 50 ms or so to a peak of between 1 and 30 mV. Typically, the transmembrane potential then decays back to resting values over the course of 20 to 60 s. Frequently, waves of depolarization are superimposed on the decay phase of the receptor potential. The in-

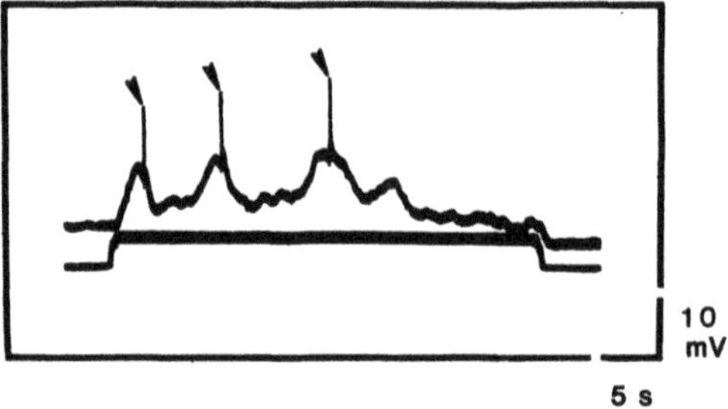

Fig. 2. Another example of a photic receptor potential which contains waves of depolarization. An action potential (arrows) is triggered by each of the waves.

itial peak of the receptor potential is often of sufficient amplitude to trigger an action potential, whose properties will be discussed later. Sometimes action potentials are triggered by each successive wave of depolarization (Fig. 2).

The time course and properties of the photic receptor potential contrast markedly with the time course and properties of most receptor potentials including receptor potentials produced by mechanical stimuli in *Stentor*. Mechanoreceptor potentials are initiated within 10 ms of the stimulus onset, rise to a sharp peak within another 20 ms, and then decay exponentially within 400 ms (Wood, 1982). These differences in time course suggest that photic and mechanical receptor potentials are produced by fundamentally different mechanisms. Theoretical formulations (Kirchhoff's Second Law) indicate that there are only two ways in which a potential change can be generated across the membrane resistance. The mechanism of receptor potential generation most commonly observed in electrophysiological analyses depends on a change in membrane resistance which selectively allows one or two species of ions to flow across the membrane and thereby changes the amount of charge on the membrane surface. In electrophysiological terminology such a mechanism is spoken of as a conductance change mechanism and is described in terms of ionic fluxes travelling through membrane channels. The second mechanism, observed much less frequently in biology, involves the activation of a transmembrane transport system which uses metabolic energy to pump ions in one direction across the membrane. Such a mechanism is referred to as an electrogenic pump.

There are a number of experimental ways of discriminating between a conductance change mechanism and an electrogenic pump mechanism.

1) Stimuli acting by means of a conductance change mechanism either cause membrane channels to open and thereby produce a large proportionate increase in membrane conductance and transmembrane current or cause previously open channels to close and thereby produce a large decrease in membrane conductance and transmembrane current. On the other hand, electrogenic pumps produce current without opening or closing ion channels; hence, the transmembrane currents they produce are generally small and the maximum amplitude of the resultant receptor potential is predictable from the resting membrane resistance. Using voltage clamp circuitry to measure transmembrane current, mechanical stimuli are observed to produce large changes in transmembrane current (5-10 nA) (Wood, 1982); whereas, photic stimuli produce current changes that are much smaller (<1 nA) (Wood, unpublished results). Both the mechanical and photic stimuli employed in these studies produced receptor potentials of between 5 and 10 mV. Passing 1 nA of current from a second intracellular microelectrode across the plasma membrane of the resting cell also produces a 10 to 20 mV change in transmembrane potential. Therefore, the photic receptor potential amplitude is predictable from the resting membrane resistance and the photic stimulation

produces only a minimal change in conductance; whereas, the mechanical stimulus produces a transient 5 to 20 fold increase in membrane conductance. These data suggest the photic receptor potential is generated by an electrogenic pump, while the mechanoreceptor potential is produced by the opening of transmembrane channels (selective for Ca^{++}) (Wood, 1982).

2) Most membrane channels allow ions to travel through them in both directions with the actual direction of ion movement through a channel being determined by the transmembrane ionic concentration gradient for the relevant ion and the transmembrane potential. For a positively charged ion such as Ca^{++}, the negative intracellular resting potential and higher extracellular concentration drive ions into the cell when Ca^{++} selective membrane channels open. Thus, mechanical stimulation of resting *Stentor* results in an inward Ca^{++} current. Using a voltage clamp, the experimenter can vary the intracellular potential and thereby vary the amount of Ca^{++} current entering the cell through mechanoreceptor channels. It is even possible to make the intracellular potential sufficiently positive that Ca^{++} ions are driven out of the cell when the mechanical stimulus opens the channels. The ability to alter the amplitude and even reverse the direction of transmembrane currents and potentials is the principle diagnostic used to characterize a current as being produced by a conductance change mechanism. For electrogenic pumps the direction of current movement is determined by the molecular structure of the pump and is generally independent of the transmembrane potential. In applying this analysis to *Stentor*, we have observed that polarizing currents which offset transmembrane potentials do not influence the amplitude of the photic receptor potential but do alter and reverse mechanoreceptor currents (Wood, 1973; 1982). Again, the photic receptor potential behaves as if it were being produced by an electrogenic pump.

3) The flux of ions through membrane channels is driven by diffusional processes and hence is only weakly temperature dependent. Further, the energy required to initiate channel opening appears to come entirely from the mechanical stimulus itself and therefore is also temperature independent. As a result, mechanoreceptor potentials are as large when recorded at 8.5° C as when recorded at 25° C and *Stentor* are equally responsive to mechanical stimuli presented at these two temperatures. On the other hand electrogenic pumping is dependent on the metabolism of high energy compounds and ionic interactions involving rather strong electrostatic forces; consequently, electrogenic pumps are temperature dependent. Correspondingly, photic receptor potentials are easy to obtain when recording at 25° C but are small and present only during the initial stages of a recording session when the cells are cooled to 8.5-10° C.

These several lines of evidence concur in demonstrating that mechanoreceptor potentials are produced by an increase in membrane conductance while photic receptor potentials are produced by activation of an electrogenic pump. Presumably this pump is activated by the light-induced increase in H^+ concentration. Possibly the pump involves an exchange of intracellular H^+ for extracellular Ca^{++} analogous to the exchange pump found in the mitochondrial membrane. If 1 H^+ were exchanged for 1 Ca^{++} then a net inward current and a depolarizing receptor potential would be produced as is observed.

We incorporate these receptor mechanisms into the electrical model by drawing a circle and arrow in the "Light" current pathway to symbolize the unidirectional net current produced by the photic electrogenic pump (Fig. 3). A switch is placed in the resistive path carrying the mechanoreceptor current to indicate that this portion of the circuit is conducting only when a mechanical stimulus closes the switch (opens the channels). Both of these receptor mechanisms are placed in the pigmented stripe area of the cell membrane as opposed to its ciliary membrane. This localization of the mechanoreceptor channels has been established by selective ligand binding and autoradiography (Wood, 1989). The photic electrogenic pump is presumed to be localized there also because of the proximity of the pigment granules to this membrane.

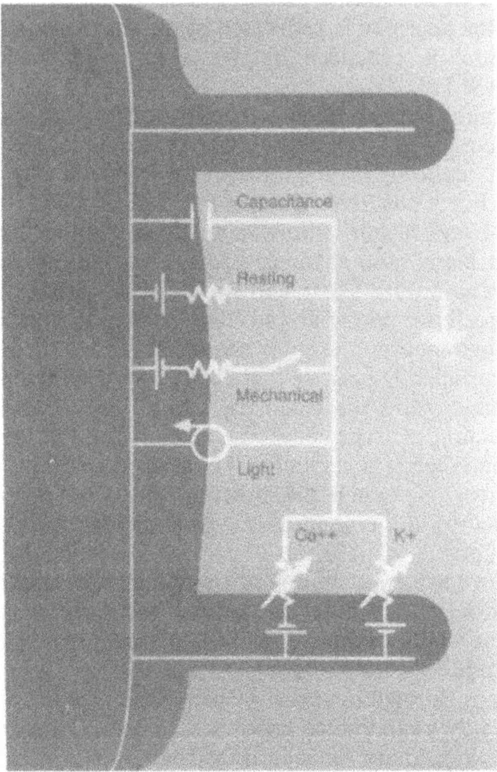

Fig. 3. The model electrical circuit which describes many electrophysiological properties of *Stentor*. The components of this model are explained in the text.

Action Potentials

Membrane depolarizations produced by photic, mechanical, or electrical stimuli elicit all-or-none action potentials from extended *Stentor* if these depolarizations exceed a threshold value of 14.7 $\pm$ 2.6 mV. When recorded from animals chilled to 8.5-10° C, these action potentials have an amplitude between 65 and 75 mV and a half amplitude duration of 194 $\pm$ 37 ms. The waveforms of action potentials produced by the three different types of stimulation are sufficiently similar to unequivocally indicate that the mechanism of their production is the same in all cases (Wood, 1970; 1973). Surprisingly, action potentials recorded from *Stentor* in the pear-shaped swimming form differ from those recorded from extended animals in that their amplitude is not all-or-none, but rather is a function of the intensity of the eliciting stimulus. Similarly, action potentials recorded from *Paramecium* are graded in amplitude and dependent on stimulus intensity (Naitoh et al., 1972). The reason for this difference between swimming and extended *Stentor* has not been studied but probably is attributable to the less negative resting potential generally observed in swimming animals (-30 to -35 mV in swimming animals vs. -48 to -56 mV in extended animals).

The mechanism of action potential production has been studied using voltage clamping and ionic substitution (Wood, 1982). Suprathreshold membrane depolarizations are observed to produce an initial influx of Ca^{++} through voltage-dependent membrane channels. The membrane conductance for K^{+} also increases at a somewhat

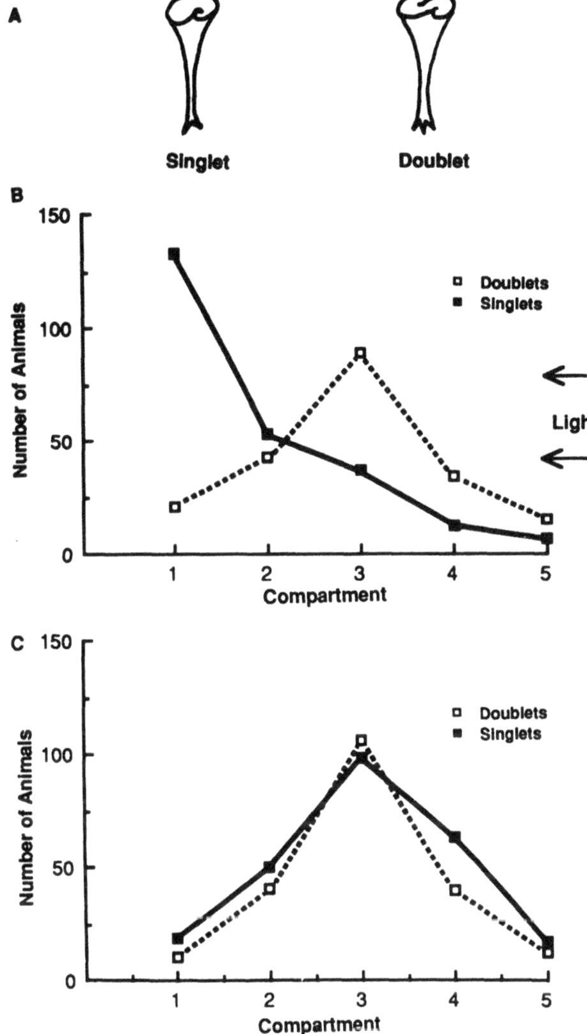

Fig. 4. A. Singlet and doublet *Stentor* diagrammatically portrayed in the extended form to show the position of the oral grooves on opposite sides of the frontal field in doublets. B. The number of doublets and singlets found in each of the 5 compartments of the photodispersal apparatus after being initially placed into the center compartment (3) and then allowed to swim freely for 12 min while being exposed to light coming from the right hand side. These data summarize the results from 5 experiments run on groups of 85 to 93 *Stentor* containing a total of 201 doublets and 242 singlets. C. The number of doublets and singlets found in each of the 5 compartments after they were run in the apparatus in the dark. These data summarize the results of 5 experiments run on groups of 86 to 98 *Stentor* for a total of 209 doublets and 244 singlets.

slower rate resulting in an efflux of K^+. In a cell which is not voltage clamped, the initial Ca^{++} influx produces the rising depolarizing phase of the action potential while the delayed K^+ efflux repolarizes the membrane to resting potential values. Assuming the presence of only these two voltage-dependent currents, we have successfully simulated *Stentor* action potentials using experimentally derived parameters and a mathematical formulation similar to that employed by Hodgkin and Huxley (1952) to simulate squid axon action potentials. Thus, while other currents may be present (eg., Ca^{++}-dependent K^+ current or a fast K^+ current) they appear to make a rather minor contribution to the waveform of an isolated action potential.

Every experimenter who has studied the electrophysiology of *Stentor* has noted the correlation between action potential production and ciliary reversal (Chen, 1972; Mergenheimer, 1970; Wood, 1973, 1976; Song et al., 1980b). This correlation is most pronounced when viewing the compound cilia which surround the frontal field and form the membranellar band. Action potentials cause all cilia around the circumference of the frontal field to flip simultaneously to a forward pointing direction. Thus, action potentials in *Stentor* produce ciliary reversal as been repeatedly observed and extensively studied in *Paramecium* and *Stylonychia* (Eckert and Naitoh, 1970; DePeyer and Machemer, 1978). It is noteworthy that the degree of ciliary reorientation and the frequency of ciliary beating recorded in *Paramecium* is a function of action potential amplitude (Machemer, 1975). Since the amplitude of action potentials in swimming *Stentor* varies as a function of stimulus intensity, this variability probably contributes to the variability observed in the duration of ciliary reversal and the degree of reorientation produced during a phobic response.

In *Paramecium* the voltage-dependent Ca^{++} channels responsible for action potential production appear to be located exclusively in the ciliary membrane. This conclusion is based on the observation that deciliation produces a loss of action potential production (Ogura and Takahashi, 1976; Dunlap, 1977) and that action potential amplitude increases in direct proportion to the length of the cilia during ciliary regeneration (Machemer and Ogura, 1979). Unfortunately, the methods which produce deciliated yet viable *Paramecia* and *Tetrahymena* (Kuznicki, 1963; Thompson et al., 1974) have killed *Stentor* in the process of deciliation so that we have been unable to reproduce this class of experiment. Nevertheless, it appears likely that voltage-dependent Ca^{++} channels are located in the ciliary membrane of *Stentor* also, since in that locus the influx of Ca^{++} would occur directly onto the molecular apparatus of the cilia. This conclusion is reinforced somewhat by the observation that ruthenium red, a Ca^{++} channel blocker, stains the cilia of the membranellar band. We also have electrophysiological evidence that voltage-dependent Ca^{++} channels are present in the membranes of internal vacuoles where they are functionally important in the production of contractions (Wood, 1990).

We incorporate these observations into our electrical model by adding two resistive current pathways and locating them in the ciliary membrane. The arrow drawn through the resistive element in these current pathways is used to signify that their resistance changes as a function of the transmembrane potential. Voltage-dependent channels located in internal membranes are ignored since it is unlikely that they contribute in the production of behavior leading to photodispersal.

Behavioral Responses which Produce Photodispersal

The electrophysiological evidence summarized above suggests that a reproducible sequence of events occurs when *Stentor* are illuminated by an effective actinic light. Initially photons of light are absorbed by stentorin located in pigment granules positioned under the membrane covering the pigmented stripe areas of the organism. The absorption of this energy causes the release of H^+ ions which directly or indirectly activate an electrogenic pump. Upon activation this pump produces a net movement of positive charge into the cell or of negative charge out of the cell. As a consequence of this charge movement, a depolarizing photic receptor potential is generated across the plasma membrane. Since the cytoplasm is a relatively good conductor, both this displaced charge and the associated receptor potential spread from their site of generation over the entire internal surface of the plasma membrane including the ciliary membrane. Depolarization of the ciliary membrane opens voltage-dependent Ca^{++} channels resulting in a Ca^{++} influx and the production of an action potential. The resulting increase in intraciliary Ca^{++} concentration triggers ciliary reversal which is the initial and most important step in the production of phobic responses. Thus, the sequence of events by which illumination elicits photophobic responses can be outlined in some detail.

However, the relationship between photophobic responses and the behaviors producing photodispersal is presently unclear. The early investigators argued that the repetitive production of photophobic responses ("avoiding responses") was sufficient to explain *Stentor*'s ability to collect in the dimmer areas of test chambers (Jennings, 1906;

Mast, 1906). In their descriptions, each photophobic response produced a reorientation in swimming direction which was randomly directed rather than being determined by the direction of the actinic light. Nevertheless, repeated photophobic responses interspersed with periods of forward swimming led by chance to a swimming orientation which was away from the light. Once this swimming direction was achieved it was maintained, since the posterior of *Stentor* was assumed to be less sensitive to light than their anterior. Therefore, exposure of the posterior of the cell to light did not lead to additional photophobic responses and the cell continued to swim away from the light.

More recently, Song and coworkers (1980a,b) have contested this rather simple and mechanistic model by arguing that *Stentor* achieve photodispersal by means of photophobic responses and a phototactic response mechanism which is in addition to and separate from the photophobic response mechanism. The initial argument supporting this point of view was based on the observation that *Stentor* will swim away from a light source even when this results in their swimming into an area of greater light intensity, in this case produced by focussing the light beam. Thus, the organisms are not responding simply to light intensity but also are sensitive to light direction. This directional sensitivity was taken as *prima-facie* evidence for the presence of a separate phototactic mechanism. Video image analysis also revealed that *Stentor* often follow curved swimming paths during their reorientation away from the light source; such curved paths could not have been produced by a pattern of photophobic responses interspersed with periods of straight swimming. Even more recently Song and colleagues have discovered a variety of experimental manipulations - adding D_2O to the medium (Iwatsuki and Song, 1985), varying K^+ and Ca^{++} concentrations (Iwatsuki and Song, 1989), and adding Ca^{++} ionophores (Prusti and Song, 1986) - which affect photophobic responses and phototactic orientation in an inverse fashion, i.e., they increase the probability of eliciting photophobic responses while simultaneously decreasing phototactic orientation through an area of focussed light. Collectively these data have served as the basis for arguing for a phototactic mechanism distinct from the photophobic response mechanism.

While we have not sought to investigate this issue directly, we have collected some data which have a bearing on it. On two separate occasions over the last 10 years, we have isolated phenotypic variants of *Stentor* which do not exhibit photodispersal and do not collect in dimly lighted areas (Wood and Marinelli, 1990).

These animals are termed "doublets" because they possess two membranellar bands and two oral grooves, one on either side of their frontal field, as opposed to the single membranellar band and single oral groove found on most *Stentor*, which will hereafter be called "singlets". Doublets are phenotypic variants rather than mutants since they can undergo an abnormal fission which produces two singlets and these singlets have not reverted to the doublet form. Their failure to exhibit photodispersal is not attributable to a disruption in their swimming behavior since they swim at speeds comparable to singlets and distribute themselves in the dark as rapidly as singlets. Doublets also produce photophobic responses and contractions when illuminated at the same fluence rates as produce these responses in singlets, hence their sensitivity to light is normal. Illumination of the frontal field of doublets produces photophobic responses at lower fluence rates than are required to produce these responses when their side or posterior is illuminated. However, singlets exhibit the same form of directional sensitivity. Doublets do differ from singlets in their ability to reorient after producing a phobic response. The average angle of reorientation produced by doublets was only 14°, whereas the average angle of reorientation produced by singlets was 57°. Such a failure to reorient is a logical consequence of the symmetrical morphology of doublets which should not lead to unbalanced propulsive forces on opposite sides of the frontal field during a phobic response.

The failure of doublets to produce both photodispersal and normal phobic responses suggests that photodispersal is heavily or entirely dependent on the normal photophobic response pattern in accord with the original proposals of Jennings (1906) and Mast (1906). Alternatively, the proposed phototactic response mechanism also may be dependent on the animals possessing an asymmetrical frontal field and cortical stripe pattern.

References

Chang, T.-H., Walker, E. B., and Song, P.-S., 1981, Effects of ionophores and ruthenium red on the phototaxis of *Stentor coeruleus* as measured by simple devices, *Photochem. Photobiol.*, 33:933.

Chen, V., 1972, "The Electrophysiology of *Stentor polymorphus*: An Approach to the Study of Behavior," University Microfilms: Ann Arbor (unpublished Ph. D. thesis).

Colombetti, G., Lenci, F., and Song, P.-S., 1982, Effects of K^+ and Ca^{2+} ions on motility and photosensory responses in *Stentor coeruleus*, *Photochem. Photobiol.*, 36:609.

DePeyer, J., and Machemer, H., 1978, Membrane excitability in *Stylonychia*: Properties of the two-peak regenerative Ca-response, *J. Comp. Physiol.*, 121:15.

Dunlap, K., 1977, Localization of calcium channels in *Paramecium caudatum*, *J. Physiol.*, 271:119.

Eckert, R., and Naitoh, Y., 1970, Passive electrical properties of *Paramecium* and problems of ciliary coordination, *J. Gen. Physiol.*, 55:467.

Feinleib, M. E., 1985, Behavioral studies of free-swimming photoresponsive organisms, *In*: "Sensory Perception and Transduction in Aneural Organisms," Colombetti, G., Lenci, F., and Song, P.- S., eds., Plenum Publishing, London, pp. 119.

Hodgkin, A. L., and Huxley, A. F., 1952, A quantitative description of membrane current and its application to conduction and excitation in nerve, *J. Physiol. (Lond.)*, 117:500.

Iwatsuki, K., and Song, P.-S., 1985, Deuterium oxide (D_2O) enhances the photosensitivity of *Stentor coeruleus*, *Biophys. J.*, 48:1045.

Iwatsuki, K., and Song, P.-S., 1989, The ratio of extracellular Ca^{2+} to K^+ ions affects the photoresponses in *Stentor coeruleus*, *Comp. Biochem. Physiol. A*, 92:101.

Jennings, H. S., 1906, "The Behavior of the Lower Organisms," Indiana University Press, Bloomington (1976 reprint).

Kim, I.-H., Prusti, R. K., Song, P.-S., Häder, D.-P., and Häder, M., 1984, Phototaxis and photophobic responses in *Stentor coeruleus*. Action spectrum and role of Ca^{2+} fluxes, *Biochim. Biophys. Acta*, 799:298.

Kuznicki, L., 1963, Recovery in *Paramecium caudatum* immobilized by choral hydrate treatment, *Acta Protozool.*, 1:177.

Machemer, H., 1975, Modification of ciliary activity by the rate of membrane potential changes in *Paramecium*, *J. Comp. Physiol.*, 101:343.

Machemer, H., and Ogura, A., 1979, Ionic conductances of membranes in ciliated and deciliated *Paramecium*, *J. Physiol.*, 296:49.

Mast, S. O., 1906, Light reactions in lower organisms. I. *Stentor coeruleus*, *J. Exp. Biol.*, 3:359.

Mergenhagen, D., 1971, Membrane potentials in *Stentor coeruleus*, *Protoplasma*, 72:359.

Møller, K. M., 1962, On the nature of stentorin, *Compt. Rend. Trav. Lab. Carlsberg*, 32:471.

Ogura, A., and Takahashi, K., 1976, Artificial deciliation causes loss of calcium-dependent responses in *Paramecium*, *Nature*, 264:170.

Naitoh, Y., Eckert, R., and Friedman, K., 1972, A regenerative calcium response in *Paramecium*, *J. Exp. Biol.*, 56:667.

Prusti, R. K., and Song, P.-S., 1986, Effects of ionomycin on the photoresponses in *Stentor coeruleus*, *Photochem. Photobiol.*, 43:84s.

Song, P.-S., Häder, D.-P., and Poff, K. L., 1980a, Phototactic orientation by the ciliate, *Stentor coeruleus*, *Photochem. Photobiol.*, 32:781.

Song, P.-S., Häder, D.-P., and Poff, K. L., 1980b, Step-up photophobic response in the ciliate, *Stentor coeruleus*, *Arch. Microbiol.*, 126:181.

Song, P.-S., Tapley, K. J., Jr., and Berlin, J. D., 1983, The photoreceptor in *Stentor coeruleus*, *In*: "The Biology of Photoreception." Society for Experimental Biology Symposium XXXVI., Cosens, D. J., and Vince-Prue, D., eds., Cambridge University Press, Cambridge, pp. 503.

Thompson, G. A., Baugh, L. C., and Walker, L. F., 1974, Non-lethal deciliation of *Tetrahymena* by a local anesthetic and its utility as a tool for studying cilia regeneration, *J. Cell Biol.*, 61:253.

Walker, E. B., Lee, T. Y., and Song, P.-S., 1979, Spectroscopic characterization of the *Stentor* photoreceptor, *Biochim. Biophys. Acta*, 587:129.

Walker, E. B., Yoon, M., and Song, P.-S., 1981, The pH dependence of photosensory responses in *Stentor coeruleus* and model system, *Biochim. Biophys. Acta*, 634:289.

Wood, D. C., 1970, Electrophysiological studies of the protozoan, *Stentor coeruleus, J. Neurobiol.*, 1:363.

Wood, D. C., 1973, Stimulus specific habituation in a protozoan, *Physiol. Behav.*, 11:349.

Wood, D. C., 1976, Action spectrum and electrophysiological responses correlated with the photophobic response in *Stentor coeruleus, Photochem. Photobiol.*, 24:261.

Wood, D. C., 1982, Membrane permeabilities determining resting, action and mechanoreceptor potentials in *Stentor coeruleus, J. Comp. Physiol.*, 146:537.

Wood, D. C., 1989, Localization of mechanoreceptors in the protozoan, *Stentor coeruleus, J. Comp. Physiol. A*, 165:229.

Wood, D. C., 1990, The functional significance of evolutionary modifications found in the ciliate, *Stentor*, *In* "Evolution of the First Nervous Systems," A. V. Anderson (ed.) Plenum Publishing, London.

Wood, D. C., and Marinelli, R., 1990, Doublet *Stentor* do not display photodispersal, *Photochem. Photobiol.*, (in press).

Action Spectroscopy

Edward D. Lipson

Department of Physics
Syracuse University
Syracuse, NY 13244-1130
U.S.A.

Introduction

Action spectroscopy - the measurement, as a function of wavelength, of the sensitivity of a particular biological response or effect - is a standard approach towards identifying receptor pigment(s) for photobiological phenomena (Jagger, 1967; Shropshire, 1972; Hartmann, 1983; Schäfer et al., 1983; Schäfer and Fukshansky, 1984; Galland, 1987). Comparison of an action spectrum with absorption spectra of known pigments often reveals the identity of the pigment, or class of pigments, involved. Some action spectra simply show the magnitude of a response as a function of wavelength, under conditions where a standard photon fluence (or fluence rate) is applied in all measurements. The difficulty with this expedient approach is that the response may well depend nonlinearly on the fluence. Such nonlinearity can introduce severe distortion into an action spectrum obtained in this way. The preferred method, instead, is to measure a separate fluence-response curve for each wavelength of interest, and then to determine the photon fluence needed to elicit a standard response level, for example 50% of maximum response. The reciprocal of the requisite photon fluence represents the sensitivity of the system, and is used as the ordinate of the action spectrum.

In this chapter, the basic methodology for action spectroscopy will be presented, and then several representative action spectra from the literature will be shown, specifically pertaining to blue-light effects in plants and microorganisms (Senger, 1980; 1984; 1987), and their elusive "cryptochrome" receptor pigments. Procedures will be described for fitting fluence-response curves to widely applicable hyperbolic saturation functions of the Michaelis-Menten type (corresponding to sigmoidal functions on semilogarithmic scales), for determining errors of the parameters, and for propagating these errors so that the action spectrum points can have error bars that are derivable from the error bars of the data points on the fluence-response curves.

Collections of representative action spectra for various blue light effects and responses have been assembled by Presti and Galland (1987) and by Galland and Senger (1988a). These typically show a peak in the near UV around 380 nm as well as major peaks in the blue region around 450 and 480 nm. Variations among these action spectra can be attributed to a number of causes, including different methodologies, different organisms, and different photoreceptor systems. To help explain the extensive variations in the near UV region, it has been suggested recently (Galland and Senger,

Biophysics of Photoreceptors and Photomovements in Microorganisms
Edited by F. Lenci *et al.*, Plenum Press, New York, 1991

1988b) that pterin chromophores may act as UV sensitizing pigments operating in association with flavin chromophores, which manage the blue sensitivity. The traditional alternative of ß-carotene, instead of flavin, has been ruled out in the case of *Phycomyces* phototropism (Presti et al., 1977), because mutants lacking ß-carotene exhibit normal phototropic sensitivity (see Figure 3 in the Chapter on Phototropism in Fungi by Lipson, this volume).

Techniques of Action Spectroscopy

The photophysical processes following light absorption by a chromophore may be represented graphically by a Jablonski diagram showing transitions among the electronic, vibrational and rotational states of the molecule (Grossweiner, 1989; see Figure 1 in the Chapter on Optical Absorption and Emission Spectroscopy of Photoreceptor Pigments (Lenci, this volume). After absorption of a photon, the chromophore is promoted from the ground state (which is usually of singlet character, and designated as S_0) to an excited electronic state (also singlet, and designated S_1, S_2, etc.); in general, the molecule also becomes excited with respect to vibrational and rotational motions (the corresponding energy spacings among these states are lower than between electronic states). The system then relaxes rapidly by internal conversion and vibrational relaxation to the lowest excited singlet level S_1 before there is time for photochemistry to take place. So, regardless of the wavelength of the photon that excited the molecule, the molecular excitation relaxes rapidly to the bottom vibrational levels based on S_1. Then the relevant photochemistry proceeds either directly from S_1 or else from the lowest triplet state, T_1, after intersystem crossing, provided that the excitation energy has not been expended meanwhile by internal conversion or fluorescence.

Considerations for Measuring and Interpreting Action Spectra

The shape of an action spectrum may be distorted significantly from the absorption spectrum of a putative chromophore. To avoid such distortions, the following conditions ideally should apply (Jagger, 1967; Shropshire, 1972):

1. The quantum yield (or quantum efficiency) - defined as the probability that a particular type of event of photobiological or photochemical interest occurs as the result of absorption of a photon - should be independent of wavelength.
2. The absorption spectrum of the receptor pigment should be the same whether measured *in vivo* or *in vitro*.
3. Screening or shading pigments, as well as scattering effects, should not cause significant wavelength-dependent distortion. In general, scattering tends to be stronger at shorter wavelength, but that is a gradual trend that normally would not obscure the detailed features of an action spectrum.

5. When an action spectrum is derived from fluence-response curves, the condition of reciprocity should be valid over the range of fluence rates and exposure times used. In other words, the response should depend on the *product* of fluence rate and exposure time, but not otherwise depend on either of these factors.
6. Light should not be totally absorbed by the sample for any wavelength under study.

Photophysical Units

In certain circumstances, fluence-response curves may be measured with broadband rather than monochromatic illumination. Among the reasons for doing so are (a) higher fluence rates are achievable with broadband filters, and (b) unless one is trying to study the wavelength dependence of a particular response or effect, it may be more representative to use broadband light covering most or all of the range of sensitivity. For example in studies of photogravitropism "threshold curves" of *Phycomyces* wild-type and mutant strains (see Figure 3 in the Chapter on Phototropism in Fungi by Lipson, this volume, and see below, Fig. 5), it is customary to employ broadband blue illumination extending over a 10-decade range of fluence rate. When broadband illumination is used, it is necessary to use energy units rather than photon units for fluence (or fluence rate). The energy units are J m^{-2} for fluence and J m^{-2} s^{-1} for fluence rate. On the other hand, in action spectroscopy, one must use monochromatic light, and in this case it is preferable to use photon units: photons m^{-2} for fluence and photons m^{-2} s^{-1} for fluence rate.

Reciprocity

For measurements employing continuous light, the strength of the stimulus is given by the fluence rate itself. When pulse light is used, however, the fluence can be adjusted by varying the exposure time (pulse width) or the fluence rate. It is preferable, for kinetic considerations, to keep the exposure time constant and vary the fluence rate, if possible. In general, though, one should establish the range of validity of reciprocity between exposure time and fluence rate.

Fundamental Relations

For action spectroscopy with pulse illumination, assume that the response R can be expressed as a function of four variables:

$$R = f(\phi\ \sigma_\lambda\ I_\lambda\ \Delta t) \quad (1)$$

where ϕ is the quantum efficiency or quantum yield, σ_λ is the cross section (see below) at wavelength λ, I_λ is the fluence rate (commonly called the "intensity", although that is strictly speaking incorrect), and Δt is the exposure time (pulse duration). The fluence is given by $F_\lambda = I_\lambda \Delta t$.

The standard procedure for measuring action spectra is first to specify a particular "criterion" response level. Then, for each wavelength, one determines the fluence (or fluence rate; see below) required to achieve that standard response level, which may be chosen, among other things. to be a) a fixed absolute level, b) a percentage (usually 50%) of the maximum response level, which itself can depend on wavelength, c) the maximum response (peak) level, for cases where the response falls off at high fluence after reaching a peak, or d) the absolute threshold (usually extrapolated downward from the rising part of the fluence-response curve, as opposed to the rather ill-defined approach of trying to find the highest fluence at which there is no perceptible response). If fluences $F_{\lambda 1}$ and $F_{\lambda 2}$ both elicit the same response level - the criterion response (noted here by R_c) -and if one assumes that Δt is the same in both experiments (alternatively, if reciprocity applies, one can correct for the different values of Δt) then the argument of the function in Eq. 1 is the same for both experiments at these two wavelengths. Further, if the quantum efficiency ϕ is assumed to be the same at both wavelengths, then

$$\frac{\sigma_{\lambda 1}}{\sigma_{\lambda 2}} = \frac{F_{\lambda 2}}{F_{\lambda 1}} \quad (2)$$

In other words, the cross section at wavelength λ is inversely proportional to the applied fluence at wavelength λ that produces the criterion response.

If, instead, the response is measured as a function of fluence rate rather than fluence, then a similar derivation starting from the equation

$$R = g(\phi \, \sigma_\lambda \, I_\lambda) \tag{3}$$

leads to the relation

$$\frac{\sigma_{\lambda 1}}{\sigma_{\lambda 2}} = \frac{I_{\lambda 2}}{I_{\lambda 1}} \tag{4}$$

Relation Between Extinction Coefficient and Cross Section

The extinction coefficient ϵ is used in conventional spectrophotometry and measured in the traditional units of $l\,mol^{-1}\,cm^{-1}$. It can be related by a conversion factor to a quantity from physics, the absorption cross section, in units of cm^2. The ratio between the intensity I transmitted through a spectrophotometric sample and the incident intensity I_0 can be derived from the following relations:

$$\frac{I}{I_0} = 10^{-\epsilon cl} = e^{-\sigma nl} \tag{5}$$

where c is the molar concentration of the pigment, l is the path length through the cuvet (usually 1 cm), σ is the absorption cross section, and n is the pigment concentration in units of molecules per cm^3. The expression in the first exponent, ϵcl, is the absorbance, A.

The extinction coefficient and the cross section are therefore interrelated by the following conversion formula:

$$\epsilon = 2.62 \cdot 10^{20} \sigma \tag{6}$$

where ϵ is in units of $l\,mol^{-1}\,cm^{-1}$ and σ is in cm^2. As an example of applying this conversion formula, consider riboflavin (or other flavin), which has an extinction coefficient of $1.25 \cdot 10^4\ l\,mol^{-1}\,cm^{-1}$ (at ~ 450 nm), then $\sigma = 4.8 \cdot 10^{-17}\ cm^2$. This represents the effective target area the chromophore presents for absorption of light. As another example, rhodopsin, with an extinction of $4 \cdot 10^4\ l\,mol^{-1}\,cm^{-1}$ (at ~ 500 nm) has $\sigma = 1.5 \cdot 10^{-16}\ cm^2$.

Derivation of Action Spectra from Fluence-Response Curves

Figure 1 shows a set of five artificial fluence-response curves to demonstrate, in an idealized situation, how action spectra may be derived from such curves. The hypothetical response is presumed to be measured as a function of the photon fluence (alternatively, fluence rate could have been chosen as the independent variable). The function $ax/(x + b)$ is plotted for five wavelengths and is graphed on two alternative scales for the abscissa, one logarithmic (a) and the other linear (b). In this example, wavelength λ_3 is clearly the most effective one, because the least amount of light is required to achieve the criterion response level (here 50% of maximum response). Of these five wavelengths, λ_5 is the least effective. So one can anticipate that, when this set

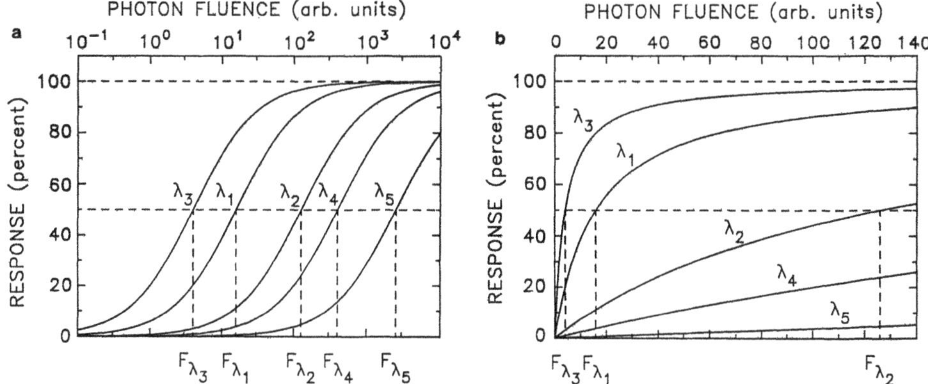

Fig. 1. Generic fluence-response curves for five wavelengths (λ_1 through λ_5) shown on logarithmic (**a**) and linear (**b**) scales for photon fluence. The curves represent the hyperbolic saturation function of the form ax/(x + b), where x is the fluence F (see top row of Table 1). On a semilogarithmic scale (**a**), the curves all have the same sigmoidal shape (see functions in rows 4 and 5 of Table 1). The fluence F needed to produce the standard criterion response (here chosen to be 50%) at each wavelength is shown by the vertical dashed lines. The designation of "arbitrary units" on the abscissa scales above each graph is meant to indicate that no special importance should be given to the absolute numbers used in this example, and, implicitly, that fluence rate could be used instead of fluence if continuous rather than pulse illumination is employed; the actual units of photon fluence would be mol m^{-2}. If fluence rate were used instead, the units would be mol m^{-2} s^{-1}.

of data is converted into an action spectrum, the ordinate representing the sensitivity, or effectiveness, will be high for λ_3 and low for λ_5.

In the semilogarithmic plot (Fig. 1a), the curves all have the same sigmoidal shape and differ only by lateral displacements. The symmetrical sigmoidal shape is a property of the hyperbolic saturation function, which is frequently used to fit fluence-response curves and other types of stimulus-response relationships in the general field of sensory physiology and in other areas of biophysics and biology (see below).

The procedure for deriving the action spectrum (Fig. 2) from these curves is straightforward. The ordinate of the action spectrum is simply the reciprocal of the fluence required to produce a criterion response as a function of wavelength. For a given application, one has to decide between plotting the action spectrum on a logarithmic or

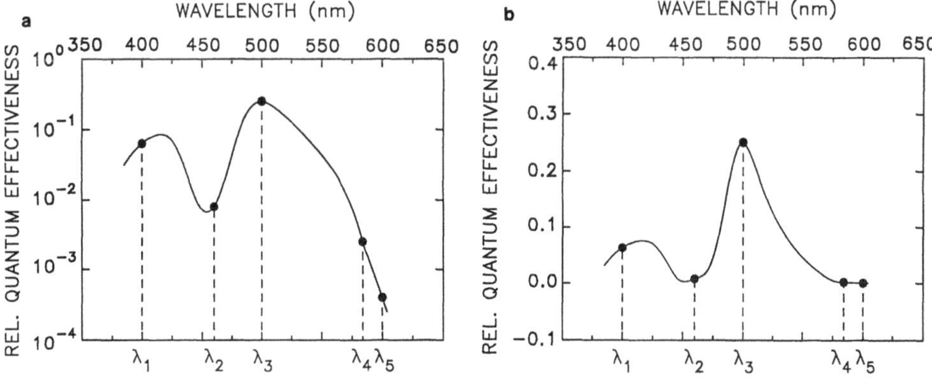

Fig. 2. Five-point action spectrum derived from the fictitious fluence-response curves in Fig. 1. The relative quantum effectiveness, shown on logarithmic (**a**) and linear (**b**) scales, is just the reciprocal of the fluence F required at each wavelength to achieve the criterion response, as indicated on Fig. 1.

linear ordinate. In *Phycomyces*, with a sensitivity range exceeding 10 orders of magnitude, it is appropriate to plot such action spectra on a logarithmic ordinate (Galland and Lipson, 1985b; 1987; Corrochano et al., 1988; Bejarano et al., 1991). This choice is particularly advantageous for comparing action spectra to one another or to absorption spectra of putative receptor pigments. Spectra that differ only by a scale factor will have identical shape on a logarithmic scale (and will be displaced by the logarithm of the scale factor). Now, if such spectra were truly identical, apart from a scale factor, they could be forced to have the same shape on a linear scale too, simply by normalizing them at any wavelength of choice (thereby superposing the spectra completely). The practical problem however is that, in the usual case when spectra are similar but not identical, choosing one wavelength or another for normalization will lead to different relative shapes. Such ambiguity and subjectivity are avoided by doing the comparison on a logarithmic scale instead. On the other hand, using a linear ordinate scale has the advantage that it is the more familiar way of viewing spectra, for example the chart records from commercial spectrophotometers.

Another consideration concerns the abscissa of the action spectrum, namely what is usually presented as a wavelength scale. On the basis of physical (i.e. quantum mechanical) principles, it would be preferable to use an energy rather than a wavelength scale; the two are related reciprocally according to the relation $E = hc/\lambda$, where E is the energy of a photon of wavelength λ, h is Planck's constant, and c is the speed of light. When an action or absorption spectrum is plotted as a function of energy, the positions of peaks can be related more readily with transitions between energy levels corresponding to molecular states (as on a Jablonski diagram; see above and Chapter on Optical Absorption and Emission Spectroscopy of Photoreceptor Pigments (Lenci, this volume). This is particularly worthwhile in the case of rhodopsin-type pigments in animals - as well as in halophilic bacteria and algae - because such pigments have spectra of standard shape except for lateral shifts in the energy of maximum absorption and vertical shifts in the absolute extinction or effectiveness. Nevertheless, most action and absorption spectra, including all the spectra shown in this chapter, are customarily plotted on a wavelength scale for the abscissa.

Error Analysis

For comparison of action spectra, and for other purposes, it is desirable to provide error bars along with the plotted values of fluence-response data. The general procedure is first to do ~ 10 repeats of each point, and then compute the mean and the standard error for each such point. An error-weighted least-squares fit is then performed to fit each set of fluence-response data to a suitable function to obtain a fluence-response curve. The parameter values and their errors (corresponding to standard errors) derived from the fit can then be used to calculate the fluence required to achieve the criterion response and the error of that fluence value. For certain functions, the criterion fluence is actually one of the parameters. In that case, its error value is available directly as a result of the fit. Otherwise, if the criterion fluence must be derived as a function of several parameters, then its error must be derived by error propagation methods (Bevington, 1969), taking into account the covariances among the estimated parameters resulting from the least-squares fits.

The method for obtaining errors during least-squares fitting (actually two related methods, the choice of which depends on whether the function being fit depends on the parameters in a *linear* or *nonlinear* fashion) is readily available in reference books on statistical and numerical methods, for example those by Hamilton et al. (1964) and by Press et al. (1985), both of which present the least squares method in an elegant, but by now standard, matrix formulation. A commercial scientific graphics program (version 4.0 of Sigmaplot from Jandel Scientific, Corte Madera, Calif.) includes such error computation in its linear and nonlinear least-squares fitting routines; however it neglects, in the current version, to provide user with the parameter covariance matrix (Hamilton, 1964) needed for error propagation when the criterion fluence depends on more than one of the fitted parameters.

The action spectrum ordinate, usually labeled as "relative quantum effectiveness", or just "effectiveness" is the inverse of the criterion fluence. According to error propagation methods, the relative error of such an inverse quantity is the same as the relative error of the quantity itself. Specifically, if we denote a criterion fluence value by F and the ordinate of the action spectrum by A, then $A = 1/F$ and their errors are related by the equation: $\sigma_A/A = \sigma_F/F$.

These error analysis methods have been applied in several recent publications (Trad and Lipson, 1987; Corrochano et al., 1988; Galland et al., 1989; Ensminger et al., 1990; Bejarano et al., 1991). Representative results from some of these papers are included below.

In Fig. 2, the units of the action spectrum ordinate are reciprocal to the fluence units, which themselves are essentially m^{-2} (apart from dimensionless entities such as photons, or moles of photons). Consequently, the action spectrum units are m^2, indicating that some area quantity is involved. This dimensionality corresponds specifically to that of a cross section (see relationship between extinction coefficient and cross section above). Some authors choose instead to label the ordinate in "relative units", for example by normalizing several action to the wavelength of maximal effectiveness of one of them. However, it is increasingly popular to retain the proper physical units of m^2 mol^{-1}. If fluence rate-response curves are used instead as the basis for the action spectra, then the corresponding reciprocal units for the "effectiveness" are m^2 s mol^{-1}.

Functions Used for Fluence-Response Curve Fits

The hyperbolic saturation function of the form $ax/(x + b)$ often arises in biophysical and biochemical applications. It is more obvious that this represents a hyperbola, if it is written in a double-reciprocal form as in a Lineweaver-Burk transformation of the Michaelis-Menten kinetics in enzymology, which employ this type of function. Other contexts where this function appears are monomolecular photochemical kinetics (Lipson and Presti, 1980) and visual physiology (Naka and Rushton, 1966; Williams and Gale, 1978). In the present context of action spectroscopy, x would stand for the fluence or, in some cases, the fluence rate. When $x = b$, the function is at the half-maximum level that is often chosen to be the criterion response. So, when one performs least-squares fits using such a function, the parameter b is the estimate of the fluence needed for the criterion response, and the effectiveness (action spectrum ordinate) is just b^{-1}.

Table 1 gives several functional forms that are useful for fitting fluence-response curves. These are all based on the hyperbolic saturation function given above and shown in the top row of the table. In certain cases, fluence-response curves have a two-component (biphasic) structure. The first component reaches a plateau as in Fig. 1, and then, at high fluence, a second component appears that may or may not approach saturation at the maximum available fluence. For such a two-component curve, if saturation is reached for the second component, then the function in the second row of Table 1 may apply, perhaps with the addition of exponents n (often attributable to cooperativity) for each component, as in the Hill function in the third row. Otherwise, if saturation is not reached at some or all wavelengths, then the function in the last row of Table 1 may be useful. The second and third terms of that function together have the shape of a hyperbola when plotted on a semilogarithmic scale (i.e. with abscissa log x). Note that this hyperbola, with a (rising) slant asymptote at high fluence, is unrelated to the hyperbola functions (for linear axes) in the first and second rows.

Table 1. Representative functions for fluence-response curve fits

Description	Function	References[a]
hyperbola	$\frac{ax}{x+b}$	1
hyperbolas plus constant[b]	$\frac{ax}{x+b} + \frac{cx}{x+d} + h$	1,2
Hill function	$\frac{ax^n}{x^n + b^n}$	3
sigmoid[c]	$\frac{a}{1 + \exp[n\,(k\text{-}u)\ln 10]}$	3
sigmoid[c]	$\frac{a}{2}\left[1 + \tanh\left(n\,(u - k)\frac{\ln 10}{2}\right)\right]$	3
Hill plus hyperbola[d]	$\frac{ax^n}{x^n + b^n} + c\log\left(\frac{x}{d}\right) + \sqrt{c^2\log^2\left(\frac{x}{d}\right) + k^2}$	4,5

[a] References: 1) Corrochano et al. (1988); 2) Galland et al. (1989); 3) Ensminger et al. (1990); 4) Trad et al. (1987); 5) Bejarano et al. (1991).

[b] This example consists of a sum of two hyperbolas plus an optional constant, h. The coefficients a and c may be positive or negative for saturation components that rise or fall, respectively.

[c] These two equivalent functions are derivable from the Hill function (so called in the familiar context of ligand binding kinetics for hemoglobin) in the third row with the substitutions $u = \log x$ and $k = \log b$. They represent the sigmoidal shape taken on by the Hill function when replotted on a semilogarithmic scale, i.e. plotted against log x instead of x. In the special case with $n = 1$, these sigmoidal functions are equivalent to the hyperbola function in the top row (see Fig. 1). In general, the Hill exponent n need not be an integer.

[d] The last two terms together comprise a hyperbola when plotted on a *logarithmic* scale for x; Note that this usage of a hyperbola is quite different from that in the top row. The hyperbola here has a horizintal asymptote for small x and a slant asymptote (with slope 2c) for large x.

In general, when these sigmoidal and other functions are fit to fluence-response curves for the purpose of action spectroscopy, there is another clear choice for criterion response besides the half-maximum (50%) level, namely a "threshold" response level. However, thresholds (defined as the highest fluence at which there is no apparent response) *per se* are difficult to measure because of signal-to-noise considerations at low response levels. A more practical measure of the threshold can be defined operationally as the fluence at which the tangent line at the inflection point (or midpoint) intersects the baseline. This level is generally somewhat higher than the actual threshold, but it has the advantages of being much better defined and easier to measure. For a fluence-response curve, or component thereof, described by the Hill function (third row of Table 1), the operational threshold fluence is smaller than the midpoint fluence by a factor of $e^{-2/n}$. In the special case when $n = 1$ (top two rows of Table 1), the factor is just e^{-2}. On a $\log_{10}$ scale, this threshold fluence is displaced to the left of the midpoint fluence by $0.869/n$ "log units".

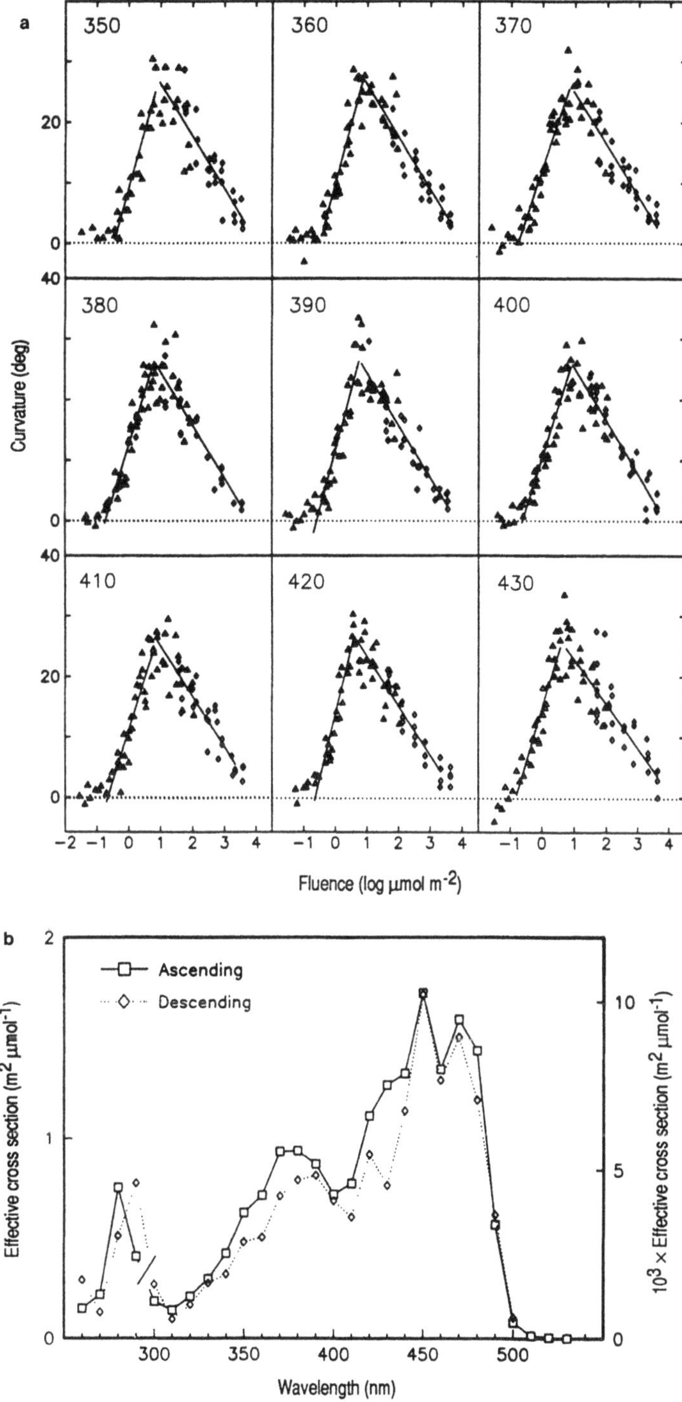

Fig. 3. Fluence-response curves (**a**) and action spectra (**b**) for first-positive phototropism in hypocotyls of alfalfa grown under red light. Only some of the fluence-response curves are shown. The two action spectra are based on a criterion response of 13° bending for the ascending and descending regions of the fluence-response curves (as indicated). After Baskin and Iino (1987).

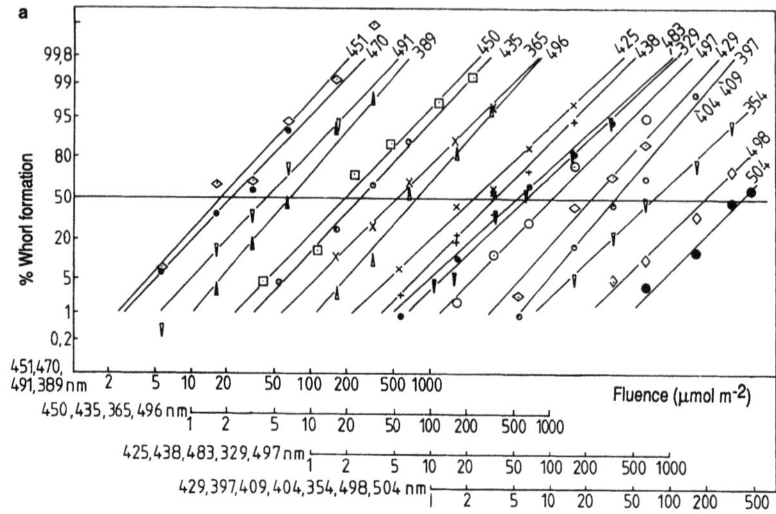

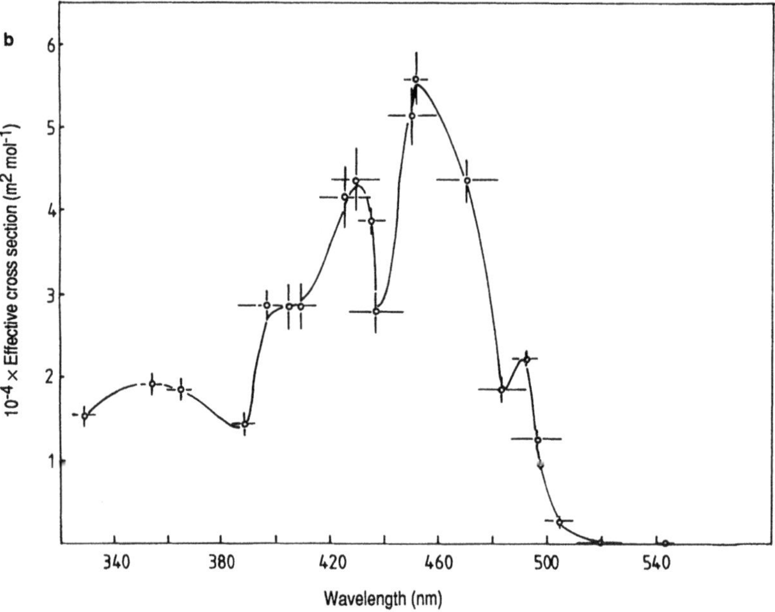

Fig. 4. Fluence-response curves (**a**) and action spectra (**b**) for light-induced formation of hair whorls in *Acetabularia mediterranea*, a photomorphogenic response controlled by blue light. **a**) Fluence-response curves for percentage formation. The ordinate is a probit scale. Wavelengths are shown at the tops of the "curves" (i.e. straight lines) and are also used to label, in groups, the appropriate logarithmic abscissa scales. For deriving the action spectrum, a criterion "response" of 50% probability was chosen. After Schmid (1987).

Representative Fluence-Response Curves and Action Spectra

Several sets of action spectra are shown below, along with the fluence-response data from which they were derived. These examples were selected, in part, because of the thorough presentations of fluence-response data in the respective publications.

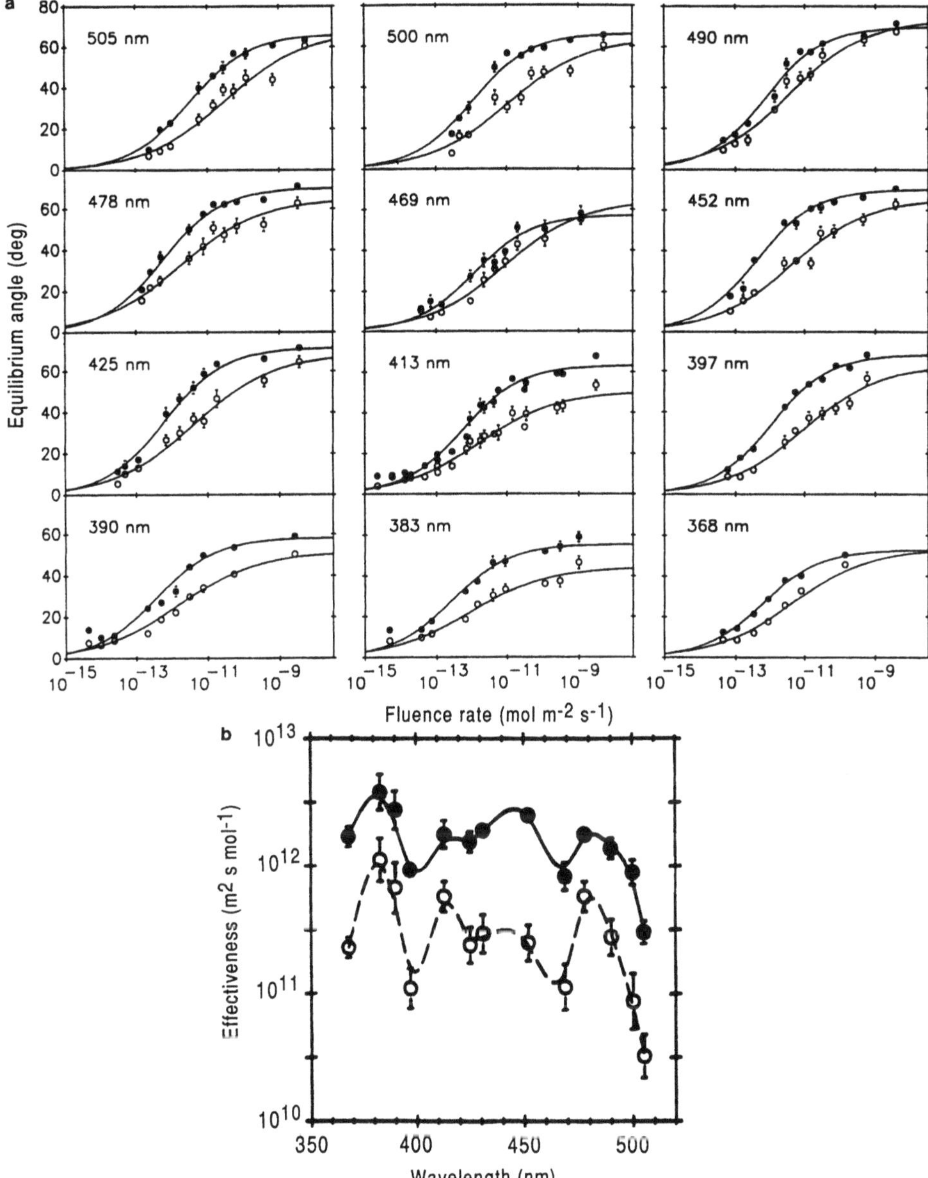

Fig. 5. Fluence-response curves (**a**) and action spectra (**b**) for photogravitropic equilibrium of *Phycomyces* sporangiophores of wild-type strain NRRL1555 (filled symbols) and mutant L150 (open symbols). Sporangiophores were exposed to unilateral light for 7 h. Note that the abscissa in (**a**) is fluence rate rather than fluence, and that the action spectrum ordinate in (**b**) correspondingly has different units than in the other figures. The curves in (**a**) were fit using the sigmoidal function in the fourth row of Table 1. After Ensminger et al. (1990).

Interested readers are referred to the papers themselves for details on the organisms, experiments, and interpretations.

The data for phototropism in alfalfa (Fig. 3) were measured at the Okazaki Large Spectrograph (Baskin and Iino, 1987). The fluence-response data (Fig. 3a) are remarkably consistent at different wavelengths; only in the UV below 300 nm do they

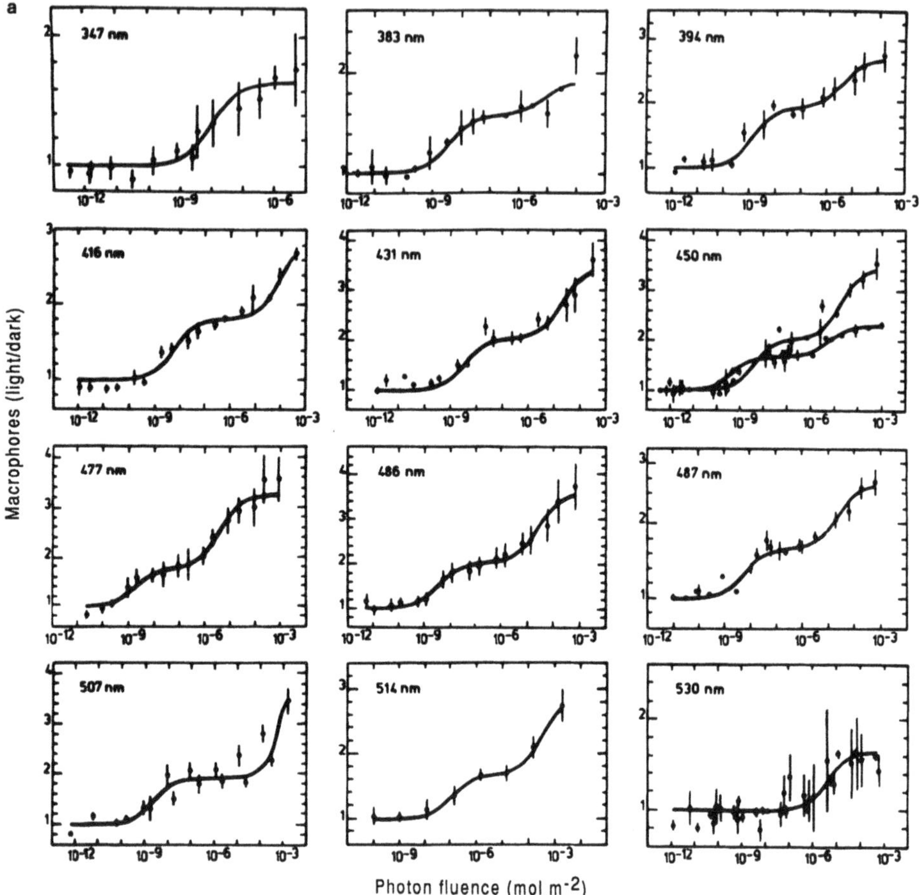

Fig. 6. Fluence-response curves (**a**) and action spectra (**b**) for light induction of sporangiophores (here designated as macrophores to distinguish them from tiny microphores formed under unfavorable conditions). This blue light-controlled photomorphogenic response is called photophorogenesis. The fluence-response curves, which were fit to functions like those in the first two rows of Table 1, are biphasic at most wavelengths, i.e. they have low-fluence and high-fluence components. The action spectra for the two components are based on a criterion response of 50% of the maximum. After Corrochano et al. (1988).

change their shape significantly (not shown here). For most wavelengths, the authors chose an absolute criterion-response level of 13° curvature for both the ascending and descending parts of the curves. In the near UV, they used instead the threshold for the ascending part and the peak for the descending part. The pair of action spectra (Fig. 3b), respectively for the ascending and descending parts of their fluence-response curves, have the typical characteristics of blue light action spectra.

Figure 4 shows fluence-response curves and action spectra for a photomorphogenic effect in the siphonaceous green alga *Acetabularia* (Schmid et al., 1987). This alga, which has been a popular model system for developmental biology, forms "hair whorls" at the growing apex in response to blue illumination. The near parallelism of the fluence-response curves, based on a logarithmic scale for photon fluence, suggests that a single blue-light photoreceptor could account for this photomorphogenesis. The shape of the action spectra itself is within the general range of action spectra found for blue light systems. Note that the error bars on the action spectrum were obtained by procedures different from those described in the present chapter.

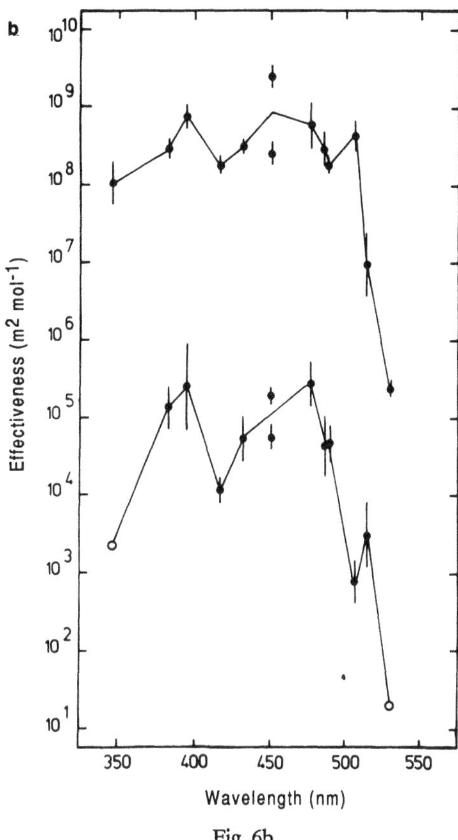

Fig. 6b

Figure 5 shows some recent "threshold curves" and derived action spectra for photogravitropism of *Phycomyces* sporangiophores (Ensminger et al., 1990). This vectorial equilibrium between phototropism and gravitropism, measured as a function of fluence rate, has been used conventionally to compare the phototropic sensitivity of the wild type and various mutants (Bergman et al., 1973; Ootaki et al., 1974; Lipson and Terasaka, 1981). This recent study, though, constitutes the first application of least-squares fitting to such threshold curves in *Phycomyces*, particularly for the generation of action spectra. The data shown here are for wild type and for one of several new mutants (here strain L150) with phototropic threshold elevated from wild type by only about 10-fold (Alvarez et al., 1989). Previously isolated night-blind mutants have displacements from about 10^3- to 10^5-fold (Bergman et al., 1973; Ootaki et al., 1974). The results for this mutant (and other related ones) show significant differences from the wild-type behavior, suggesting that the mutation may affect the photoreceptor complex, as found previously for certain other night-blind mutants (Galland and Lipson, 1985b).

Figure 6, like Figure 4, shows fluence-response curves and an action spectrum for a blue-light controlled photomorphogenic response, in this case for sporangiophore development in *Phycomyces* (Corrochano et al., 1988). That paper also presents results for a complementary response, namely photoinhibition of "microphores", miniature sporangiophores produced under stress conditions. Blue light suppresses microphores and promotes the occurrence of "macrophores", the normal sporangiophores. At most wavelengths, the fluence-response curves are biphasic with low- and high-fluence components. One therefore obtains two action spectra for macrophorogenesis. For the photoinhibition of microphores, which also has biphasic fluence-response curves, there are two more action spectra. All four spectra, while showing general blue light sensitivity, differ significantly from one another. This allows for the possibility that there may be up to four photoreceptors mediating these responses.

An analogous study was performed on carotenogenesis in *Phycomyces* (Bejarano et al., 1991; results not shown here). The fluence response curves again were biphasic, with low- and high-fluence components; however only the low-fluence component reached saturation at the maximal fluence employed. Therefore the function in the bottom row of Table 1 was used for the analysis of the fluence-response data in that study. The two action spectra so obtained show distinct blue light sensitivities, suggesting the possible participation of more than one photosystem.

An earlier application of the function in the bottom row of Table 1 was in analyzing the fluence-response curves for a particular light-induced absorbance change (LIAC) studied *in vivo* on samples from *Phycomyces* (Trad and Lipson, 1987). For a general discussion of this LIAC approach for spectroscopic analysis of blue light systems *in vivo* and *in vitro*, see Chapter on LIAC Activity in Higher Plants (Caubergs, this volume).

Another way to measure action spectra is to employ a balance or null method, in which the criterion response is zero, but without the need to work at absolute threshold levels. Two lights, one a fixed "reference" light and the other a variable "test" light are applied from opposite directions or in temporal alternation, in such a way that no net response occurs. Recent examples for action spectroscopy on *Phycomyces* are balance experiments on phototropism (Galland and Lipson, 1985a) and null experiments on the light-growth response (Ensminger et al., 1991; Ensminger and Lipson, 1991).

This chapter has emphasized action spectra for blue light effects, believed to be mediated by flavin receptor pigments (Galland and Senger, 1988a) with likely contributions by pterin chromophores that absorb in the near UV (Galland and Senger, 1988b). In closing, I will mention the action spectroscopy work of Foster and coworkers that has led to the discovery of a rhodopsin in the alga *Chlamydomonas* that serves as photoreceptor for phototaxis (Foster et al., 1984) and for photocarotenogenesis (Foster et al., 1988b). The *Chlamydomonas* system is discussed in the Chapter on Photoreception in *Chlamydomonas* (Hegemann, this volume). As discussed in the Chapter on Color Discriminating Pigments in *Halobacterium halobium* (Spudich, this volume) rhodopsins have been studied extensively in bacteria, specifically in *Halobacterium halobium* (see Spudich, this volume; also see Stoeckenius et al., 1988 for recent action spectroscopy).

The occurrence of rhodopsin in *Chlamydomonas* was actually predicted by Foster and Smyth (1980) based on their reanalysis of the action spectra of Nultsch et al. (1971) on *Chlamydomonas* which included full fluence-response curve analysis. To overcome screening effects (1980; 1988), Foster and Smyth chose threshold as the criterion response for their analysis.

Subsequently Foster and coworkers (Foster et al., 1984; 1990; Foster et al., 1988a; 1989) carried out action spectroscopy studies of phototaxis in which retinal analogs were added to a mutant that does not produce ß-carotene (metabolic precursor of retinal) in the dark. When retinal or various analogs were added, phototactic sensitivity could be restored, leading to the conclusion that the receptor pigment for phototaxis in *Chlamydomonas* is a rhodopsin.

In related work, it was shown that a rhodopsin is also the receptor pigment for controlling synthesis of ß-carotene (Foster et al., 1988b); thus *Chlamydomonas* rhodopsin evidently autoregulates the synthesis of its retinal chromophore. This result is surprising, in that light-induced carotene synthesis in most organisms studied is believed to be mediated by conventional blue-light photoreceptors rather than rhodopsin pigments. For example, in *Phycomyces* the phototransduction pathway for light-induced carotene synthesis seems to share the same photoreceptors as the other blue light responses (including phototropism, which has normal threshold sensitivity in caroteneless mutants (see Fig. 4 and text of Chapter Phototropism in Fungi by Lipson, this volume). Further, the action spectra for light-induced carotene synthesis in *Phycomyces* (Bejarano et al., 1991) suggest typical blue light photoreceptors.

This brief chapter on action spectroscopy has given primary emphasis to applications in the field of blue light photoreception. It has not delved into action spectroscopy for photochromic receptors, such as phytochrome, which is dealt with in many of the references on action spectroscopy cited in the introduction. Some attention has also been given to formal data analysis techniques to encourage readers to adopt such methods for analyzing experimental results from action spectroscopy and other studies.

Acknowledgements

Work in this laboratory has been supported by grants from the National Institutes of Health, the National Science Foundation, the United States-Spain Joint Committee for Scientific and Technological Cooperation and the United States-Israel Binational Science Foundation. I am grateful to David Durant for assistance in preparing Figures 1 and 2.

References

Alvarez, M. I., Eslava, A. P., and Lipson, E. D., 1989, Phototropism mutants of *Phycomyces blakesleeanus* isolated at low light intensity, *Exp. Mycol.*, 13:38.

Baskin, T. I., and Iino, M., 1987, An action spectrum in the blue and ultraviolet for phototropism in alfalfa, *Photochem. Photobiol.*, 46:127.

Bejarano, E. R., Avalos, J., Lipson, E. D., and Cerdá-Olmedo, E., 1991, Photoinduced accumulation of carotene in *Phycomyces*, *Planta*, in press.

Bergman, K., Eslava, A. P., and Cerdá-Olmedo, E., 1973, Mutants of *Phycomyces* with abnormal phototropism, *Mol. Gen. Genet.*, 123:1.

Bevington, P. R., 1969, "Data Reduction and Error Analysis for the Physical Sciences," McGraw-Hill, New York.

Corrochano, L. M., Galland, P., Lipson, E. D., and Cerdá-Olmedo, E., 1988, Photomorphogenesis in *Phycomyces*: fluence response curves and action spectra, *Planta*, 174:315.

Ensminger, P. A., Chen, X., and Lipson, E. D., 1990, Action spectra for photogravitropism of *Phycomyces* wild type and three behavioral mutants (L150, L152, and L154), *Photochem. Photobiol.*, 51:681.

Ensminger, P. A., and Lipson, E. D., 1991, Action spectra of the light-growth response in three behavioral mutants of *Phycomyces*, *Planta*, in press.

Ensminger, P. A., Schaefer, H. R., and Lipson, E. D., 1991, Action spectra of the light-growth response of *Phycomyces*, *Planta*, in press.

Foster, K., Saranak, J., Derguini, F., Jayathirtha, V., Zarrilli, G., Okabe, M., Fang, J.-M., Shimizu, N., and Nakanishi, K., 1988a, Rhodopsin activation: a novel view suggested by *in vivo Chlamydomonas* experiments, *J. Am. Chem. Soc.*, 110:6588.

Foster, K., Saranak, J., Derguini, F., Zarilli, G., Johnson, R., Okabe, M., and Nakanishi, K., 1989, Activation of *Chlamydomonas* rhodopsin *in vivo* does not require isomerization of retinal, *Biochemistry*, 28:819.

Foster, K. W., Saranak, J., and Dowben, P. A., 1990, Spectral sensitivity, structure, amd activation of eukaryotic rhodopsins: activation spectroscopy of rhodopsin analogs in *Chlamydomonas*, *J. Photochem. Photobiol. B:Biol.*, in press.

Foster, K. W., Saranak, J., Patel, N., Zarilli, G., Okabe, M., Kline, T., and Nakanishi, K., 1984, A rhodopsin is the functional protoreceptor for phototaxis in the unicellular eukaryote *Chlamydomonas*, *Nature*, 311:756.

Foster, K., Saranak, J., and Zarrilli, G., 1988b, Autoregulation of rhodopsin in *Chlamydomonas reinhardtii*, *Proc. Natl. Acad. Sci. USA*, 85:6379.

Foster, K. W., and Smyth, R. D., 1980, Light antennas in phototactic algae, *Microbiol. Rev.*, 44:572.

Galland, P., 1987, Action spectroscopy, *in:* "Blue Light Responses: Phenomena and Occurrence in Plants and Microorganisms," Senger, H., ed., CRC Press, Boca Raton, Florida, p. 37.

Galland, P., and Lipson, E. D., 1985a, Action spectra for phototropic balance in *Phycomyces blakesleeanus*: dependence on reference wavelength and intensity range, *Photochem. Photobiol.*, 41:323.

Galland, P., and Lipson, E. D., 1985b, Modified action spectra of photogeotropic equilibrium in *Phycomyces blakesleeanus* mutants with defects in genes *madA*, *madB*, *madC*, and *madH*, *Photochem. Photobiol.*, 41:331.

Galland, P., and Lipson, E. D., 1987, Blue-light reception in *Phycomyces* phototropism: evidence for two photosystems operating in low- and high-intensity ranges, *Proc. Natl. Acad. Sci. USA*, 84:104.

Galland, P., Orejas, M., and Lipson, E. D., 1989, Light-controlled adaptation kinetics in *Phycomyces*: evidence for a novel yellow-light absorbing pigment, *Photochem. Photobiol.*, 49:493.

Galland, P., and Senger, H., 1988a, The role of flavins as photoreceptors, *J. Photochem. Photobiol. B:Biol.*, 1:277.

Galland, P., and Senger, H., 1988b, The role of pterins in the photoreception and metabolism of plants, *Photochem. Photobiol.*, 48:811.

Grossweiner, L. I., 1989, Photophysics, *in:* "The Science of Photobiology," Smith, K. C., ed., Plenum Press, New York, p. 1.

Hamilton, W. C., 1964, "Statistics in Physical Science," Ronald Press, New York.

Hartmann, K. M., 1983, Action spectroscopy, *in:* "Biophysics," Hoppe, W., Lohmann, W., Markl, H., and Ziegler, H., eds., Springer-Verlag, Berlin, Heidelberg, New York, p. 115.

Jagger, J., 1967, "Introduction to Research in Ultraviolet Photobiology," Prentice-Hall, Englewood Cliffs, NJ.

Lipson, E. D., and Presti, D., 1980, Graphical estimation of cross sections from fluence-response data, *Photochem. Photobiol.*, 32:383.

Lipson, E. D., and Terasaka, D. T., 1981, Photogeotropism in *Phycomyces* double mutants, *Exp. Mycol.*, 5:101.

Naka, K. I., and Rushton, W. A. H., 1966, S-potentials from colour units in the retina of fish (Cyprinidae), *J. Physiol.*, 185:536.

Nultsch, W., Throm, G., and von Rimscha, I., 1971, Phototaktische Untersuchungen an *Chlamydomonas reinhardtii* Dangeard in homokontinuierlicher Kultur, *Arch. Microbiol.*, 90:47.

Ootaki, T., Fischer, E. P., and Lockhart, P., 1974, Complementation between mutants of *Phycomyces* with abnormal phototropism, *Mol. Gen. Genet.*, 131:233.

Press, W. H., Flannery, B. P., Teukolsky, S. A., and Vetterling, W. T., 1985, "Numerical Recipes," Cambridge University Press, Cambridge, England.

Presti, D. E., and Galland, P., 1987, Photoreceptor biology of *Phycomyces*, *in:* "*Phycomyces*," Cerdá-Olmedo, E., and Lipson, E. D., eds., Cold Spring Harbor Laboratory, Cold Spring Harbor, New York, p. 93.

Presti, D., Hsu, W. J., and Delbrück, M., 1977, Phototropism in *Phycomyces* mutants lacking ß-carotene, *Photochem. Photobiol.*, 26:403.

Schäfer, E., and Fukshansky, L., 1984, Action spectroscopy, *in:* "Techniques in Photomorphogenesis," Smith, H., and Holmes, M. G., eds., Academic Press, London, p. 109.

Schäfer, E., Fukshansky, L., and Shropshire, W., Jr., 1983, Action spectroscopy of photoreversible pigment systems, *in:* "Encyclopedia of Plant Physiology", New Series, Vol. 16 A,B, "Photomorphogenesis," Shropshire, W., Jr., and Mohr, H., eds., Springer-Verlag, Berlin, Heidelberg, New York, p. 39.

Schmid, R., Idziak, E.-M., and Tünnermann, M., 1987, Action spectrum for the blue-light-dependent morphogenesis of hair whorls in *Acetabularia mediterranea*, *Planta*, 171:96.

Senger, H., ed., 1980, "The Blue Light Syndrome," Springer-Verlag, Berlin, Heidelberg, New York.

Senger, H., ed., 1984, "Blue Light Effects in Biological Systems," Springer-Verlag, Berlin, Heidelberg, New York.

Senger, H., ed., 1987, "Blue Light Responses: Phenomena and Occurrence in Plants and Microorganisms," Vol. I and II, CRC Press, Boca Raton, Florida.

Shropshire, W., Jr., 1972, Action spectroscopy, *in:* "Phytochrome," Mitrakos, K., and Shropshire, W., Jr., eds., Academic Press, London, p. 162.

Smyth, R. D., Saranak, J., and Foster, K. W., 1988, Algal visual systems and their photoreceptor pigments, *Prog. Phycol. Res.*, 6:225.

Stoeckenius, W., Wolff, E., and Hess, B., 1988, A rapid population method for action spectra applied to *Halobacterium halobium*, *J. Bact.*, 170:2790.

Trad, C. H., and Lipson, E. D., 1987, Biphasic fluence-response curves and derived action spectra for light-induced absorbance changes in *Phycomyces* mycelium, *J. Photochem. Photobiol. B:Biol.*, 1:169.

Williams, T. P., and Gale, J. G., 1978, "Compression" of retinal responsivity: V-log I functions and increment thresholds, *in:* "Visual Psychophysics and Physiology," Armington, J. C., ed., Academic Press, New York, p. 129.

Phototropism in Fungi

Edward D. Lipson

Department of Physics
Syracuse University
Syracuse, NY 13244-1130
U.S.A.

Introduction

Phototropism, the curvature of a growing part of a plant or fungus toward or away from light, is a well known, but not particularly well understood, phenomenon. The general topic of phototropism has been extensively reviewed (Foster, 1977; Dennison, 1979; Hertel, 1980; Gressel and Horwitz, 1982; Pohl and Russo, 1984; Briggs and Baskin, 1988; Firn, 1990; Iino, 1990). This chapter will focus on two Zygomycete fungi, *Phycomyces* and *Pilobolus*, both of which are currently under active investigation. A recent review (Galland, 1990) compares phototropism in *Phycomyces* with that in higher plants. For a more extensive coverage of the classical work on fungi, see the review article by Page (1968). A more recent phototropism review with some coverage on fungi is that by Pohl and Russo (1984). Page (1968) has listed fungi with known phototropism and offered the generalization that, although phototropism occurs widely in terrestrial fungi, it does not seem to occur in aquatic fungi.

Models for Phototropism

In higher plant phototropism, there are two traditional, competing models, popularly known as the Cholodny-Went and the Blaauw models (see phototropism reviews cited above). Briefly, the Cholodny-Went model assumes that phototropism is governed by redistribution of growth-controlling substances. The favorite candidate for this in higher plants has been the auxin indoleacetic acid (IAA), although controversies remain on this and other aspects of phototropism (Briggs and Baskin, 1988; Hasegawa et al., 1989; Firn, 1990). According to the Blaauw model, phototropism is presumed to be generated as a superposition of localized light-growth responses. Both types of models have been considered in fungi as well as in plants. The most recent serious resurrection of the Blaauw model in fungi was the rotation model of Dennison and Foster (1977) in *Phycomyces*. However, there have been a number of contrary results (Iino and Schäfer, 1984; Galland et al., 1985), in particular that the kinetics for the light-growth response (see definition below) often differ considerably from the kinetics for phototropism. For a discussion of other criticisms of the rotation model, see Hertel (1980).

The results for *Phycomyces* overall favor the redistribution of growth effectors rather than superposition of localized light-growth responses (Galland et al., 1985).

Phototropism in Phycomyces

Both the sporangiophores and the mycelium of *Phycomyces* exhibit a number of responses to blue light stimulation. Most attention has been given to the sporangiophore responses: phototropism and the related light-growth response (transient change of the elongation rate in response to a change in fluence rate, usually applied symmetrically). The mycelial photoresponses include light-induced carotene synthesis (photocarotenogenesis) and light-induced sporangiophore initiation (photophorogenesis). These mycelial phenomena do not absolutely require light, but can be promoted by illumination. The responses and other general areas of research on *Phycomyces* are reviewed in a recent monograph (Cerdá-Olmedo and Lipson, 1987b), which includes a comprehensive bibliography (Shropshire, 1987) of the *Phycomyces* literature over the past 100 years.

Figure 1 shows two examples of phototropic responses of *Phycomyces* sporangiophores. In the experiment for the left curve, the fluence rate was maintained the same after a transition from symmetrical to unilateral light. Under this condition, the latency has a minimal value of several minutes. If the fluence rate is reduced 500-fold during the transition to unilateral illumination, then the latency increases to almost 50 min, because the system needs time to adapt to this lower light level. This protocol has been exploited for investigating the adaptation kinetics of *Phycomyces* phototropism (Galland and Russo, 1984; Galland et al., 1984; 1989a,b), which are generally similar to

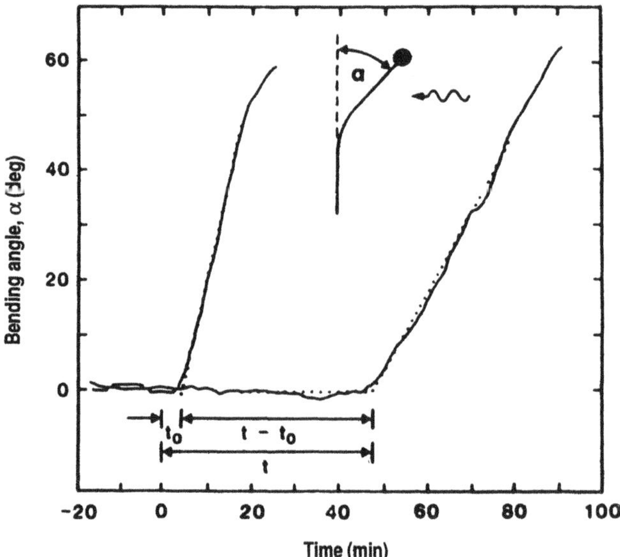

Fig. 1. Time course of phototropism in *Phycomyces* sporangiophores. For the curve at left, a sporangiophore was adapted for 45 min by exposure to a horizontal light beam at a fluence rate of 1.3 x 10^{-2} W m^{-2} (452 nm), by rotating the sporangiophore about its axis at 10 rpm. This rotation simulates the condition of light impinging uniformly from all azimuthal directions. The phototropic stimulus is initiated at time zero simply by stopping the rotation. Bending ensues after a latent interval (t_0) of several minutes. For the curve on the right, the fluence rate was reduced by a factor of 500 at time zero. The increase of the latency (t) is a manifestation of the adaptation process underlying phototropism. After Galland (1990).

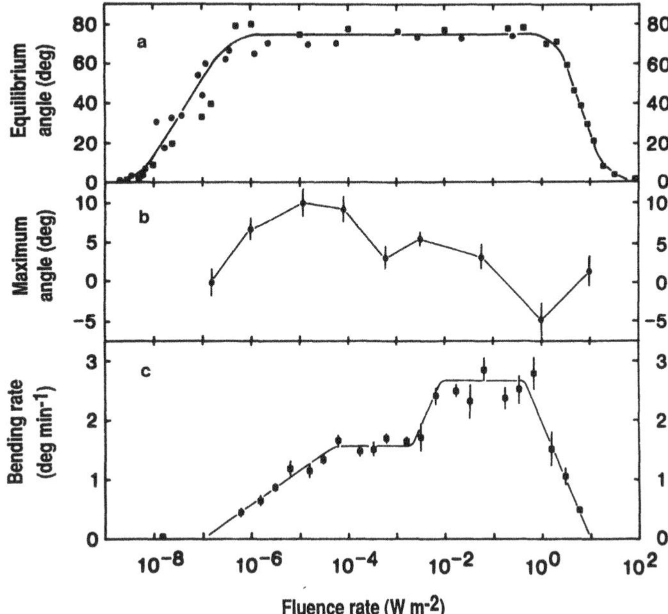

Fig. 2. Fluence-rate dependence of three aspects of phototropism in *Phycomyces* sporangiophores under equilibrium and nonequilibrium conditions. **a.** Photogravitropic equilibrium , as in Fig. 3 (WT curve). **b.** Maximal angle of transient bending in response to 30 s pulses (450 nm), following 3 h of dark adaptation. **c.** Bending rate, determined as indicated by the dotted line on the left curve of Fig. 1. After Galland and Lipson (1987b).

adaptation kinetics studied in the light-growth response (Delbrück and Reichardt, 1956; Lipson and Block, 1983). For general reviews, see Galland and Lipson (1987b) and Galland (1989) and, for practical information on photobiological methods for *Phycomyces*, see Lipson and Galland (1987) and Galland and Lipson (1987a). Kinetic measurements of phototropism can be facilitated by a semi-automatic method using time-lapse video equipment and an electronic protractor (Lipson and Häder, 1984; used to obtain the records shown in Fig. 1) or else by computerized image analysis techniques (Popescu et al., 1989b).

Phycomyces phototropism and light-growth response have an absolute sensitivity range of 10 orders of magnitude (Fig. 2), similar to the range of visual sensitivity. In our vision there are two sets of photoreceptor cells, namely the rods and the cones, to manage this enormous range. However, the *Phycomyces* sporangiophore, a single-celled structure, can somehow handle this whole range by itself. As in vision, a sophisticated adaptation process permits range adjustment of sensitivity (Delbrück and Reichardt, 1956; Lipson and Block, 1983; Galland and Russo, 1984; Galland, 1989).

Three different measures of phototropism as a function of fluence rate are shown in Fig. 2. The equilibrium between phototropism and gravitropism provides a convenient assay for comparing the behavioral mutants of *Phycomyces*, specifically for their phototropic threshold and operating range (Fig. 3; see below and see Chapter on Action Spectroscopy (Lipson, this volume). The normal threshold is $\sim$1 nW m^{-2}. Figure 2b shows a fluence-response curve for phototropic responses to 30 s pulses of light. The "maximal angle" is plotted because, for *Phycomyces*, the bending response to such pulses is transient. In Fig. 2b, the rate of steady-state bending (slope of curves as in Fig. 1) of sporangiophores is shown as a function of fluence rate. A general comparison of phototropism in *Phycomyces* to that in higher plants has been presented recently by Galland (1990).

Phototropism Mutants: Threshold Curves and Classification Scheme

A collection of well characterized behavioral mutants is available in *Phycomyces*, including a large number of phototropism (*mad*) mutants, affected in nine genes (*madA* to *madI*). Those affected in genes *madA*, *madB* and *madC* are greatly reduced in sensitivity for both phototropism and the light-growth response (Fig. 3). Their thresholds are elevated by over 10^3-fold for *madA* and by about 10^5-fold for *madB* and *madC* mutants. Phototropism mutants have recently been isolated with only about a 10-fold loss of sensitivity (Alvarez et al., 1989; see curve for strain L150 in Fig. 3, and see Fig. 5 of Chapter on Action Spectroscopy (Lipson, this volume); some of these have been assigned to the new gene *madI* (Campuzano et al., 1990).

The "stiff" mutants (*madD*, *madE*, *madF*, and *madG*) have slow, weak bending responses; they are called stiff because of their sluggish bending responses to all types of stimuli. Conversely, the hypertropic (*madH*) mutants (Lipson et al., 1980; Lipson and Terasaka, 1981) have generally enhanced bending responses, including in the threshold region. In the bottom half of Fig. 3 is shown a threshold curve for a piloboloid (*pil*) mutant (Koga and Ootaki, 1983a; b), so-called because such mutants have a subsporangial swelling resembling that of *Pilobolus* (see below). The *pil* mutants have thresholds similar to wild type, but they exhibit negative phototropism under the normal light conditions where other mutants and the wild type have positive phototropism (Koga et al., 1984).

Figure 4 presents a genetic classification scheme based on phenotypic analyses (behavioral and biochemical) of *Phycomyces* mutants (Bergman et al., 1973; Cerdá-Olmedo and Lipson, 1987a). In the present context of phototropism, the pathway from blue light to tropisms (and growth responses) is of special interest. The discovery that *madB* and *madC* mutants have altered action spectra for phototropism (Galland and

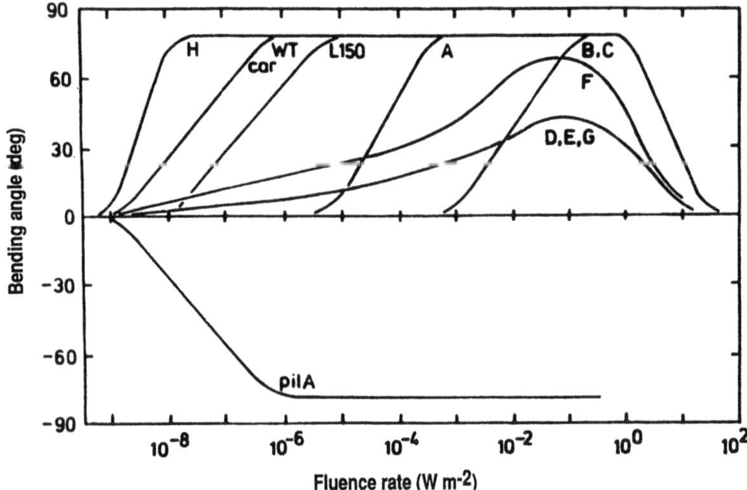

Fig. 3. Photogravitropic equilibrium threshold curves (data points not shown; see Fig. 4a for typical data) for sporangiophores of *Phycomyces* wild type (WT; strain NRRL1555) and representative mutants. Sporangiophores were exposed in a "threshold box" to 8-10 h of unilateral blue light (broadband, 450 nm, or 488 nm). The letters A-H stand for the phototropism genes *madA* through *madH*. The mutant L150 has yet to be assigned to a gene (probably *madI*; see text). Carotene (*car*) mutants lacking the bulk pigment ß-carotene have similar threshold behavior to the wild type strain. The curve in the bottom half is for a representative piloboloid (*pil*; see text) mutant that exhibits negative phototropism. After Galland and Lipson (1987b).

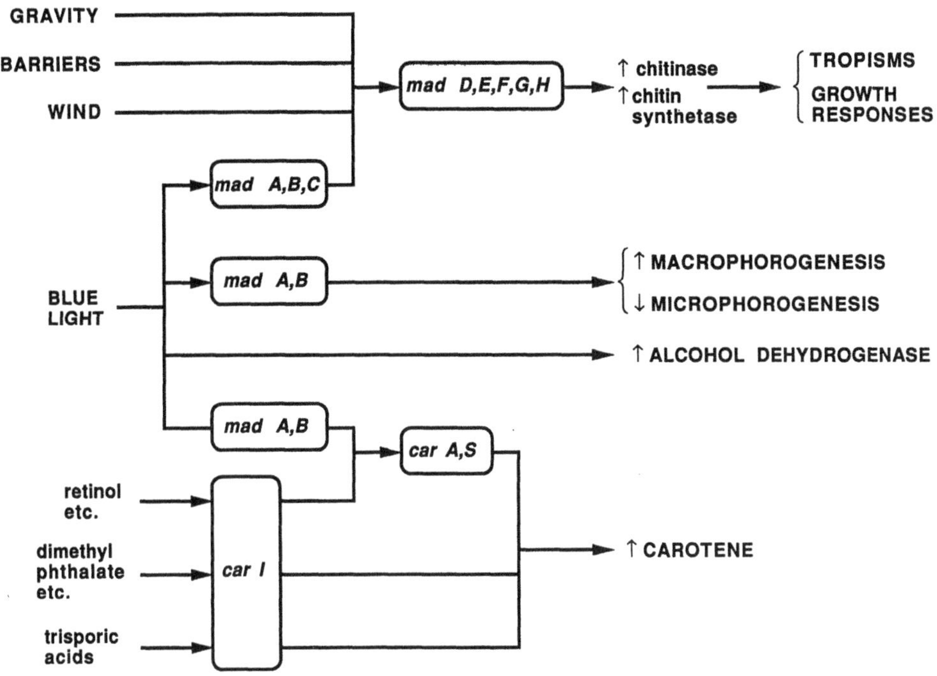

Fig. 4. Schematic representation of putative sensory transduction pathways in *Phycomyces* inferred from analysis of phototropism (*mad*) mutants and carotene (*car*) mutants. After Cerdá-Olmedo and Lipson (1987a).

Lipson, 1985) suggests that they are photoreceptor mutants and therefore deserving special attention in the search for blue light photoreceptors in *Phycomyces* (see below).

Light-Growth Response

Transient changes in elongation rates induced by light stimuli occur in many organisms that show phototropism. Analysis of *Phycomyces* mutants isolated for abnormal phototropism has shown that phototropism and the light-growth response share photoreceptor systems and operating range (Bergman et al., 1973). However, recent experiments testing whether superposition of localized light-growth responses might account for phototropism (Blaauw model) have given negative results; specifically, the two types of photoresponses tend to be kinetically dissimilar (Cosgrove, 1985; Galland et al., 1985; Kataoka, 1987).

System analysis methods, based on the Wiener theory of nonlinear system identification, have been applied extensively to the light-growth response of *Phycomyces* (reviewed by Galland and Lipson, 1987b; Lipson and Pratap, 1988). Comparative system-analysis studies of the light-growth response of single and double *mad* mutants and the wild type have revealed extensive mutual interactions among the *mad* gene products. These results suggest that there may be a molecular complex for sensory transduction (located most likely in the plasma membrane of the growing zone, which is in the upper 3 mm of the sporangiophore beneath the 0.5 mm diameter spherical sporangium), instead of a sequential pathway suggested by the formal genetic scheme in Fig. 4. Nevertheless, it is clear from Fig. 4 and other results that phototropism and the light-growth response share many features in common, including photosystems and adaptation mechanisms (see above).

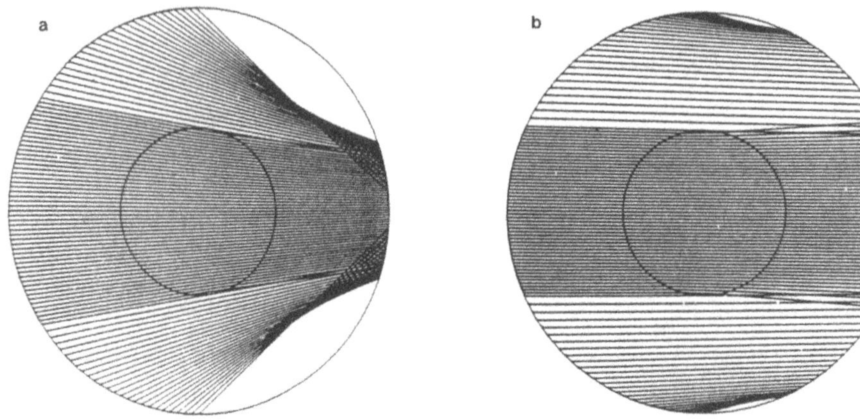

Fig. 5. Computer generated ray-tracing diagrams for light of wavelength 510 nm incident on a *Phycomyces* sporangiophore in air (**a**) and in a medium of refractive index n = 1.306 (**b**). Collimated light is assumed to be incident from the left. Effects of multiple internal reflection are not taken into account. After Fukshansky and Steinhardt (1989).

Optical Aspects of Phototropism

Consideration of optical effects is crucial for understanding of phototropism. When a *Phycomyces* sporangiophore is exposed to unilateral blue light, the light is refracted at the cell surface and becomes concentrated in a bright band on the distal side. The azimuthal light profile around the sporangiophore somehow leads to the differential growth that underlies phototropism. To a first approximation, the sporangiophore acts optically like a cylindrical lens. Other optical phenomena need to be considered, besides refraction, including multiple internal reflection, absorption, scattering, and diffraction. Unlike a cylindrical glass lens, the sporangiophore has a number of radial regions with differing refractive indices, notably the vacuole, tonoplast, cytoplasm, cell wall, and plasma membrane. Several investigators have done computer calculations taking these cellular and optical factors into account to varying extents (Foster, 1977; Steinhardt and Fukshansky, 1985; Tsuru et al., 1988). For a summary of earlier work, including confirmation of the lens effect, see Galland and Lipson (1987b).

Sophisticated optical calculations (Steinhardt and Fukshansky, 1985), employing a diffusion approximation to Maxwell's equations of electromagnetism and applying the formalism of differential geometry, have provided detailed azimuthal profiles for comparison to experimental results on *Phycomyces* phototropism under various conditions (Fukshansky and Steinhardt, 1987; Steinhardt et al., 1987; Steinhardt et al., 1989), for example a) sporangiophores immersed in media of different refractive indices to abolish or invert the lens effect and b) sporangiophores illuminated bilaterally with lights of different quality.

An important application of the theory has been to consider where the photoreceptor molecules for phototropism are located. Experiments on the light-growth response with polarized light (Jesaitis, 1974) showed that the photoreceptors in *Phycomyces* are oriented and dichroic. Based on this result and other considerations, it has been widely assumed that the photoreceptors are likely to reside in the plasma membrane within the growing zone. That is presumably where the enzymatic machinery resides for controlling cell wall growth. The main polymer in the cell wall is chitin (Gamow et al., 1987), and the major growth enzymes are believed to be chitin synthetase and chitinase; the former has been found to be regulated by light (Herrera-Estrella and Ruiz-Herrera, 1983; Ruiz-Herrera et al., 1990). However, the tonoplast membrane has also been proposed as a possible site of a the photoreceptors (Meistrich et al., 1970).

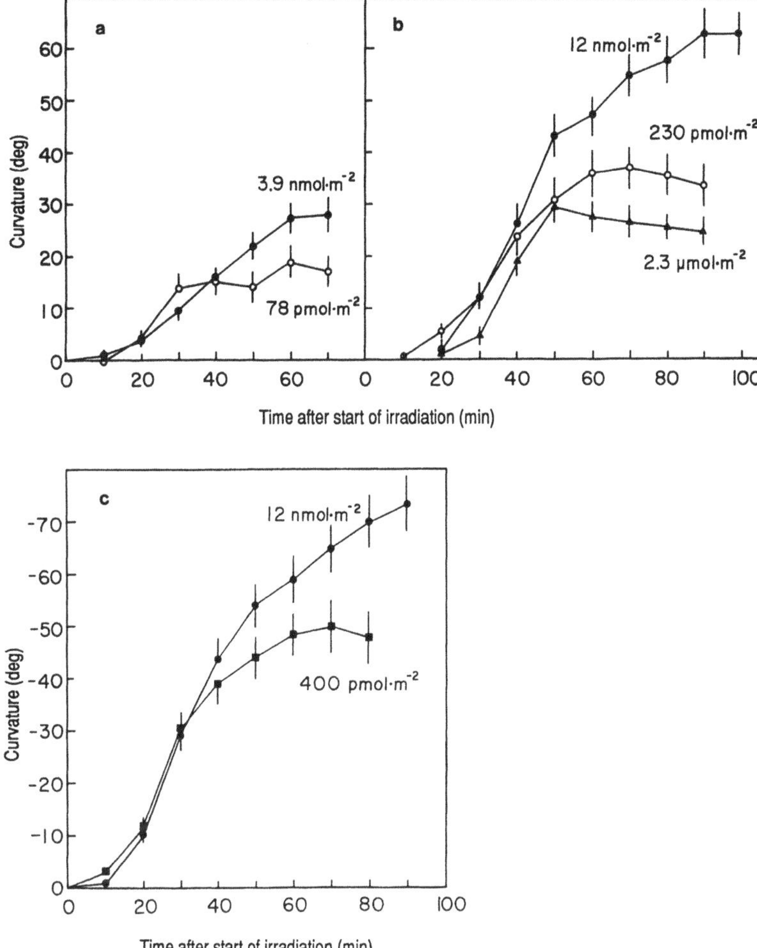

Fig. 6. Time course of phototropism in *Pilobolus* sporangiophores at various fluences, as indicated, of 450 nm blue light for 1 min (**a**); 30 min (**b**); and 280 nm UV-B radiation at a fluence rate of 6.7 pmol m^{-2} s^{-1} for 1 min or 30 min (**c**). After Kubo and Mihara (1988).

The computations of Steinhardt et al. (1989) favor the plasma membrane instead of the vacuole as the site of the photoreceptors. The reason is evident in Fig. 5: the contrast between the proximal and distal sides is far greater at the plasma membrane than at the vacuole under normal illumination conditions (Fig. 5a). Under the artificial condition with the sporangiophore immersed in a medium of neutral refractive index (Fig. 5b), the contrast between the two sides is virtually abolished at the plasma membrane site. The illumination contrast at the vacuole appears to be negligible under both conditions.

Besides the location of the photoreceptor molecules, the question remains as to their orientation. Application of the optical theory of Steinhardt and Fukshansky (1985) favors a transverse, tangential orientation of the chromophore transition dipoles (Steinhardt et al., 1989), consistent with the conclusions of Jesaitis (1974) based on light-growth response experiments with polarized light. The main application of the optical calculations will, in due course, be to develop a full theory of phototropism. To that end, Fukshansky and Steinhardt (1987) and Steinhardt et al. (1987) have extended their calculations of azimuthal light profiles, taking into account the presumed transition dipole orientation, to compute so-called excitation profiles.

In other recent work, Fukshansky and coworkers have reinvestigated the UV-B sensitivity of phototropism in *Phycomyces*. From earlier work, it had been generally accepted that the negative phototropism of *Phycomyces* in the UV region around 300

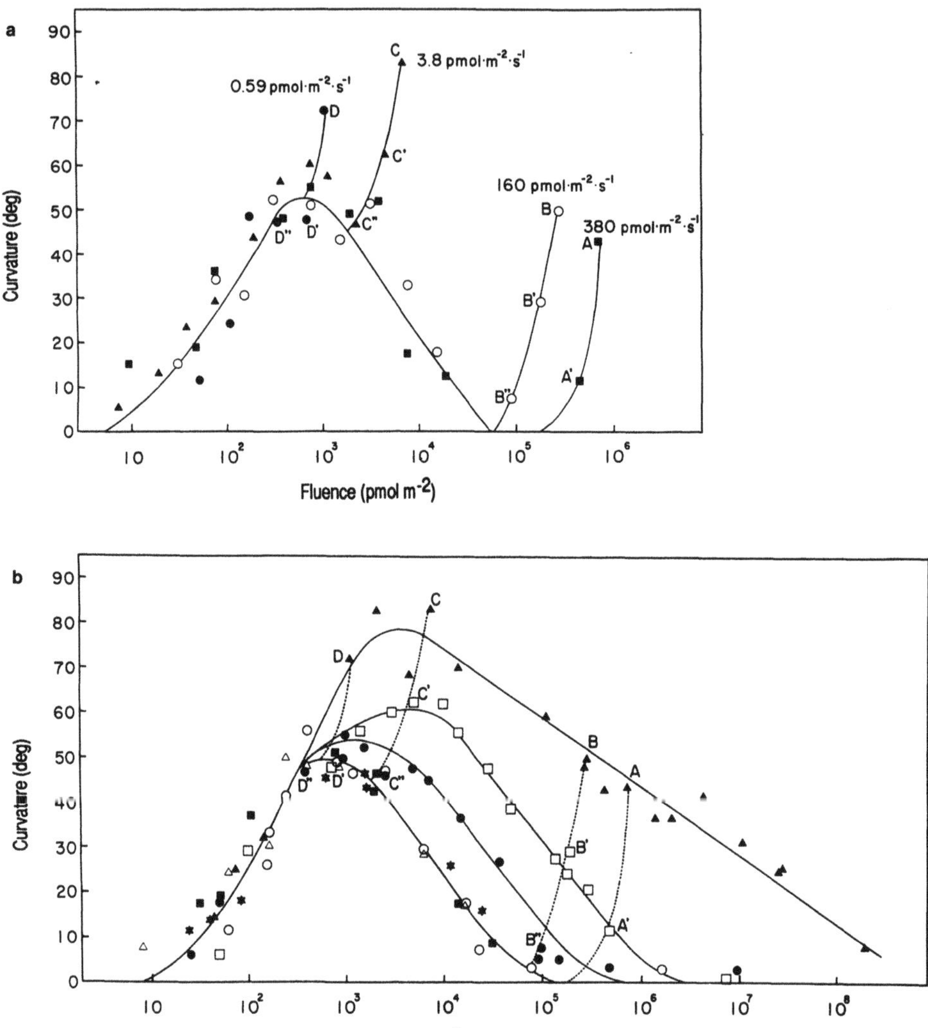

Fig. 7. Fluence-response curves for phototropism in *Pilobolus* obtained with blue light (450 nm). **a.** For each of the four curves shown, the fluence was varied by changing the exposure time. As indicated for the four data sets, the fluence rate was 0.59 pmol m^{-2} s^{-1} (filled circles) 3.8 pmol m^{-2} s^{-1} (triangles), 160 pmol m^{-2} s^{-1} (open circles), or 380 pmol m^{-2} s^{-1} (squares). The curved segments extending to the upper right from the common bell-shaped curve represent second-positive phototropism. **b.** Fluence was varied by changing the fluence rate. For the seven series shown, the exposure time was set at 0.125 s (filled squares), 1 s (open triangles), 10 s (stars), 100 s (open circles), 10 min (filled circles), 20 min (open squares), or 30 min (filled triangles). The broken curves correspond to the second-positive part of the fluence-response curves in **a.** Curvature was measured 60 min after the end of irradiation. After Kubo and Mihara (1988).

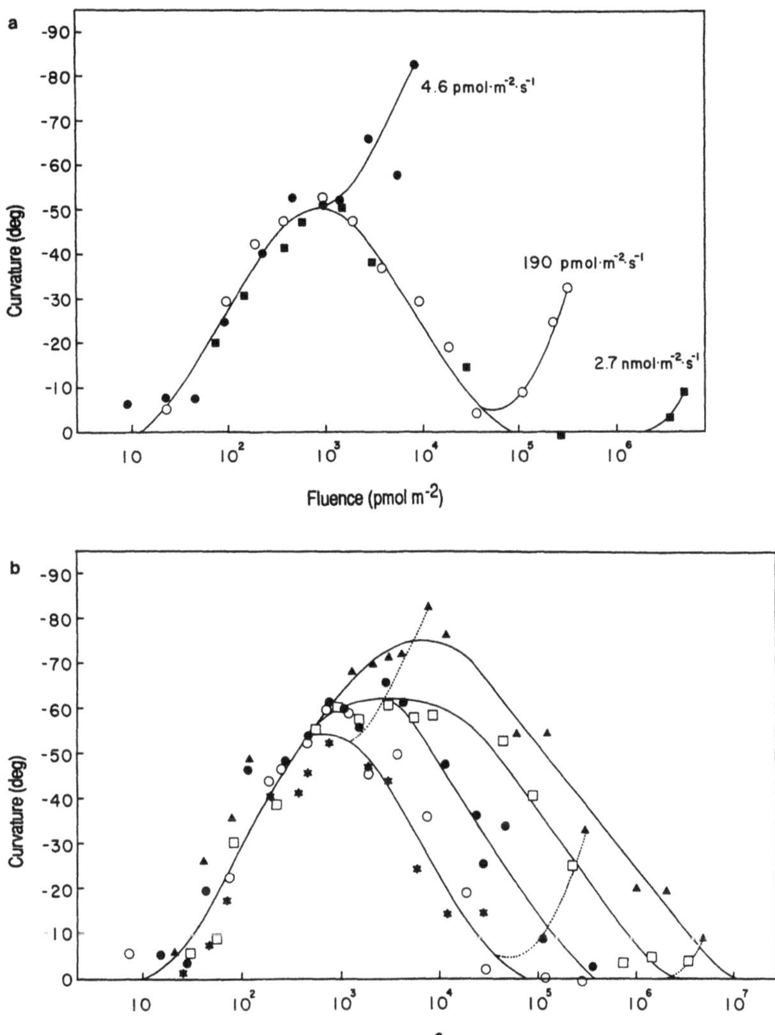

Fig. 8. Fluence-response curves for phototropism in *Pilobolus* obtained with UV-B radiation (280 nm). **a.** Fluence was varied by changing the exposure time for the three fluence rates as indicated. **b.** Fluence was varied by changing the fluence rate. The exposure time was set at 10 s (stars), 100 s (open circles), 10 min (filled circles), 20 min (open squares), or 30 min (filled triangles). The broken curves correspond to the second-positive part of the fluence-response curves in **a.** After Kubo and Mihara (1988).

nm was attributable to attenuation by the UV-absorbing gallic acid in the vacuole (Delbrück and Shropshire, 1960). Popescu et al. (1989a) have found, interestingly, that phototropism in this UV region is not purely negative. Rather, it depends critically on the absolute fluence rate. For each UV wavelength, there is a neutral point, or "compensation point", a fluence rate above which phototropism is negative and below which it is positive. This behavior has been interpreted in terms of their optical theories. The main conclusion from this work is that there seems to be a separate UV-B photoreceptor in *Phycomyces*.

Phototropism in Pilobolus

The genus *Pilobolus*, including *P. kleinii* and *P. crystallinus*, is within the class Zygomycetes and the order Mucorales, like *Phycomyces*. *Pilobolus* is characterized by a subsporangial swelling and has the remarkable capability of ejecting its sporangium for distances of over one meter, in particular after its has bent toward a light source.

The phototropism of *P. kleinii* had been studied earlier this century (Page and Curry, 1966; Page, 1968). Physiological studies of phototropism in *P. crystallinus* have been carried out recently in Kyoto, and some aspects of that work are summarized below (Kubo and Mihara, 1988; 1989). The main thrust of the recent work has been to measure fluence-response curves for UV-B and blue light.

Figure 6 shows the time course of phototropism in *Pilobolus* in response to 1 min and 30 min exposures to 450 nm blue light and 280 nm UV-B. Illumination with UV-B produces negative phototropism (Fig. 6c), as occurs in *Phycomyces* (see above). Reciprocity between fluence rate and exposure time (see Chapter on Action Spectroscopy by Lipson, this volume) was valid for fluences below 400 pmol m^{-2}. Fluence-response curves are shown for blue light (Fig. 7) and UV-B (Fig. 8). In response to blue light pulses, the curves exhibit the first-positive and second-positive regions as found in phototropism of higher plants (Briggs, 1960; Blaauw and Blaauw- Jansen, 1970). The curves for UV-B and blue light are almost superposable, especially for the common bell-shaped feature in all four curves (Figs. 7 and 8). This result suggests that a common photoreceptor system may be used for both UV-B and blue light.

Recent experiments on *Pilobolus* have shown that blue preirradiation can shift fluence response curves to higher fluence, and furthermore that the phototropic sensitivity can recover during a period in darkness following the preirradiation (Kubo and Mihara, 1989). After preirradiation at high fluence (~ 1 μmol m^{-2}), the shift of the fluence-response curves to lower fluence is biphasic. The transient loss of sensitivity is an adaptation phenomenon, like that in *Phycomyces* phototropism (see above). The specific details of this behavior, notably the biphasic aspect, suggest that there may be multiple photoreceptors acting in *P. crystallinus*, as in *Phycomyces* (Galland and Lipson, 1987b; Kubo and Mihara, 1989).

Photoreceptors

The efforts towards identifying blue light photoreceptors (sometimes termed "cryptochrome(s)" on a tentative basis) in *Phycomyces* have been reviewed by Presti and Galland, 1987. As mentioned above, the primary contenders for chromophores are flavins with possible participation of pterins as UV sensitizers (Galland and Senger, 1988a; b). For a current review of photoreceptors in general, including blue light photoreceptors, see Lipson and Horwitz (1991). The major obstacles that have restricted the specific identification of flavin and flavoprotein photoreceptors are a) that there are a vast number of flavoproteins involved in general metabolism in every type of living cell and b) that flavins and flavoproteins are all very similar spectrally. In *Phycomyces*, the discovery that *madB* and *madC* mutants have altered action spectra for phototropism (Galland and Lipson, 1985) suggests that they may be photoreceptor mutants, and may thereby offer a genetic assay for the receptor pigments in conjunction with biochemical and spectroscopic studies (Horwitz et al., 1985; Pollock et al., 1985a,b; Trad et al., 1987).

In plants, two main photoreceptor systems control phototropism: (a) the blue light receptor pigments tentatively called "cryptochrome(s)" and the well-studied plant photomorphogenic pigment, phytochrome. Phytochrome is a photochromic (photoreversible) pigment; its interconvertible forms, called Pr and Pfr, have absorption peaks in the red and near infrared region, respectively at ~ 660 and ~ 730 nm. Cryptochrome action spectra typically show peaks around 370, 450 and 480 nm and a sharp cutoff beyond 500 nm. Fungi appear to share the blue-light photoreceptors with plants, but there is only limited evidence for the occurrence for phytochrome in fungi. In particular, a recent report suggests that phytochrome may be present in the Ascomycete *Aspergillus nidulans* (Mooney and Yager, 1990).

Traditionally, the chief candidates for cryptochromes have been flavins and carotenes, which show similar absorption in the blue region of the spectrum. They differ however in the near UV region, where ß-carotene shows very weak absorption, while flavins show substantial absorption. Flavins are generally preferred as photoreceptor candidates compared to carotene (Presti et al., 1977; Galland and Senger, 1988a). The considerable variation of UV peaks in action spectra for blue-light effects may be due to the participation of pterin pigments as accessory chromophores or sensitizing pigments (Galland and Senger, 1988b).

Carotene (*car*) mutants have very similar threshold behavior to wild type, even though some are completely lacking in ß-carotene. That has been one of the main pieces of evidence that ß-carotene is unlikely to serve as the photoreceptor for phototropism in *Phycomyces* (Presti et al., 1977). In addition, photophysical considerations speak against ß-carotene as a primary photoreceptor (rather than an accessory pigment), notably the very high quantum efficiency for internal conversion, a process which competes with photochemistry (Song and Moore, 1974; De Fabo, 1980; Song, 1984).

Signal Transduction and Internal Messengers

The role of internal messengers and other cellular effectors for signal transduction has been worked out much more extensively in animal systems than in plants and fungi (Horwitz, 1989; Lipson and Horwitz, 1991). In a recent study of phototropism in maize coleoptiles, it was found that cytoplasmic Ca^{++} and H^{+} concentrations increase in cells on the shaded (more rapidly growing) side (Gehring et al., 1990). These and other results support the classical Cholodny-Went hypothesis, namely that auxin redistribution is fundamental to phototropism. Protein phosphorylation has also been implicated recently in the sensory transduction processes associated with phototropism (Short and Briggs, 1990; see Chapter by Short et al., this volume). This effect offers a novel approach towards isolation of blue-light receptors and identification of transduction steps for phototropism. In *Phycomyces*, there have been only a few studies to test agents commonly involved in signal transduction phenomena. For example, cyclic AMP (dibutyryl form) has been shown to affect sporangiophore growth transiently (Cohen, 1974) and, in more recent work, calmodulin effectors and microtubular agents have been implicated in *Phycomyces* photoresponses (Valenzuela and Ruiz-Herrera, 1989; Ruiz-Herrera et al., 1990). In order to introduce substances into sporangiophores that are normally in air, special experimental techniques are required (viz. immersion in solutions, or microinjection). Notwithstanding such technical obstacles, this general approach deserves considerably more attention to help reveal the cellular components and mechanisms underlying phototropism.

Acknowledgements

Work in this laboratory has been supported by grants from the National Institutes of Health, the National Science Foundation, the United States-Spain Joint Committee for Scientific and Technological Cooperation and the United States-Israel Binational Science Foundation.

References

Alvarez, M. I., Eslava, A. P., and Lipson, E. D., 1989, Phototropism mutants of *Phycomyces blakesleeanus* isolated at low light intensity, *Exp. Mycol.*, 13:38.

Bergman, K., Eslava, A. P., and Cerdá-Olmedo, E., 1973, Mutants of *Phycomyces* with abnormal phototropism, *Mol. Gen. Genet.*, 123:1.

Blaauw, O. H., and Blaauw-Jansen, G., 1970, The phototropic responses of *Avena* coleoptiles, *Acta Bot. Neerl.*, 19:755.

Briggs, W. R., 1960, Light dosage and phototropic responses of corn and oat coleoptiles, *Plant Physiol.*, 35:951.

Briggs, W. R., and Baskin, T. I., 1988, Phototropism in higher plants - controversies and caveats, *Botan. Acta*, 101:133.

Campuzano, V., Díaz Mínguez, J. M., Eslava, A. P., and Alvarez, M. I., 1990, A new gene (*madI*) involved in the phototropic response of *Phycomyces*, *Mol. Gen. Genet.*, 223:148.

Cerdá-Olmedo, E., and Lipson, E. D., 1987a, A biography of *Phycomyces*, *in:* "*Phycomyces*," Cerdá-Olmedo, E., and Lipson, E. D., eds., Cold Spring Harbor Laboratory, New York, p. 7.

Cerdá-Olmedo, E., and Lipson, E. D., eds., 1987b, "*Phycomyces*," Cold Spring Harbor Laboratory, Cold Spring Harbor, New York.

Cohen, 1974, Cyclic AMP levels in *Phycomyces* during a response to light, *Nature*, 251:144.

Cosgrove, D. J., 1985, Kinetic separation of phototropism from blue-light inhibition of stem elongation, *Photochem. Photobiol.*, 42:745.

De Fabo, E., 1980, On the nature of the blue light photoreceptor: still an open question, *in:* "The Blue Light Syndrome," Senger, H., ed., Springer-Verlag, Berlin, p. 187.

Delbrück, M., and Shropshire, W., Jr., 1960, Action and transmission spectra of *Phycomyces*, *Plant Physiol.*, 35:194.

Delbrück, M., and Reichardt, W., 1956, System analysis for the light growth reactions of *Phycomyces*, *in:* "Cellular Mechanisms in Differentiation and Growth," Rudnick, D., ed., Princeton University Press, Princeton, New Jersey, p. 3.

Dennison, D. S., 1979, Phototropism, *in:* "Encylopedia of Plant Physiology," New Series Vol. 7, "Physiology of Movements," Haupt, W., and Feinleib, M. E., eds., Springer-Verlag, Berlin, Heidelberg, New York, p. 506.

Dennison, D. S., and Foster, K. W., 1977, Intracellular rotation and the phototropic response of *Phycomyces*, *Biophys. J.*, 18:103.

Firn, R. D., 1990, Phototropism - the need for a sense of direction, *Photochem. Photobiol.*, 52:255.

Foster, K. W., 1977, Phototropism of coprophilous *Zygomycetes*, *Annu. Rev. Biophys. Bioeng.*, 6:419.

Fukshansky, L., and Steinhardt, A. R., 1987, Spatial factors in *Phycomyces* phototropism: analysis of balanced responses, *J. Theor. Biol.*, 129:301.

Galland, P., 1989, Photosensory adaptation in plants, *Botan. Acta*, 102:11.

Galland, P., 1990, Phototropism of the *Phycomyces* sporangiophore: a comparison with higher plants, *Photochem. Photobiol.*, 52:233.

Galland, P., and Lipson, E. D., 1985, Modified action spectra of photogeotropic equilibrium in *Phycomyces blakesleeanus* mutants with defects in genes *madA*, *madB*, *madC*, and *madH*, *Photochem. Photobiol.*, 41:331.

Galland, P., and Lipson, E. D., 1987a, Light calibrations and radiometric units, *in:* "*Phycomyces*," Cerdá-Olmedo, E., and Lipson, E. D., eds., Cold Spring Harbor Laboratory, Cold Spring Harbor, New York, p. 375.

Galland, P., and Lipson, E. D., 1987b, Light physiology of *Phycomyces* sporangiophores, *in:* "*Phycomyces*," Cerdá-Olmedo, E., and Lipson, E. D., eds., Cold Spring Harbor Laboratory, Cold Spring Harbor, New York, p. 49.

Galland, P., and Russo, V. E. A., 1984, Light and dark adaptation in *Phycomyces* phototropism, *J. Gen. Physiol.*, 84:101.

Galland, P., and Senger, H., 1988a, The role of flavins as photoreceptors, *J. Photochem. Photobiol. B:Biol.*, 1:277.

Galland, P., and Senger, H., 1988b, The role of pterins in the photoreception and metabolism of plants, *Photochem. Photobiol.*, 48:811.

Galland, P., Pandya, A., and Lipson, E. D., 1984, Wavelength dependence of dark-adaptation in *Phycomyces* phototropism, *J. Gen. Physiol.*, 84:739.

Galland, P., Palit, A., and Lipson, E. D., 1985, *Phycomyces*: phototropism and light-growth response to pulse stimuli, *Planta*, 165:538.

Galland, P., Corrochano, L. M., and Lipson, E. D., 1989a, Subliminal light control of dark adaptation kinetics in *Phycomyces* phototropism, *Photochem. Photobiol.*, 49:485.

Galland, P., Orejas, M., and Lipson, E. D., 1989b, Light-controlled adaptation kinetics in *Phycomyces*: evidence for a novel yellow-light absorbing pigment, *Photochem. Photobiol.*, 49:493.

Gamow, R. I., Ruiz-Herrera, J., and Fischer, E. P., 1987, The cell wall of *Phycomyces*, *in:* "*Phycomyces*," Cerdá-Olmedo, E., and Lipson, E. D., eds., Cold Spring Harbor Laboratory, Cold Spring Harbor, New York, p. 223.

Gehring, C. A., Williams, D. A., Cody, S. H., and Parish, R. W., 1990, Phototropism and geotropism in maize coleoptiles are spatially correlated with increases in cytosolic free calcium, *Nature*, 345:528.

Gressel, J., and Horwitz, B., 1982, Gravitropism and phototropism, *in:* "The Molecular Biology of Plant Development," Smith, H., and Grierson, D., eds., ("Molecular Biology of Plant Development", Vol. 18), Blackwell Scientific Publications, Oxford, p. 405.

Hasegawa, K., Sakoda, M., and Bruinsma, J., 1989, Revision of the theory of phototropism in plants: a new interpretation of a classical experiment, *Planta*, 178:540.

Herrera-Estrella, L., and Ruiz-Herrera, J., 1983, Light response in *Phycomyces blakesleeanus*: evidence for roles of chitin biosynthesis and breakdown, *Exp. Mycol.*, 7:362.

Hertel, R., 1980, Phototropism of lower plants, *in:* "Photoreception and Sensory Transduction in Aneural Organisms," Lenci, F., and Colombetti, G., eds., Plenum Press, New York and London, p. 89.

Horwitz, B. A., 1989, The potential for second messengers in light signaling, *in:* "Second Messengers in Plant Growth and Development," Boss, W. F., and Morre, D. J., eds., Alan R. Liss, Inc., New York, p. 289.

Horwitz, B. A., Trad, C. H., and Lipson, E. D., 1985, Differential spectrophotometry of *Phycomyces* mutants with abnormal photoresponses, *Photochem. Photobiol.*, 44:207.

Iino, M., 1990, Phototropism - mechanism and ecological implications, *Plant Cell Environ.*, 13:633.

Iino, M., and Schäfer, E., 1984, Phototropic response of the stage I *Phycomyces* sporangiophore to a single pulse of blue light, *Proc. Natl. Acad. Sci. USA*, 81:7103.

Jesaitis, A. J., 1974, Linear dichroism and orientation of the *Phycomyces* photopigment, *J. Gen. Physiol.*, 63:1.

Kataoka, H., 1987, The light-growth response of *Vaucheria*. A *conditio sine qua non* of the phototropic response?, *Plant Cell Physiol.*, 28:61.

Koga, K., and Ootaki, T., 1983a, Complementation analysis among piloboloid mutants of *Phycomyces blakesleeanus*, *Exp. Mycol.*, 7:161.

Koga, K., and Ootaki, T., 1983b, Growth of the morphological, piloboloid mutants of *Phycomyces blakesleeanus*, *Exp. Mycol.*, 7:148.

Koga, K., Sato, T., and Ootaki, T., 1984, Negative phototropism of the piloboloid mutants of *Phycomyces blakesleeanus* Bgff, *Planta*, 162:97.

Kubo, H., and Mihara, H., 1988, Phototropic fluence-response curves for *Pilobolus crystallinus* sporangiophore, *Planta*, 174:174.

Kubo, H., and Mihara, H., 1989, Blue-light-induced shift of the phototropic fluence-response curve in *Pilobolus* sporangiophores, *Planta*, 179:288.

Lipson, E. D., and Block, S. M., 1983, Light and dark adaptation in *Phycomyces* light-growth response, *J. Gen. Physiol.*, 81:845.

Lipson, E. D., and Galland, P., 1987, Measurement of *Phycomyces* light responses, *in:* "*Phycomyces*," Cerdá-Olmedo , E., and Lipson, E. D., eds., Cold Spring Harbor Laboratory, New York, p. 367.

Lipson, E. D., and Häder, D.-P., 1984, Video data acquisition for movement responses in individual organisms, *Photochem. Photobiol.*, 39:437.

Lipson, E. D., and Horwitz, B. A., 1991, Photosensory reception and transduction, *in:* "Sensory Receptors and Signal Transduction," Spudich, J., ed., ("Modern Cell Biology", Vol. 7), Academic Press, New York, in press.

Lipson, E. D., and Pratap, P., 1988, System analysis of *Phycomyces* light-growth response with Gaussian white noise and sum-of-sinusoids test stimuli, *Ann. Biomed. Eng.*, 16:95.

Lipson, E. D., and Terasaka, D. T., 1981, Photogeotropism in *Phycomyces* double mutants, *Exp. Mycol.*, 5:101.

Lipson, E. D., Terasaka, D. T., and Silverstein, P. S., 1980, Double mutants of *Phycomyces* with abnormal phototropism, *Mol. Gen. Genet.*, 179:155.

Meistrich, M. L., Fork, R. L., and Matricon, J., 1970, Phototropism in *Phycomyces* as investigated by focused laser radiation, *Science*, 169:370.

Mooney, J. L., and Yager, L. N., 1990, Light is required for conidiation in *Aspergillus nidulans*, *Genes Dev.*, 4:1473.

Page, R. M., 1968, Phototropism in fungi, *in:* "Photophysiology," Giese, A. C., ed., Academic Press, New York, p. 65.

Page, R. M., and Curry, G. M., 1966, Studies on phototropism of young sporangiophores of *Pilobolus kleinii*, *Photochem. Photobiol.*, 5:31.

Pohl, U., and Russo, V. E. A., 1984, Phototropism, *in:* "Membranes and Sensory Transduction," Colombetti, G., and Lenci, F., eds., Plenum, New York, p. 231.

Pollock, J. A., Lipson, E. D., and Sullivan, D. T., 1985a, Analysis of microsomal flavoproteins from *Phycomyces* sporangiophores: candidates for the blue-light photoreceptor, *Planta*, 163:506.

Pollock, J. A., Lipson, E. D., and Sullivan, D. T., 1985b, Electrophoretic analysis of proteins from night-blind mutants of *Phycomyces*, *Biochem. Genet.*, 23:377.

Popescu, T., Roessler, A., and Fukshansky, L., 1989a, A novel effect in *Phycomyces* phototropism - positive bending and compensation spectrum in far UV, *Plant Physiol.*, 91:1586.

Popescu, T., Zangler, F., Sturm, B., and Fukshansky, L., 1989b, Image analyzer used for data acquisition in phototropism studies, *Photochem. Photobiol.*, 50:701.

Presti, D., Hsu, W. J., and Delbrück, M., 1977, Phototropism in *Phycomyces* mutants lacking ß-carotene, *Photochem. Photobiol.*, 26:403.

Presti, D. E., and Galland, P., 1987, Photoreceptor biology of *Phycomyces*, *in:* "*Phycomyces*," Cerdá-Olmedo, E., and Lipson, E. D., eds., Cold Spring Harbor Laboratory, Cold Spring Harbor, New York, p. 93.

Ruiz-Herrera, J., Martinez-Cadena, G., Valenzuela, C., and Reyna, G., 1990, Possible roles of calcium and calmodulin in the light-stimulation of wall biosynthesis in *Phycomyces*, *Photochem. Photobiol.*, 52:217.

Short, T. W., and Briggs, W. R., 1990, Characterization of a rapid, blue light-mediated change in detectable phosphorylation of a plasma membrane protein from etiolated pea (*Pisum sativum* L.) seedlings, *Plant Physiol.*, 92:179.

Shropshire, W., Jr., 1987, *Phycomyces* bibliography, *in:* "*Phycomyces*," Cerdá-Olmedo, E., and Lipson, E. D., eds., Cold Spring Harbor Laboratory, Cold Spring Harbor, New York, p. 381.

Song, P. S., and Moore, T. A., 1974, On the photoreceptor pigment for phototropism and phototaxis: is a carotenoid the most likely candidate?, *Photochem. Photobiol.*, 19:435.

Song, P.-S., 1984, Photophysical aspects of blue light receptors: the old question (flavins versus carotenoids) re-examined, *in:* "Blue Light Effects in Biological Systems," Senger, H., ed., Springer-Verlag, Berlin, Heidelberg, New York, p. 75.

Steinhardt, A., Popescu, T., and Fukshansky, L., 1989, Is the dichroic photoreceptor for Phycomyces phototropism located at the plasma membrane or at the tonoplast?, *Photochem. Photobiol.*, 49:79.

Steinhardt, A. R., and Fukshansky, L., 1985, Diffusion approximation for scattering in a cylinder: optics of phototropism, *J. Opt. Soc. Am. A*, 2:1725.

Steinhardt, A. R., Shropshire, W., Jr., and Fukshansky, L., 1987, Invariant properties of absorption profiles in sporangiosphores of Phycomyces blaskesleeanus under balancing bilateral illumination, *Photochem. Photobiol.*, 45:515.

Trad, C. H., Horwitz, B. A., and Lipson, E. D., 1987, Light-induced absorbance changes in extracts of *Phycomyces* sporangiophores: modifications in night-blind mutants, *J. Photochem. Photobiol. B:Biol.*, 1:305.

Tsuru, T., Koga, K., Aoyama, H., and Ootaki, T., 1988, Optics in *Phycomyces blakesleeanus* sporangiophores relative to determination of phototropic orientation, *Exp. Mycol.*, 12:302.

Valenzuela, C., and Ruiz-Herrera, J., 1989, Inhibition of phototropism in *Phycomyces* sporangiophores by calmodulin antagonist and antimicrotubular agents, *Curr. Microbiol.*, 18:11.

High-Fluence Rate Monochromatic Light Sources, Computerized Analysis of Cell Movements, and Microbeam Irradiation of a Moving Cell: Current Experimental Methodology at the Okazaki Large Spectrograph

Masakatsu Watanabe

National Institute for Basic Biology
Okazaki National Research Institutes
Okazaki, Aichi 444
Japan

The Okazaki Large Spectrograph (OLS) and the Tunable Laser Irradiation System

The Okazaki Large Spectrograph (OLS)

Previous biological spectrographs (Parker et al., 1946; Monk and Ehret, 1956; Jacques et al., 1964; Balegh and Biddulph, 1970; Watanabe and Furuya, 1974) which enabled simultaneous irradiation of many samples with different wavelengths have been very useful tools for action spectroscopy, especially with whole plant systems (Borthwick et al., 1952; Shen-Miller et al., 1969; Wada and Furuya, 1974; Inoue and Furuya, 1975).

The Okazaki Large Spectrograph (OLS) (Fig. 1), built in 1980 at the National Institute for Basic Biology (NIBB), Okazaki, Japan, (Watanabe et al., 1982) is the world's largest and most powerful biological spectrograph and has been extensively used for photobiological action spectroscopy under the NIBB Cooperative Research Program for The Use of the OLS by the members of the Institute and visiting scientists (Watanabe, 1985, 1988).

It provides a gigantic and powerful spectrum covering ultraviolet, visible and infrared regions (250-1000 nm in wavelength) onto a 10 m-long x 20 cm-high focal curve (with a linear dispersion of about 0.8 nm cm^{-1}), where samples are placed for simultaneous irradiation, with monochromatic photon fluence rates (10^{16} photons s^{-1} cm^{-2} + 20 % at each wavelength between 400-700 nm; $5x10^{14}$ and $5x10^{15}$ photons s^{-1} cm^{-2} at 250 and 300 nm, respectively, with a maximum entrance slit width of 5 cm giving a spectral full width at half maximum of about 5.5 nm) which is more than twice as much as that of the tropical sunlight at noon if compared at each wavelength of the spectrum. This performance of the OLS was realized by the combined action of a 30 kW xenon short-arc lamp and a 90 x 90 cm compound diffraction grating which is double-blazed at 250 and 500 nm. Two types of threshold boxes, which can be placed at 45 nm or 10 nm intervals, enable simultaneous irradiation with decreasing series of seven fluence rates at each wavelength (Saji et al., 1982, 1983; Ohki et al., 1982; Lumsden et

Biophysics of Photoreceptors and Photomovements in Microorganisms
Edited by F. Lenci *et al.*, Plenum Press, New York, 1991

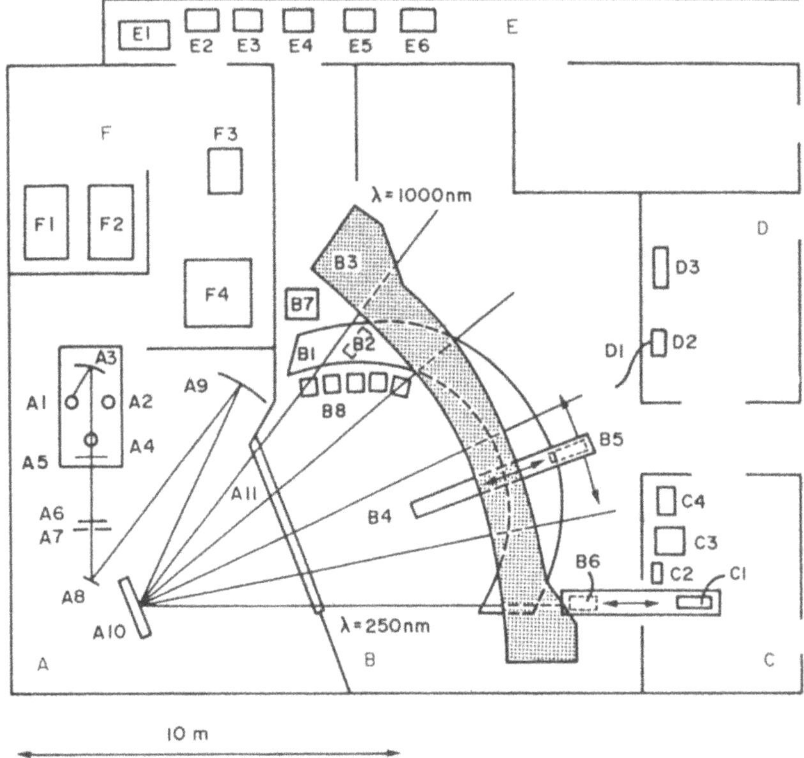

Fig. 1. Spatial arrangement of the Okazaki Large Spectrograph. A1, 30 kW Xe lamp; A7, entrance slit; A9, condensing mirror; A10, plane grating (double-blazed at 250 and 500 nm); B1, focal curve stage; sample box; B3-6 and C2, box transfer robot; C2, control panel; C3, CRT terminal; D1, optical fiber bundle; D2, its outlet; E1, host microcomputer. (Watanabe et al., 1982).

al., 1987; Yoshida et al., 1989). An optical fiber bundle system is available as an irradiation light guide for electrophysiological experiments (Arikawa and Aoki, 1982).

Extension of Action Spectroscopy into the Ultraviolet Region for the Blue/Near-Ultraviolet Light Effects in Plants and Fungi

The chemical identity of the photoreceptor pigment(s) involved in the so-called blue/near-ultraviolet light effects controlling various biological processes in plants and fungi are still an open question (Galland and Senger, 1988a,b). Flavins (or flavoproteins) and carotenoids (or carotenoproteins) are the commonly suggested candidates because the absorption spectra of both of them in the near-ultraviolet and visible regions are similar to the action spectra for the blue/near-ultraviolet light effects. Taking into consideration the fact that free flavins have a very pronounced absorption peak at about 260-280 nm whereas free carotenoids do not have such a pronounced peak there, several projects have been performed at the NIBB using the OLS to extend the action spectra for the blue/near-ultraviolet effects into the ultraviolet region in various fungal and plant systems, e.g., photoinduction of sexual reproduction in fungi (Fig. 2a, b), photoinhibition of red-light induced spore germination in ferns (Fig. 2c, d), phototropic bending in a dicotyledonous plant (Fig. 2e), and photoavoidance in a slime mold (Fig. 2f). All of the above 6 ultraviolet action spectra show a very pronounced peak in the ultraviolet region at about 260-280 nm, in 5 cases of which (Fig. 2a-c,e, and f) higher

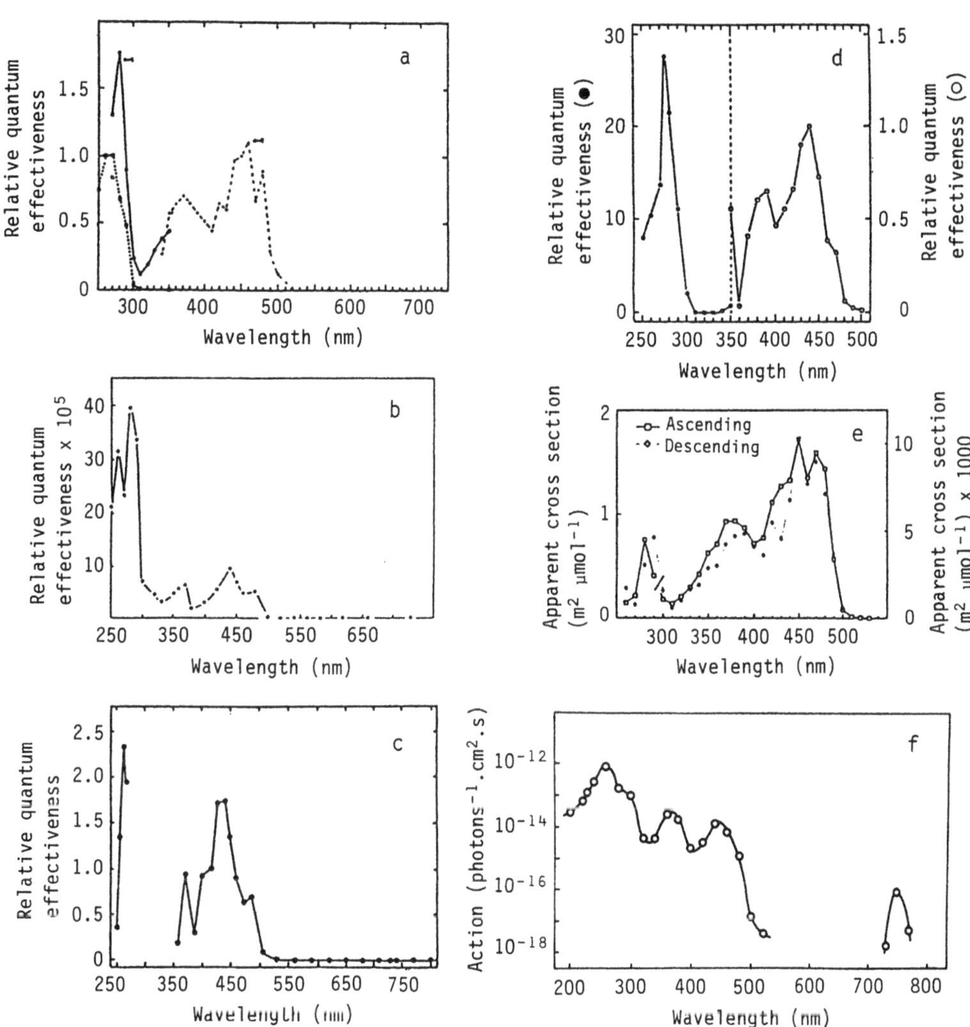

Fig. 2. Action spectra for (a) photo-induction of perithecial formation in *Gerasinospora* (a-1, a-2) (Inoue and Watanabe, 1984), (b) primodial initiation of *Coprinus* (Durand and Furuya, 1985), (c) photoinhibition of red-light-induced germination of spores in *Pteris*. (Sugai et al., 1984), (d) photoinhibition of red-light-induced germination of spores in *Adiantum* (Sugai and Furuya, 1985), (e) phototropism in alfalfa (Baskin and Iino, 1987) and (f) photoavoidance in *Physarum* (Ueda et al., 1988).

than the peak at about 460 nm in the blue region. These action spectra resemble the absorption spectra of flavins rather than those of carotenoids, thus suggesting the former as the more probable candidates for the photoreceptor pigments than the latter, although other possibilities are not yet excluded, e.g. that the energy of the ultraviolet light is absorbed by the aromatic amino acid residues of the protein moiety and then transferred to the chromophoric moiety of carotenoproteins, or participation of other pigments (Watanabe, 1988).

Action Spectrum for Resetting the Circadian Phototaxis Rhythm in Chlamydomonas

Kondo et al. (1991) studied the photobiology of phase-resetting the *Chlamydomonas* clock by brief (3 s to 15 min) light pulses administered during a 24 h dark period, making full use of the advantage of simultaneous irradiation at many wavelengths enabled by the Okazaki Large Spectrograph. Its action spectrum exhibited two prominent peaks, at 520 nm and 660 nm. The pigment that is participating in resetting the clock by green light (520 nm) is unknown. The phase shift by red light (660 nm) was not diminished by subsequent administration of far-red light (730 nm), even if the red light pulse was as short as 0.1 s. This constitutes the first report of a regulatory action by red light in *Chlamydomonas.*

The Tunable Two-Wavelength CW Laser Irradiation System

A tunable two-wavelength CW laser irradiation system (Fig. 3) has been constructed as a complementary light source to the OLS to be used in irradiation experiments which specifically require ultra-high fluence rates (Iino, 1988; Ogura et al., 1985, 1990) as well as ultra-high spectral, temporal (Choi et al., 1989), and spatial resolutions. It is composed of a high-power Ar ion laser (Coherent, Innova 20) (336.6-528 nm, 20 W output), two CW dye lasers (Coherent, CR-599-01) (420-930 nm, 250-1000 mW output), Acoustooptic (A/O) modulators (up to 40 MHz) to chop the laser beam, a beam expander, and a tracking microbeam irradiator (MIMO scope, described below) with an infra-red phase contrast observation system.

The Tracker-Cell Movement Analyzer System (TCMA)

The TCMA System

In the study of tactic and/or phobic responses such as algal and bacterial photomovements, recording and analysis of the movements at the individual cell level are essentially important (Hand and Davenport, 1970). For this purpose, photograhic (Feinleib and Curry, 1967; Watanabe and Furuya, 1982), video (Boscov and Feinleib, 1979), computerized video (Davenport et al., 1970), and computerized video, automatic analysis (Takahashi and Kobatake, 1982; Häder and Lebert, 1985) methods have been devised and used.

We have made up a new computerized video-automatic analysis method [the Tracker-Cell Movement Analyzer (TCMA)] that enables a more flexible and more detailed time-related tracking and analyses of cell movements, especially behavioral responses to light stimuli (Kondo et al., 1988). It is composed of two parts: The Tracker system (Fig. 4), which is a computerized videosystem to digitally record and display the swimming tracks of the moving cells in consecutive time segments, and the Cell Movement Analyzer Program (CMA), which is a software to automatically analyze the data provided by the Tracer through thinning, tracking, and statistical analysis processes. The major unique points of this method are as follows:

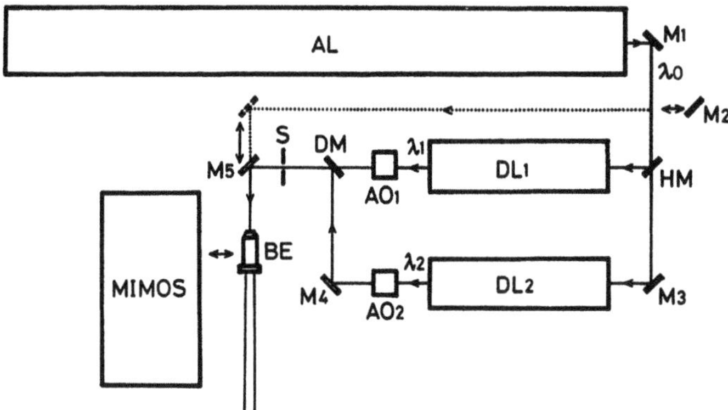

Fig. 3. Block diagram of the tunable two-wavelength CW laser irradiation system. AL, Ar ion laser; M1-5, mirrors; HM, half-mirror; DL1-2, dye lasers; AO1-2, acoustooptic modulators; DM, dichroic mirror; S, slit, BE, beam expander; MIMOS, MIMO scope.

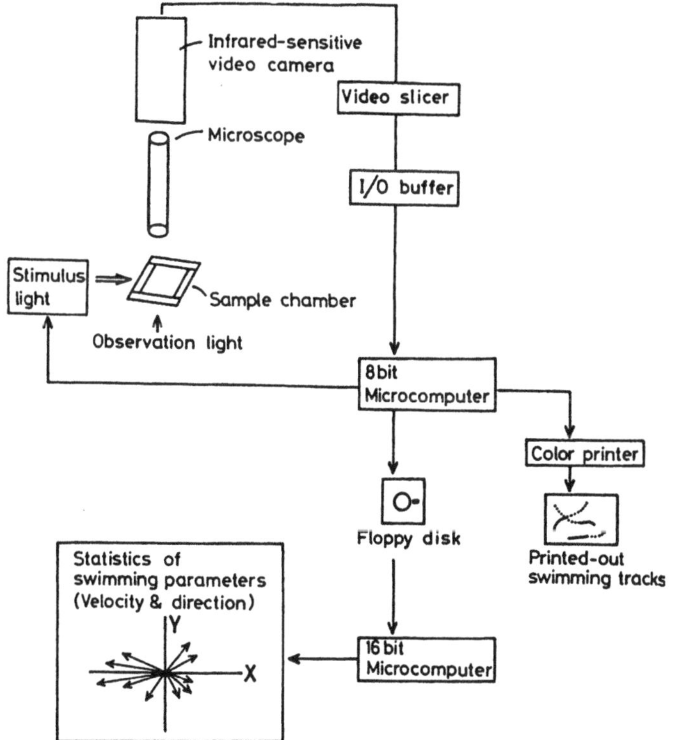

Fig. 4. Block diagram of the Tracker system: a 1-inch silicon vidicon camera (C1000-02, Hamamatsu Photonics) with an I/O buffer (M998, Hamamatsu Photonics) and a video slicer (M1005, Hamamatsu Photonics); an 8-bit microcomputer system (BIWAK DMS-808, Digi-Tech) with a Z80A CPU, two 64 kByte static RAM boards, two FDD for 1 MByte floppy disks and a BIWAC Bus line (Kondo et al., 1988).

i) it can display the swimming tracks of all the individual cells in the microscopic field in consecutive time segments of variable length, with different colors for different time segments;

ii) it can automatically do measurement of the change with consecutive time segments in swimming direction of each cell as the function of the initial swimming direction, as well as statistical analysis of the swimming directions and speeds of the cell populations in each time segment. It provides new possibilities for many experimental analyses of cellular behavioral responses to external stimuli. It is provided for visiting users, as well as inside users, as an attachment to the Okazaki Large Spectrograph.

Circadian Rhythms of Phototactic Orientation and Motility in Chlamydomonas

As an example of its many possible applications, an analysis of the circadian rhythms of phototactic orientation and motility in *Chlamydomonas* was performed (Kondo et al., 1988). It was thus clearly demonstrated that the velocity of motile cells, the phototactic orientation and the percentage of motile cells, changed rhythmically in a circadian manner, all in the same phase, i.e. the maximum occurred at the "subjective dawn" and the minimum at the "subjective dusk". These facts can quantitatively account for the observed circadian change in the photoaccumulation activity (Fig. 5).

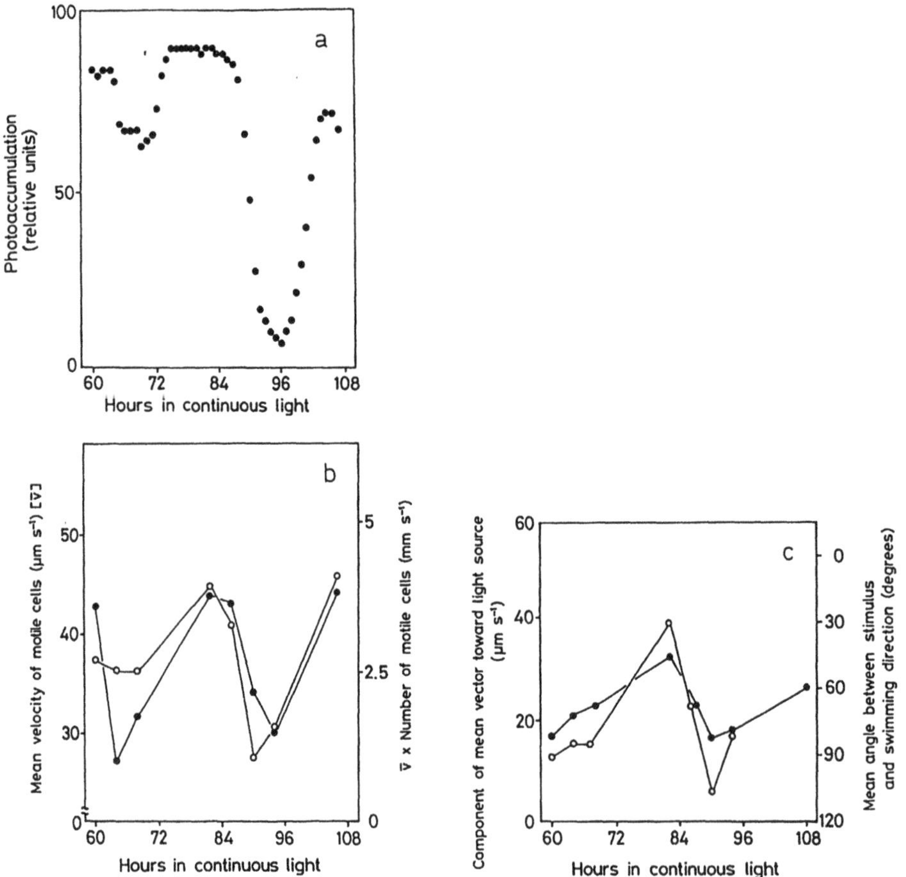

Fig. 5. Circadian changes in the photoaccumulation ability, motility and phototactic orientation (Kondo et al., 1988).

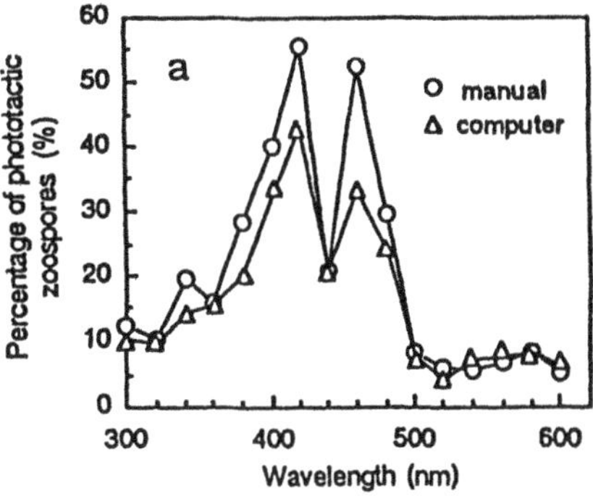

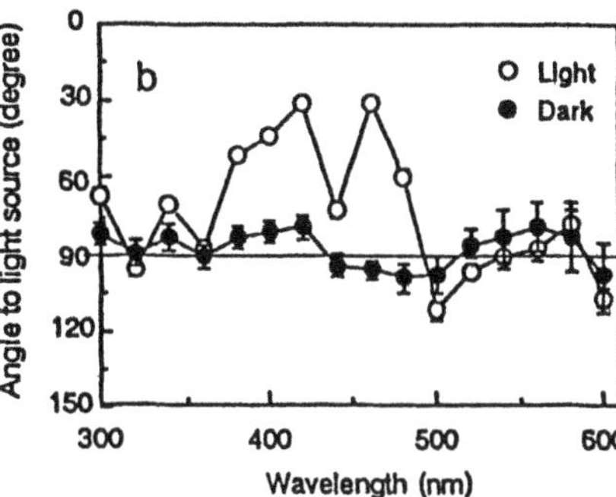

Fig. 6. Action spectrum for phototactic orientation in zoospores of *Pseudochorda* (a) based on the percentage of zoospores of which the acute angle between the swimming tracks and the optical axis of the stimulus light were within 15 degrees. Circles, data obtained manually; triangles, data obtained by CMA program. (b) based on the average angle of swimming tracks to the direction of light at each wavelength. Open circles, in the presence of stimulating light; closed circles, 4.5 s dark period before irradiation (dark control) (Kawai et al., 1991).

Action Spectra for Phototaxis in Zoospores of the Brown Alga Pseudochorda

Most planktonic algae are able to detect the orientation and intensity of light, and to move towards or to escape from it. These phototactic responses are mostly so-called blue-light responses but the exact mechanism and photoreceptive pigments involved have yet to be elucidated (Nultsch and Häder, 1988; Lenci and Ghetti, 1989).

Brown algae are mostly sedentary benthic macroalgae. However, they have flagellated reproductive cells which often show obvious phototaxis. Recently, a flavin-like substance has been reported to occur in the posterior flagella of brown algal swarmers, and it has been suggested that this substance is involved in photoreception for the phototactic system (Müller et al., 1987; Kawai, 1988). Action spectra for phototaxis in zoospores of the brown alga, *Pseudochorda gracilis* (Laminariales), were examined in the wavelength range between 300 and 600 nm using the OLS and the TCMA. The action spectrum reported here (Fig. 6) has two major peaks, one at ca. 420 nm and the other at ca. 460 nm. Although the latter peak coincides well with the one observed in most blue-light reactions, the former peak is obviously different in wavelengh from another major peak (at ca. 380 nm) found in most blue-light reactions, including phototaxis in *Euglena* (Galland and Senger, 1988a, b; Nultsch and Häder, 1988; Watanabe 1988; Lenci and Ghetti, 1989; Sugai and Furuya, 1990).

In this connection, it is noteworthy that Kawai et al. (1990) have reported a very similar action spectrum, with major peaks at ca. 430 and 450 nm, for phototaxis in male gametes of another brown alga *Ectocarpus siliculosus*. It is quite interesting to note that another strikingly similar action spectrum has been reported for hair whorl formation in the gigantic unicellular green alga, *Acetabularia* (Schmid, 1984).

The *in vivo* action spectrum for photoreactivation in *Streptomyces* (Jagger et al., 1970) and the *in vitro* action spectrum of the purified enzyme (Eker, 1978) have a peak at ca. 445 nm and lack the obvious near-UV peak (at about 380 nm) typical for absorption spectra of riboflavin and flavoenzymes with FMN or FAD as chromophores (Galland and Senger, 1988a). The chromophore of the *Streptomyces* DNA photolyase is a 8-hydroxy-5-deazaflavin (Eker et al., 1981). The free 8-hydroxy-5-deazaflavins from *Streptomyces* have absorption spectra with a peak at 420 nm and also no near-UV peak, at pH 8.3 (Eker et al., 1981). Therefore, it is a tempting hypothesis that the action spectra for the brown algal phototaxis and for hair whorl formation in *Acetabularia*

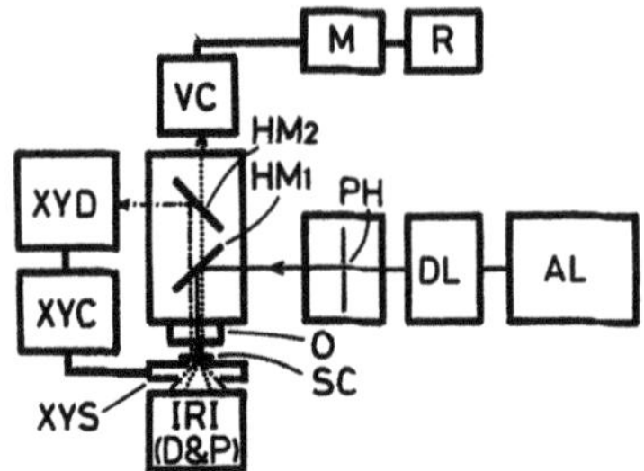

Figure 7. Block diagram of the MIMO scope. IRI(D&P), infrared illuminator (dark-field at 750-950 nm and phase-contrast at 650-750 nm); XYS, xy-stage; SC, sample chamber; O, objective lens, HM1-2, half-mirrors; AL, Ar ion laser; DL, dye laser, PH, pinhole; XYD, xy-detector; XYC, xy-controller; VC, video camera; M, video monitor; R, video recorder.

might be explained by combining the absorption spectra of free (for the 420 nm peak) and protein-bound (for the 450-460 nm peak) 8-hydroxy-5-deazaflavins.

Action Spectra for Photophobic Responses and Phototactic Response in Dunaliella

We determined the action spectra of the photophobic responses as well as the phototactic response in the green alga *Dunaliella salina* (Volvocales) using both single cells and populations (Wayne et al., 1991). The action spectra of the photophobic responses have maxima at 510 nm, the spectrum for phototaxis has a maximum at 450-460 nm. These action spectra are not compatible with the hypothesis that flavoproteins are the photoreceptor pigments, and we suggest that carotenoproteins or rhodopsins act as the photoreceptor pigments. We also conclude that the phototactic response in *Dunaliella* is an elementary response, quite independent of the step-up and step-down photophobic responses.

The Microsope for Microbeam Irradiation of a Moving Organism (MIMO Scope; MIMOS)

Microbeam Irradiation of a Moving Organism

The intracellular localization of photoreceptor pigments involved in light-stimulated developmental processes and motile responses have been determined with the aid of microbeam irradiation (Haupt et al., 1969; Wada and Furuya, 1978; Furuya et al., 1980). Recently, microbeam irradiation of rapidly swimming single cells has become possible with the development of the MIMO scope (Fig. 7) (Watanabe et al., in prep.). This technique enables us to locate the photoreceptor pigments involved in light-stimulated motile responses including phototaxis, the step-down and the step-up photophobic responses.

Localization of the Photoreceptive Site for the Step-up Photophobic Response in the Green Alga Dunaliella salina

The intracellular photoreceptive site involved in the step-up photophobic response in the green alga *Dunaliella salina* was investigated by partially irradiating a cell, using the MIMO scope (Kadota et al., 1988, in prep.). The swimming algal cells were irradiated with 514.5 nm light which was supplied from an Ar ion laser connected to the microscope. The movements of the cells were observed with infrared light upon irradiation of a whole cell with the stimulus light. The cell instantaneously changed its direction of movement and stopped for approximately 100 ms, after which the cell resumed swimming. This results in a slowdown in the forward velocity. Using the instantaneous turn of more than 90 degrees as a criterion of the response, a fluence rate-response curve was obtained. The response was roughly linear to the log of the fluence rate up to 2.8 x 10^{16} photons cm^{-2} s^{-1} (=110 W m^{-2}). Partial irradiation of the cell was conducted using a fluence rate of 1.4 x 10^{16} photons cm^{-2} s^{-1} (=54 W m^{-2}), which represents the ascending part of the curve. The results showed that the photophobic reaction was induced only after irradiation of the anterior half but not after light exposure to the posterior half. These data indicate that the photoreceptive site responsible for the step-up photophobic response is localized in the anterior region of the cell. The nature of the photoreceptor pigment involved in the step-up photophobic response in *Dunaliella* remains unknown.

Acknowledgements

Most of the work described here have been supported by the NIBB Cooperative Research Program for the Use of the Okazaki Large Spectrograh. I thank many collaborators especially Dr. Takao Kondo and Mr. Mamoru Kubota (National Institute for Basic Biology); Dr. Hiroshi Kawai (Hokkaido University): Dr. Akeo Kadota (Tokyo Metropolitan University); Prof. Randy Wayne (Cornell University). I thank Mrs. Junko Watanabe (National Institute for Basic Biology) for drawing some of the illustrations.

References

Arikawa, K., and Aoki, K., 1982, Response characteristics and occurrence of extraocular photoreceptors on lepidopteran genitalia, *J. Comp. Physiol.*, 148:483.

Balegh, S. E., and Biddulph, O., 1970, The photosynthetic action spectrum in the bean plant, *Plant Physiol.*, 46:1.

Baskin, T. I., and Iino, M., 1987, An action spectrum in the blue and ultraviolet for phototropism in alfalfa, *Photochem. Photobiol.*, 46:127.

Borthwick, H. A., Hendricks, S. B., Parker, M. W., Toole, E. H., and Toole, V. K., 1952, A reversible photoreaction controlling seed germination, *Proc. Natl. Acad. Sci. U. S. A.*, 38:662.

Boscov, J. S., and Feinleib, M. E., 1970, Phototactic response of *Chlamydomonas* to flashes of light. II. Response of individual cells, *Photochem. Photobiol.*, 30:499.

Choi, K.-S., Watanabe, M., and Furuya, M., 1989, Effects of long-term storage on phytochrome-mediated germination in lettuce seeds, *Bot. Mag. Tokyo*, 102:181.

Davenport, D., Culler, G. J., Greaves, J. O. B., Forward, R. B., and Hand, W. G., 1970, The investigation of the behavior of microorganisms by computerized television, *IEEE Trans. Biomed. Engineer.*, 17:230.

Durand, R., and Furuya, M., 1985, Action spectra for stimulatory and inhibitory effects of UV and blue light on fruitbody formation in Coprinus congregatus, *Plant Cell Physiol.*, 26:1175.

Eker, A. P. M., 1978, Some properties of a DNA photoreactivating enzyme from *Streptomyces griseus*, *in*: "DNA repair mechanisms," P. C. Hanawalt, E. C. Friedberg, and C. F. Fox, eds., Academic Press, New York, p. 129.

Eker, A. P. M., Dekker, R. H., and Berends, W., 1981, Photoreactivation enzyme from *Streptomyces griseus*. IV. On the nature of the chromophoric cofactor in *Streptomyces griseus* photoreactivating enzyme, *Photochem. Photobiol.*, 33:65.

Feinlieb, M. E. H., and Curry, G. M., 1967, Methods for measuring phototaxis of cell populations and individual cells, *Physiol. Plant.*, 20:1083.

Furuya, M., Wada, M., and Kadota, A., 1980, Regulation of cell growth and cell cycle by blue light in *Adiantum* gametophytes, *in*: "Blue Light Syndrome," H. Senger, ed., Springer-Verlag, Berlin, p. 119.

Galland, P., and Senger, H., 1988a, The role of flavins as photoreceptors, *J. Photochem. Photobiol. B: Biol.*, 1:277.

Galland, P., and Senger, H., 1988b, The role of pterins in the photoreception and metabolism of plants, *Photochem. Photobiol.*, 48:811.

Häder, D.-P., and Lebert, M., 1985, Real time computer-controlled tracking of motile microorganisms, *Photochem. Photobiol.*, 42:509.

Hand, W. G., and Davenport, D., 1970, The experimental analysis of phototaxis and photokinesis in flagellates, *in*: "Photobiology of Microorganisms," P. Halldal, ed., Wiley, London, p. 253.

Haupt, W., Mörtel, G., and Winkelnkemper, I., 1969, Demonstration of different dichroic orientation of phytochrome Pr and Pfr, *Planta*, 88:183.

Iino, M., 1988, Pulse-induced phototropisms in oat and maize coleoptiles, *Plant Physiol.*, 88:823.

Inoue, Y., and Furuya, M., 1975, Perithecial formation in *Gerasinospora reticulispora*. IV: Action spectra for the photoinduction, *Plant Physiol.*, 55:1098.

Inoue, Y., and Watanabe, M., 1984, Perithecial formation in *Gerasinospora reticulispora*. VII: Action spectra in UV region for the photoinduction and the photoinhibition of photoinductive effect brought by blue light, *Plant Cell Physiol.*, 25:107.

Inoue, Y., 1984, Re-examination of action spectroscopy in blue/near-UV light effects, *in*: "Blue Light Effects in Biological Systems," H. Senger, ed., Springer-Verlag, Berlin, p. 110.

Jacques, R., Chabbal, R., Couard, P., and Jacquinot, P., 1964, Mise au point d'un illuminateur spectral a usage biologique, *C. R. Acad. Sci. Paris*, 259:1581.

Jagger, J., Takebe, H., and Snow, J. M., 1970, Photoreactivation of killing in *Streptomyces*: Action spectra and kinetic studies, *Photochem. Photobiol.*, 12:185.

Kadota, A., Wayne, R., Kubota, M., and Watanabe, M., Localization of the photoreceptive site for the step-up photophobic response in a green alga, *Dunaliella salina* as analyzed by the partial irradiation of a cell, *Photochem. Photobiol.*, in prep.

Kawai, H., 1988, A flavin-like autofluorescent substance in the posterior flagellum of golden and brown algae, *J. Phycol.*, 24:114.

Kawai, H., Müller, D. G., Fölster, E., and Häder, D.-P., 1990, Phototactic responses in the gametes of the brown alga, *Ectocarpus siliculosus.*, *Planta*, 182:292.

Kawai, H., Kubota, M., Kondo, T., and Watanabe, M., 1991, Action spectra for phototaxis in zoospores of the brown alga *Pseudochorda gracilis, Protoplasma*, in press.

Kondo, T., Kubota, M., Aono, Y., and Watanabe, M., 1988, A computerized video system to automatically analyze movements of individual cells and its application to the study of circadian rhythms in phototaxis and motility in *Chlamydomonas reinhardtii, Protoplasma, Suppl.* 1:185.

Kondo, T., Johnson, C. H., and Hastings, J. W., 1991, Action spectrum for resetting the circadian phototaxis rhythm in the CW 15 strain of *Chlamydomonas*. I. Cells in darkness, *Plant Physiol.*, in press.

Lenci, F., and Ghetti, F., 1989, Photoreceptor pigments for photomovement of microorganisms: some spectroscopic and related studies, *J. Photochem. Photobiol. B: Biol.*, 3:1.

Lumsden, P. J., Saji, H., and Furuya, M., 1987, Action spectra confirm two separate actions of phytochrome in the induction of flowering in *Lemna paucicostata* 441, *Plant Cell Physiol.*, 28:1237.

Monk, G. S., and Ehret, C. F., 1956, Design and performance of a biological spectrograph, *Radiat. Res.*, 5:88.

Müller, D. G., Maier, I., and Müller, H., 1987, Flagellum autofluorescence and photoaccumulation in heterokont algae, *Photochem. Photobiol.*, 46:1003.

Nultsch, W., and Häder, D.-P., 1988, Photomovement in motile microorganisms. II, *Photochem. Photobiol.*, 47:837.

Ogura, T., Yoshikawa, S., and Kitagawa, T., 1985, Resonance Raman spectra of catalytic intermediates of cytochrome c oxydase detected with a mixed flow transient apparatus, *Biochem. Biophys. Acta*, 832:220.

Ogura, T., Takahashi, S., Shinzawa-Ito, K., Yoshikawa, S., and Kitagawa, T., 1990, Observation of the $Fe^{4+}=O$ stretching Raman band for cytochrome oxidase compound B at ambient temperature, *J. Biol. Chem.*, 265:14721.

Ohki, K., Watanabe, M., and Fujita, Y., 1982, Action of near UV and blue light on the photocontrol of phycobiliprotein formation; a complementary chromatic adaptation, *Plant Cell Physiol.*, 23:651.

Parker, M. W., Hendricks, S. B., Borthwick, H. A., and Scully, N. J., 1946, Action spectrum for the photoperiodic control of floral initiation of short-day plants, *Bot. Gaz.*, 108:1.

Saji, H., Furuya, M., and Takimoto, A., 1982, Spectral dependence of night-break effect on photoperiodic floral induction in *Lemna paucicostata* 441, *Plant Cell Physiol.*, 23:623.

Saji, H., Vince-Prue, D., and Furuya, M., 1983, Studies on the photoreceptors for the promotion and inhibition of flowering in dark-grown seedlings of *Pharbitis nil* Choisy, *Plant Cell Physiol.*, 24:1183.

Schmid, R., 1984, Blue light effects on morphogenesis and metabolism in *Acetabularia*, *in*: "Blue Light Effects in Biological Systems," H. Senger, ed., Springer-Verlag, Berlin, p. 419.

Shen-Miller, J., Cooper, P., and Gordon, S. A., 1969, Phototropism and photoinhibition of basipolar transport of auxin in oat coleoptiles, *Plant Physiol.*, 44:491.

Sugai, M., Tomizawa, K., Watanabe, M., and Furuya, M., 1984, Action spectrum between 250 and 800 nanometers for the photoinduced inhibition of spore germination in *Pteris vittata*, *Plant Cell Physiol.*, 25:205.

Sugai, M., and Furuya, M., 1985, Action spectrum in ultraviolet and blue light region for the inhibition of red-light-induced spore germination in *Adiantum capillus-veneris* L., *Plant Cell Physiol.*, 26:953.

Takahashi, T., and Kobatake, Y., 1982, Computer-linked automated method for measurement of the reversal frequency in phototaxis of *Halobacterium halobium*, *Cell Struct. Funct.*, 7:183.

Ueda, T., Mori, Y., Nakagaki, T., and Kobatake, Y., 1988, Action spectra for superoxide generation and UV and visible light photoavoidance in plasmodia of *Physarum polycephalum*, *Photochem. Photobiol.*, 48:705.

Wada, M., and Furuya, M., 1974, Action spectrum for the timing of photoinduced cell division in *Adiantum* gametophytes, *Physiol. Plant.*, 32:377.

Wada, M., and Furuya, M., 1978, Effects of narrow-beam irradiations with blue and far-red light on the timing of cell division in *Adiantum* gametophytes, *Planta*, 126:85.

Watanabe, M., and Furuya, M., 1974, Action spectrum of phototaxis in a cryptomonad alga, *Cryptomonas* sp., *Plant Cell Physiol.*, 15:413.

Watanabe, M., and Furuya, M., 1982, Phototactic behavior of individual cells of *Cryptomonas* sp. in response to continuous and intermittent light stimuli, *Photochem. Photobiol.*, 35:559.

Watanabe, M., Furuya, M., Miyoshi, Y., Inoue, Y., Iwahashi, I., and Matsumoto, K., 1982, Design and performance of the Okazaki Large Spectrograph for photobiological research, *Photochem. Photobiol.*, 36:491.

Watanabe, M., 1985, The Okazaki Large Spectrograph and its application to action spectroscopy, *in*: "Photobiology 1984," J. W. Longworth, J. Jagger, and W. Shropshire, Jr., eds., Praeger, New York, p. 37.

Watanabe, M., 1988, The Okazaki Large Spectrograph and the extension, into the ultraviolet region, of the action spectra for the signaling effects of blue and near-ultraviolet light in plants and fungi, *Photomed. Photobiol.*, 10:83.

Watanabe, M., Kubota, M., Kadota, A., Morita, K., and Matsubara, M., A tracking microscope for microbeam irradiation of a moving organism, *Photochem. Photobiol.*, in prep.

Wayne, R., Kadota, A., Watanabe, M., and Furuya, M., 1991, Photomovement in *Dunaliella salina*: Fluence rate-response curves and action spectra, *Planta*, in press.

Yoshida, K., Takimoto, A., and Sasaki, Y., 1989, Fluence-response relationship for photogene expression in etiolated pea seedlings, *Photochem. Photobiol.*, 50:121.

Participants

Members of the Advisory and Organizing Committees indicated by an asterisk.

AKYILDIZ, Selda
Hacettepe University
Dept. of Molecular Biology
Ankara
Turkey

AKYÜZ, Sevim
Ondokuz Mayis University
Department of Physics
Fen-edebiyat Faculty
Samsun
Turkey

AKYÜZ, Tanil
Mineral Research and Exploration
Institute of Turkey (M.T.A.)
Spectro-Chemistry Lab.
MAT Dairesi
Balgat
Ankara
Turkey

ARISOY, Münevver
Hacettepe University
Faculty of Science
Department of Biology
06532 Beytepe
Ankara
Turkey

ARMITAGE, Judith P.
University of Oxford
Microbiology Unit
Department of Biochemistry
South Parks Road
Oxford OX1 3QU
UK

*BALDOCCHI, Maria Antonia
CNR Institute of Biophysics
via San Lorenzo 26
56127 Pisa
Italy

BANERJEE, Milan
University of Poona
Department of Physics
Pune 411 007
India

BARSANTI, Laura
CNR Institute of Biophysics
via San Lorenzo 26
56127 Pisa
Italy

BELOZERSKAYA, Tatiana A.
A.N. Bach Inst. of Biochemistry
33 Leninsky Prospekt
Moscow 117071
USSR

BERND, Meta
Max-Planck-Inst. for Biochemistry
D-8033 Martinsried bei München
Germany

BRIGGS, Winslow
Carnegie Institution of Washington
Department of Plant Biology
290 Panama Street
Stanford CA 94305-1297
USA

BRODHUN, Bonita
Friedrich-Alexander University
Botany and Pharm. Biology Inst.
Staudtstrasse 5
D-8520 Erlangen
Germany

BRYL, Krzysztof
Univ. of Agriculture and Technology
Dept. of Physics and Biophysics
10-957 Olsztyn
Poland

CAUBERGS, Roland
Leerstoel Voor Algemene Plantkunde
Rijkuniversitair Centrum
Groenenborgerlaan 171
B-2020 Antwerpen
Belgium

*CHECCUCCI, Giovanni
Florence University
Animal Biology and Genetics Dept.
via Romana 17
50125 Firenze
Italy

CIHANGIR, Nilufer
Hacettepe University
Faculty of Science
Department of Biology
06532 Beytepe
Ankara
Turkey

*COLOMBETTI, Giuliano
CNR Institute of Biophysics
via San Lorenzo 26
56127 Pisa
Italy

COTUGNO, Antonio
CNR Institute of Cybernetics
Via Toiano 6
80072 Arco Felice (NA)
Italy

CUBEDDU, Rinaldo
Polytechnics of Milano
Institute of Physics
Piazza Leonardo da Vinci 32
20133 Milano
Italy

DENIZLI, Adil
Hacettepe University
Chemical Engineering Department
06532 Beytepe
Ankara
Turkey

DOUGHTY, Michael
University of Waterloo
Faculty of Sciences
School of Optometry
Waterloo Ontario N2L 3G1
Canada

FARRENS, Dave
University of Nebraska-Lincoln
Department of Chemistry
Lincoln NE 68588-0304
USA

FERRANDO, Elisa
Max-Planck-Inst. of Biochemistry
Dept. of Membrane Biochemistry
D-8033 Martinsried bei München
Germany

FORCHIASSIN, Flavia
Buenos Aires University
Biological Sciences Department
Buenos Aires
Argentina

GERBER, Sabine
Friedrich-Alexander University
Botany and Pharm. Biology Inst.
Chair of Botany I
Staudtstrasse 5
D-8520 Erlangen
Germany

*GHETTI, Francesco
CNR Institute of Biophysics
via San Lorenzo 26
56127 Pisa
Italy

*GIOFFRE', Domenico
CNR Institute of Biophysics
Via San Lorenzo 26
56127 Pisa
Italy

GUALTIERI, Paolo
CNR Institute of Biophysics
via San Lorenzo 26
56127 Pisa
Italy

*HÄDER, Donat-Peter
Friedrich-Alexander University
Botany and Pharm. Biology Inst.
Chair of Botany I
Staudtstrasse 5
D-8520 Erlangen
Germany

HÄDER, Maria
Friedrich-Alexander University
Botany and Pharm. Biology Inst.
Staudtstrasse 5
D-8520 Erlangen
Germany

HARZ, Hartmann
Max-Planck-Inst. for Biochemistry
Dept. of Membrane Chemistry
D-8033 Martinsried bei München
Germany

HASER, Harald
Friedrich-Alexander University
Botany and Pharm. Biology Inst.
Staudtstrasse 5
D-8520 Erlangen
Germany

HAUPT, Wolfgang
Friedrich-Alexander University
Botany and Pharm. Biology Inst.
Staudtstrasse 5
D-8520 Erlangen
Germany

HEGEMANN, Peter
Max-Planck-Inst. of Biochemistry
D-8033 Martinsried bei München
Germany

HERRMANN, Heike Kornelia
Friedrich-Alexander University
Botany and Pharm. Biology Inst.
Staudtstrasse 5
D-8520 Erlangen
Germany

HILDEBRAND, Eilo
Jülich Research Center
Institute for Processing
Biological Information
Postfach 1913
D-5170 Jülich
Germany

JOHNSSON, Anders
The University of Trondheim
Department of Physics
N-7055 Dragvoll
Norway

KJELDSTAD, Berit
The University of Trondheim
Department of Physics
N-7055 Dragvoll
Norway

KÖNIG, Karsten
Ballhausgasse 6
DDR-6900 Jena
East Germany

KRAH, Monika
Max-Planck-Inst. of Biochemistry
Dept. of Membrane Biochemistry
D-8033 Martinsried bei München
Germany

KRZYZANOWSKI, Roman
Friedrich-Alexander University
Botany and Pharm. Biology Inst.
Staudtstrasse 5
D-8520 Erlangen
Germany

*LENCI, Francesco
CNR Institute of Biophysics
via San Lorenzo 26
56127 Pisa
Italy

LIPSON, Edward
Syracuse University
Department of Physics
201 Physics Building
Syracuse NY 13244-1130
USA

LUCIA, Sabina
University of Genova
Institute of Biophysics
via Giotto 2
16135 Genova
Italy

LYUDNIKOVA, Tamara A.
A.N. Bach Inst. of Biochemistry
33 Leninsky Prospekt
Moscow 117071
USSR

MALEC, Przemyslaw
Jagiellonian University
Zurzycki Inst. of Molecular Biology
Al. Mickiewicza 3
31-120 Krakow
Poland

MARANGONI, Roberto
University of Genova
Institute of Biophysics
Salita superiore della Noce 10
16100 Genova
Italy

MARINO, Michael
University of Pittsburgh
Dept. of Behavioral Neuroscience
Pittsburgh, PA 15260
USA

MARWAN, Wolfgang
Max-Planck-Inst. of Biochemistry
D-8033 Martinsried bei München
Germany

MIEDZIEJKO, Ewa
Agricultural University
Dept. of Physics
Wojska Polskiego 38/42
PL 60-637 Poznan
Poland

MUSIO, Carlo
CNR Institute of Cybernetics
Via Toiano 6
80072 Arco Felice (NA)
Italy

NAVARATNAM, Suppiah
The North East Wales Institute
Faculty of Research and Innovation
Deeside Clwyd CH5 4BR
UK

NULTSCH, Wilhelm
Philipps University of Marburg
Faculty of Biology
Botany Inst.
D-3550 Marburg/Lahn
Germany

OLCAY, Murat
Hacettepe University
Chemical Engineering Department
06532 Beytepe
Ankara
Turkey

PASSARELLI, Vincenzo
CNR Institute of Biophysics
via San Lorenzo 26
56127 Pisa
Italy

PETRACCHI, Donatella
CNR Institute of Biophysics
Via San Lorenzo 26
56127 Pisa
Italy

PFAU, Jürgen
Philipps University of Marburg
Faculty of Biology
Botany Inst.
D-3550 Marburg/Lahn
Germany

PHILLIPS, Glyn O.
The North East Wales Institute
Deeside Clwyd CH5 4BR
UK

POSUDIN, Yuri
Ukrainian Agricultural Academy
Kiev 252041 Goloseevo
USSR

RAGOT, Richard
CNRS LENA - URA 654
Applied Electrophysiology and
Neurophysiology Laboratory
47 Bd. de l'Hopital
75651 Paris Cedex 13
France

RÜFFER, Ursula
Philipps University of Marburg
Faculty of Biology
Botany Inst.
D-3550 Marburg/Lahn
Germany

SAYIN, Deniz
Hacettepe University
Chemical Engineering Department
06532 Beytepe
Ankara
Turkey

SCHARF, Birgit
Max-Planck-Institut
für Ernährungsphysiologie
Rheinlanddamm 201
D-4600 Dortmund 1
Germany

SCHIMZ, Angelika
Jülich Research Center
Institute for Processing
Biological Information
Postfach 1913
D-5170 Jülich
Germany

SCHUCHART, Hartwig
Philipps University of Marburg
Faculty of Biology
Botany Inst.
D-3550 Marburg/Lahn
Germany

SHORT, Timothy W.
Carnegie Institution of Washington
Department of Plant Biology
290 Panama Street
Stanford CA 94305-1297
USA

SINESHCHEKOV, Oleg
M.V. Lomonosov Moscow State Univ.
Biology Department
Moscow 119899
USSR

*SONG, Pill-Soon
University of Nebraska-Lincoln
Department of Chemistry
Hamilton Hall
Lincoln NE 68588-0304
USA

SPONZA, Delia
Dakuz Eylül Universitesi
Nimaelik-Nükendislit Fak.
Gerze Nukendislit Bolüma
Bornova Izmir
Turkey

SPUDICH, John L.
Yeshiva University
Albert Einstein College of Medicine
Anatomy and Structural Biol Dept.
1300 Morris Park Avenue
Bronx, NY 10461
USA

TAKAHASHI, Tetsuo
Hokkaido University
Faculty of Pharmaceutical Sciences
Department of Biophysics
Sapporo 060
Japan

TARONI, Paola
CNR Quantum Electronics Center
Inst. of Physics of Polytechnics
Piazza Leonardo da Vinci 32
20133 Milano
Italy

*TROVATO, Raffaella
CNR Institute of Biophysics
Via San Lorenzo 26
56127 Pisa
Italy

UHL, Reiner
University of Washington
School of Medicine
Department of
Physiology and Biophysics
SJ-40
Seattle, WA 98195
USA

UNAL, Yesim
Hacettepe University
Chemical Engineering Department
06532 Beytepe
Ankara
Turkey

VORNLOCHER, Hans-Peter
Friedrich-Alexander University
Botany and Pharm. Biology Inst.
Staudtstrasse 5
D-8520 Erlangen
Germany

WALNE, Patricia L.
Univ. of Tennessee
Dept. of Botany
Program in Cellular
Molecular and
Developmental Biology
Knoxville, TN 37996-1100
USA

WATANABE, Masakatsu
Okazaki National Research Inst.
National Inst. for Basic Biology
Okazaki Aichi 444
Japan

WOOD, David
University of Pittsburgh
Dept. of Behavioral Neuroscience
Pittsburgh PA 15260
USA

YAN, Bing
Physiological laboratory
University of Cambridge
Downing street
Cambridge CB2 3EG
UK

YANG, Hanjing
University of Wisconsin-Madison
College Agricultural and Life Sci.
Department of Biochemistry
420 Henry Mall
Madison, WI 53706-1569
USA

YARGICOGLU, Piraye
Akdeniz University
Dept. of Physiology and Biophysics
Antalya
Turkey

Index

GPSR Compliance
The European Union's (EU) General Product Safety Regulation (GPSR) is a set of rules that requires consumer products to be safe and our obligations to ensure this.

If you have any concerns about our products, you can contact us on

ProductSafety@springernature.com

In case Publisher is established outside the EU, the EU authorized representative is:

Springer Nature Customer Service Center GmbH
Europaplatz 3
69115 Heidelberg, Germany

www.ingramcontent.com/pod-product-compliance
Ingram Content Group UK Ltd.
Pitfield, Milton Keynes, MK11 3LW, UK
UKHW051127260726
13967UKWH00010B/2906

* 9 7 8 1 4 6 8 4 5 9 8 9 0 *